HABITATS
of the
WORLD

HABITATS
of the
WORLD

A Field Guide *for* Birders, Naturalists *and* Ecologists

Iain Campbell | **Ken Behrens**
Charley Hesse | **Phil Chaon**

Special contributors
Sam Woods, Pablo Cervantes,
and **Anais Campbell**

PRINCETON UNIVERSITY PRESS
PRINCETON AND OXFORD

Published by Princeton University Press
41 William Street, Princeton, New Jersey 08540
6 Oxford Street, Woodstock, Oxfordshire OX20 1TR

press.princeton.edu

All Rights Reserved

Library of Congress Cataloging-in-Publication Data
Names: Campbell, Iain, 1969- author. | Behrens, Ken,
    author. | Hesse, Charlie, author. | Chaon, Philip, author.
Title: Habitats of the world / Iain Campbell with Ken
    Behrens, Charlie Hesse, Philip Chaon.
Description: Princeton : Princeton University Press, [2021] |
    Includes bibliographical references and index.
Identifiers: LCCN 2021008539 (print) | LCCN 2021008540
    (ebook) | ISBN 9780691197562 (cloth) |
    ISBN 9780691225968 (ebook)
Subjects: LCSH: Habitat (Ecology) | Biogeography.
Classification: LCC QH541 .C343 2021 (print) | LCC QH541
    (ebook) | DDC 577--dc23
LC record available at https://lccn.loc.gov/2021008539
LC ebook record available at https://lccn.loc.
    gov/2021008540

British Library Cataloging-in-Publication Data is available

This book has been composed in Cambay Devanagari

Printed on acid-free paper. ∞

Printed in Italy

10 9 8 7 6 5 4 3 2 1

Editorial: Robert Kirk and Abigail Johnson
Production Editorial: Ellen Foos
Jacket Design: Wanda España
Production: Steven Sears
Publicity: Matthew Taylor and Caitlyn Robson
Copyeditor: Amy K. Hughes
Typeset and Design: D & N Publishing, Wiltshire, UK

Cover art: (front) © Christine Elder, (back) © Iain Campbell

Section opener photos:

Australasia
Termite mounds. © Iain Campbell, Tropical Birding
Bush Thick-knee, © Iain Campbell, Tropical Birding

Neotropics
Puma. © Keith Barnes, Tropical Birding
Resplendent Quetzal. © Pablo Cdfrvantes, Tropical Birding

Indo-Malaysia
Vishnu statue. © Iain Campbell, Tropical Birding

Afrotropics
African lake. © Iain Campbell, Tropical Birding
Lesser Masked Weaver. © Iain Campbell, Tropical Birding

Palearctic
Przewalski's Horse. © Keith Barnes, Tropical Birding

Nearctic
Grand Canyon. © Ben Knoot, Tropical Birding

CONTENTS

List of Figures and Sidebars   8

INTRODUCTION   9

Genesis of the Book   9

What Do We Cover as a Distinct Habitat
in This Book?   9

Zoogeographic Regions   10

Habitat Nomenclature   12

Common Canopy Leaf Types and the Forests
Where You May Find Them   13

Climate Descriptions and Graphs   13

Habitat Key   17

Global Biomes with Latitude and
Precipitation   21

Taxonomy   22

Endemic Bird Areas   22

About This Book   23

Abbreviations   24

**HABITATS OF AUSTRALASIA
(Australia, New Zealand, and New Guinea)**   25

■ Gondwanan Conifer Rainforest   26

■ Dune and Rocky Spinifex   28

■ Chenopod and Samphire Shrubland   30

■ New Zealand Beech Forest   33

■ Australasian Tropical Lowland Rainforest   35

■ Australasian Subtropical and
Montane Rainforest   38

■ Australasian Temperate Rainforest   41

■ Australasian Tropical Semi-Deciduous Forest   43

■ Australasian Monsoon Vine Forest   45

■ Australasian Sandstone Escarpments   48

■ Open Eucalypt Savanna   50

■ Tetrodonta Woodland Savanna   53

■ Brigalow and *Callitris* Woodlands   57

■ *Melaleuca* Savanna   59

■ Sheoak Woodland   61

■ Australasian Tussock Grassland   63

■ Australasian Lowland Heathland   66

■ Australasian Alpine Heathland   69

■ Wet Sclerophyll Forest   72

■ Dry Sclerophyll Forest   75

■ Temperate Eucalypt Woodland   77

■ Mallee Woodland and Scrubland   81

■ Mulga Woodland and Acacia Shrubland   83

■ Australasian Temperate Wetland   86

■ Australasian Tropical Wetland   87

■ Australasian Mangrove   89

■ Australasian Rocky Coastline and Sandy Beach   92

■ Australasian Sandy Cays   94

■ Australasian Tidal Mudflat and Salt Marsh   95

■ Australasian Pelagic Waters   97

■ Australasian Large-Scale Farming   98

**HABITATS OF THE NEOTROPICS
(Central and South America)**   101

■ Valdivian Rainforest and Neotropical
Mixed Conifer Forest   102

■ Neotropical Desolate Desert   104

■ Galápagos Desert and Scalesia   108

■ Neotropical Thornscrub   112

■ Puna   114

■ Patagonian Steppe   116

■ Neotropical Semidesert Scrub   118

■ Monte   120

■ Neotropical Pine-Oak Woodland   123

■ Neotropical Lowland Rainforest   124

■ Neotropical Semi-Evergreen Forest   129

■ Neotropical Cloud Forest   131

■ Yungas   135

■ Elfin and Stunted Cloud Forest   138

■ Magellanic Rainforest   141

■ Neotropical Dry Deciduous Forest   143

■ Caatinga   146

■ Chaco Seco and Espinal   149

■ Cerrado   152

Pampas and Campo 156

Matorral Sclerophyll Forest and Scrub 159

Paramo 161

Antarctic Tundra and Tussock Grass 165

Neotropical Flooded Grassland and Wetland 168

Igapó and Várzea Flooded Forest 171

Neotropical Mangrove 174

Neotropical Tidal Mudflat 175

Neotropical Rocky Coastline and Sandy Beach 177

Neotropical and Antarctic Pelagic Waters 179

Neotropical Cropland 181

Neotropical Grazing Land 183

HABITATS OF INDO-MALAYSIA
(Southeast Asia and India) **185**

Indo-Malayan Pine Forest 186

Indo-Malayan Thornscrub 189

Indo-Malayan Tropical Lowland Rainforest 192

Indo-Malayan Semi-Evergreen Forest 196

Indo-Malayan Tropical Montane Rainforest 199

Indo-Malayan Subtropical Broadleaf Forest 202

Indo-Malayan Peat Swamp Forest 206

Kerangas 209

Indo-Malayan Limestone Forest 211

Indo-Malayan Moist Deciduous Forest 213

Indo-Malayan Dry Deciduous Forest 217

Indo-Malayan Seasonally Flooded Grassland 219

Indo-Malayan Montane Grassland 223

Indo-Malayan Freshwater Swamp Forest 225

Indo-Malayan Freshwater Wetland 227

Indo-Malayan Mangrove Forest 229

Indo-Malayan Tidal Mudflat and Salt Pan 231

Indo-Malayan Rocky Coastline and Sandy Beach 233

Indo-Malayan Pelagic Waters 234

Indo-Malayan Offshore Islands 235

Paddy Fields and Other Cropland 237

Indo-Malayan Cities and Villages 239

HABITATS OF THE AFROTROPICS
(Sub-Saharan Africa) **241**

Afrotropical Desert 243

Karoo 246

Malagasy Spiny Forest 249

Dragon Blood Semidesert 253

Afrotropical Lowland Rainforest 255

Afrotropical Monsoon Forest 260

Afrotropical Swamp Forest 264

Afrotropical Montane Forest 267

Indian Ocean Rainforest 273

Malagasy Dry Deciduous Forest 278

Guinea Savanna 281

Miombo Woodland 285

Gusu Woodland 289

Mopane Savanna 292

Afrotropical Dry Thorn Savanna and Thornscrub 296

Afrotropical Moist Mixed Savanna 301

Inselbergs, Koppies, and Cliffs 305

Afrotropical Tropical Grassland 308

Afrotropical Montane Grassland 312

Fynbos 316

Afrotropical Montane Heath 320

Afrotropical Freshwater Wetland 322

Afrotropical Salt Pan 325

Afrotropical Mangrove 327

Afrotropical Tidal Mudflat and Salt Marsh 329

Afrotropical Rocky Coastline and Sandy Beach 330

Afrotropical Pelagic Waters 331

Afrotropical Offshore Islands 333

Afrotropical Cropland 334

Afrotropical Grazing Land 336

Afrotropical Cities and Villages 337

HABITATS OF THE PALEARCTIC
(Europe, Northern Asia, and North Africa) **339**

Eurasian Spruce-Fir Taiga 340

Siberian Larch Forest 343

■ Eurasian Montane Conifer Forest   346
■ Mediterranean and Dry Conifer Forest   348
■ Beringian Taiga Savanna   350
■ Palearctic Hot Desert   352
■ East Asian Cold Desert   355
■ Central Asian Cold Desert   357
■ Temperate Desert Steppe   359
■ Palearctic Hot Shrub Desert   361
■ Palearctic Semidesert Thornscrub   364
■ East Asian Moist Mixed Forest   366
■ European Moist Mixed Forest   369
■ East Asian Temperate Bamboo Forest   373
■ Colchic Deciduous Rainforest   375
■ European Temperate Deciduous Forest   377
■ East Asian Temperate Deciduous Forest   380
▨ Palearctic Forest-Steppe   382
▨ Palearctic Subtropical Savanna   385
■ Mediterranean Oak Forest   387
■ European Heathland and Moorland   389
■ Maquis   391
■ Garrigue   393
▨ Western Flower Steppe   395
▨ Eastern Grass Steppe   397
■ Arctic Polar Desert   399
■ Eurasian Rocky Tundra   401
■ Eurasian Boggy Tundra   405
■ Eurasian Alpine Tundra and
  Himalayan Montane Desert   407
■ Palearctic Temperate Wetland   411
■ Palearctic Tidal Flat   414
■ Palearctic Rocky Coastline and Sandy Beach   416
■ Palearctic Pelagic Waters   418
■ Palearctic Cropland   418
■ Palearctic Grazing Land   420

**HABITATS OF THE NEARCTIC
(North America)**   **421**

■ Nearctic Spruce-Fir Taiga   422
■ Nearctic Montane Spruce-Fir Forest   425
■ Montane Mixed-Conifer Forest   428

■ Longleaf Pine Savanna   431
■ High-Elevation Pine Woodland   434
■ Lodgepole Pine Forest   437
■ Nearctic Temperate Rainforest   440
■ Pinyon-Juniper Woodland   444
■ Madrean Pine-Oak Woodland   447
■ Chihuahuan Desert Shrubland   450
■ Columnar Cactus Desert   453
■ Salt Desert Shrubland   456
■ Nearctic Desert Grassland   458
■ Sagebrush Shrubland   461
■ Mesquite Brushland and Thornscrub   464
■ Bald Cypress–Tupelo Forest   468
■ Nearctic Temperate Deciduous Forest   471
■ Nearctic Temperate Mixed Forest   475
■ Western Riparian Woodland   479
■ Aspen Forest and Parkland   482
▨ Nearctic Cloud Forest   484
■ Nearctic Tropical Dry Forest   486
▨ Cedar Savanna   488
▨ Nearctic Oak Savanna   491
▨ Florida Oak Scrub   494
▨ Shortgrass Prairie   496
▨ Tallgrass Prairie   499
■ Pacific Chaparral   502
■ Nearctic Rocky Tundra   505
■ Nearctic Boggy Tundra   507
■ Nearctic Alpine Tundra   509
■ Nearctic Freshwater Wetland   512
■ Nearctic Mangrove   514
■ Nearctic Tidal Mudflat   516
■ Nearctic Salt Marsh   517
■ Nearctic Rocky Coastline and Sandy Beach   519
■ Nearctic Pelagic Waters   521
■ Nearctic Offshore Islands   523
■ Nearctic Cropland and Grazing Land   525

PLANT NAMES   **527**
INDEX   **541**

8

LIST OF FIGURES AND SIDEBARS

FIGURES

Fig. 1. Zoogeographic regions used in this book   10

Fig. 2. Zoogeographic boundaries between Asia and Australasia   12

Fig. 3. Leaf types   13

Fig. 4. Sample climate graphs   14

Fig. 5. Biomes of the world   17

Fig. 6. Influence of latitude, precipitation, and temperature on habitats   21

Fig. 7. How biomes blend into one another   22

SIDEBARS

Sidebar 1.1. Reg: It May Appear Desolate, but Look Closer   32

Sidebar 1.2. Forest Canopies and Undergrowth Can Change at Different Rates   74

Sidebar 1.3. Habitat Transition through New South Wales, Australia, with Climate Graphs   80

Sidebar 2.1. The Effects of World Ocean Currents on Climate   107

Sidebar 2.2. Hot-Spot Volcanism and Plate Movements   111

Sidebar 2.3. Tree-Line and Krummholz Forests   137

Sidebar 2.4. What Makes a Habitat?   151

Sidebar 2.5. Duricrusts and Desertification   155

Sidebar 2.6. Subduction and the Andes   164

Sidebar 3.1. Habitats across India with Climate Graphs   186

Sidebar 3.2. Island Arc Collisions: Building Archipelagos   195

Sidebar 3.3. Why Are There Clouds in the Cloud Forest?   202

Sidebar 3.4. Why There Are No Volcanoes in the Himalayas: The Continental-Continental Collisions   205

Sidebar 3.5. What Is a Monsoon?   222

Sidebar 4.1. Habitat Transition in Senegal, Africa   242

Sidebar 4.2. Refugia within Expanding and Contracting African Rainforest   259

Sidebar 4.3. Africa's Great Rift Valley   266

Sidebar 4.4. Albertine Rift: The Hub of Africa's Montane Forest   271

Sidebar 4.5. African Sky Islands   272

Sidebar 4.6. Sambirano Rainforest   277

Sidebar 4.7. Africa's Lost Mammals   305

Sidebar 5.1. Extinction of People, Plants, and Animals of the Sahara   354

Sidebar 5.2. Habitats in Flux   363

Sidebar 5.3. Desert Refugia   366

Sidebar 5.4. The Forest Recolonization of Europe   369

Sidebar 5.5. All Those Beeches   379

Sidebar 6.1. Superlative Trees   437

Sidebar 6.2. Covering the Caribbean   468

Sidebar 6.3. Habitat Use by Migratory Birds   474

Sidebar 6.4. Mountains Can Be Two-Faced   478

INTRODUCTION

GENESIS OF THE BOOK

All the authors of this field guide have had a lifelong fascination with biogeography and wildlife habitats. Like the vast majority of other passionate traveling naturalists, we are most consistently interested in birds and larger mammals, while also paying some attention to reptiles, amphibians, butterflies, and other groups, especially in places where they're conspicuous. We have all been frustrated by the approach to habitat classification used in most books and the complete absence of habitat information in many field guides. An understanding of habitats is fundamental to becoming a knowledgeable traveling naturalist, but gaining this understanding has often required slowly piecing things together yourself. What we're hoping to do in this book is present our view of global wildlife habitats, in order to allow others to understand them far more readily than we were able to.

There are innumerable lenses through which planet Earth's habitats can be assessed. Geology, geography, and botany are all critically important. But we don't view any of them as the final word on habitats, and much of what these models prioritize is of little immediate relevance to traveling naturalists. A specialist in entomology or herpetology will also apply a different, and fascinating, lens to the world. A recent study of Illinois divided that US state into nearly 100 habitats based on their distinct assemblages of insects! None of these lenses is invalid, and all of them reveal fascinating things about this planet's biogeography. Our reason for prioritizing a larger mammal and bird "lens" is that we look at the world primarily through this lens, and so do the vast majority of the world's traveling naturalists. A handful of specialists seek out chameleons in Tanzania's Eastern Arc Mountains, whereas millions of tourists visit the Serengeti and Masai Mara to see big mammals and glamorous birds. A few people venture to the Amazon to seek out its incredible diversity of insects, but masses visit rainforest lodges in search of monkeys, Hoatzin, and an overall "jungle" experience. So our approach in this book might lack the academic purity of a cleanly geological or botanical approach to the world's habitats, but we think it has far greater utility to most world travelers than any other previous perspective on habitats.

In its attempt to cover the wildlife habitats of the entire globe, this is an ambitious book, in which hard decisions had to be made about what to include and exclude. We freely admit that habitats like wetlands, anthropogenic environments, and the oceans, are all worthy of far more detailed coverage. People who know their local area well may be frustrated by a lack of information about "their patch." Please remember that this book is about giving people a broad view. It also offers a sort of "virtual travel"; the first thing all the authors do when they find out they're headed somewhere new on the globe is conduct a bit of research about the local wildlife habitats. Deciding to "split" or "lump" some habitats was very tricky, and some of our decisions could be argued endlessly. But condensing a huge amount of research into a finite number of pages, and simply finishing this project, required a certain brutality. Our approach is certain to alienate some, but we firmly believe it will be both enjoyable and useful to other global naturalists like ourselves.

WHAT DO WE COVER AS A DISTINCT HABITAT IN THIS BOOK?

We evaluate habitats based on two main criteria: 1) Their visual distinctiveness, which can be easily assessed by a casual observer. 2) Their assemblage of wildlife, primarily meaning larger mammals and birds, since that is our lens throughout this book. Most of the listed habitats are easily validated by a moderate score in both categories. An example is African miombo woodland, which is quite distinct in appearance from adjacent savanna habitats and supports a fairly distinctive set of wildlife, including quite a few species restricted to this habitat. But in some cases, one or the other criterion is of predominant importance. Except to the eye of a trained botanist, Indian Ocean rainforest is not very different from other humid broadleaf forests around the world. But it has almost a continent's worth of diversity for many groups and virtually no overlap with any other place on earth. So, it is considered a distinct habitat. An example of the opposite case is African Mopane savanna. This habitat is characterized by the dominance of the Mopane tree, which is highly distinctive and easily recognizable. So the Mopane savanna qualifies strongly for the first criterion, even as it lacks a cohesive set of wildlife, having, rather, a subset of the wildlife of surrounding habitats.

ZOOGEOGRAPHIC REGIONS

The book is organized by zoogeographic regions (fig. 1). These are similar to the conventional continents used by geographers and the floral kingdoms recognized by botanists but with some important differences.

There is a chapter for each zoogeographic region (except Antarctica, covered briefly in the Neotropics chapter), which contains individual accounts for all of its habitats. An alternative way to organize the book would have been by habitat category (see the Habitat Key below), but a continental approach seemed more useful, especially to travelers. The broad habitat categories, which are color-coded on the maps, and the "Habitat Affinities" section at the beginning of each account, are ways of cross-referencing similar habitats across zoogeographic regions. These are our zoogeographic regions:

— **Australasia** is everything east and south of Sulawesi and Bali. The exact placement of this line is often debated; all the islands between Borneo and New Guinea are a transition zone in the region referred to as Wallacea (fig. 2). Deer occur on the Lesser Sundas and Sulawesi but no farther east. Monkeys reach Sulawesi, the Lesser Sundas, and a few other islands but extend no farther east.

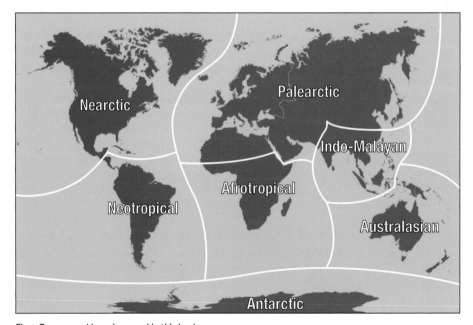

Fig. 1. Zoogeographic regions used in this book.

Meanwhile, marsupials are far more common in Australasia than Wallacea. The presence of eucalypts (*Eucalyptus* spp.) has been used as evidence that Sulawesi is part of Australasia, but they were an insignificant part of the Australian flora until very recently, so are not a valid indicator of what constitutes Australasia. The Lesser Sundas are included in Australasia because they have savanna and monsoon forest that are the same as habitats in Australia, with significant bird species overlap. Oceania (lands of the c. and s. Pacific Ocean including Micronesia, Melanesia, and Polynesia) is included as part of Australasia but doesn't receive extensive coverage because wildlife distribution is determined more by each island's remoteness rather than by the habitats it supports. Most species in Oceania tolerate a broad range of habitats.

— The **Neotropics** runs from Central America south through all of South America. It should be noted that despite this name, much of s. South America has a nontropical climate. The limit with the Nearctic is a political one of convenience along the Guatemala-Mexico border. Antarctica is included in this section.

— **Indo-Malaysia** region, or the Asian tropics, includes the Indian subcontinent and southeast Asia up to Sulawesi and Bali (excluding the rest of Wallacea and the Lesser Sundas). In the western and central regions, it is clearly demarcated in the Thar Desert along the India-Pakistan border and the heights of the Himalayas. In e. China, the division is messier. It is based on the transition from the predominantly tropical evergreen and semi-evergreen forests, with a monsoonal climate, to the predominately deciduous and coniferous and non-monsoonal areas of the Palearctic. This line runs, approximately, west from Shanghai along the Yangtze River.

— The **Afrotropics** is all of Africa south of the Sahara, with a transitional area in the southwestern part of the Arabian Peninsula, which has overlap with Africa. The Sahara and the Arabian Desert effectively divide the Afrotropics from the Palearctic. Northernmost Africa has much more in common biologically with Europe than with the rest of the continent and is included in the Palearctic region.

— The vast **Palearctic** region consists of n. Africa, Europe, most of the Middle East, and n. and c. Asia. Although there is great diversity within this region, it all shares much in common biologically and contains no clear divides. The conventional geographical divide between Europe and Asia, at the Ural Mountains, is of little biological importance.

— The **Nearctic** includes all of North America, running to Mexico's southern border. Biologically, this southern boundary is messy, leaving the political boundary as good as any option. But it is supported by the fact that there is only limited crossover of habitats, such as outliers lowland rainforest and cloud forest from the Neotropics, and dry conifer forest from the Nearctic.

"Rival" Lines and the Arbitrary Divide Between Asia and Australasia

Where should we draw the line between Indo-Malaysia and Australasia? The best-known boundary is Wallace's Line, named after the famous biologist and explorer Alfred Russel Wallace (fig. 2). It follows the edge of the Sunda Shelf, an area of submerged continental shelf under shallow seas averaging about 330 ft. (100m) deep, that connects the islands of Sumatra, Borneo, Java, and Bali with continental se. Asia. These islands have been connected to the mainland during recent ice ages when sea levels were as much as 390 ft. (120m) lower than today. East of Bali and Borneo is a deep ocean trench, which has been the edge of the Sunda Shelf for around 50 million years. Despite the short distance between the islands of Bali and Lombok, many birds and mammals simply haven't crossed this short stretch of ocean.

Biologist Max Carl Wilhelm Weber suggested another line based on mammal data. To the west of Weber's Line is Sulawesi and the Lesser Sundas (from Lombok to Babar), and to the east are the

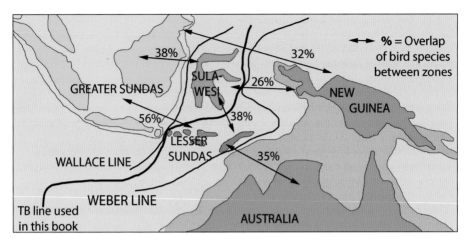

Fig. 2. Zoogeographic boundaries between Asia and Australasia. This illustration shows Weber's and Wallace's Lines, the two most famous zoographic boundaries between Asia and Australasia, and the line used in this book. Also indicated are percentages of overlap in bird species between the various regions.

Moluccas, including Halmahera, Buru, Seram, and the Tanimbar Islands. So which line do we use? Wallace's Line is based on bird data, whereas Weber's Line is based on mammal data. The area between Wallace's and Weber's Lines, known as Wallacea, is essentially a transitional zone containing a mixture of species of both Indo-Malayan and Australasian origin. These species must have arrived over the sea, and this process certainly favored species with greater dispersal ability. Any line we draw through Wallacea to separate the Indo-Malayan and Australasian regions is going to be rather subjective.

Our combined bird-and-mammal approach has led us to place the Lesser Sundas in Australasia, and Sulawesi in the Indo-Malayan zoogeographic region.

HABITAT NOMENCLATURE

Within the zoogeographic regions explained above, there are individual habitats, which are divided into the broad categories explained in the Habitat Key section. This section includes some diagrams and definitions that will help the reader to understand some of the most important terms used in naming and defining habitats. These are terms that appear over and over in the book.

A Few Important Terms to Start

Although wherever possible we used non-technical English throughout this book, there is a variety of unavoidable terms, jargon, and concepts that need fuller, and often quite detailed, explanation. To fully explain these terms, we have constructed an extensive online glossary and concepts page on this volume's sister website: www.habitatsoftheworld.org.

Desert. Very dry and either unvegetated or sparsely vegetated habitat.
Grassland. Habitat dominated by grasses, with few shrubs or trees.
Halophytic. Refers to a plant that can grow in highly saline environments.

Heath/Heathland. Shrubland dominated by fine-leaved evergreen members of the erica family (Ericaceae). Moorland is a moist type of heath.

Rainforest. Lush forest that receives abundant moisture.

Savanna. A lightly wooded or treeless tropical grassland with a prominent wet and dry season.

Steppe and Prairie. Grassland in areas with a cold-winter/warm-summer climate.

Taiga/Boreal Forest. Forest of spruce and fir trees that grows in harsh northern climates, all the way around the North Pole.

Tundra. Open, treeless habitat of extreme environments that are covered in snow for most of the year.

Wetland. Habitat that is frequently or permanently flooded.

Xeric. Refers to a dry environment with little moisture.

Xerophytic. Refers to a plant that needs very little water and can grow in xeric conditions.

COMMON CANOPY LEAF TYPES AND THE FORESTS WHERE YOU MAY FIND THEM

Fig. 3 presents the most common leaf types used in describing different types of forest canopy and some of the habitats where they are prominent. This does not take into account the many types of leaves of understory plants such as grasses, sedges, ferns, euphorbias, and cacti.

CLIMATE DESCRIPTIONS AND GRAPHS

Throughout the book, the habitat descriptions include a brief overview of the climate. In sidebars sprinkled throughout the book, we provide deeper looks into the relationships between vegetation due to latitude and climate (fig. 6), the interaction of general climate with fire, the influence of drought (sidebar 2.4), and anthropogenic effects on habitat modification (sidebar 5.2). Looking at habitat distributions and their relationships to not only temperature and rainfall but also distribution of rainfall through the year, it became apparent that this rainfall distribution is often a more important factor in vegetation type than average precipitation alone. To help illustrate these variations through the year, we have created climate graphs for each habitat, based on the original work of Walter and Lieth, though we have heavily modified them to make them easier to read and interpret.

LEAF SHAPE	LEAF NAME	HABITATS
	Conifer Needle Thin linear leaves. Usually evergreen.	Taiga, dry conifer forests
	Deciduous Broadleaf Broad, thin leaves that grow quickly and last one season.	Temperate deciduous forests, wet/dry deciduous forests
	Evergreen Broadleaf Broad, thin, often with drip tips. They last a long time.	Rainforests, cloud forests
	Sclerophyllous Evergreen Thick, leathery leaves resist transpiration and fires. Often contain toxins. They last a long time.	Eucalypt forests, sclerophyll forests, heathlands, maquis, fynbos, mallee, Mulga, matorral, cerrado, chaparral
	Microphyllous Small Leaves that resist transpiration.	Acacia savanna, thornscrub, Chaco seco, desert scrubs

Fig. 3. Leaf types.

Reading these graphs may seem intimidating at first, but when their relevance is explained, they become more scrutable (see fig. 4). When temperature and precipitation are plotted together, and where each 20mm (0.8 in.) of precipitation is compared to each 10 degrees Celsius (50°F), for average daily temperatures, some really interesting patterns appear. When the precipitation plot drops below the temperature plot, the area is in a time of drought, and plants are stressed. We have colored these periods in orange. When the precipitation plot is above the temperature line, the area has a surplus of water, and plant growth is strong; these periods are colored light blue. However, once the precipitation exceeds 100mm (4 in.) a month, there is an extreme surplus of water, regardless of the temperature, and most runs off and is not used by plants; we have colored these periods in dark blue. Because the whole method makes sense only when used with the metric system, we have included average daily temperature only in Celsius and rainfall in millimeters on the graphs.

Fig. 4. Sample climate graphs.

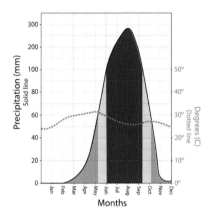

SAVANNA

> Temperature hot throughout the year
> Drought conditions in winter
> Very intense monsoonal summer rains

LOWLAND RAINFOREST

> Temperature hot throughout the year
> Excessive precipitation throughout the year

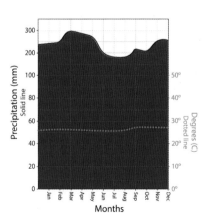

MEDITERRANEAN SCRUB
(INCLUDING CHAPARRAL)

> Cold winters, warm summers
> Wet winters
> Moderately dry summers

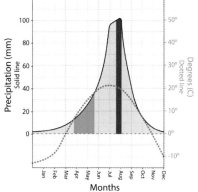

STEPPE AND PRAIRIE

> Very cold winters, hot summers
> Wetter in summer than in winter

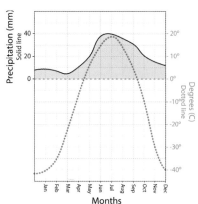

TAIGA

> Very cold winters, cool summers
> Moderate precipitation throughout
  the year

Fig. 4. Sample climate graphs *(continued)*

TEMPERATE DECIDUOUS FOREST

> Cold winters, cool summers
> Precipitation throughout the year, but more in summer

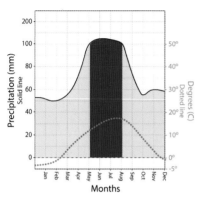

WARM DESERT

> Dry and hot throughout the year

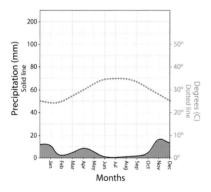

HABITAT KEY

The broad habitat categories and subcategories used in this book are briefly explained in this section; the example habitats listed can be located in the table of contents. Refer to "Habitat Nomenclature," above, for further explanation of most of the terms used here. Note that the color-coding used in the world map (fig. 5) corresponds with that used in the maps throughout the book. These descriptions broadly correspond to the major ecological community types known as "biomes."

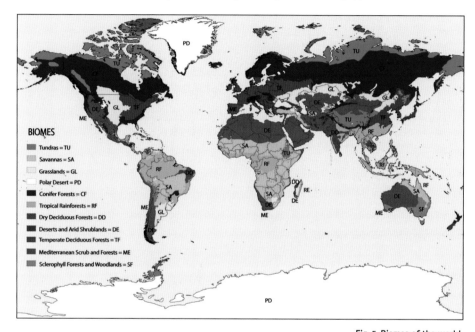

Fig. 5. Biomes of the world.

CONIFER FORESTS: Forests made up of coniferous trees (which generally don't seasonally lose their leaves, with the exception of larches and few others).

- **Taiga (Boreal Forest):** Spruce and fir forests of the far north.
  Example: **Eurasian Spruce-Fir Taiga**
- **Temperate Conifer Forest:** Humid forests (usually) in temperate areas, including mountains.
  Example: **Nearctic Temperate Rainforest**
- **Dry Conifer Forest:** Forests and woodlands in dry temperate and tropical areas.
  Example: **Nearctic Montane Mixed-Conifer Forest**

DESERTS AND ARID SCRUBS: Arid areas with little plant growth.

- **Barren Desolate Desert:** Harshest of deserts, where little grows, with areas of rock and sand dunes.
  Example: **Neotropical Barren Desert (e.g., Atacama)**
- **Hot Desert:** Tropical/subtropical deserts with hot summers and cool or warm winters.
  Example: **Palearctic Hot Desert (e.g., Sahara)**

- **Cold Desert:** Temperate deserts with hot/warm summers and cold winters.
  Example: **Palearctic East Asian Cold Desert (e.g., Gobi)**
- **Desert and Semidesert Shrubland:** Open arid areas with small shrubs with generally
  small leaves; cacti or euphorbias can be present and may be large.
  Example: **Nearctic Columnar Cactus Desert**
- **Desert Steppe:** Very arid areas where grass growth is sparse and ephemeral.
  Example: **Palearctic Temperate Desert Steppe**
- **Desert and Semidesert Thornscrub:** Arid areas with thickets of tall thorn bushes and little
  grass growth.
  Example: **Palearctic Semidesert Thornscrub**
- **Arctic Polar Desert:** Extremely cold and dry areas where almost nothing grows.

TEMPERATE DECIDUOUS FORESTS: Forests that are either entirely broadleaf deciduous
or are mixed with evergreen conifers.

- **Temperate Deciduous Forest:** Temperate broadleaf deciduous forests where most trees lose their
  leaves in winter.
  Example: **Nearctic Temperate Deciduous Forest**
- **Temperate Moist Mixed Forest:** Broadleaf forests of deciduous trees that also include evergreen
  broadleaf trees and/or conifers.
  Example: **Palearctic European Moist Mixed Forest**

TROPICAL HUMID FORESTS: Quintessential warm and wet rainforest-type environments.

- **Lowland Rainforest:** Wet, evergreen tall forest with thick full canopy cover and open undergrowth.
  Example: **Neotropical Lowland Rainforest (e.g., Amazon Basin)**
- **Semi-Evergreen Forest:** Generally humid forests with near-complete canopy cover in which a minority
  of the trees lose their leaves.
  Example: **Indo-Malayan Semi-Evergreen Forest**
- **Montane/Subtropical Evergreen Forest:** Warm, wet forests with almost closed canopy
  of evergreen or partially deciduous trees.
  Example: **Neotropical Cloud Forest**

DRY DECIDUOUS FORESTS: Warm forests that lose most of their leaves in dry periods.

- **Closed Deciduous Forest:** Closed-canopy, dry deciduous forests that in summer appear lush,
  but many trees lose their leaves in winter. Fire-intolerant.
  Example: **Indo-Malayan Moist Deciduous Forest**
- **Open Deciduous Forest:** Open-canopy forests where canopy trees can lose leaves in the dry
  season and the undergrowth consists of small-leaved shrubs, cacti, and/or euphorbias. Fire-intolerant.
  Example: **Neotropical Caatinga**

SAVANNAS AND SEASONALLY MOIST SHRUBLANDS: Habitats with an open
canopy, lots of grass or shrubbery, and a strongly seasonal (usually wet-summer/dry-winter)
climate. Most habitats in this category are heavily influenced by fire.

- **Open Broadleaf Woodland:** Non-sclerophyllous woodlands that can have tall trees and open
  canopy cover. Can have many deciduous trees. Fire-tolerant.
  Example: **Afrotropical Guinea Savanna**
- **Thorn Savanna:** Open (tall or short) woodlands with lots of grass cover. Trees are often dominated
  by acacias (of various genera), many of them with spines.
  Example: **Neotropical Chaco Seco and Espinal**

■ **Broadleaf Sclerophyllous Savanna:** Low shrublands or forests that can be thick. Tend to not have as much grass as other savannas. Plants have thick leathery leaves, thick bark. Fire-tolerant.
Example: **Neotropical Cerrado**

■ **Mixed Shrub Savanna:** Thickets and low-canopy dry forest that has grass cover and a mix of coniferous, sclerophyllous, and broadleaf shrubs or trees. Fire-tolerant.
Example: **Australasian Brigalow Woodland**

GRASSLANDS AND STEPPES: Habitats that are dominated by grasses with or without shrubs and flowers, and few or no trees. Fire-dependent.

■ **Temperate Grassland:** Grasslands that have moderate or warm summers but cold winters. Can receive precipitation as snow in winter or rain in summer, but growing season is usually restricted to spring and summer. Fire-tolerant
Example: **Palearctic Western Flower Steppe**

■ **Tropical Grassland:** Grasslands that have moderate winters and hot summers. Grasses can grow throughout the year but are dependent on rain. Fire-tolerant.
Example: **Australasian Tussock Grassland**

■ **Montane Grassland:** Grasslands in highlands, often receiving orographic rainfall. Fire-tolerant.
Example: **Afrotropical Montane Grassland (e.g., Highveld)**

■ **Flooded Grassland:** Grasslands that spend most of the year as lush grasslands but turn into huge wetlands during the wet season.
Example: **Neotropical Flooded Grassland and Wetland (e.g., Pantanal)**

MEDITERRANEAN FORESTS, WOODLANDS, AND SHRUBLANDS: Thick scrub in areas with climates with cold, wet winters and dry, hot summers.

■ **Maquis, Chaparral, and Matorral:** Low shrubland that can be either closed or open. Dominated by fire and grazing. Plants are often similar to those of nearby forests.
Example: **Palearctic Maquis**

■ **Mediterranean Heathland:** Low heathlands with nearly 100% ground cover of sclerophyllous bushes and forbs. Fire-dependent.
Example: **Afrotropical Fynbos**

SCLEROPHYLLOUS FORESTS AND WOODLANDS: Forests and woodlands where the majority of the canopy trees are eucalypts and/or have small leathery leaves.

■ **Wet Sclerophyll Forest:** Tall, straight-trunked trees that form a canopy where the tree branches touch each other. Understory is lush and wet. Canopy trees are fire-tolerant, but the understory is fire-intolerant.
Example: **Australasian Wet Sclerophyll Forest**

■ **Dry Sclerophyll Forest:** Tall, straight-trunked canopy trees that form a canopy where the tree branches can touch each other but are too widely spaced to form a closed canopy. Understory is grass- and shrub-dominated, with plants not found in neighboring rainforests. Fire-tolerant.
Example: **Australasian Dry Sclerophyll Forest**

■ **Eucalypt Woodland:** Open, spaced woodland with a thin canopy, dominated by eucalypts with short, crooked trunks.
Example: **Mallee Woodland**

■ **Australasian Acacia Woodlands:** Open spaced woodlands and shrublands, often with a thick canopy, dominated by acacias with short trunks.
Example: **Mulga Woodland and Acacia Shrubland**

TUNDRAS: Very low vegetation dominated by mosses and many lichens. Found at extreme latitudes or elevations, where temperatures, snow cover, or exposure to wind prohibit the growth of trees.

Example: **Eurasian Rocky Tundra**

FRESHWATER HABITATS: Habitats whose most important aspect is their inundation with fresh water.

- **Swamp Forest:** Forested habitats that are seasonally or permanently inundated.
  Example: **Neotropical Igapó and Várzea Flooded Forest**
- **Freshwater Wetland:** Nonforested habitats whose most important aspect is that they are seasonally or permanently flooded.
  Example: **Australasian Tropical Wetland**

SALT-DOMINATED HABITATS: Habitats where the dominant force is the presence of high levels of salt in the water or soil.

- **Salt Pan:** Areas in which evaporation or volcanic activity have produced extremely high salt concentrations in the soil. Mostly unvegetated, though algae grows quickly when floods occur.
  Example: **Afrotropical Salt Pan**
- **Mangrove:** A specialized forest that grows in tidally flooded coastal areas.
  Example: **Australasian Mangrove**
- **Salt Marsh:** Salt-tolerant marsh vegetation that grows in sheltered coastal areas that are periodically flooded with seawater.
  Example: **Nearctic Salt Marsh**
- **Tidal Mudflat:** Nutrient-rich areas of mud that are frequently flooded with seawater, usually in estuaries.
  Example: **Nearctic Tidal Mudflat**
- **Rocky Coastline and Sandy Beach:** Nutrient-poor sandy and rocky beaches, cliffs, and other coastline types.
  Example: **Indo-Malayan Rocky Coastline and Sandy Beach**
- **Pelagic Waters:** Marine environments with deep water.
  Example: **Australasian Pelagic Waters**
- **Offshore Islands:** Small islands that are well offshore and that support a low growth of grass and/or shrubs.
  Example: **Afrotropical Offshore Islands**

ANTHROPOGENIC HABITATS: The primary force in shaping these habitats is the presence of humans.

- **Grazing Lands:** Areas that are heavily grazed by domestic animals.
- **Cultivated Lands:** Areas cultivated by humans for the production of crops.
- **Human Habitation:** Areas directly inhabited by humans.

GLOBAL BIOMES WITH LATITUDE AND PRECIPITATION

The graph in fig. 6 shows which habitats are most likely at any one latitude and precipitation level. What is interesting, and at first glance counterintuitive, is that habitats such as savannas and steppes cover massive areas of the planet but exist in narrow climatic bands, while other habitats such as temperate and subtropical rainforests are not very common across the planet but occur over wide ranges of precipitation. The other takeaway from the diagram is that some habitats, such as tundras, are heavily latitude- (and by inference temperature-) influenced yet can exist over a wide range of precipitation levels. Other habitats, such as Mediterranean scrub and semi-evergreen (including dry deciduous) forests, occur in a narrow precipitation range but exist over a wide range of latitudes (and again, by inference, temperatures). Across the world, precipitation of 1,200mm (48 in.) a year, or more importantly 100mm (4 in.) per month, seems to be a division between evergreen forests and those that undergo some stress due to (even short) periods of drought (sidebar 2.4). This diagram does not take into account variations due to elevation (sidebar 3.1) or continentality (sidebar 2.1). Please note that the size of the habitat shown is not indicative of the actual extent of that habitat over the planet.

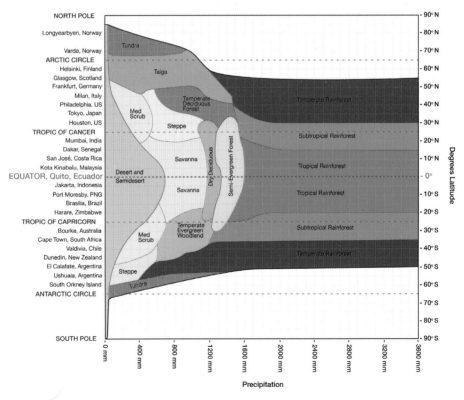

Fig. 6. Influence of latitude, precipitation, and temperature on habitats.

Transition through West African biomes from desert (A) to rainforest (C), showing habitats in between with their climate data.

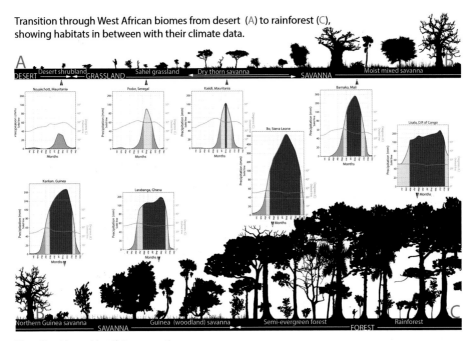

Fig. 7. How biomes blend into one another.

TAXONOMY

For birds we follow the eBird-Clements taxonomy. It is up to date and carefully maintained, and is the most popular taxonomy for American birders. For mammals, reptiles, amphibians, insects, and plants, we mainly follow Wikipedia. This is sure to shock some scientists and purists, but in writing this book, we found Wikipedia to be accurate and up to date for the groups that we know intimately well, giving us confidence that other groups are covered similarly well. In an effort to keep the text flowing, we have avoided using scientific names except to avoid confusion. Plant taxonomy is far less defined than that of birds or mammals so we have included a plant list, so every plant mentioned in the text can be cross-referenced with its scientific name.

ENDEMIC BIRD AREAS

Endemic Bird Areas (EBAs), first identified in 1987 by Birdlife International, are defined as areas that contain two or more bird species with restricted ranges of less than 19,300 sq. mi. (50,000km²). A Secondary EBA is one that contains the range of a single species. Range-restricted species are defined as those with a breeding range of less than 19,300 sq. mi., as recorded historically, i.e., since 1800. If at some point after 1800, the species had a breeding range larger than this, it is not considered a range-restricted species. The size of each EBA is flexible and is dictated by the ranges of the species contained therein.

While the identification of an EBA is a valuable tool in pinpointing areas of endemism, it doesn't show the whole picture, especially for the habitat-based approach used in this book. In the EBA designation, there is a natural bias toward island species, the ranges of which are intrinsically restricted. The same holds true for continental species restricted to montane environments. Despite the fact that a bird species is restricted to a single continental lowland habitat, it can have a large distribution that is more extensive than the measurable threshold of a range-restricted species, and the area in which it lives may not count as an EBA, even though it is a major area of endemism. Examples of such excluded habitats include the monte of the Neotropics and the Mulga of Australia. In addition, this book uses a dual bird and larger mammal lens, which obviously is different from that of the purely bird-driven EBA concept.

ABOUT THIS BOOK

Each of the habitat accounts includes the following sections, which are briefly explained here:

In a Nutshell: A succinct explanation of what makes a habitat distinctive and worthy of separation from other habitats.

Habitat Affinities: Habitats from other zoogeographic regions that are structurally similar. This is a way of cross-referencing similar habitats across the book. Perhaps one of the habitats mentioned will be familiar to you, helping you to understand the unfamiliar habitat covered.

Species Overlap: The habitats that have the most similar assemblages of larger mammals and birds. These are ranked from the most similar habitat down. The vast majority of these are habitats within the same zoogeographic region as the habitat covered.

Habitat Silhouette: These silhouettes are designed to give a quick visual snapshot of a habitat, showing some of its distinctive plant shapes and its overall height and structure. They include a human silhouette for scale. These diagrams are obviously simplifications, especially in the case of habitats with a huge range of variation, such as Africa's dry thorn savanna.

Range Map: These are visual representations of a habitat's occurrence within a given zoogeographical region. Dark shading is used for areas where the habitat is the predominant habitat, or one of the predominant habitats. In some maps, pale shading is used to indicate areas where the habitat is found only locally.

Description: This section explains what makes a habitat distinctive and how it works. Some of the commonly included information is the height and composition of the various layers of vegetation, the overall "feel" and accessibility, local temperature and rainfall, and a brief discussion of some of the conservation challenges. The accompanying climate graphs are discussed in "Climate Descriptions and Graphs," above. In these descriptions, we have purposefully chosen not to always include exactly the same information, or to put it in the same order. This is both to allow us to stress what is most important about a given habitat and simply to vary these sections to keep them interesting for readers.

Wildlife: This section may be the most interesting for a typical reader. Beyond the nuts and bolts of what makes a habitat distinctive, and what makes it work, most visitors are keen to learn about and to find its wildlife. Throughout this book, when considering wildlife, larger mammals and birds are our primary focus. Species that are restricted to a certain habitat (endemics) are given special weight, as finding these will be the priority for many visitors. A species that is an indicator species for that habitat has "(IS)" beside its name.

Endemism: This section talks about endemic hotspots within a given habitat, including Endemic Bird Areas, or "EBAs" (see above). Some habitats, especially montane ones or those occurring on islands, host many distinct nodes of endemism and are given their own Endemism section. For more general habitats, if there are no major zones of endemism, the details are described in the main description section or are incorporated into the wildlife text.

Distribution: This section, and the accompanying range map, indicate where the habitat occurs within a given zoogeographic region. The elevations at which it is found are sometimes mentioned, though this information may also be in the Description.

Where to See: These are places that you can visit to experience a given habitat. In general, these are the most readily or frequently visited places, in the most accessible country or countries.

Photos: Photos are included that illustrate both the habitat itself and some of its charismatic wildlife.

Sidebars: Throughout the book there are boxes or sidebars that discuss aspects of a habitat, biome, or region—in some cases these discussions are slightly more tangential, in others more in-depth. Many of these are about geology, ecology, and climate. We have chosen to place this sort of information in side boxes to make it more accessible and relevant (rather than in long and dry introductory sections that are likely to be ignored by most readers!).

ABBREVIATIONS

Directions (north, south, east, west, central) are abbreviated only when they directly precede a geographical place name.

aka	also known as	km	kilometer	s.	south/southern
c.	central	km²	square kilometer	sc.	south-central
cm	centimeter	lb.	pound	se.	southeastern
e.	east/eastern	m	meter	sp.	species (singular)
EBA	Endemic Bird Area	Ma	million years ago	spp.	species (plural)
ec.	east-central	mi.	mile	sq. mi.	square mile
ft.	foot/feet	mm	millimeter	sw.	southwestern
ha	hectare	n.	north/northern	w.	west/western
in.	inch/inches	nc.	north-central	wc.	west-central
IS	indicator species	ne.	northeastern	YBP	years before present
kg	kilogram	nw.	northwestern		

HABITATS OF AUSTRALASIA
(AUSTRALIA, NEW ZEALAND, AND NEW GUINEA)

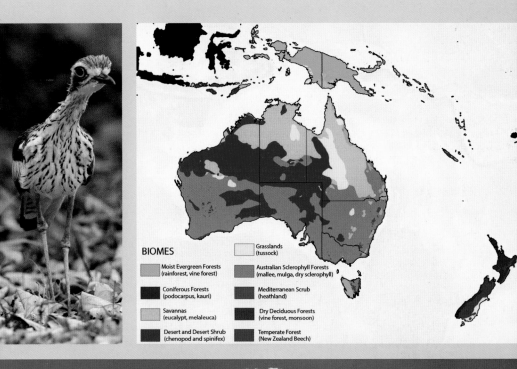

BIOMES

Grasslands
(tussock)

Moist Evergreen Forests
(rainforest, vine forest)

Australian Sclerophyll Forests
(mallee, mulga, dry sclerophyll)

Coniferous Forests
(podocarpus, kauri)

Mediterranean Scrub
(heathland)

Savannas
(eucalypt, melaleuca)

Dry Deciduous Forests
(vine forest, monsoon)

Desert and Desert Shrub
(chenopod and spinifex)

Temperate Forest
(New Zealand Beech)

GONDWANAN CONIFER RAINFOREST

IN A NUTSHELL: Moist, mixed conifer and broadleaf-evergreen rainforests that once covered much of New Zealand; these have more conifers than Australian or Asian rainforests. **Habitat Affinities:** VALDIVIAN RAINFOREST AND NEOTROPICAL MIXED CONIFER FOREST. **Species Overlap:** NEW ZEALAND BEECH FOREST.

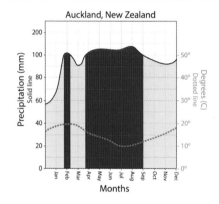

DESCRIPTION: Gondwanan conifer rainforest encompasses a variety of mixed conifer and broadleaf-evergreen rainforests in New Zealand, which are distinguished mostly by the relative composition and dominance of podocarp and kauri conifers (mainly Black and New Zealand Kauri). These distinctions occur along a continuum, with little effect on widespread wildlife distribution, and are best treated as microhabitats. The *Podocarpus*-dominated conifer-broadleaf forest in temperate zones of New Zealand is the country's tallest and most complex forest and is the New Zealand equivalent of the TEMPERATE RAINFOREST of se. Australia. These forests are a mixture of podocarp conifers such as Rimu and Kahikatea, both of which reach over 150 ft. (45m), and smaller broadleaf evergreens, which together form a dense canopy. The understory is a mix of shrubs, ferns, and tree ferns. At higher elevations, before these forests blend into NEW ZEALAND BEECH FOREST, the trees tend to be shorter and the habitat floristically poorer than in the lowlands. The kauri-dominated coniferous and broadleaf evergreen forests, which are warm temperate rainforests, are found farther north and are the New Zealand equivalent of the SUBTROPICAL RAINFOREST of se. Australia. In these forests, broadleaf trees and Nikau Palms form a canopy at around 100 ft. (30m), with emergent kauri trees towering to 150 ft. (45m) tall. When kauri conifers are small, they have a "Christmas tree" shape, but as mature trees, they have very straight trunks with no branches until near the very top, where the branches form a very compact crown, giving the tree a lollipop appearance. The understory of this forest includes tree ferns, shrubs, and orchids.

WILDLIFE: Since before the start of the Pleistocene (at least since around 2.5 million years ago), and before the arrival of humans, the wildlife of New Zealand evolved without mammal predators or herbivores; the only native terrestrial mammals were three species of bats. The introduction of

Tāne Mahuta, the world's largest known kauri tree, Waipoua Forest, North Island, New Zealand. © SCOTT WATSON, TROPICAL BIRDING TOURS

Above: **The Tui lives in New Zealand in mixed forests often comprising both beech and coniferous species such as *Podocarpus*.** © LISLE GWYNN, TROPICAL BIRDING TOURS

Left: **The Kea inhabits wooded valleys in the mountains of South Island, New Zealand.** © KEITH BARNES, TROPICAL BIRDING TOURS

herbivores such as sheep and rabbits has altered habitats, but it was predation from introduced rats and cats from Europe that initially had the most devastating toll on birdlife. A later introduction from Australia, the omnivorous Common Brushtail Possum, has caused massive damage by overbrowsing native vegetation and causing death to some canopy trees, by competition for other food sources, and by predation of native bird nests. The possum population has not yet stabilized, and overall final effects on the ecology of the forests have not been determined.

Since the native birdlife of Gondwanan conifer rainforest has been destroyed by these introduced mammals, most birds seen in these forests are species introduced from Australia, such as Australian Magpie; or Europe, such as Eurasian Blackbird and Song Thrush. Native birds that do cling on here include the Kea, Tomtit, New Zealand Bellbird, Tui, and Whitehead. Nonpasserine birds include the three species of brown kiwis, Yellow-crowned Parakeet, and Kakapo, a critically endangered flightless parrot.

DISTRIBUTION: Historically, these forests covered most of the North Island of New Zealand and most of the western half of the South Island. The Maori people deforested around half of the native forest prior to European arrival, and now less than 15% remains.

WHERE TO SEE: Waipoua Forest, North Island, New Zealand.

DUNE AND ROCKY SPINIFEX

IN A NUTSHELL: A spiky, low grassland that can be near impenetrable. **Habitat Affinities:** There are no grasslands like spinifex. **Species Overlap:** Little overlap with AUSTRALASIAN TUSSOCK GRASSLAND.

DESCRIPTION: The word "spinifex" usually refers to a genus of grasses (*Spinifex*) found in coastal areas around the world, but in Australia "spinifex" is the name of a desert habitat that includes hummock grasses of the genera *Triodia* and, to a lesser extent, *Plectrachne*. The spinifex landscape described here is the quintessential habitat of the arid inland and, in various forms, covers large tracts of desert Australia. In c. Australia are large areas of longitudinal sand dunes, distinct from the crescent-shaped sand dunes of Africa and Asia. In this geomorphological environment and on the vast sandy colluvial plains, spinifex forms the main sand-binding habitat. Spinifex is also found in rocky terrains, and although the grass systems are similar, dune and rocky spinifex areas have different suites of animals. In both types, the

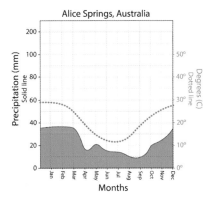

Alice Springs, Australia

Spinifex habitat at the base of Uluru (formerly called Ayers Rock), Northern Territory, Australia. © IAIN CAMPBELL, TROPICAL BIRDING TOURS

Spinifex Pigeon, a handsome resident of rocky spinifex areas, Northern Territory, Australia.
© BEN KNOOT, TROPICAL BIRDING TOURS

grasses are usually widely spaced and have a curious and unique growth form: large, low hummocks that are usually not very high but may be up to 5 ft. (1.5m) wide, with each stem growing from the same node as the root, so that the stems have independent sources for nutrients. The roots are stiff, and the stems of many grasses of the spinifex habitats are infused with silica at the tips, which makes the plants very rigid.

WILDLIFE: The grasses of spinifex habitats are spiny and nutrient-poor and so are unattractive to grazing mammals such as the kangaroos so common through most of Australia. The few mammals that once lived in this habitat, such as the Greater Bilby or Rufous Hare-Wallaby, have been decimated, not only by a change in the fire regime of the region but also the introduction of the food competitor European Rabbit and the predatory feral cat. Without large herbivores and their predators, the main herbivores are termites, and the food web builds from there. Although of limited use as a food source, spinifex cover forms a very important refuge for sheltering birds, lizards, and mammals.

Birds typical of rocky spinifex areas include most of the grasswrens, such as Pilbara Grasswren (IS) and Rusty Grasswren (IS). Other birds using the spinifex for cover include the Rufous-crowned Emuwren (IS), Painted Firetail, Spinifex Pigeon (IS), and the newly rediscovered Night Parrot. Overhead, this is the favorite habitat of the rarest of all Australian raptors, Gray Falcon. The sandy spinifex lands support a variety of rare birds that are limited to it, including Striated Grasswren (IS) and Eyrean Grasswren.

The odd Thorny Devil, one of Australia's most recognizable reptiles, occupies spinifex areas.
© CHRIS WATSON

The populations of many lizards fluctuate greatly in a cyclical manner, and even species cohabiting the same habitat vary in their microhabitat preferences, which are regulated by fire regimes and rain cycles. This means that species described in field guides as uncommon may be abundant in some years, and vice versa. Some of the lizards you can hope to see here are the Thorny Devil; Pygmy Desert and Short-tailed Monitors; Military, Rusty, and Blue-lined Dragons; and Mesa, Bearded, and Pale Knob-tailed Geckos. Snakes present in this habitat include the Desert Death Adder, Narrow-banded Shovel-nosed Snake, Moon Snake, and Desert Banded Snake.

DISTRIBUTION: Spinifex habitat is common from extreme sw. Queensland all the way across the country to the tropical Western Australia coast. Any visitor to Australia's arid interior is likely to come across spinifex, e.g., on the Birdsville and Strzelecki Tracks (South Australia), around Alice Springs and Uluru (Northern Territory), and near Mt. Isa (Queensland).

WHERE TO SEE: MacDonnell Ranges, Northern Territory, Australia.

CHENOPOD AND SAMPHIRE SHRUBLAND

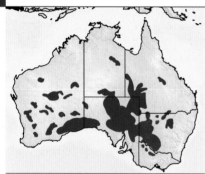

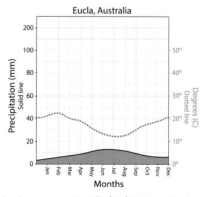

IN A NUTSHELL: Very low, sparse shrubland tolerant of salt and aridity. **Habitat Affinities:** PALEARCTIC SEMIDESERT THORNSCRUB. **Species Overlap:** DUNE AND ROCKY SPINIFEX; ACACIA SHRUBLAND.

DESCRIPTION: This desert environment is defined by the scarcity of plant life; it is an open habitat where trees struggle to take hold, and its vegetation mainly consists of low, open shrubs, which can create a canopy up to 6 ft. (2m) in height but are usually 1–3 ft. (0.3–1m) high. Very scattered emergent trees, generally Mulga, Desert Sheoak, and River Red Gum, are sometimes present. Samphires, the most common type of chenopods, including succulent shrubs such as sea asparagus, occur in more saline environments, at the margins of semipermanent and ephemeral salt lakes, and sometimes in waterlogged environments. On extensive clay pans and aeolian sand plains, *Atriplex* saltbushes dominate. One species, Oldman Saltbush, can grow to 9 ft. (3m) high and looks very similar to saltcedar (tamarisk) stands of the African Sahel.

Shrublands of gibber peneplains, aka "reg" (sidebar 1.1), are the extreme of this environment; the vegetation is lower and more open than in other chenopod and samphire shrublands, and the habitat appears barren, except for very sparse perennial plant growth and, after infrequent rains,

Above: **Samphire shrublands in New South Wales, Australia, home to Plains-wanderer, White-winged Fairywren, and Red Kangaroo.** © SAM WOODS, TROPICAL BIRDING TOURS

Right: **Plains-wanderer amid blooming chenopod and samphire shrublands, after a season of heavy rains, in New South Wales, Australia.** © SAM WOODS, TROPICAL BIRDING TOURS

grasses. The small pebbles and cobbles (gibbers) that cover the ground are exposed and become varnished with silica, giving them a shiny, smooth texture, and they form an almost impenetrable layer on top of the fine red sands below. Because gibber plains are so harsh, limited human impact has been possible, but the chenopod shrublands have been seriously overgrazed and modified in the Hay Plain of New South Wales, Australia.

WILDLIFE: These shrublands appear to support few birds and mammals; lizards are the most abundant animals. Amid the samphire and chenopod shrubs, expect to find lizards like the Saltbush Slender Blue-tongue, Claypan Dragon, and Nullarbor Earless Dragon. In the gibber peneplains, there are even more specialized lizards like the Gibber Gecko, Gibber Dragon, Smooth-snouted Earless Dragon, and the bizarre looking Pebble-mimic Dragon.

Those birds that occur are special species highly restricted to gibber shrubland habitats, including Gibber Chat (IS), Thick-billed Grasswren (IS), and Inland Dotterel (IS). Very sparsely distributed and highly nomadic, they are very hard to find, even within this open habitat. Chenopod and samphire specialists include Chestnut-breasted Whiteface, Gibber Chat, Orange Chat, Plains-wanderer (IS), Naretha Bluebonnet, Rufous Fieldwren, Slender-billed Thornbill, White-fronted Chat, and Nullarbor Quail-thrush.

Most of the mammals that exist here are small marsupial mice or bats. The larger mammals are the Euro Wallaroo, Red Kangaroo, and Dingo. Unfortunately, feral cats have thrived in this habitat and seriously depleted the native mammal and reptile population.

DISTRIBUTION: Chenopod and samphire habitats occur across arid s. Australia on the Nullarbor Plain, the lower Lake Eyre basin, and the Hay Plain. Gibber shrublands occur along the Strzelecki and Birdsville Tracks (South Australia), around Coober Pedy (South Australia), and in far w. New South Wales.

WHERE TO SEE: North and south of Booroorban, New South Wales, Australia; Sturt National Park, New South Wales, Australia; n. South Australia.

An Inland Dotterel spotlighted at night in chenopod and samphire shrublands in New South Wales, Australia, after a season of low rainfall. © SAM WOODS, TROPICAL BIRDING TOURS

SIDEBAR 1.1   REG: IT MAY APPEAR DESOLATE, BUT LOOK CLOSER

Reg, called "gibber" in Australia and "serir" in the e. Sahara desert of n. Africa, is regarded as the most desolate desert environment of the tropics and mid-latitudes. The rocks on the surface are sandblasted by harsh winds and develop a shiny coating, earning the descriptive term "desert varnish." When vegetation is cleared, burned, or overgrazed, or dunes move, the smallest particles (or fine fraction) of clays, silts, and sands are blown away, and the larger pebbles are winnowed to a point that they form interlocking pavement. Over many years, the silica and iron in what little water moves through the area concentrates on the surface of the stones as a siliceous glass with iron oxides. Although for the vast majority of the time this environment looks completely lifeless, there are nutrients under this pavement containing a seed bank. When the rains do come, the reg comes alive with ephemeral flowers and grasses, which die off in a few short weeks. When humans directly irrigate very small holes in the substrate, these soils can be quite productive, as the varnished pavement reduces evaporation from the soil around the plant.

NEW ZEALAND BEECH FOREST

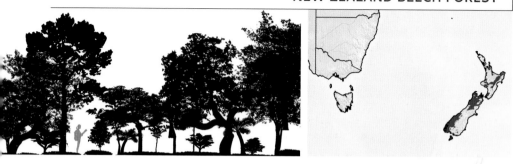

IN A NUTSHELL: Remnants of ancient Gondwanan cold, wet forests. **Habitat Affinities:** NEOTROPICAL MAGELLANIC RAINFOREST. **Species Overlap:** GONDWANAN CONIFER RAINFOREST.

DESCRIPTION: This habitat includes many tree species, but beeches usually dominate the canopy. New Zealand beech forests have abundant gnarled branches and a moss-laden forest floor. The canopy usually reaches around 60–100 ft. (20–30m) high and is typically uninterrupted, making the forest well shaded and therefore fairly open at ground level and easy to walk through. Typically shorter than the AUSTRALASIAN TEMPERATE RAINFOREST and Northern Hemisphere deciduous forests, these forests are also more uniform than Australian temperate rainforests and lack emergent trees. The five beech species that characterize this habitat are Hard Beech, Black Beech, Red Beech, Silver Beech,

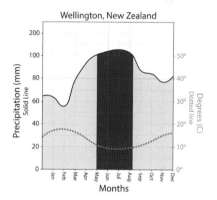

Wellington, New Zealand

Moss-covered trees on Mt. Climie, North Island, New Zealand. © PSEUDOPANAX AT EN.WIKIPEDIA, CC BY-SA 4.0, HTTPS://CREATIVECOMMONS.ORG/ LICENSES/BY-SA/4.0, VIA WIKIMEDIA COMMONS

and Mountain Beech. The New Zealand beeches (*Nothofagus*) are not related to the beeches (*Fagus*) of the Northern Hemisphere, but they appeared similar to early white settlers and botanists. Generally, the forests are dominated by one or two of the beech species—which species depends on rainfall, temperature, and underlying rock type. Beech stands are found within temperate rainforests in a few locations on mainland Australia and Tasmania, but they are never in patches sufficient enough to really support a different faunal assemblage from the temperate rainforests. Beech forests usually seed en masse (called masting) every four to six years.

WILDLIFE: Within these beech forests, native scale insects produce a liquid called honeydew, which is an important food resource for native birds and insects. Introduced German and Common Wasps have become very numerous in some areas, and exploit the honeydew, at the expense of the bird populations, while also predating on native insects in the forests, and so present a major peril to native bird populations. New Zealand Bellbird, Tui, and other native nectivorous birds feed on honeydew-secreting scale insects. Other native bird species occurring in this habitat include kiwis, Yellow-crowned Parakeet, Kaka, Tomtit, Pipipi, Yellowhead, and Rifleman.

New Zealand has a very small native mammal population, consisting of several bat species. At least two of these, New Zealand Long-tailed Bat and New Zealand Lesser Short-tailed Bat, occur in this habitat but are not exclusive to it. All non-bat, nonmarine mammals are introduced. The most obvious of the introduced mammal species in this habitat is the Common Brushtail Possum, which seems ubiquitous and is a significant predator of native bird populations. Other pest mammals are Fallow Deer, Short-tailed Weasel, and Brown Rat. When the mass seeding occurs, it produces a huge amount of food and a resulting population explosion of rats and mice. This in turn dramatically affects the population of native animals when the seed supply is exhausted and they become the food source for plague numbers of introduced predators.

Endemism: Yellowhead, an endangered passerine, appears to be the only faunal species completely confined to this habitat; Red Beech forests are now its primary habitat (it occurred in other habitats historically). New Zealand beech forest is the strongly preferred habitat of Yellow-crowned Parakeet, although this species not exclusive to *Nothofagus* forests.

South Island and North Island Saddlebacks have been successfully reintroduced into native beech forests, after introduced predators decimated their numbers.
© LISLE GWYNN, TROPICAL BIRDING TOURS

DISTRIBUTION: Approximately one-quarter of all native forest cover remains in New Zealand, and two-thirds of this is beech forest, covering some 10 million acres (4 million ha). It is the dominant forest type on South Island (where it is particularly plentiful in the west) and on North Island is found mostly around the mountains and hills.

WHERE TO SEE: Fiordland National Park, Victoria Forest Park, and Wakatipu, South Island, New Zealand; Remutaka Forest Park, North Island, New Zealand.

AUSTRALASIAN TROPICAL LOWLAND RAINFOREST

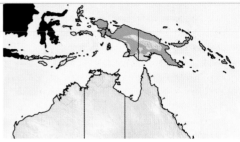

IN A NUTSHELL: Tall, very humid forests found at low elevations. **Habitat Affinities:** NEOTROPICAL LOWLAND RAINFOREST; INDO-MALAYAN TROPICAL LOWLAND RAINFOREST; AFROTROPICAL LOWLAND RAINFOREST. **Species Overlap:** AUSTRALASIAN SUBTROPICAL AND MONTANE RAINFOREST.

DESCRIPTION: Tropical lowland rainforest is the dominant vegetation in low-lying, wet (but not submerged) tropical environments of New Guinea, surrounding islands, and ne. Australia. It is characterized by a mega-diverse plant community, with a large variety of evergreen trees that form a thick canopy and block out the sun, resulting in a relatively open understory. The forest canopy here is high, around 130 ft. (40m). In light gaps created by tree falls or landslides, there are massive bursts of seedlings of larger trees and smaller succession trees. These plants fight for the new available light and form a different microenvironment that has a very thick undergrowth, until one or two trees win the battle and smother out the others. Australasian lowland rainforest differs from similar habitats in Africa (AFROTROPICAL LOWLAND RAINFOREST) and Asia (INDO-MALAYAN TROPICAL LOWLAND RAINFOREST)

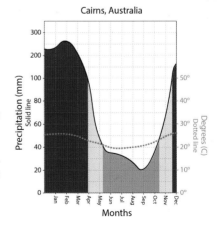

Cairns, Australia

in having figs and sclerophyllous plants (often eucalypts) present within the canopy. It further varies from Asian rainforest in that dipterocarps are not the dominant family. The lowland rainforests of New Guinea and ne. Australia are remnants of forests that were once more widespread.

WILDLIFE: Although this tropical lowland rainforest appears similar to INDO-MALAYAN TROPICAL LOWLAND RAINFOREST, the bird assemblages are very different. Unlike the se. Asian forests, Australasian lowland rainforest has only one hornbill species, Blyth's Hornbill; instead, the ecological niche is filled by many large parrot and pigeon species. The paradise-kingfishers fill the niche taken by forest-dwelling kingfishers in Asia. Other groups of birds typical of this Australasian forest type are birds-of-paradise, bowerbirds, fig-parrots, fruit-doves, and ground birds like the jewel-babblers that fill the niche of the Malaysian Rail-babbler in Malaysia or the *Picathartes* rockfowl in Africa. This rainforest is much more extensive in New Guinea, where the many mountain ranges separate forest blocks and bird species, and the New Guinean rainforest thus supports many more bird species than its Australian counterpart. Nonetheless, the lowland forests of n. Australia and s. New Guinea share many species, such as Southern Cassowary, Buff-breasted Paradise-Kingfisher (IS), Papuan Pitta (IS), Eclectus Parrot (IS), Palm Cockatoo, White-faced Robin, and Magnificent Riflebird.

Australasia does not have native primates, ungulates, or bovines, so the rainforest canopy here is populated by marsupials such as possums, Common Spotted Cuscus, and tree-kangaroos, as well as a large number of smaller bats and fruit bats. The forest floor is dominated by pademelons, bandicoots, and native rats. A night walk through lowland rainforest of ne. Australia might reveal Musky Rat-Kangaroo, Red-legged Pademelon, and Northern Brown Bandicoot, all scurrying through the undergrowth, and the gorgeous black-and-white Striped Possum and enigmatic Bennett's Tree-Kangaroo in the canopy. Spectacled and Little Red Flying Foxes and Northern Blossom Bat soar around the canopy. New Guinean Quoll, though smaller than a domestic cat, is the largest modern

Lowland rainforest in the wet tropics of ne. Queensland, Australia, home to Southern Cassowary and Victoria's Riflebird. © IAIN CAMPBELL, TROPICAL BIRDING TOURS

mammal predator of these forests in New Guinea. Tree-kangaroos, which have to be seen to be believed, with their thick tails for balance and strong forearms for grip, reach their highest diversity in New Guinea, where they are the largest of the mammals found in the lowland rainforests.

Herps are well represented in lowland rainforests, where *Litoria* species such as White-lipped, Orange-thighed, and Green-eyed Tree Frogs all occur in the canopy. Terrestrial frogs include Northern Barred Frog, Australian Wood Frog, and Fry's Frog. Snakes typical of this environment include the Green Tree and Scrub Pythons. Some geckos are much more common in rainforest environments than in surrounding eucalypt forests, including Northern Velvet Gecko and McIlwraith and Northern Leaf-tailed Geckos. Unfortunately, the introduced Cane Toad has had a massive impact on this environment in Australia and has decimated populations of the less competitive terrestrial frogs.

Endemism: Because the islands north of Australia are so mountainous, they harbor many distinct areas of endemism; the main areas are the north and south sides of the main mountain range in New Guinea and the island of Halmahera (Moluccas). The isolated rainforests of the ne. Australian wet tropics are also areas of endemism.

DISTRIBUTION: Tropical rainforest is the dominant lowland habitat of New Guinea, where large swaths of it still exist, and surrounding islands of Halmahera, New Britain, New Ireland, and Bougainvillea, and most of the Solomon Islands in Oceania. Few large areas are protected by national parks, but the human population is still small in much of this region. Clearing by logging companies and for African Oil Palm plantations poses the greatest threat to these habitats north of Australia.

In Australia, the vast majority of this habitat was cleared many decades ago, and the remaining tracts are well protected by the national parks system. In the Iron Range on the Cape York Peninsula (Queensland), large tracts of lowland rainforest remain untouched and feel truly wild. Here you can look for birds whose Australian range is limited to lowland tropical rainforest, such as Magnificent Riflebird, White-faced Robin, Northern Scrub-Robin, and Green-backed Honeyeater.

WHERE TO SEE: Daintree National Park, Queensland, Australia; Kiunga, Papua New Guinea; Weda, Halmahera, Indonesia; Nimbokrang, West Papua, Indonesia.

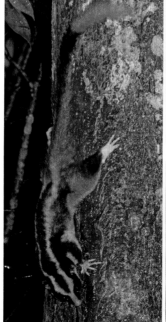
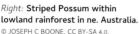

Right: **Striped Possum within lowland rainforest in ne. Australia.** © JOSEPH C BOONE, CC BY-SA 4.0, HTTPS://CREATIVECOMMONS.ORG/ LICENSES/BY-SA/4.0, VIA WIKIMEDIA COMMONS

Far right: **Australia's heaviest bird, Southern Cassowary lives within lowland rainforest in Australia and New Guinea.** © SAM WOODS, TROPICAL BIRDING TOURS

AUSTRALASIAN SUBTROPICAL AND MONTANE RAINFOREST

IN A NUTSHELL: Closed-canopy humid forest with many epiphytes and extensive understory. **Habitat Affinities:** NEOTROPICAL CLOUD FOREST; AFROTROPICAL MONTANE FOREST.

Species Overlap:
AUSTRALASIAN
TROPICAL LOWLAND
RAINFOREST;
WET SCLEROPHYLL
FOREST.

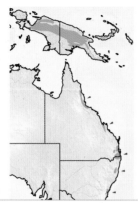

DESCRIPTION: Subtropical forests of Australasia are very similar to the cloud forests of Africa (AFROTROPICAL MONTANE FOREST) and South America (NEOTROPICAL CLOUD FOREST). They grow in the cooler and more seasonal southern parts of Australia and farther north at elevations higher than AUSTRALASIAN TROPICAL LOWLAND RAINFOREST. These forests have slightly lower tree diversity, more epiphytes, and a much more extensive understory than lowland rainforests. The canopies of the subtropical forests of n. New South Wales, Australia, are dominated by Strangler Fig, Red Cedar, Blackbean, and Hoop Pine. The understory has plants like Stinging Tree (to be avoided, as the name implies) and Bangalow Palm.

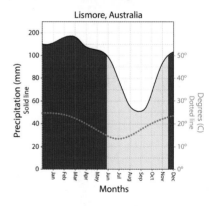

Lismore, Australia

In the north of Australia and New Guinea, lowland tropical rainforests merge into these montane forests at around 3,300 ft. (1,000m). These forests have low canopies, at 45–75 ft. (15–25m), and the undergrowth is very thick. At lower elevations, canopy species such as New Guinea Oak are dominant, and Hoop Pine and Klinki Pine are the dominant conifers. Higher up, the forest changes, and beech, podocarp conifers, and myrtles become dominant.

WILDLIFE: The mountain rainforests of New Guinea are extensive and are different enough from the surrounding AUSTRALASIAN TROPICAL LOWLAND RAINFOREST that a vast array of endemic birds and bird families specific to this habitat has developed. In New Guinea, this is the zone where birds-of-paradise reach their highest abundance, with species such as sicklebills and astrapias. Other types of birds typical of the New Guinean montane forests include Wattled Ploughbill (IS),

Crested Satinbird in montane rainforest near Mt. Hagen, Papua New Guinea. © SAM WOODS, TROPICAL BIRDING TOURS

Subtropical forest canopy in Lamington National Park, near Brisbane, Queensland, Australia. © IAIN CAMPBELL, TROPICAL BIRDING TOURS

Regent Bowerbird in montane forest at O'Reilly's resort, within Lamington National Park, near Brisbane, Queensland, Australia. © IAIN CAMPBELL, TROPICAL BIRDING TOURS

tiger-parrots, Snow Mountain Robin complex (IS), berrypeckers, Chestnut Forest-Rail, Papuan Whipbird, Lorentz's Whistler, Black Sittella, Belford's Melidectes (IS), Smoky Honeyeater (IS) and Friendly Fantail. Among several bird species limited to the montane rainforests of n. Queensland, Australia, are Tooth-billed and Golden Bowerbirds, Mountain Thornbill, Atherton Scrubwren, Fernwren, Bridled Honeyeater, Gray-headed Robin (IS), Bower's Shrikethrush, and Chowchilla. Birds found in subtropical rainforest at more southerly latitudes include Marbled Frogmouth, Regent Bowerbird (IS), Australian Logrunner (IS), Green Catbird, and Paradise Riflebird (IS).

There are many mammals in this habitat, though they are devilishly hard to see. Those on New Guinea and surrounding islands include Mountain Cuscus, Calaby's Pademelon, and Doria's Tree-Kangaroo. A visit to the subtropical rainforests of Australia's Lamington National Park (Queensland) could yield Red-necked and Red-legged Pademelons, Sugar Glider, Short-eared Brushtail Possum, and Common Ringtail Possum.

Endemism: The highlands of the islands north of Australia are so riddled with endemic bird areas and centers of plant diversity that almost any mountainous area is part of one. The major endemic areas of New Guinea are the mountains of the Vogelkop (or Bird's Head) Peninsula; the island's main range (with a significant difference between its east and west sides); the Adelbert Mountains; and

Lumholtz's Tree-Kangaroo in montane forest within the Atherton Tableland of ne. Australia. © SAM WOODS, TROPICAL BIRDING TOURS

the mountains of the Huon Peninsula. In Australia the endemic areas are the Border Ranges near Brisbane, the Atherton Tableland near Cairns, and the Lockhart River in Cape York Peninsula, all in Queensland. On New Caledonia, the suite of endemics includes New Caledonian Imperial-Pigeon, New Caledonian Parakeet, and New Caledonian Whistler.

DISTRIBUTION: On the islands of New Guinea, New Ireland, New Britain, and the Moluccas, montane rainforest habitat is widespread above 3,300 ft. (1,000m), reaching up to 13,000 ft. (4,000m) on higher passes. In many areas it is hardly touched, due to the incredibly rugged topography. Subtropical and montane rainforests are very limited in Australia, found in two main areas: In ne. Queensland, montane rainforests occur in the cooler upland areas, generally above 2,000 ft. (600m). Farther south, subtropical rainforests are generally found at lower elevations, from the coast to the lower reaches of the mountains. This habitat also occurs in New Caledonia, mainly on the east coast and the southern part of the island.

WHERE TO SEE: Mt. Hypipamee National Park, Queensland, Australia; Mt. Lewis, Queensland, Australia; Tari Valley, Papua New Guinea; Kumul Lodge, Papua New Guinea.

The Eastern whipbird is a common and highly vocal bird of Eastern Australia. © IAIN CAMPBELL, TROPICAL BIRDING TOURS

AUSTRALASIAN TEMPERATE RAINFOREST

IN A NUTSHELL: Cold wet forests of se. Australia that have a closed coniferous or broadleaf canopy.
Habitat Affinities: AFROTROPICAL MONTANE FOREST; NEOTROPICAL ELFIN FOREST; NEW ZEALAND BEECH FOREST.
Species Overlap: AUSTRALASIAN SUBTROPICAL RAINFOREST.

DESCRIPTION: Temperate rainforests in Australia are characterized by lower plant diversity than AUSTRALASIAN TROPICAL LOWLAND RAINFOREST or SUBTROPICAL RAINFOREST and a simpler structure, with fewer large vines and more ferns and hanging mosses. At the northern limit of this forest in s. Queensland, there are Antarctic Beech trees at the tops of ranges surrounded by subtropical rainforest. In Tasmania, the canopy is made up of Myrtle Beech, Southern Sassafras, and other trees. The structure can range from a tall, dense forest with a canopy at 130 ft. (40m) and little light reaching the understory, to a shorter forest with a canopy at 115 ft. (35m), many gaps in the canopy, and a thick understory of Mother Shield Fern and Soft Tree Fern.

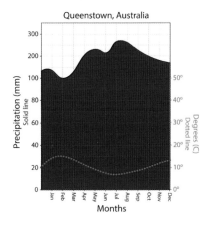

Queenstown, Australia

WILDLIFE: In Australia, the only bird endemic to this habitat is the Rufous Scrub-bird (IS), though Bassian Thrush, Satin Flycatcher, Olive Whistler, and Pink Robin are all more common in this habitat than in surrounding areas.

Mammals of this habitat in Australia include Eastern Ringtail Possum, Eastern Pygmy Possum, Eastern Quoll, and Red-necked and Rufous-bellied Pademelons (Tasmania only for the lattermost).

Endemism: The endemic Rufous Scrub-bird is found in se. Queensland and ne. New South Wales. Most of the endemic birds of Tasmania use this habitat, though none of them are endemic to it.

DISTRIBUTION: These rainforests grow down to sea level in Tasmania but above 3,300 ft. (1,000m) in se. Queensland. They are found in isolated pockets in higher elevations of Lamington National Park (Queensland) and in e. New South Wales and e. Victoria. They are most common in the western half of Tasmania, where they are the dominant forest type.

Above: **Temperate rainforests in Tasmania, Australia are home to a series of bird species endemic to the island. This includes a set of handsome robins, such as this Pink Robin** *(left).* PHOTOS © IAIN CAMPBELL, TROPICAL BIRDING TOURS

Below: **The unique Platypus occurs in creeks and rivers in the rainforests of Australia.** © IAIN CAMPBELL, TROPICAL BIRDING TOURS

WHERE TO SEE: Barrington Tops, New South Wales, Australia; Toolangi State Forest, Victoria, Australia; Mt. Field National Park, Tasmania, Australia.

AUSTRALASIAN TROPICAL SEMI-DECIDUOUS FOREST

IN A NUTSHELL: Thick, scrubby semi-deciduous forests of island areas with a prolonged dry season. **Habitat Affinities:** INDO-MALAYAN SEMI-EVERGREEN FOREST. **Species Overlap:** AUSTRALASIAN TROPICAL LOWLAND RAINFOREST; AUSTRALASIAN MONSOON VINE FOREST.

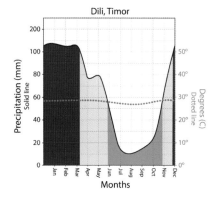

Dili, Timor

DESCRIPTION: "Australasian tropical semi-deciduous forest" is a catchall phrase for a variety of relatively dry microhabitats found on the islands to the north of Australia and arcing around to New Caledonia, east of Australia. There are slight variations in the percentage of deciduous trees, though they are always a minor component of these dry forests. Thornscrub also occurs within this suite but again is a microhabitat within the broader suite of closed woodland. Usually, these dry forests have a canopy between 15 ft. (5m) and 45 ft. (15m) high, are dense, and contain many vines. In Timor (Lesser Sunda Islands), which has the oldest and best developed of these semi-deciduous forests, they are stands of the evergreen eucalypt Timor Mountain Gum, mixed with an understory of casuarinas and smaller deciduous trees along with large-leaved mid-story broadleaf evergreen trees that seem incongruous with the casuarinas. This habitat may contain a few palms but rarely has cycads in any number, and walking around in it is easy. In many areas, these forests represent an ecotone between savanna and AUSTRALASIAN MONSOON VINE FOREST or TROPICAL LOWLAND RAINFOREST and therefore share most of their species with these surrounding habitats. For example, none of New Caledonia's five endemic plant families are represented in dry forest, suggesting that it is much younger habitat than surrounding wet forests. One alarming feature is that these forest types are dry or monsoonal but not fire tolerant. The increasing, human-caused fire regime is rapidly turning this habitat into a savanna with a grass understory in many areas.

WILDLIFE: In the Lesser Sundas, many of the same bird groups represented in the MONSOON VINE FOREST occur in semi-deciduous forests of the mountains, such as parrots, white-eyes, flowerpeckers, honeyeaters, and gerygones. Many montane species are restricted to higher-elevation forests (above 2,625 ft./800m), including Timor Cuckoo-Dove, Timor Imperial-Pigeon, Sunda Cuckoo, Flores Scops-Owl, Pygmy Cupwing, White-browed White-eye, Mountain Leaf Warbler, Yellow-breasted Warbler, Brown-capped Fantail, Little Pied Flycatcher, Bare-throated Whistler,

The magnificent Komodo Dragon. © SAM WOODS, TROPICAL BIRDING TOURS

and Mount Mutis Parrotfinch. Timor Leaf Warblers are also more abundant at higher elevations. Some of the species that are found in this same habitat on multiple islands in the Lesser Sundas are Black-backed Fruit-Dove, Mountain Leaf and Yellow-breasted Warblers, Mountain White-eye, Sunda Bush Warbler, Chestnut-backed Thrush, and Blood-breasted Flowerpecker.

Komodo Dragon, the world's largest lizard, reaching 10 ft. (3m) long and 150 lb. (70kg), inhabits tropical semi-deciduous forests on a handful of islands in Indonesia.

Endemism: Where this habitat has been isolated on islands for a long geological time period, it has given rise to high levels of avian endemism, particularly in the Lesser Sundas, where many of the islands have a distinct set of dozens of endemic or near-endemic birds of their own, including pigeons and doves, leaf warblers, fantails, honeyeaters, owls, flycatchers, and finches. For example, on Timor, the montane semi-deciduous forests are home to Timor Imperial-Pigeon, Olive-headed and Iris Lorikeets, Yellow-eared Honeyeater, Plain Gerygone, and Timor Leaf Warbler, while in this habitat on Flores, Leaf Lorikeet, Brown-capped Fantail, White-browed and Dark-crowned White-eyes, and Bare-throated Whistler are found. Montane semi-deciduous forest is also home to a vocally distinct, local Flores form of White-browed Shortwing, which many consider an endemic species, "Flores Shortwing." This bird is largely confined to this habitat.

DISTRIBUTION: These habitats encompass small areas in an arc from Lombok, Indonesia, eastward to the French territory of New Caledonia but occur as large areas on the west side of New Caledonia, east side of Timor, lowlands of the Lesser Sundas, lowlands of Sumba, and the south and southwest coast of New Guinea.

WHERE TO SEE: Western New Caledonia; Moluccas.

This tropical semi-deciduous forest on Gunung Mutis, a mountain on the Indonesian island of Timor, is home to a number of regional endemic bird species. © SAM WOODS, TROPICAL BIRDING TOURS

AUSTRALASIAN MONSOON VINE FOREST

IN A NUTSHELL: The jungle of n. Australia and the islands to the north, with well-defined wet and dry seasons and lots of semi-deciduous trees and shrubs. **Habitat Affinities:** INDO-MALAYAN MOIST DECIDUOUS FOREST. **Species Overlap:** AUSTRALASIAN TROPICAL LOWLAND RAINFOREST.

DESCRIPTION: This rainforest-like habitat is both a wet and a dry habitat, depending on the season. The difference between this forest and true rainforest does not seem to be overall rainfall but rather the duration and intensity of the dry season. Australasian monsoon vine forest is not fire-tolerant and is readily transformed to savanna when exposed to increased fire frequency and/or intensity. Much of the canopy consists of large fig trees. Many of the other trees are seasonally deciduous, losing leaves in the dry season but forming a very thick canopy and bearing fruit in the wet season. The understory is very open, with lots of vines. Palms are common and form a distinct sub-canopy. The ground cover is thick during the wet season and sparse during the dry season, with dry leaves on the ground creating a leaf-litter layer around 3 in. (7.5cm) deep.

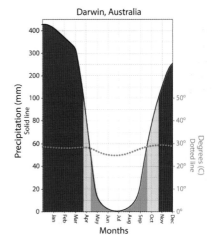

Darwin, Australia

WILDLIFE: In the Lesser Sundas, including Timor, the bird assemblage includes Flores Green-Pigeon, Timor Green-Pigeon, Timor Bushchat, Cinnamon-banded Kingfisher, Fawn-breasted Whistler, and Timor White-eye. As monsoon vine forest is widespread across the Lesser Sundas, and these islands are rich in endemic species, many island endemics occur within this habitat. However, most species appear not to be dependent on monsoon forest and (like many island species) are able to exist in a variety of local habitat types. Only a few species occur widely across multiple islands in this habitat in the Lesser Sundas, such as Indonesian Honeyeater, Cinnamon-banded Kingfisher, Red-cheeked Parrot, Elegant Pitta, and Broad-billed Flycatcher; most of the other species on these islands are endemic to the one island. On Flores, some other monsoon species include Flores White-eye and Black-fronted Flowerpecker. On mainland Australia and New Guinea, this habitat abuts extremely different savanna habitat, so the changes in bird assemblages can be immediate. In n. Australia, these forests are characterized by Black-banded Fruit-Dove (IS), Rainbow Pitta (IS), and Green-

Monsoon forest showing the abundance of vines, in Cape York Peninsula, Queensland, n. Australia. © IAIN CAMPBELL, TROPICAL BIRDING TOURS

backed Gerygone (IS). Large-tailed Nightjar also occurs here, though, interestingly, in the rest of its range it is associated more with savanna terrains.

There are few mammals on the islands of the Lesser Sundas in general, and no conspicuous ones in monsoon forests, birds being the most visible (nonhuman) natural component of this habitat. In Australia, typical reptiles include Rough-scaled Python and Ring-tailed, Mourning, and Northern Velvet Geckos. Most mammals are much more catholic in habitat choice, occurring both in this forest and nearby rainforest, with examples including Common Brushtail Possum, Rock Ringtail Possum, and Black-footed Tree-rat.

The gorgeous Rainbow Pitta, photographed in monsoon forest near Darwin, Northern Territory, Australia. © IAIN CAMPBELL, TROPICAL BIRDING TOURS

Endemism: In Australia the monsoon vine forest of the Northern Territory supports many strict endemics, which are absent from adjacent habitats, such as Rainbow Pitta, Arafura Fantail, and Black-banded Fruit-Dove. The various islands that are dominated by this habitat are rich in localized endemics, which have probably evolved more because of the isolated nature of the islands than because of habitat preferences. For example, on the island of Timor, the birds occurring in this habitat include Slaty Cuckoo-Dove, Timor Cuckoo-Dove, Timor Green-Pigeon, Olive-shouldered Parrot, Timor Blue Flycatcher, Timor White-eye, Buff-banded Bushbird, Black-banded Flycatcher, Timor Stubtail, Timor Bushchat, Timor Friarbird, Plain Gerygone, Orange-banded Thrush, and Tricolored Parrotfinch, which

Above: **Yellow-breasted Boatbill occurs in monsoon vine forest and tropical lowland forest in Australia and New Guinea.** © IAIN CAMPBELL, TROPICAL BIRDING TOURS

Right: **Black Flying Fox in riparian monsoon vine forest in Northern Territory, Australia.** © SAM WOODS, TROPICAL BIRDING TOURS

are all island endemics or near endemics shared with some smaller islands. On Komodo, a relict population of Yellow-crested Cockatoo occurs in this forest habitat; this species was formerly more widespread in the islands prior to extensive hunting.

DISTRIBUTION: Throughout its range, Australasian monsoon vine forest is found mainly in close proximity to the coast, often on sandy soils. It is the dominant habitat in much of the Lesser Sundas, the Moluccas, and on small Pacific islands eastward to New Caledonia. In the *MELALEUCA* SAVANNA zone of far s. New Guinea, this habitat is found in moist areas where the trees are better protected from fire. Across n. Australia, monsoon vine forest may be found along dry creek beds, sheltered gullies, and along edges of sandstone escarpments. There are scattered small patches in e. Australia, including around the Iron Range on Cape York Peninsula and Inskip Point in Queensland.

WHERE TO SEE: East Point, Fogg Dam, and Howard Springs, Northern Territory, Australia; Manupeu Tanah Daru National Park, Sumba, Indonesia; Oelnasi, Timor, Indonesia.

AUSTRALASIAN SANDSTONE ESCARPMENTS

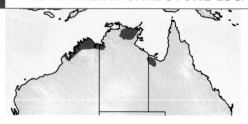

IN A NUTSHELL: A blend of habitats where escarpments holding SPINIFEX grasslands, other grasses, and sparse trees meet MONSOON VINE FOREST or OPEN EUCALYPT SAVANNA below them. **Habitat Affinities:** Gullies in NEOTROPICAL CAATINGA. **Species Overlap:** OPEN EUCALYPT SAVANNA.

Left: **Wilkins' Rock-Wallaby alongside Aboriginal rock art at Ubirr, in Kakadu National Park, Northern Territory, Australia.** © SAM WOODS, TROPICAL BIRDING TOURS

Below: **Sandstone escarpments are home to Chestnut-quilled Rock-Pigeon and Sandstone Shrikethrush, while the monsoon forest at their base is critical habitat for Black-banded Fruit-Dove.** © NICK ATHANAS, TROPICAL BIRDING TOURS

Black-banded Fruit-Dove in monsoon forest at
the base of a sandstone escarpment in Kakadu
National Park, Northern Territory, Australia.
© NICK ATHANAS, TROPICAL BIRDING TOURS

DESCRIPTION: Rather than a specific habitat, this is a mélange of microhabitats in a very small area with a small suite of birds that prefers them. Quartzites, quartz sandstones, and a few arkosic sandstones form massif plateaus with dramatic escarpments in n. Australia. Because these rock types are so resistant to chemical weathering, they form a mesa, and mechanical weathering and erosion take place on the plateau edges, leading to spectacular cliffs (scarps) that appear to rise out of the surrounding savanna. The rocky slopes at the bottom of the escarpments usually support AUSTRALASIAN MONSOON VINE FOREST, where water is concentrated, or ROCKY SPINIFEX in drier areas. The top of the escarpments may support the ground cover of rocky spinifex, with occasional, very stunted eucalypts. In the plateau-top swales, where sandy and sometimes even shallow clay soils have had a chance to develop, a scattered OPEN EUCALYPT or *MELALEUCA* SAVANNA may develop.

WILDLIFE: The ROCKY SPINIFEX and eucalypt assemblage at the tops of escarpments is home to some of Australia's most range-restricted bird species, such as Black Grasswren (IS) and White-quilled Rock-Pigeon (IS) of the Kimberley region of Western Australia, and White-throated Grasswren (IS) and Chestnut-quilled Rock-Pigeon (IS) of Arnhem Land in Northern Territory. White-lined Honeyeater and Sandstone Shrikethrush (IS) are typical of the escarpment edge. The monsoon forest habitat at the bottom of Nourlangie Rock, Northern Territory, is the base for Australia's most range-restricted pigeon, Black-banded Fruit-Dove.

This habitat is also a great place to look for some very range-restricted mammals, such as Rock Ringtail Possum, Black Wallaroo, and Short-eared Rock-Wallaby. Reptiles, even more specialized in this area than birds, include Giant Cave Gecko, Marbled Velvet Gecko, Northern Banded Knob-tailed Gecko, Arnhem Land Spotted Dtella, Olive Python, and the very range-restricted Oenpelli Python. Frog-watchers can find Rockhole and Masked Frogs here.

DISTRIBUTION: This habitat is restricted to two bands in tropical n. Australia: the western plateaus, in a band from the Kimberley in n. Western Australia east toward sw. Northern Territory; and an eastern plateau in Arnhem Land, in ne. Northern Territory. The most accessible eastern escarpments occur in Kakadu National Park, at sites like Nourlangie Rock, which is accessible for wheelchair users, and Gunlom, which requires a strenuous walk from the monsoon forest to the top of the escarpment. The western escarpments are quite remote, but places like Mitchell Falls (Western Australia) are accessible by four-wheel-drive vehicles.

WHERE TO SEE: Nourlangie Rock and Ubirr, Kakadu National Park, Northern Territory, Australia.

OPEN EUCALYPT SAVANNA

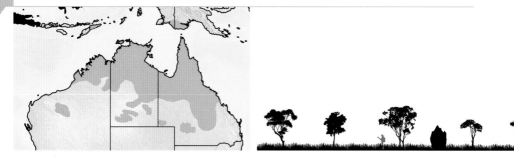

IN A NUTSHELL: A very open savanna with broadly spaced trees, mainly eucalypts, and many large termite mounds. **Habitat Affinities:** AFROTROPICAL MOIST MIXED SAVANNA. **Species Overlap:** TETRODONTA WOODLAND SAVANNA.

DESCRIPTION: This is the Australasian equivalent of the East African savannas. In the semiarid but well-drained parts of n. Australia, s. New Guinea, and surrounding islands, where there is a distinct summer rain regime (i.e., a monsoonal climate), savannas are the norm. On mainland Australia they can take on a variety of structures, from savanna woodland with an open canopy, similar to the TETRODONTA WOODLAND SAVANNA, to tree savanna characterized by trees with crooked trunks scattered randomly across the landscape

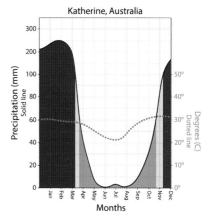

Open eucalypt savanna at the southern end of the Cape York Peninsula in Queensland, Australia. © IAIN CAMPBELL, TROPICAL BIRDING TOURS

(where tree cover varies from 30% to 5%), which transitions to either Mitchell Grass tussock grassland savanna or spinifex-type grassland.

In contrast to the flora of many of the better-wooded savannas of Africa or Asia, the dominant tree species of Australian savannas are restricted to this habitat and are not also found in the surrounding deciduous or evergreen forests. The *Eucalyptus* trees in open eucalypt savannas are generally shorter (15–35 ft./5–12m tall) than those in the tetrodonta woodland savanna, and the canopy is rounded and is not normally dominated by just one species. Most of the trees are evergreen, though some do lose their leaves during very dry periods. Boxes are the most common eucalypts in Western Australia, mixed with Pindan, a wattle species. In Queensland, ironbarks dominate, and some boxes are mixed with Pindan. The extent of grass in the understory varies greatly with rainfall and soil. Speargrass is dominant in the wetter savannas, Mitchell Grass in the drier areas, and the grasses of DUNE AND ROCKY SPINIFEX in the driest terrains and on rocky ridges. In New Guinea and the Moluccan islands, this habitat can consist of a tree layer of a few species with straight trunks and a canopy around 100 ft. (30m) high. In less fertile soils, the eucalypts tend to be less than 60 ft. (20m) tall and have a crooked shape. Both canopy types tend to have a ground layer of head-height grasses.

The most characteristic feature of this environment is the near-annual fires toward the start of the wet season, when lightning strikes are common but the ground is not yet very humid. The natural fire regime is augmented by anthropogenic fires, as Aboriginal Australians regularly burn this environment to maintain the parkland landscape. In the same way that temperate environments often vary drastically in winter and summer, these habitats seem very different during the wet and dry seasons. After a burn-off, the ground is bare, with exposed clay, sand, or many ironstone pisoliths, and patches of unburnt dry grass scattered across the landscape. In the wet season, the grasses are very thick, and many areas have ephemeral lakes. Rivers through this habitat often have thick stands of grass that extend out into the dry savanna.

WILDLIFE: Wildlife is remarkably similar across the range of this habitat, with animal distribution changing from north to south rather than east to west. Many bird species move over large areas in

Waterholes in open eucalypt savanna in Australia can attract concentrations of birds, including parrots and finches, like this rare Gouldian Finch (photo: Northern Territory, Australia).

Right: **Purple-crowned Fairywren is a localized specialist of riparian canegrass within open eucalypt savanna in n. Australia.**
© IAIN CAMPBELL, TROPICAL BIRDING TOURS

Below: **Antilopine Wallaroo is widespread within the open eucalypt savannas of n. Australia.** © SAM WOODS, TROPICAL BIRDING TOURS

search of blossoming trees or seeding grasses. Blossom nomads to be expected when trees are in bloom include Varied Lorikeet, Banded Honeyeater, Yellow-tinted Honeyeater, Gray-headed Honeyeater, and Golden-backed Honeyeater (IS). Seedeaters feature heavily in the grassy understory, including a range of finches such as Gouldian (IS), Long-tailed, Masked, and Double-barred Finches. Black-tailed Treecreeper, Great Bowerbird, Australian Pratincole, and Partridge Pigeon all occur here, with the latter two species preferring recently burned areas. Range-restricted species within this habitat include Hooded (IS) and Golden-shouldered Parrots (IS), which need large terrestrial termite mounds in which to nest. Wetlands and waterholes in open eucalypt savanna become a refuge for many birds, including most seedeaters as well as insectivores and nectivores, and during the dry season, waterholes are the best place to find most bird and mammal species. Around rivers, the range-restricted Purple-crowned Fairywren is a major target for birders.

Both Agile Wallaby and Antilopine Wallaroo inhabit and overlap in habitat, with the Agile preferring slightly more closed savanna and the Antilopine preferring grassier, more open areas, where it feeds with Northern Nail-tail Wallaby and Euro Wallaroo. This habitat is also one of the best for finding the Dingo, which is the main native predator of the smaller marsupials. The main large bats present in this habitat are Little Red and Black Flying Foxes, which, rather than being blossom nomads, concentrate on fruiting trees in the better-watered areas. Open eucalypt savanna in the Lesser Sundas is very low in mammal diversity, with no native mammals, other than some nonforest bat species, such as Northern Blossom Bat.

DISTRIBUTION: This habitat is widespread across n. Australia from Derby, Western Australia, on the west coast, to the western edge of the Atherton Tableland around Mareeba, Queensland. It is the dominant habitat through the northern quarter of the country. This savanna is also extensive along the south coast of New Guinea, interspersed with patches of monsoon forest along drainage lines and other moister areas. Savannas in the intermontane valleys and along the north coast of New Guinea may have been created by humans. Eucalypt savanna is also found on many islands of the Lesser Sundas and the Moluccas.

WHERE TO SEE: Mt. Carbine, Queensland, Australia; Victoria River, Northern Territory, Australia.

TETRODONTA WOODLAND SAVANNA

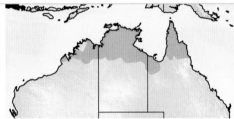

IN A NUTSHELL: Savanna woodland with a very uniform canopy, where the trees are tall-trunked and show some canopy overlap. **Habitat Affinities:** AFROTROPICAL MIOMBO WOODLAND. **Species Overlap:** OPEN EUCALYPT SAVANNA.

DESCRIPTION: This habitat, which looks like an open forest at a distance, has a canopy height of up to 100 ft. (30m), though it is usually around 50 ft. (15m). The canopy is almost monotypic, with the vast majority of trees being Darwin Stringybark (*Eucalyptus tetrodonta*), mixed with a minor component of Narrow-leaved Ironbark, Woollybutt, Melville Island Bloodwood, and Moly Redbox. There is not really a substantial sub-canopy or brushy understory; the ground cover is almost entirely tropical grasses. This grass is high and thick in the wet season and at the beginning of the dry season, but by the end of the dry season much of it is burned off, exposing the savanna floor of bauxite pisoliths and ironstone pisolites, and the habitat takes on a park-like feel.

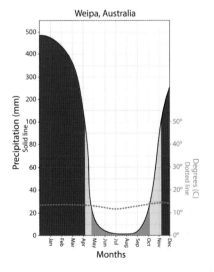

Weipa, Australia

WILDLIFE: Most of the birds in this habitat are found across n. Australia, such as Rufous-banded, Rufous-throated, and Banded Honeyeaters. Some species are more restricted, like Black-tailed Treecreeper and Northern Shrike-tit in Northern Territory, and Black-backed Butcherbird of Cape York Peninsula, Queensland.

Agile Wallaby is the most common, or at least the most conspicuous, mammal in this environment, outnumbering the less common Antilopine Wallaroo. Northern Brown Bandicoot moves into this habitat from nearby monsoon forest during the wet season when the grass grows and terrestrial insects, worms, and seeds become abundant.

Reptiles can be conspicuous in this habitat, including species such as Chameleon Dragon, Spotted Tree Monitor, Australian Green Tree Snake, Greater Black Whipsnake, and the extremely dangerous Coastal Taipan, which is 40 times more venomous than the diamondback rattlesnakes of North America. During the wet season, besides the ubiquitous invasive Cane Toad, native frogs such as Little Collared Frog, Tawny Rocket Frog, Marbled Frog and Mimic Toadlet can be found in this habitat. The introduction of the poisonous Cane Toad has had a disastrous effect on mammal and reptile predator populations of this habitat and has decimated populations of the Northern Quoll.

DISTRIBUTION: This habitat is widespread in the more humid, tropical bi-seasonal regions of Australia from the Kimberley region, Western Australia, across to Cape York Peninsula, Queensland. It is found mainly on well-drained, iron-rich, bauxite and lateritic plateaus. Poorly drained depressions within these habitats are often filled with AUSTRALASIAN MONSOON VINE FOREST or *MELALEUCA* SAVANNA.

WHERE TO SEE: Extensive stands occur around the town of Weipa in w. Cape York Peninsula Queensland, Australia.

Right: **Squatter Pigeons are largely terrestrial residents of this habitat, often best found by checking the verges of dirt roads.** © IAIN CAMPBELL, TROPICAL BIRDING TOURS

Burns are a natural part of the cycle of tetrodonta woodland savanna habitat, attracting Black Kites on the hunt for prey escaping the inferno. © IAIN CAMPBELL, TROPICAL BIRDING TOURS

BRIGALOW AND *CALLITRIS* WOODLANDS

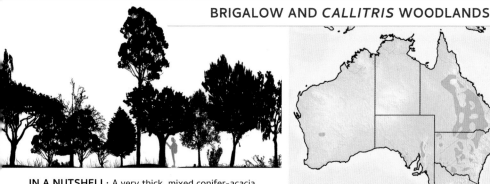

IN A NUTSHELL: A very thick, mixed conifer-acacia semiarid woodland with thick undergrowth, which in some regions is dominated by Brigalow (*Acacia harpophylla*) and in others by *Callitris* (cypress-pines). **Habitat Affinities:** No other habitats have this mix of small conifers and acacias. **Species Overlap:** MULGA WOODLAND AND ACACIA SHRUBLAND; OPEN EUCALYPT SAVANNA.

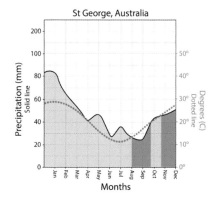

St George, Australia

DESCRIPTION: Biome classifications generally consider Brigalow woodlands to be a form of summer-rain tropical savanna woodland, and *Callitris* woodlands to be winter-rain Mediterranean woodlands. We combine them because they not only have remarkably similar canopy and undergrowth structure and botany but also, more importantly, share so many bird, mammal, and reptile species. This habitat occurs in e. and sw. Australia between the WET and DRY SCLEROPHYLL FOREST of the Great Dividing Range and subcoastal areas, and the truly arid habitats of the interior. Brigalow woodland can have a very thick canopy, usually 30–45 ft. (10–15m) high. Two acacias, Brigalow and Gidgee, are the most common canopy trees, along with Black-butt, boxes, Coolabah, sheoaks, and *Callitris* species such as Black, White, and Baily's Cypress-Pines. In slightly wetter areas, this habitat can have understory thickets that are actually lush. Despite being a nonhumid habitat, Brigalow woodland was described by early settlers as a form of dry rainforest and referred to as "scrub" in the same manner they described AUSTRALASIAN SUBTROPICAL RAINFOREST. The *Callitris* woodlands are slightly more open in the canopy and denser in the understory, and have a similar tree assemblage, with fewer acacias, more *Callitris*, and the same numbers of eucalypts. They are much more difficult to walk through than the surrounding OPEN EUCALYPT SAVANNA.

WILDLIFE: Typical bird species are Turquoise Parrot, Glossy Black-Cockatoo, Gilbert's Whistler (IS), Rufous Whistler, Speckled Warbler, Yellow Thornbill, Red-capped Robin, Noisy Friarbird, Striped Honeyeater, Apostlebird, White-winged Chough, Plum-headed Finch, and Diamond Firetail.

   The mammals present in this habitat include the endangered Bridled Nail-tail Wallaby and Northern Hairy-nosed Wombat, as well as the more common Whiptail Wallaby, Eastern Gray Kangaroo, and Red-necked Wallaby.

Reptiles include Golden-tailed Gecko, Common Death Adder, Brigalow Scaly-foot, Ornamental Snake, and Dunmall's Snake.

DISTRIBUTION: This habitat is found in e. Australia from south of Townsville, Queensland (inland of the Great Dividing Range, west of Brisbane), south to Leeton, s. New South Wales. The northern expression of this habitat is the Brigalow woodland of Queensland and n. New South Wales. The *Callitris* woodlands occur farther south in New South Wales and Victoria. There are very small stands of *Callitris* in c. Western Australia, which are thought to be remnant stands from the Miocene. This habitat has been extensively cleared in the southern areas because it is so productive for agriculture. In the northern part of the range, those areas not cleared for crops have been extensively grazed and altered, and few large stands of Brigalow woodland still exist.

WHERE TO SEE: Binya State Forest, New South Wales, Australia; Durikai State Forest, Queensland, Australia.

Red-capped Robin is particularly abundant in Brigalow woodland.
© IAIN CAMPBELL, TROPICAL BIRDING TOURS

Brigalow habitat in s. Queensland, Australia. © IAIN CAMPBELL, TROPICAL BIRDING TOURS

MELALEUCA SAVANNA

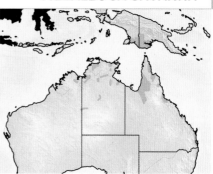

IN A NUTSHELL: A seasonally flooded savanna dominated by *Melaleuca* trees. **Habitat Affinities:** There is no similar global habitat. **Species Overlap:** AUSTRALASIAN MONSOON VINE FOREST; OPEN EUCALYPT SAVANNA.

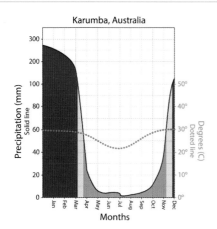

Karumba, Australia

When flowers are in bloom and abundant nectar is available, Red-collared Lorikeets visit this habitat.

© IAIN CAMPBELL, TROPICAL BIRDING TOURS

DESCRIPTION: This thick savanna dominated by *Melaleuca* (paperbark or tea trees) is found in marshy environments near the coast or in hilly terrain as shrub growth. Some occurs in similar climatic zones to OPEN EUCALYPT SAVANNA, but *Melaleuca* savanna tends to dominate in low-lying areas that are flooded in the wet season and where fires are frequent. Here it replaces the acacias that do not tolerate fire and the eucalypts that prefer better-drained environments. *Melaleuca* groves tend to be of one or two species, the common ones in wetter parts of n. Australia being Broad-leaved and Weeping Paperbarks. Most of the trees have a characteristic bark that exfoliates in large sheets, and the trees usually have off-white to cream trunks that are burned black at the base. The height of *Melaleuca* savanna ranges from 10 ft. (3m) in shrublands to 100 ft. (30m) in woodlands, with the same species varying in height depending on water conditions. In flood-prone areas there is little understory, and the ground can be bare, with clay exposed. Away from rivers, the density of trees decreases, transitioning from woodland with a heath or grass understory to wooded savanna with a tussock-grass understory and even to grassland savanna with few trees.

Above: **Melaleuca savanna near Darwin, Northern Territory, n. Australia.** © SAM WOODS, TROPICAL BIRDING TOURS

Right: **Finches, such as the Double-barred Finch, can be found feeding in the grassy understory of Melaleuca savanna.** © IAIN CAMPBELL, TROPICAL BIRDING TOURS

WILDLIFE: These *Melaleuca* stands are prone to mass flowering, and during these times they can be pumping with birds. Nectivorous birds include Red-collared Lorikeet, Bar-breasted Honeyeater (IS), Brown-backed Honeyeater (IS), and Scarlet Myzomela. Other birds of the canopy include Silver-crowned Friarbird, Little Shrikethrush, Northern Fantail, and Paperbark Flycatcher. The grassy understory is the domain of numerous finches, and the Red-backed Fairywren.

While terrestrial mammals, such as the Agile Wallaby, are few, the dominant groups in n. Australia are flying foxes and other bats, such as Black Flying Fox and Northern Blossom Bat.

DISTRIBUTION: This habitat occurs throughout tropical and subtropical ne. and nc. Australia, New Guinea, and New Caledonia. It is widespread in the Trans-Fly region of New Guinea, as well as in n. Australia, especially around the Gulf Country (along the Gulf of Carpentaria), and extends through the tropical parts of Australia. There are small patches of *Melaleuca* savanna in coastal areas of s. Queensland and n. New South Wales.

WHERE TO SEE: Lakefield National Park, Queensland, Australia; Fogg Dam, Northern Territory, Australia; Wasur National Park, West Papua, Indonesia.

Agile Wallaby is one of the few terrestrial mammals to occur regularly in *Melaleuca* savanna.
© IAIN CAMPBELL, TROPICAL BIRDING TOURS

SHEOAK WOODLAND

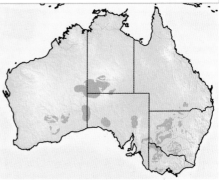

IN A NUTSHELL: An arid to semiarid tree savanna to open woodland in which the main canopy trees, although angiosperms, appear very similar to conifers. **Habitat Affinities:** PALEARCTIC SUBTROPICAL SAVANNA has similar structure. **Species Overlap:** MULGA WOODLAND AND ACACIA SHRUBLAND.

DESCRIPTION: This arid woodland appears similar, at a distance, to the other open woodlands of temperate e. Australia. It often occurs with an understory of grasses, such as SPINIFEX hummock grasses or tussock-forming Mitchell Grass, and small shrubs, but the most obvious vegetation is widely spaced sheoak trees, usually *Allocasuarina* species, such as Desert Sheoak. Young trees look very different from mature trees in that they have a spiky tangle of short branches the whole way up the trunk, resisting grazers. When the tree grows to about 10 ft. (3m) tall, and the taproot hits the water table, the crown grows out wider (like African acacias), and the tangle along the trunk dies off. The leaves of the tree are narrow and almost pine-needle-like, but they droop down to form a thick

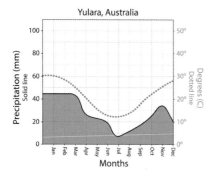

Yulara, Australia

mass of shady cover. In coastal Australia, this habitat is dominated by members of the *Casuarina* genus, such as Beach Sheoak, which closely resemble the arid sheoaks and grow in areas of high salinity.

Sheoak woodland and the Kata Tjuta rock formations, also known as the Olgas, in Uluru–Kata Tjuta National Park, Northern Territory, Australia. © IKIWANER, LICENSED UNDER CC BY-SA 3.0, HTTPS://CREATIVECOMMONS.ORG/LICENSES/BY-SA/3.0/

WILDLIFE: Although no birds are restricted to sheoak woodland, it is the preferred habitat of the rare and elusive Princess Parrot (IS). Other birds that occur here are Purple-backed Fairywren, Black-faced Woodswallow, Ground Cuckooshrike, Pink Cockatoo, and Hooded Robin.

Mammals of this habitat include Red Kangaroo, Euro Wallaroo, and Greater Bilby. Introduced mammals play a greater role in this environment, where Dromedary (One-humped Camel) and Feral Horse are both common.

Splendid Fairywren occurs in sheoak woodland.
© IAIN CAMPBELL, TROPICAL BIRDING TOURS

Numerous reptiles live in sheoak woodlands, though none are exclusive to it. The most notable species are Western Hooded Scaly-foot, Great Desert Skink, Broad-banded Sand-swimmer, Centralian Blue-tongued Lizard, Black-collared Dragon, and Central Netted Dragon.

DISTRIBUTION: Much of this habitat is inaccessible, existing in remote, rarely visited lands like the Great Sandy Desert (Western Australia) and the Great Victoria Desert (Western Australia–South Australia), though it exists in smaller patches in far w. New South Wales and far e. South Australia.

WHERE TO SEE: It is possible to access this habitat in Northern Territory, Australia, southwest of Alice Springs, including around Watarrka (Kings Canyon) and Uluru–Kata Tjuta National Parks, and also at Newhaven Sanctuary northwest of Alice Springs.

Red-backed Kingfisher occupies wooded habitats away from water, including sheoak woodland.
© IAIN CAMPBELL, TROPICAL BIRDING TOURS

AUSTRALASIAN TUSSOCK GRASSLAND

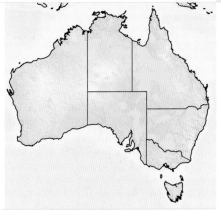

IN A NUTSHELL: Almost treeless grassy plains.
Habitat Affinities: AFROTROPICAL TROPICAL GRASSLAND. **Species Overlap:** SHEOAK WOODLAND; DUNE SPINIFEX; CHENOPOD SHRUBLAND.

DESCRIPTION: This grassland habitat is the Australian equivalent of the TROPICAL GRASSLAND of Africa, the TALLGRASS PRAIRIE of North America, or the PAMPAS of South America. The dense, tufted perennial grasses of Australasian tussock grassland grow to 2–3 ft. (0.6–1m) in height, with a crown usually around 2 ft. (0.6m) wide. The tussocks themselves are well spaced, usually about 3 ft. (1m) apart, so there is often bare earth exposed between grass clumps. Long-lived, with a dual root system for absorbing both smaller rainfalls and deeper subsoil moisture, these grasses can withstand drought and then burst into life after a good rain, giving the landscape the look of a well-planted lawn. Mitchell Grass is the most common species of tussock grass in n. Australia. Few trees can grow in the cracking clay soils or withstand the near-annual fire regime, so the habitat has a very expansive, open, rolling-downs feel to it. At a distance it looks like DUNE SPINIFEX but up close has a much softer appearance, and with many more nutrients, it is much more productive for birds and mammals.

WILDLIFE: Tussock grasslands are an important habitat for Flock Bronzewing (IS), which, along with Letter-winged Kite (IS), has a boom-and-bust life cycle, with numbers fluctuating dramatically. The kite's numbers are dependent on the population of its main food source, the Long-haired Rat. The Flock Bronzewing was heading toward extinction in the same manner as the North American Passenger Pigeon; flocks of hundreds of thousands in the 1800s were reduced to mere hundreds by the early 1900s. With strict protection, numbers have rebounded, and flocks of thousands are again being reported. Many other species of birds use these grasslands, including Australian Bustard, Australian Pratincole, and Inland Dotterel. The very rare Night Parrot is found in some areas, and it has been speculated that it may be found in this habitat and DUNE SPINIFEX throughout c. Australia.

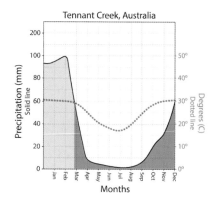

A pair of Brolga and a keen young photographer within tussock grasslands in w. Queensland, Australia. © IAIN CAMPBELL, TROPICAL BIRDING TOURS

The star mammal in tussock grassland is Greater Bilby, an odd marsupial that was brought back from near extinction and still requires management. Larger grazers such as Red Kangaroo and Common Wallaroo are much more abundant in this habitat than in dune spinifex, and these large herbivores provide a food source for carrion eaters such as Dingo, Wedge-tailed Eagle, Black-breasted Kite, and Black Kite.

Numerous lizards occur, such as the Lined Earless Dragon and the impressive Spencer's Monitor. Snakes include Fierce Snake, Speckled Brown Snake, and Ingram's Brown Snake.

DISTRIBUTION: Australasian tussock grasslands occur on more fertile soils than DUNE AND ROCKY SPINIFEX, mainly in inland ne. Australia, particularly nw. Queensland and s. Northern Territory. Although this habitat is widespread, most of the areas where it occurs are relatively remote, including the Barkly Tableland of s. Northern Territory and the Diamantina region of Queensland. Because tussock grassland is so much more productive than spinifex, it has been almost completely exploited for wide-scale cattle grazing. Overall, the habitat seems to be preserved, but overgrazing is a serious issue in drought periods and has caused a reduction in native animal numbers.

WHERE TO SEE: Connells Lagoon, Barkly Tableland, Northern Territory, Australia.

Tussock grasslands are prime breeding habitat for the Australian Bustard.
© IAIN CAMPBELL, TROPICAL BIRDING TOURS

AUSTRALASIAN LOWLAND HEATHLAND

IN A NUTSHELL: Low sclerophyllous scrub in a winter-rainfall and fire-prone environment. **Habitat Affinities:** AFROTROPICAL FYNBOS. **Species Overlap:** AUSTRALASIAN ALPINE HEATHLAND; MALLEE WOODLAND AND SCRUBLAND.

DESCRIPTION: This habitat occurs in areas with nutrient-poor soils that are very shallow, heavily leached, and sand-dominated, in areas with harsh, often highly saline winds. These conditions create an extremely arid microclimate, so the plants that grow here are much hardier than those of surrounding environments. This habitat is found in areas with a Mediterranean climate, such as the south coasts of New South Wales and Western Australia, but also occurs in some mountainous areas on poor soils. In general appearance and structure, coastal heathlands are similar across Australia. The floral makeup of any small area is successional due to fire and in constant flux. When areas are freshly burned, the heathland is very short (1–2 ft./30–60cm tall) and made up of many species of wildflowers and other forbs, reaching maximum floral diversity around four years after burning. At this stage, the habitat is easy to walk through. After about five years, the herbaceous growth is gradually replaced by shrubs, typically *Banksia* and *Melaleuca* (paperbark) species, along with some others. These shrubs are normally very thick and difficult to walk through, forming a canopy with a height around 6 ft. (2m), from which emerge a few small mallee-form (with most branches sprouting from the tree base) eucalypts. The vegetation at this stage, with a profusion of highly productive flowers around head height, is very attractive to wildlife. After about 10 years, most species of *Banksia*, *Grevillea*, and *Hakea* have been outcompeted by a few larger banksias, such as Heath-leaved Banksia and Saw Banksia, which can grow to 50 ft. (15m) tall, though the heath canopy is usually around 25–30 ft. (8–10m). In this final stage, the habitat is nowhere near as productive as in earlier successions.

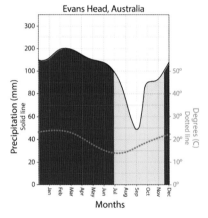

Evans Head, Australia

Coastal heathlands in New Zealand have a similar growth pattern, usually dominated by Kumeraho, which is 6–10 ft. (2–3m) tall in final succession. However, in New Zealand this habitat appears to be a product of human removal of the original kauri forest (see GONDWANAN CONIFER RAINFOREST) rather than a naturally occurring habitat. In stark contrast to heathlands of Australia

Above: **Coastal lowland heathland in Royal National Park, near Sydney, New South Wales, Australia.**
© SAM WOODS, TROPICAL BIRDING TOURS

or South Africa, the heathlands of New Zealand are extremely poor in endemic plant species, with most plants being part of the wet forest assemblage. In the absence of the original forest canopy layer, Kumeraho and an association of heath shrubs and sedges grow in a now-open environment. In some areas of the South Island there is a gradual change from these heathlands to forests, where forest trees like the Kamahi grow in a shrub form.

WILDLIFE: Birds that may be spotted in early to mid-succession are the shy and difficult-to-see Southern Emuwren (IS) and Chestnut-rumped Heathwren (IS). One of the main target birds for nature observers in e. Australia, the semi-nocturnal Ground Parrot (IS), is best found flying across heathland at dusk, at a height of 3–4 ft. (1–1.2m). It is, however, the nectivorous

Right: **The handsome Western Spinebill is restricted to Western Australia, where it is commonly encountered in heathland.** © SAM WOODS, TROPICAL BIRDING TOURS

Above: **New Holland Honeyeater is very abundant in coastal lowland heathland (photo: New South Wales, Australia).** © IAIN CAMPBELL, TROPICAL BIRDING TOURS

Below: **A Honey Possum drawing nectar on a heathland banksia in Western Australia.** © SAM WOODS, TROPICAL BIRDING TOURS

birds that are most obvious in this habitat through all successions. Little and Western Wattlebirds and New Holland, White-cheeked, and Tawny-crowned Honeyeaters compete noisily for the highly productive flowers, particularly in spring when the heath bursts into spectacular flower.

Mammals abound but are mostly nocturnal species that are not easily seen. A night walk around Two Peoples Bay in Western Australia could turn up Honey Possum, Western Pygmy Possum, Southern Brown Bandicoot and the newly rediscovered Gilbert's Potoroo. The diurnal kangaroos include Quokka and Western Gray Kangaroo.

Reptiles are frequently found in this type of habitat. Red-bellied Black Snake and Common Death Adder are both common in n. New South Wales, along with Lace Monitor, one of the biggest lizards in the world.

DISTRIBUTION: Variations of this heathland are found from the Moluccas through New Guinea to New Caledonia. In New Zealand it is found in far n. North Island around Spirits Bay (Piwhane), on the west side of the South Island, and on Stewart Island. In New Guinea it is very limited (too limited to be mapped) and always associated with rare areas of infertile soil. The vast majority of Australasian lowland heathland habitat is found on mainland Australia, mostly in the subtropical and temperate regions from Shark Bay in the west, around the south coast, and up the east coast to Cape York Peninsula. It is very common in sw. Western Australia, Tasmania, and along n. New South Wales coast.

WHERE TO SEE: Broadwater National Park, near Evans Head, New South Wales, Australia; Two Peoples Bay, Western Australia.

AUSTRALASIAN ALPINE HEATHLAND

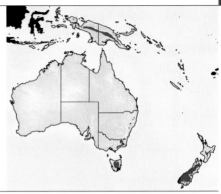

IN A NUTSHELL: Low sclerophyllous scrublands in areas that are very exposed to strong winds, have infertile soils, and are prone to fire. **Habitat Affinities:** AFROTROPICAL FYNBOS. **Species Overlap:** AUSTRALASIAN LOWLAND HEATHLAND; TEMPERATE EUCALYPT WOODLAND.

DESCRIPTION: These alpine heathlands look similar to NEARCTIC ALPINE TUNDRA, with small plants that manage to survive in extremely cold, windy, and dry conditions, often growing in infertile soil or rock fields. These tropical and mid-latitude alpine heathlands are subject to extremes in temperature and water availability but do not have the many months of darkness and the months of continuous sunshine experienced by Arctic tundra. The plants in New Guinea, Australia, and New Zealand have very little in common with those of the Northern Hemisphere tundra, and Australasian alpine heathland complexes are far more floristically rich than their northern counterparts, supporting a mixture of endemic shrubs, grasses, and mosses. Small-leaved shrubs, herbs, and tussock grasses dominate the vegetation, which is usually less than 6 ft. (2m) tall. Areas with poorer drainage have cushion bogs and grass bogs that resemble the wet PARAMO of the n. Andes. Generally, it is easy to move through this environment, though the ground is often uneven underfoot. Species richness decreases with elevation, and the makeup of the heath changes to few shrubs mixed with cushion plants and bare rock. The lower edges of these alpine heathlands have low trees (<10 ft./3m tall), which are not stunted, windswept versions of larger trees but often their smaller close relatives that are better suited to this environment. The tree species at the tree line are beeches in New Zealand, eucalypts in Australia, and rhododendrons and tree ferns in New Guinea.

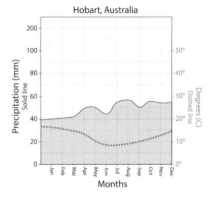

WILDLIFE: Although the alpine heathlands in Australia are floristically rich, they do not hold many endemic vertebrate species. Characteristic birds during summer include Flame Robin and Striated Fieldwren. Alpine heathland is far more interesting in New Zealand and seems to be the least disturbed of the country's habitats. Any short visit to these heathlands on the South Island of New Zealand should result in encounters with Kea, a large, comical-looking alpine parrot that, besides berries and plants, eats birds, mammals, and carrion, and has even been recorded attacking sheep. Another bird that thrives in this environment is New Zealand Rockwren (IS), aka South Island Wren, which is found in alpine heathlands of the South Island.

Snow Gums can form loose groves in protected areas of alpine heathland (photo: near Hobart, Tasmania).
© SAM WOODS, TROPICAL BIRDING TOURS

The alpine heathlands of New Guinea, sometimes referred to as subalpine shrublands, are less diverse than habitats at lower elevations but have a very distinct avifauna, mainly comprising such groups as honeyeaters, robins, and munias. These heaths adjoin grasslands where local species like Snow Mountain Quail (IS) (Snow Mountains, West Papua, Indonesia) and Alpine Munia (se. Papua New Guinea) occur, as well as the more widespread Alpine Pipit and Papuan Marsh-Harrier. In the skies overhead, the wide-ranging Mountain Swiftlet is seen regularly, while shrubbery in this area is home to Brown Quail, Archbold's Nightjar, Painted Tiger-Parrot, and a series of high-living honeyeaters: Gray-streaked and Orange-cheeked Honeyeaters, and Long-bearded, Short-bearded, and Sooty Melidectes. Robins are less conspicuous, but Subalpine and Snow Mountain Robins occur in the highlands of West Papua, along with finches such as the wide-ranging Mountain Firetail and the more local Snow Mountain Munia. The most striking of all the inhabitants of this habitat is the Crested Berrypecker, from an endemic New Guinean family. None of the heath species appear to be restricted to the habitat, except perhaps the grassland species (e.g., Snow Mountain Quail, Alpine Pipit); most others range into the nearby grasslands or the edges of the upper montane forests.

Mammals that occur in this habitat include Eastern Long-beaked Echidna, Long-tailed Pygmy Possum, and Subalpine Woolly Rat, while Calaby's Pademelon, Striped and Raffray's Bandicoots, Habbema and Speckled Dasyures, and Coppery Ringtail Possum occur in nearby subalpine grasslands.

DISTRIBUTION: In mainland Australia, this habitat is limited, occurring along the Great Dividing Range and in the Snowy Mountains of New South Wales. It is common throughout the highlands

Flame Robin is conspicuous in Australian alpine heathlands in springtime, when it sings from the tops of shrubs and trees. © IAIN CAMPBELL, TROPICAL BIRDING TOURS

Short-beaked Echidna is often encountered foraging on roadside verges in heathland areas of Australia.
© BEN KNOOT, TROPICAL BIRDING TOURS

of Tasmania. It is very localized but widely distributed in the highest elevations of the New Guinea highlands. In New Zealand, alpine heathlands are found throughout highland areas, particularly on the South Island, where they also extend down to sea level.

WHERE TO SEE: Mt. Wellington, Tasmania, Australia; Snow Mountains, West Papua, Indonesia.

WET SCLEROPHYLL FOREST

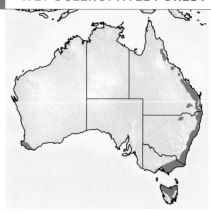

IN A NUTSHELL: A forest with a eucalypt canopy and an understory of plants usually associated with rainforest. **Habitat Affinities:** Fills the niche of semi-evergreen forests but not related to others globally. **Species Overlap:** AUSTRALASIAN SUBTROPICAL RAINFOREST; DRY SCLEROPHYLL FOREST.

DESCRIPTION: Wet sclerophyll forest has an open canopy dominated by tall eucalypts and forms in high-rainfall, high-fertility areas. These are the tallest forests in Australia and among the tallest in the world, reaching 180 ft. (55m), with some trees as tall as 260 ft. (80m). The dominant trees are Rose Gum and Sydney Blue Gum in e. Australia and Karri in Western Australia. Although wet sclerophyll forest may look like rainforest from a distance, the small leaves still allow significant light through the canopy, resulting in a dense understory of rainforest-type plants, so walking through this habitat is very difficult and wandering is near impossible. The transition between wet sclerophyll and rainforest can be clinal and depends on fire frequency. Where the forests occur beside each other, rainforest grows in wet gullies, and wet sclerophyll dominates the more exposed slopes. There is no equivalent forest type in other parts of the world.

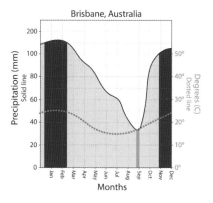

Brisbane, Australia

WILDLIFE: This is the tale of two habitats (see sidebar 1.2). The similarity of the understory to rainforest, both physically and in terms of plant species composition, leads to significant overlap in bird species. Birds of the understory include Superb Lyrebird, Crescent Honeyeater, Eastern Spinebill, Satin Flycatcher, Pilotbird, Eastern Whipbird, and Satin Bowerbird. In contrast, the canopy is very similar to that of DRY SCLEROPHYLL FOREST, so the canopy bird species are typical of that habitat

Above: **Black-headed Honeyeater in wet sclerophyll habitat in Tasmania.** © IAIN CAMPBELL, TROPICAL BIRDING TOURS

Right: **The threatened Koala is still locally abundant in a few wet sclerophyll forests.** © SAM WOODS, TROPICAL BIRDING TOURS

Below: **Wet sclerophyll forest near Melbourne, Australia.** © IAIN CAMPBELL, TROPICAL BIRDING TOURS

and include Crimson Rosella, Australian King-Parrot, Red-browed Treecreeper, Bell Miner (IS), and Spotted Pardalote.

Besides the rainforest mammals, like Red-legged Pademelon, that occur in the wet sclerophyll understory, this habitat holds some habitat-restricted mammals, such as Northern Bettong, Yellow-bellied Glider, Parma Wallaby, Squirrel Glider, and Australian Swamp Rat. Around Sydney, wet sclerophyll forest is home to Eastern Gray Kangaroo, Swamp Wallaby, Common Ringtail Possum, Common Brushtail Possum, and Eastern Pygmy Possum, along with the occasional Feathertail Glider. Koala was once a common mammal of wet sclerophyll forests but has suffered significant population declines.

Reptiles typical of this habitat, though rare, include Pale-headed Snake, Broad-headed Snake, and Rosenberg's Goanna.

DISTRIBUTION: In the tropics of e. Australia, wet sclerophyll forests occur patchily along the western margins of rainforests. In the subtropics, they are a buffer between higher-elevation rainforests and DRY SCLEROPHYLL FOREST. In temperate areas, such as sw. Western Australia and the Blue Mountains of New South Wales, these forests are more extensive, forming large stands. This is not a habitat native to New Zealand, but with planted eucalypt forests extending over 125,000 acres (50,000ha), it is now more common than many native forests, though it can be regarded as an ecological desert, devoid of animal life.

WHERE TO SEE: Blue Mountains (e.g., Mt. Wilson, Wollemi National Park), New South Wales, Australia; Bruny Island, Tasmania, Australia.

SIDEBAR 1.2	**FOREST CANOPIES AND UNDERGROWTH CAN CHANGE AT DIFFERENT RATES**

Where habitats merge, the plant and bird assemblages of different parts of the habitat do not necessarily change at the same time. In s. Queensland, you could walk from AUSTRALASIAN SUBTROPICAL RAINFOREST into WET SCLEROPHYLL FOREST without noticing a change in the undergrowth; the plants and undergrowth birds seem to be the same. It is only when you look up that you notice that the canopy trees have changed and the corresponding canopy birds have changed. Continue walking from the wet sclerophyll to the DRY SCLEROPHYLL FOREST, and the canopy birds will remain the same, but the understory species change dramatically, as the undergrowth transitions from shrubs and small broadleaf trees to grasses and more broadly spaced shrubs.

DRY SCLEROPHYLL (EUCALYPT) FOREST	WET SCLEROPHYLL (EUCALYPT) FOREST	SUBTROPICAL RAINFOREST

Eucalypt canopy animal assemblage: e.g., Spotted Pardalote, Red-browed Treecreeper, White-throated Honeyeater, Koala.

Rainforest canopy animal assemblage: e.g., Paradise Riflebird, Regent Bowerbird.

Grassland animal assemblage: e.g., Spotted Quail-thrush, Pretty-faced Wallaby.

Rainforest undergrowth animal assemblage: e.g., Eastern Whipbird, Southern Logrunner, Albert's Lyrebird, Rufous Fantail, Red-necked Pademelon, Red-legged Pademelon.

DRY SCLEROPHYLL FOREST

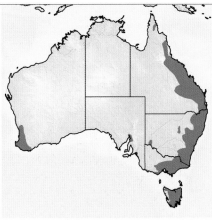

IN A NUTSHELL: Eucalypt forest that has a nearly closed canopy and an open understory with significant amounts of grass. **Habitat Affinities:** No similar non-Australian habitats. **Species Overlap:** WET SCLEROPHYLL FOREST; TEMPERATE EUCALYPT WOODLAND.

DESCRIPTION: This is an open forest of medium to low height, generally lower than 100 ft. (30m), with closely growing but regularly spaced trees that form a linked, flat, but still sparse canopy of a few dominant eucalypt species, notably Jarrah in Western Australia and Spotted Gum in New South Wales. The large amount of light that reaches the understory allows for a thin mid-canopy of acacias and a complete ground cover. The understory can vary greatly. In some areas, there is thick shrub, resembling the understory of WET SCLEROPHYLL FOREST, dominated by Native Cherry (a member of the sandalwood family) and Golden Wattle. In less fertile or drier areas, the undergrowth is very different, mainly consisting of tussock grasses, grass trees, casuarinas, and heath plants such as small banksias. This is a fire-dependent habitat, and most of the understory shrubs are adapted to regenerate after fire, while most canopy trees rapidly regrow leaves.

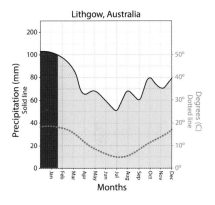

Lithgow, Australia

WILDLIFE: The dry sclerophyll forests of se. Australia share most bird species with either the adjoining TEMPERATE EUCALYPT WOODLAND or the adjoining WET SCLEROPHYLL FOREST. Typical birds of the canopy in e. Australia include Little Lorikeet (IS), Red-browed Treecreeper, White-naped Honeyeater, and White-throated Treecreeper. Western Yellow Robin is common in Western Australia, and Yellow-throated Honeyeater is common in Tasmania. The undergrowth is the favorite haunt of Blue-breasted Fairywren, Spotted Quail-thrush (IS), White-browed Scrubwren, Striated Thornbill, and Painted Buttonquail.

Sunset in dry sclerophyll forest close to Lamington National Park, near Brisbane, Queensland, Australia. © SAM WOODS, TROPICAL BIRDING TOURS

As with most Australian mammals, the vast majority found in the dry sclerophyll are nocturnal, including Rufous Bettong, Eastern Pygmy Possum, Spotted-tailed Quoll, Southern Brown Bandicoot, and Squirrel Glider. Larger ground grazers here include Black-striped Wallaby, Pretty-faced Wallaby, and Brush-tailed Rock-Wallaby. This is also one of the main habitats for Koala.

Many endangered reptiles are found in these forests; notable species include Zigzag Velvet Gecko, Border Thick-tailed Gecko, and Broad-headed Snake.

DISTRIBUTION: Dry sclerophyll forest occurs widely in e. Australia, from the west of the Atherton Tableland (Queensland), through Capertee Valley and Hunter Valley (New South Wales), to much of e. Victoria and ne. Tasmania. It is extensive in s. Western Australia but has been extensively cleared for agriculture.

WHERE TO SEE: Capertee Valley, New South Wales, Australia; Lamington National Park, Queensland, Australia; Wungong Regional Park, Western Australia.

Parrots are conspicuous in Australia, including Scaly-breasted Lorikeet (shown here), which inhabits a variety of eucalypt-dominated habitats. © IAIN CAMPBELL, TROPICAL BIRDING TOURS

Pretty-faced Wallaby forages in the grassy understory of dry sclerophyll, Queensland, Australia.
© GABRIEL CAMPBELL, TROPICAL BIRDING TOURS

TEMPERATE EUCALYPT WOODLAND

IN A NUTSHELL: A eucalypt woodland with an open canopy and a shrubby or grassy ground cover. **Habitat Affinities:** NEOTROPICAL MATORRAL SCLEROPHYLL FOREST AND SCRUB. **Species Overlap:** DRY SCLEROPHYLL FOREST; MALLEE WOODLAND AND SCRUBLAND; MULGA WOODLAND AND ACACIA SHRUBLAND.

DESCRIPTION: This habitat type encompasses a variety of distinctive woodlands with different canopy trees, including the once-extensive ironbark and box eucalypt woodlands of inland se. Australia, the *Dryandra* eucalypt woodlands of w. Australia (with Wandoo, Powderbark Wandoo, and Salmon White Gum), and the White Gum woodlands of s. Australia. Botanical classifications combine these habitats with tropical woodlands such as TETRODONTA WOODLAND SAVANNA, and some canopy species such as Narrow-leaved Ironbark occur in both habitats. However, we split these habitats based on their different climates, their different understories, and most importantly, their vastly different bird assemblages.

The canopy of temperate eucalypt woodland is usually around 60–75 ft. (20–25m) high and made up of eucalypts: ironbarks, boxes, and Darwin Stringybark. The canopy is typically quite open. The mid-canopy, at around 25–30 ft. (8–10m), is dominated by wattles, such as Golden Wattle. Woodlands on floodplains generally have a grassy understory, while those in elevated areas support a shrubbier understory. Walking through this habitat is normally easy.

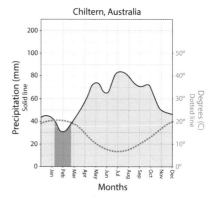

Chiltern, Australia

WILDLIFE: Many of the canopy bird species that use this habitat are blossom nomads, which will move north and south along with the flowering of eucalypts and/or the emergence of insects associated with those flowerings. The two most notable and sought-after bird species are Painted Honeyeater and Regent Honeyeater (IS). Others include Fuscous and Yellow-faced Honeyeaters and White-browed Woodswallow. Ground passerines, which tend to be more sedentary, include White-winged Chough; Apostlebird; Jacky-winter; Yellow, Inland, and Yellow-rumped Thornbills; Rufous Whistler; and Hooded Robin. This habitat is favored by Variegated Fairywren in the southeast and Blue-breasted Fairywren in the southwest. This zone also has the highest density of parrot/cockatoo species anywhere in the world, with 22 species found on the western slopes of New South Wales alone. Some of the most spectacular of these are Musk Lorikeet (IS) and Swift (IS), Superb, and Turquoise Parrots.

Ironbark and box woodlands, like this example in Victoria, are a distinctive type of temperate eucalypt woodland, favored by the critically endangered Regent Honeyeater. © IAIN CAMPBELL, TROPICAL BIRDING TOURS

Above: **Temperate eucalypt woodlands, such as Dryandra State Forest, south of Perth, Western Australia, are widespread in sw. Australia.** © IAIN CAMPBELL, TROPICAL BIRDING TOURS

Below: **Blue-breasted Fairywren inhabits temperate eucalypt woodland in Western Australia.** © IAIN CAMPBELL, TROPICAL BIRDING TOURS

Echidnas are relatively common in this kind of habitat, which has enough termites for food and is open enough to allow these mammals to move around freely. Kangaroos that live here include Eastern Gray Kangaroo and Swamp Wallaby. Wombats can be found in many of the more extensive reserves, though do not adapt to farming lands as well as kangaroos. Most of Australia's smaller marsupial populations are very endangered; however, large areas of this habitat are now being fenced off, so some bettongs (very small kangaroos), such as the Woylie Bettong and the Burrowing Bettong, as well as the Western Barred Bandicoot, are again findable.

DISTRIBUTION: These woodlands characterize the regions in e. and sw. Australia between the DRY SCLEROPHYLL FOREST of the Great Dividing Range and subcoastal areas, and the truly arid habitats of the interior (see sidebar 1.3). These habitats have taken the brunt of clearing for the wheat fields of New South

Common Wombat occurs in temperate eucalypt woodland.
© IAIN CAMPBELLL, TROPICAL BIRDING TOURS

Wales and Western Australia and have been almost completely cleared in New South Wales. What remain are small remnants and a series of human-maintained forest corridors along country roads. Although fragmented, these remnants form the backdrop to the iconic "country" of New South Wales, Victoria, and Western Australia. Well-known birding locations with examples of this habitat include Weddin Mountains National Park (New South Wales), Binya State Forest (New South Wales), and Chiltern (Victoria).

WHERE TO SEE: Weddin Mountains National Park, New South Wales; Chiltern, Victoria, Australia.

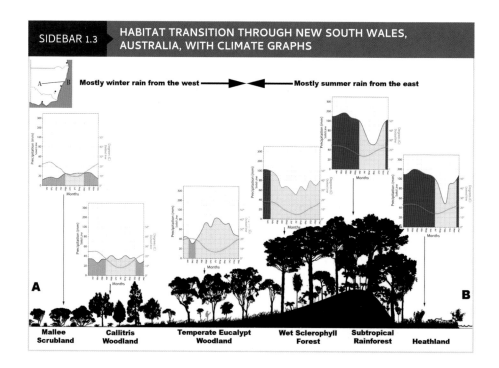

SIDEBAR 1.3

HABITAT TRANSITION THROUGH NEW SOUTH WALES, AUSTRALIA, WITH CLIMATE GRAPHS

Mostly winter rain from the west ➤◄ Mostly summer rain from the east

A

B

Mallee Scrubland

Callitris Woodland

Temperate Eucalypt Woodland

Wet Sclerophyll Forest

Subtropical Rainforest

Heathland

MALLEE WOODLAND AND SCRUBLAND

IN A NUTSHELL: A low, dry, uniform woodland of "mallee-form" eucalypts—with most branches sprouting from the tree bases—and very little undergrowth. **Habitat Affinities:** NEOTROPICAL CHACO SECO AND ESPINAL. **Species Overlap:** MULGA WOODLAND AND ACACIA SHRUBLAND.

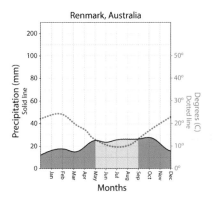

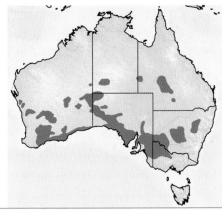

DESCRIPTION: Mallee is a scrubby woodland dominated by several species of eucalypt, such as White Mallee, Red Mallee, Yorrell, Gray Mallee, Fruit-ridged Mallee, and Soap Mallee, that form a distinctive structure. Found in areas of s. Australia with a Mediterranean climate, mallee woodland and scrubland generally grows on nutrient-poor, sandy soils on flat, open ground or gentle dunes. The lack of nutrients and low rainfall restrict the growth of the eucalypts, which develop in a coppice form (all branches sprouting from the low base), usually around 10 ft. (3m) high and rarely exceeding 20 ft. (6m) high. Mallee is usually open below, with little ground cover, but has a closed canopy formed by the interlocking of the multi-stemmed eucalypts. When ground cover exists, it is usually in the form of SPINIFEX grasses (*Triodia*) or sparse shrubs. There is usually a covering of leaf litter. Mallee is a fire-resistant habitat that undergoes rapid resprouting in the wet season after a fire.

Mallee is the primary habitat for Yellow-plumed Honeyeater. © IAIN CAMPBELL, TROPICAL BIRDING TOURS

Mallee in sw. New South Wales, showing the distinctive mallee-form trees, with coppice-like, multi-stemmed trunks. © IAIN CAMPBELL, TROPICAL BIRDING TOURS

WILDLIFE: Mallee areas are great places to go herp hunting, hosting numerous species of geckos, such as Thick-tailed Gecko and Beaded Gecko, many species of legless lizards, and Burton's Snake-Lizard. Notable skinks include the Western Blue-tongued Lizard and the flat Shingleback (or Stumpy-tailed) Lizard.

Large mammals include Red and Western Gray Kangaroos and Common Wombat and, in a few locations, Southern Hairy-nosed Wombat.

Shingleback Lizard sunning itself on a road cutting through the mallee in Western Australia.
© SAM WOODS, TROPICAL BIRDING TOURS

Birding in mallee habitat can be extremely varied because many of the characteristic species are blossom nomads, such as White-fronted, Purple-gaped (IS),Yellow-plumed (IS), and Black Honeyeaters, which are found in the area only when their food plants are in bloom. There are also resident species restricted to mallee, such as Chestnut Quail-thrush (IS), Red-lored Whistler (IS), Malleefowl, and Shy Heathwren. **Endemism:** Birds endemic to mallee include Black-eared Miner (IS), Mallee Emuwren (IS), Red-lored Whistler (IS), and Southern Scrub-Robin (IS).

DISTRIBUTION: Mallee woodland occurs through much of semiarid s. Australia. A large portion of the original mallee cover has been cleared, though some large tracts occur in Stirling Range National Park (Western Australia), Little Desert National Park (Victoria), and Gluepot Reserve (South Australia). However, many narrow corridors exist, and long strips of mallee habitat follow roads in many areas.

WHERE TO SEE: Round Hill Nature Reserve, New South Wales, Australia; Gluepot Reserve, South Australia; Stirling Range National Park, Western Australia.

MULGA WOODLAND AND ACACIA SHRUBLAND

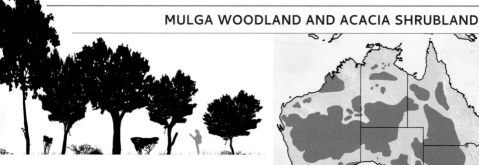

IN A NUTSHELL: Open acacia woodlands and shrublands of interior and s. Australia with a grassy or sclerophyllous ground cover. **Habitat Affinities:** NEOTROPICAL CHACO SECO AND ESPINAL. **Species Overlap:** TEMPERATE EUCALYPT WOODLAND; BRIGALOW WOODLAND.

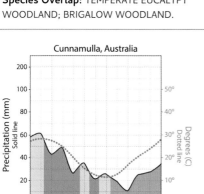

Cunnamulla, Australia

Precipitation (mm) Solid line

Degrees (C) Dotted line

Months

DESCRIPTION: Mulga woodland can be described as the non-fire-resistant acacia-dominated sister habitat to the fire-resistant eucalypt-dominated MALLEE WOODLAND AND SCRUBLAND. It gets its name from its dominant tree, an acacia called Mulga. In more arid areas, it occurs as shrubland, which is open, with stands of Mulga generally less than 15 ft. (5m) high, often interspersed with other shrubs, such as *Eremophila* (figwort family), and occasional eucalypts, such as Morrel in the west and Silver-leaved Ironbark in the east. In more hostile environments, Mulga woodland gradually gives way to CHENOPOD SHRUBLAND in areas of higher salinity, and to

Above: **Much of the remaining Mulga woodland is privately owned and overgrazed, as shown here, resulting in the absence of a significant understory.** © IAIN CAMPBELL, TROPICAL BIRDING TOURS

Left: **Pink Cockatoo is most abundant in Mulga areas.**
© BEN KNOOT, TROPICAL BIRDING TOURS

SPINIFEX habitat in drier areas, with the remaining Mulga and acacia shrubland decreasing to a shrub-steppe type of habitat.

In some areas, generally where drainage is impeded, Mulga woodlands can form a dense monotypic canopy, rarely up to 45 ft. (15m) tall, which makes this habitat look like a forest from more humid areas. Generally though, the canopy of Mulga trees is interspersed with Poplar Box, Gidgee, and Brigalow trees in the east and Myall in the southeast. The understory and ground cover are made up of shrubs and grasses, which vary markedly depending on subtle changes in substrate, so while the canopy of Mulga woodland may seem uniform from above, the understory can vary dramatically.

WILDLIFE: Many of the bird and mammal species found here range widely from one side of Australia to the other, and there is extensive overlap with MALLEE SCRUBLAND and TEMPERATE EUCALYPT WOODLAND. Birds that are widespread through the Mulga woodland of c. Australia include White-browed Treecreeper, Bourke's Parrot (IS), Redthroat (IS), Spiny-cheeked Honeyeater, Mulga Parrot, Chestnut-breasted Quail-thrush (IS), and Splendid Fairywren. Hall's Babbler (IS) is restricted to dense Mulga woodlands, while Gray Honeyeater and Slaty-backed Thornbill (IS) prefer acacia shrubland.

Mulga is one of the preferred habitats for the larger kangaroos, and Red, Western Gray, and Eastern Gray Kangaroos and Common Wallaroo are all found here in good numbers. Arboreal mammals are far fewer in number; the most common species is the Common Brushtail Possum.

Geckos feature heavily in Mulga woodlands, including velvet geckos, Box-patterned Gecko, Burrow-plug Gecko, Western Beaked Gecko, and Spiny-tailed Gecko. A large number of other lizards live here, including the Yakka Skink. Snakes typical of this environment include the Mulga Snake, Red-naped Snake, Dwyer's Snake, Mud Adder, Ringed Brown Snake, and Yellow-naped Snake.

Endemism: Hall's Babbler is a Mulga woodland endemic from sc. Queensland.

DISTRIBUTION: Mulga woodland and acacia scubland habitat is widespread and often encountered across Australia's arid interior, from Kalgoorlie, Western Australia, through Alice Springs and Uluru–Kata Tjuta National Park, Northern Territory, across to wc. Queensland.

WHERE TO SEE: Bowra Station, sw. Queensland, Australia.

The world's largest macropod, Red Kangaroo is common in Mulga woodland. © IAIN CAMPBELL, TROPICAL BIRDING TOURS

AUSTRALASIAN TEMPERATE WETLAND

IN A NUTSHELL: The billabongs and wetlands of s. Australia and New Zealand, which are much like the freshwater marshes of the Northern Hemisphere. **Habitat Affinities:** PALEARCTIC TEMPERATE WETLAND. **Species Overlap:** AUSTRALASIAN TROPICAL WETLAND.

DESCRIPTION: Wetlands can consist of permanent depressions, river floodplains, oxbow lakes (billabongs), human-made lakes, sewage works, and the ephemeral lakes of desert regions. Temperate wetlands generally lack the mats of floating vegetation that characterize the AUSTRALASIAN TROPICAL WETLANDS of the north, and they attract a very different suite of animals. All Australasian temperate wetlands are subject to the vagaries of the climate, and while most are permanent, they still experience fluctuations in water levels, filling up after rain and slowly evaporating during times of drought. The wetlands of New Zealand are far less affected by drought and therefore remain more or less permanent, with only minor water fluctuations. The lakes of far inland Australia, like Lake Eyre, are empty most of the time and may fill up only every few years. More permanent but shallow lakes can be thickly vegetated with Cumbungi (cattail or bulrush), which provides thick cover for shyer species. Some wetlands include flooded woodland, and often there are stands of dead trees with many tree cavities. Since these are surrounded by water, limiting access by land predators (mainly mammals), they are havens for hole-nesting birds such as parrots.

WILDLIFE: Bird populations in temperate wetlands fluctuate along with water levels, and while some species stay on, others are nomadic in nature, roaming great distances. Therefore, some species may be present at certain sites year-round for years at a time, but they may also be absent for similar periods in response to changing water availability. Sporadically flooded wetlands are important nesting areas for birds such as Australian Pelican and Banded Stilt. Permanent lakes attract secretive species like Australian, Spotless, and Baillon's Crakes; Black-tailed Nativehen (IS); and Australasian Bittern. In more coastal areas, the open waters are often permanent and attract species like Hardhead, Pink-eared, Blue-billed (IS), and Musk Ducks. The reedy edges are home to

A temperate wetland hosting Australian Pelicans, Pied Stilts, and Black Swans in se. Australia.
© IAIN CAMPBELL, TROPICAL BIRDING TOURS

Straw-necked Ibis is an abundant Australian wetland species (photo: s. Australia). © IAIN CAMPBELL, TROPICAL BIRDING TOURS

passerines such as Australian Reed Warbler and Little Grassbird (IS).

DISTRIBUTION: Temperate wetlands in New Zealand and Australia are widespread, and most cities and small towns have a sewage pond or town weir that holds some wetland birds. Some good examples of accessible permanent wetlands with a nice diversity of species are Whangamarino Wetland, North Island, New Zealand; and in Australia, Fivebough Wetlands, near Leeton, New South Wales; Gum Swamp near Forbes, New South Wales; Werribee Sewage Plant near Melbourne, Victoria; and Herdsman Lake in Perth, Western Australia.

WHERE TO SEE: Fivebough Wetlands, New South Wales, Australia; Werribee Sewage Plant, Victoria, Australia.

AUSTRALASIAN TROPICAL WETLAND

IN A NUTSHELL: Permanent or ephemeral wetlands, often surrounded by very arid landscapes. **Habitat Affinities:** AFROTROPICAL FRESHWATER WETLAND. **Species Overlap:** AUSTRALASIAN TEMPERATE WETLAND.

DESCRIPTION: Tropical wetlands are expanses of water that exist as part of river floodplains, billabongs (oxbow lakes), or even man-made structures such as dams, and these waters combine with swamp savanna, open marsh, herbaceous swamp, and flooded grasslands to form an incredibly species-rich mosaic of microhabitats. These natural wetlands can range from many feet deep in billabongs and near dams to extremely shallow on floodplains, with large swaths of exposed mud. Because most of these wetlands are in areas of monsoonal climate, they change dramatically through the year. Some of the shallow wetlands are ephemeral, filling up during the wet season (December–April), when the water can rise 15 ft. (5m), before drying out over the rest of the year, perhaps completely, before the rain arrives again. A feature of these tropical wetlands, particularly in shallow areas, is mats of floating vegetation. The wetlands are sometimes bordered by reeds or exposed mudflats in shallow areas and by *Pandanus* (screw pine) where there are steeper banks. Water lilies are often a major component and at times can cover a whole body of water.

WILDLIFE: The water-lily habitat is important for birds like Comb-crested Jacana and Green Pygmy-Goose and may also support massive congregations of Magpie Geese and whistling-ducks.

Yellow Water billabong in Kakadu National Park, Australia.
© IAIN CAMPBELL, TROPICAL BIRDING TOURS

At the end of the dry season, these wetlands may hold thousands of birds, including Black-necked Stork, Magpie Goose, Cotton and Green Pygmy-Geese (IS), Plumed and Wandering Whistling-Ducks, Pied and Little Pied Cormorants, a profusion of ibises, spoonbills, herons, and egrets, and masses of shorebirds, including Black-fronted and Red-kneed Dotterels and Australian Pratincole. In the dry season, these wetlands are also an important refuge for land birds such as pigeons, parrots, and finches, which come to drink.

Two of Australia's most sought-after reptiles, the monstrous and exceptionally aggressive Saltwater (or Estuarine) Crocodile and the much smaller and more passive Freshwater Crocodile both live in tropical wetlands. Northern Snake-necked Turtle and Northern Snapping Turtle compete with Merten's Water Monitor for basking positions on the water's edge, though the arrival of the Cane Toad has resulted in a serious population decline of the water monitor. The Keelback, a snake that does not suffer from the Cane Toad's toxin, can be found in good numbers around tropical wetlands, along with Australian Green Tree Snake, which hunts in brush overhanging the water's edge.

The water itself does not hold many mammals compared with temperate wetlands, but Agile Wallaby and Antilopine Wallaroo can often be found around wetland edges.

Comb-crested Jacana, a characteristic inhabitant of tropical wetlands, in n. Australia. © IAIN CAMPBELL, TROPICAL BIRDING TOURS

DISTRIBUTION: Tropical wetlands are found across far n. and e. Australia in areas with a monsoonal climate and in the lowlands of New Guinea, Halmahera (Moluccas), and the Lesser Sundas. The wetlands of the Sepik region and the northern slope of New Guinea are less monsoon-dependent and savanna-oriented, being surrounded by rainforest and swamp forest, while those of the island's Trans-Fly region are surrounded by savanna and are closely associated with the wetlands of mainland Australia. The most famous wetland in Australia is the enormous Yellow Water in Kakadu National Park (Northern Territory). The Yellow Water Cruise offers one of the best opportunities in the world for waterbird photography from a boat. In ne. Queensland, Lake Mitchell and Mareeba Wetlands on the Atherton Tableland are good examples of tropical wetlands.

WHERE TO SEE: Yellow Water, Kakadu National Park, Northern Territory, Australia; Wasur National Park, West Papua, Indonesia.

AUSTRALASIAN MANGROVE

IN A NUTSHELL: Mangroves along tropical coasts take on many different structures, from thin strips to full-blown forests. **Habitat Affinities:** INDO-MALAYAN MANGROVE FOREST (mangroves from se. Asia are closely related). **Species Overlap:** AUSTRALASIAN TROPICAL LOWLAND RAINFOREST.

DESCRIPTION: Mangroves are trees that are adapted to grow in a flooded environment. Most grow in intertidal conditions, often in large inlets and tidal river systems, though freshwater species do exist. Within our region, species diversity is highest in New Guinea and Cape York Peninsula, Australia, where diverse mangrove forests also support several salt-tolerant plant species that are not classified as mangroves, such as Mangrove Palm and Mangrove Lily. These mangrove forests can also have plants such as orchids from surrounding rainforests, which grow as epiphytes on the trunks and branches of mangrove trees. In the most developed of these forests, which are generally bordered on the land side by AUSTRALASIAN TROPICAL LOWLAND RAINFOREST, the change from mangrove forest to rainforest can be imperceptible from the air, and it is only at ground level that the boundary between habitats is obvious. There is considerable structural variation, both within and between species of mangroves, and different species tend to grow in slightly different conditions. Some large mangrove trees have straight trunks going into the mud, while most *Rhizophora* species have prop roots. These roots branch out spider-like from the bottom quarter of the trees and give the mangrove forests an eerie feel. On the ocean side, mangrove stands are protected from the sea by smaller fringe mangroves (described overleaf).

Moving south from northernmost Australia and New Guinea, the species diversity of mangroves drops off rapidly, from forests with some 40 species of mangrove, to just six species in far n. New South Wales. In s. New South Wales, New Zealand, and s. Western Australia, there is only one species, Gray (aka White) Mangrove. The southern monotypic stands tend to be shorter, less than 15 ft. (5m) tall, and lack the structure and species variety of the northern forests.

In both tropical and temperate areas, the fringe mangroves closest to the sea face more tidal action and wave action than more distal stands. The shrubs and trees growing here have to be most resistant to salty water, and they tend to be more stunted than more protected stands. Generally, fringe trees have aerial roots, which resemble pencils sticking out of the ground. The shrubs and trees here are generally 5–10 ft. (1.5–3m) tall and thicker than the protected stands. They gradually extend out into either the sandy beaches or tidal mudflats on the shoreline.

WILDLIFE: In each of the zones in our region, there are animals restricted to the mangrove forests. In n. Australia, mangrove bird species include Chestnut Rail (IS), Mangrove Robin (IS), Dusky Gerygone, Red-headed Myzomela, Australian Yellow White-eye, and Mangrove Fantail (IS). There are very few mammals restricted to n. Australian mangroves, with the exception of the False Water Rat, found from Northern Territory to se. Queensland. Due to the inaccessibility of this habitat to terrestrial mammal predators, bats such as the Black and Spectacled Flying Foxes use mangroves as roosting sites around n. Australia. Snakes present include the White-bellied Mangrove Snake, Banded Sea Krait, and Little File Snake. The most notable reptile of the northern mangroves as far south as c. Queensland is the Saltwater Crocodile, the apex predator of this environment. Lizards such as Mangrove Monitor and Rusty Monitor also live in these forests.

Farther south, the mangrove stands of se. Queensland hold Mangrove Honeyeater and Mangrove Gerygone. In this area, Australian Mudlark (aka Magpie-Lark) preferentially nests in mangrove stands; these birds lay deep-red eggs, in contrast to the off-cream eggs of birds of this species living outside the mangroves. Gray-headed Flying Foxes use these mangroves as roosting sites.

From the Moluccas to the Solomon Islands, the equatorial and tropical mangrove forests form an important habitat for a very distinctive bird assemblage. The most obvious group are the kingfishers, with Beach Kingfisher and Collared Kingfisher having strong associations with mangroves. Feeding on the mud at the base of the mangroves are more-terrestrial species, such as Chestnut Rail and Red-billed Brushturkey, as well as roosting Beach Thick-knee. Although the canopies do have passerines such as White-bellied Pitohui, Island Whistler, and Olive-crowned Flowerpecker, these mangrove forests are richest in parrots and pigeons, such Red Lory, Brown Lory, Blue-capped Fruit-Dove, Wallace's Fruit-Dove, and Yellowish Imperial-Pigeon.

Mangrove boardwalk, Cairns, Queensland, Australia.
© IAIN CAMPBELL, TROPICAL BIRDING TOURS

Australia's most fearsome predator, Saltwater Crocodile often loafs at the edges of mangroves.
© IAIN CAMPBELL, TROPICAL BIRDING TOURS

Beach Thick-knee foraging beside mangroves in ne. Australia.
© SAM WOODS, TROPICAL BIRDING TOURS

Even within a mangrove area, the distribution of some species of birds is highly stratified. For example, in Broome, Western Australia, the White-breasted Whistler prefers the fringe mangroves closer to the water's edge, while the sympatric Black-tailed Whistler prefers taller mangrove forests farther inland.

Endemism: Mangroves in Australasia, more so than any other continent, hold many species that are not found in other habitats. Because of the large extent of mangroves around the Australian coast, most mangrove birds are not considered range-restricted species, because the criteria for this designation require they be found in an area smaller than 19,300 sq. mi. (50,000km²). However, if the criteria took into account just how narrow the band of mangroves really is over the vast distances in which the habitat occurs, many of the Australian mangrove species would be regarded as endemic. With the criteria the way they are now, Mangrove Honeyeater (IS), from ne. Queensland and ne. New South Wales, is not regarded as range-restricted. This is a severe limitation of the "range-restricted" concept.

DISTRIBUTION: Extensive mangrove forest systems can be found in Australia around Cairns airport, Queensland; Darwin, Northern Territory; and Broome, Western Australia. Stands of Gray Mangrove are found around Brisbane airport, Queensland, and Ballina, New South Wales. Most of the Moluccan islands, New Britain, and the Solomon Islands, are ringed with mangrove forests. Large mangrove forests exist on the Fly and Kikori Rivers of New Guinea. In New Zealand, there are mangroves on the northern part of the North Island.

WHERE TO SEE: Mangrove forest: Jack Barnes Mangrove Boardwalk, Cairns, Queensland, Australia. Mangrove stand: Boondall Wetlands, Queensland, Australia.

AUSTRALASIAN ROCKY COASTLINE AND SANDY BEACH

IN A NUTSHELL: The typical sandy beaches and rocky headlands of s. Australia. Many have been developed, but there are still thousands of miles of undeveloped coastline. **Habitat Affinities:** NEOTROPICAL ROCKY COASTLINE AND SANDY BEACH. **Species Overlap:** AUSTRALASIAN TIDAL MUDFLAT.

Australia has a profusion of parrots seemingly adapted to every environment. Rock Parrot specializes in beach habitats. © SAM WOODS, TROPICAL BIRDING TOURS

DESCRIPTION: Most of coastal s. Australia and New Zealand is dominated by erosional headlands and rocky coastlines interspersed with depositional sandy beaches. The rocky areas can be in the form of sea cliffs, such as the many around s. Tasmania; rocky flats formed by lava flow, such as Flat Rock, in Ballina, New South Wales; or a wave-cut platform of sedimentary sandstones, such as Long Reef in New South Wales. The sandy beaches are often not simply one beach with foredune but a series of relict dunes paralleling the beach and representing paleo-shorelines, often containing saline lagoons that are used by wetland birds and shorebirds. In s. Australia, the slope of the beach changes markedly, from

A Pied Oystercatcher rests on a sandy beach. On rockier coastlines, this species can be seen alongside Sooty Oystercatcher. © PABLO CERVANTES DAZA, TROPICAL BIRDING TOURS

Hooded Plover nests on sandy beaches in s. Australia.
© SAM WOODS, TROPICAL BIRDING TOURS

very flat through the summer months to much steeper during the winter months, when rougher seas remove much of the sand that was deposited in the warmer season.

WILDLIFE: These habitats tend to be the domain of shorebirds and marine species. Rocky headlands are the preferred habitat of shorebirds like Sooty Oystercatcher, Whimbrel, Black-faced Cormorant, and Ruddy Turnstone. Sandy beaches provide habitat for Pied Oystercatcher, Hooded Plover, Sanderling, most tern species, and gulls. Penguins, including Little and Yellow-eyed Penguins, come ashore to Australian beaches late in the afternoon and nest in dunes.

Mammals that frequent these shores include marine mammals, such as Australian and New Zealand populations of Australasian Fur Seal, while Common Bottlenose Dolphins are often seen close to the shoreline. Non-marine mammals are rare in this habitat, though Dingoes can sometimes be found on remote beaches of the Australian mainland.

DISTRIBUTION: These habitats are found all around Australia, though more prominently in the temperate areas. Some great wild beaches are found in Broadwater National Park, New South Wales. Headlands popular for land-based sea-watching include Magic Point in Sydney, New South Wales; Bass Point near Shellharbour, New South Wales; Cape Nelson near Portland, Victoria; Eaglehawk Neck and Bruny Island, Tasmania; and Cape Naturaliste near Albany, Western Australia.

WHERE TO SEE: Bruny Island, Tasmania, Australia; Bass Point, New South Wales, Australia; Evans Head, New South Wales, Australia.

AUSTRALASIAN SANDY CAYS

IN A NUTSHELL: In the waters of n. Australia and the islands to the north and east, bare or sparsely vegetated sandy cays form over coral reefs. **Habitat Affinities:** INDO-MALAYAN OFFSHORE ISLANDS. **Species Overlap:** AUSTRALASIAN SANDY BEACH.

DESCRIPTION: In shallow tropical seas, large coral barrier-reef systems can develop. In some localities within these large systems, the sands derived from the breakdown of coral become concentrated to form sandy cays. In many areas these small islands are covered at high tide, but some are high enough to support sparse vegetation. Usually the vegetation is hardy grasses and shrubs, but on some islands, small, stunted trees can develop.

Right: **Tropical fish, like this Moorish Idol, abound on the barrier reefs of e. Australia.** © BEN KNOOT, TROPICAL BIRDING TOURS

Below: **A typical scene during breeding season on Michaelmas Cay, Queensland, Australia. Brown Boobies, Brown Noddies, and Sooty Terns dominate.** © IAIN CAMPBELL, TROPICAL BIRDING TOURS

WILDLIFE: These remote cays are very important for tropical seabirds, as they provide predator-free breeding areas. Some species are widespread and catholic in their nesting sites, breeding on nearly all the sandy cays. Species breeding on the more open, sandy, and sparsely grass-dominated islands include Brown and Black Noddies (IS), and Great Crested, Lesser Crested, Little, Bridled, Roseate, Black-naped (IS), and Sooty Terns (IS). Other species, like Red-footed Boobies and Lesser and Great Frigatebirds, require trees for nesting, while burrow-nesting species like shearwaters and petrels nest on the larger islands that have significant ground cover. These seabirds are joined by less pelagic species such as Pacific Reef-Herons and migrants such as Ruddy Turnstone and Sanderling.

DISTRIBUTION: These coral reefs and corresponding coral cays surround n. and ne. Australia (with the exception of the Gulf of Carpentaria), New Guinea, the Moluccas, and other islands of the sw. Pacific east to New Caledonia, wherever the seas are shallow enough to support reef growth. The biggest swath of these habitats is in the Great Barrier Reef, which extends down the e. Australian coast from n. Cape York Peninsula in ne. Queensland south to c. Queensland. The reef system is very accessible, with numerous tourist resorts strung along its entire length. The most accessible sandy cay is Michaelmas Cay, a 4-acre (1.5ha) island, about 20 mi. (30km) offshore from the coastal city of Cairns, which is widely considered the premier cay for visiting birders. It can hold up to 200,000 breeding seabirds.

WHERE TO SEE: Michaelmas Cay, Queensland, Australia; Green Island, Queensland, Australia.

AUSTRALASIAN TIDAL MUDFLAT AND SALT MARSH

IN A NUTSHELL: The muddy areas that are periodically exposed either through fluctuating tidal waters or movement of water by wind. **Habitat Affinities:** INDO-MALAYAN TIDAL MUDFLAT. **Species Overlap:** AUSTRALASIAN SANDY BEACH.

DESCRIPTION: Mudflats are usually exposed layers of estuarine sands, silts, and clays, and are found in bays and lagoons, on shallow, mud-dominated beaches, and along edges of slow-flowing rivers. In tidal mudflats, the sediment is deposited in the intertidal zone, where the swaths of mud are exposed and submerged twice a day. In other very shallow areas, wind can cause significant water-level change, and the silts are exposed or submerged based on wind direction. At depth, the muds are anaerobic, due to the small sediment size, so only a very specialized assemblage of animals is able to exist just below the mud surface. The regularly exposed silts and clays are extremely bioturbated by massive numbers of small organisms, including clams, Saltwater Yabbies, and worms. Crabs abound in this habitat, spending time both under the mud and feeding on the exposed flats when the waters recede. Few plants can grow

Red-breasted Dotterel is confined to New Zealand, where it forages on tidal flats during winter, in some areas moving inland to breed.
© LISLE GWYNN, TROPICAL BIRDING TOURS

in these harsh environments, but the mudflats are ringed by mangroves in tropical and subtropical areas. In more temperate regions, they are ringed with salt-marsh grasses that are inundated with each tide or, higher up, are inundated a few times a month. These grasses are able to survive in hypersaline conditions even more exteme than in the mudflat closer to the water. Walking on the dryer or sandy edges of these flats is easy, but care is required as some areas have very soft mud and are difficult to navigate. Other, rarer areas have quicksand, such as Homebush Bay, Sydney, where, as the author can attest from personal experience as a 15-year-old, one can sink thigh-deep into the mud, become very worried about the incoming tide, and need to be rescued by local police!

WILDLIFE: Mudflats and salt marshes are important roosting and feeding areas for many shorebirds. The vast majority of species are migratory waders, which breed in Siberia or Alaska and are therefore present only during the austral summer (October–March), but some species are resident, such as Beach Thick-knee (IS), Pied Oystercatcher, and Red-capped Plover. Mudflats also provide important resting and feeding areas for terns and gulls. Tidal flats vary in structure, and some species prefer flats with a sandy substrate (e.g., Sanderling and Lesser and Greater Sand-Plovers), while others prefer the muddy substrates of estuaries. Different shorebirds also feed in different areas of the flats, with the species segregating on the shoreline depending on their leg and bill structures, which dictate the depth of water in which they can feed. It can be difficult to see shorebirds at low tide, as they are often quite a distance away. The best viewing opportunities are generally as the tide comes in and birds are forced up the shore by the incoming tide. Once the tide has covered the feeding areas, the birds will travel to high-tide roosts, where thousands of waders may gather to wait for the next low tide. This can be the best opportunity to view and photograph them, when the many birds in one place allow direct comparisons of different species.

DISTRIBUTION: Tidal mudflats are found throughout Australasian coastal regions. They are most prevalent around estuaries and sheltered bays and, in Australia, also beaches in the north of the country where the seas are calm.

WHERE TO SEE: Roebuck Bay near Broome, Western Australia; Lee Point near Darwin, Northern Territory, Australia; the Esplanade in Cairns, Queensland, Australia; Richmond River estuary, Ballina, New South Wales, Australia; Port Phillip Bay, Victoria, Australia; Wasur National Park, West Papua, Indonesia; Biak Island, Indonesia; Miranda, North Island, New Zealand.

Wrybill breeds along rivers in New Zealand but forms feeding flocks on tidal flats during the winter months.
© LISLE GWYNN, TROPICAL BIRDING TOURS

AUSTRALASIAN PELAGIC WATERS

IN A NUTSHELL: The coastal, continental, and deep-sea waters throughout the region. **Habitat Affinities:** AFROTROPICAL PELAGIC WATERS. **Species Overlap:** None.

Chatham Albatross, which breeds on Chatham Island (off New Zealand's main islands), mainly forages in Australasian waters. © LISLE GWYNN, TROPICAL BIRDING TOURS

DESCRIPTION: In many tropical areas, the edge of the continental shelf (aka continental break), where the ocean rapidly becomes much deeper, is usually far offshore; all the waters between Australia and New Guinea are shallow. Where the edge of the continental shelf is close to shore, as it is along the southeast and southwest coasts of Australia (but not the Great Australian Bight), and the east and southwest coasts of New Zealand, upwellings of nutrient-rich, colder waters produce substantial concentrations of sea life, from microscopic organisms right up the food chain to large pelagic fish. This continental break is always more productive than the shallow waters coastward and the very deep oceans seaward.

WILDLIFE: New Zealand and Australia have a diverse pelagic bird and cetacean fauna and are widely considered some of the best seabird-viewing areas of the world. They attract a great diversity of pelagic birds, including shearwaters, petrels, storm-petrels, prions, albatrosses, and gannets. Seasonality is very important to pelagic bird movements: generally the austral winter (June–August) is best for pelagic groups such as prions, albatrosses, and some of the more southerly breeding petrel species (e.g., Cape Petrel, Northern Giant-Petrel, and storm-petrels). The austral summer (October–February) is better for groups such as shearwaters, northerly breeding petrels (e.g., Providence, Great-winged, and most other *Pterodroma* petrel species), and tropical-breeding species such as tropicbirds and boobies. Generally, the calmer the sea, the duller the birding and the better the cetacean-watching, as many bird species require updrafts produced from waves to fly and feed.

The cetaceans found in the pelagic waters of the region include Southern Right, Long-finned Pilot, Pygmy Right, Blue, and Antarctic Minke Whales.

DISTRIBUTION: Around most of s. Australia, regardless of the width of the continental shelf, the shelf break (the edge of the shelf) and continental slope are very steep, dropping to the deep and flat abyssal plain. Most of New Zealand also has a wide continental shelf, with a gradual continental slope. The notable exceptions are in sw. South Island, where the continental shelf to the abyssal plain is extremely steep; the sw. tip of the North Island; and a small area in ne. South Island where a trough and sea canyon cut through the continental shelf and the continental slope. This canyon, near Kaikoura, South Island, is debatably the most spectacular pelagic seabird location in the world, a place where many pelagic birds and cetaceans can be seen from shore, though there are many short boat trips available to get very close to the action. Most Australian pelagic tours depart from southern ports and take a few hours to get out to the continental break; regular trips leave from Sydney and Wollongong, New South Wales; Eaglehawk Neck, Tasmania; Portland, Victoria; and Albany and Perth, Western Australia.

WHERE TO SEE: Wollongong, New South Wales, Australia; Eaglehawk Neck, Tasmania, Australia; Kaikoura, South Island, New Zealand.

AUSTRALASIAN LARGE-SCALE FARMING

IN A NUTSHELL: The farmland that has replaced most of the forests of New Zealand and se. Australia. **Habitat Affinities:** AFROTROPICAL GRAZING LAND; NEARCTIC CROPLAND. **Species Overlap:** Overlap with surrounding habitats when small corridors are preserved.

DESCRIPTION: Large-scale crop farming, and especially wheat farming, is common throughout s. Australia. The wheat farms of New Zealand are the most productive in the world, recording almost 7.5 short tons per acre (17 metric tons/ha). This farming has drastically altered massive tracts of natural habitat throughout inland se. and sw. Australia and the South Island of New Zealand. This has resulted in the severe fragmenting of habitats such as MALLEE WOODLAND AND SCRUBLAND in New South Wales, Victoria, and Western Australia. While some species have lost out within this equation, the new croplands and permanent water sources have benefited others. Country roads of

Red-rumped Parrots often feed on farmlands and perch on their perimeter fences in small groups (photo: s. Australia). © BEN KNOOT, TROPICAL BIRDING TOURS

Cockatiels are native to Australia and may gather in large flocks around bores on Outback farmlands (photo: n. Australia). © SAM WOODS, TROPICAL BIRDING TOURS

inland Australia in these wheat belts have become established strongholds of birds such as Rufous and Brown Songlarks, Australasian Bushlark, Australian Kestrel, Black-shouldered Kite, and a host of parrot species, including Australian Ringneck, Greater Bluebonnet, and Red-rumped Parrot. Conversely, species such as Malleefowl, Gilbert's Whistler, and Turquoise and Superb Parrots (among others) have lost much of their range to the expansion of these agricultural operations.

Grazing lands for beef cattle, dairy cattle, and sheep in New Zealand have a lot in common with the dairy-cattle grazing lands of e. and sw. Australia, where clear-felling of the original forest has resulted in the typical Western European image of cleared fields with hedgerows and an occasional stand of trees for shade. In the traditionally forested regions of New Guinea and elsewhere in the sw. Pacific, pig and goat grazing, coconut plantations, and cassava fields result in more of a polyculture environment.

Widespread beef-cattle grazing occurs in the savanna and open woodland parts of lowland New Guinea and has occurred in some form across nearly the whole of the Australian continent, even the arid inland. The intensity of this grazing depends on the suitability of local conditions. While the effect has not been significant on some inland areas where grazing is light due to lack of

Western Gray Kangaroo grazing on agricultural land.
© IAIN CAMPBELL, TROPICAL BIRDING TOURS

nutrients, such as in DUNE AND ROCKY SPINIFEX habitat, more productive areas, such as Mitchell Grass expanses and most of e. Australia, have been heavily impacted. Much of the inland beef-cattle grazing occurs in unfenced private concessions with open-range, low-density operations (reminiscent of the vast ranches of the American Southwest), and the original savanna or woodland is intact or only slightly disturbed here. These private holdings appear much wilder and less obviously human-modified than the landscape of the south of the country, which has undergone dramatic and conspicuous modification. In the north, the bores and dams associated with cattle farms can be highly productive for birds, which gather to drink in significant numbers, especially in late afternoon and at dusk. Indeed, this can be one of the best ways to find many northern birds. Examples are cockatiels, Gouldian Finch as well as other finches, and a variety of honeyeaters. Agile Wallaby and Antilopine Wallaroo occur in low-intensity grazing lands, as do Sand Goanna and Frilled Lizard.

In New Zealand, sheep farming is concentrated in the center of the North Island and the east side of the South Island. In se. South Island, where the native grasslands were easily converted to highly productive pastures, sheep vastly outnumber humans by over 40 to 1. In Australia, intense sheep farming is concentrated in the southern rangelands (i.e., the Great Dividing Range and the slopes to its west, or inland), where large swaths of natural open woodlands and MALLEE WOODLAND AND SCRUBLAND have been cleared, though here much more of the original vegetation is left in corridors, as compared with dairy farms in coastal areas of Australia or sheep farms in New Zealand. Most sheep farming, however, is on more marginal land, such as expanses of w. New South Wales and inland Western Australia, where comparatively little of the original MULGA WOODLAND AND ACACIA SHRUBLAND has been cleared, and the sheep feed mainly on native grasses, shrubs, and chenopod plants. Southern grazing lands provide habitat for Emu, songlarks, Banded Lapwing, buttonquails, and even Plains-wanderer in some specific areas of New South Wales and Victoria. Some of the larger kangaroos, such as Eastern Gray, Western Gray, and Red, have thrived in these southern grazing lands and should be easily found if you are within their range.

BIOMES

- Pine-Oak Woodlands
- Dry Deciduous Forests
- Temperate Grasslands
- Evergreen and Semi-Evergreen Forests
- Coniferous Forests
- Savannas
- Paramo and Tundras
- Deserts and Arid Shrublands
- Mediterranean Woodlands
- Neotropical Regions

HABITATS OF THE NEOTROPICS

(CENTRAL AND SOUTH AMERICA)

VALDIVIAN RAINFOREST AND NEOTROPICAL MIXED CONIFER FOREST

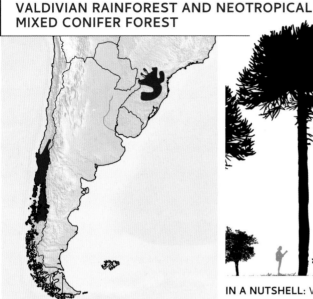

IN A NUTSHELL: Very wet broadleaf and coniferous forests that have evolved separately in Chile and se. Brazil. **Habitat Affinities:** NEARCTIC TEMPERATE RAINFOREST. **Species Overlap:** MAGELLANIC RAINFOREST.

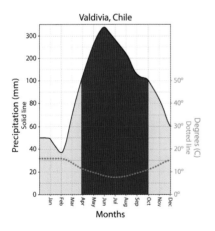

Valdivia, Chile

DESCRIPTION: This very broad habitat category contains many localized habitats, some of which have highly distinctive dominant canopy species. Because they have broadly similar forest structure and wildlife assemblages, they are all considered together here. In the s. Brazilian highlands, the broadleaf Atlantic forests blend with conifers to form a mixed conifer-broadleaf forest similar to the forest of this type in Chile, though the rainfall comes from the mid-Atlantic rather than the s. Pacific Ocean. In Chile south of Santiago, coastal areas are covered in wet broadleaf evergreen, broadleaf deciduous, and wet conifer forests. In the south of this region, Valdivian forests are extremely wet, receiving up to 5 ft. (1.5m) of precipitation per year. Here, the southern beeches (*Nothofagus*) of the MAGELLANIC RAINFOREST farther south are mixed with broadleaf trees and stands of Patagonian Cypress. Sometimes the forest is dominated by evergreen *Nothofagus* that has a layered two-canopy structure; this is referred to as *siempre verde* forest. In other areas, the Patagonian Cypress forms soaring, homogenous stands, and this forest is referred to as *alerce* forest. On the Andean side of this zone, in areas of high precipitation, low temperatures, and poor soils, grows a very distinctive *Araucaria* conifer known as the Monkey Puzzle Tree. This relict species from the Jurassic period

Striking Valdivian rainforest in Nahuelbuta National Park, Chile.
© NICK ATHANAS, TROPICAL BIRDING TOURS

forms towering forests 130 ft. (40m) tall. At the top of the daunting straight trunk, the canopy forms an umbrella shape, similar to an oversize Dragon Blood Tree of Socotra (Yemen) or the Umbrella Pine of the Mediterranean basin. The trunks of these trees are not buttressed as they are in NEOTROPICAL LOWLAND RAINFOREST. The canopy usually has a wide range of tree heights and is rather open, allowing more light, relative to other rainforests, to reach the understory; hence the understory is extremely thick with bamboo and ferns. Climbing plants are common in these forests and form a significant proportion of the mid-canopy.

WILDLIFE: Because the wet habitats of s. Chile are surrounded by desert and semidesert, a very unique animal life has developed in these forests. Although diversity is low, in comparison to the moister forests in the tropics, the percentage of endemics is far higher. Notable bird species include Black-throated Huet-huet (IS), Slender-billed Parakeet (IS), Chucao Tapaculo, Ochre-flanked Tapaculo, and Fire-eyed Diucon. Mammal diversity is low in these Chilean forests, but they host some very local species, such as Monito del Monte, a member of an ancient order of marsupials; Southern Pudu, the world's smallest deer; and Kodkod, a rare cat.

Several large and impressive tapaculos occur in Valdivian rainforest, including Chucao Tapaculo, of c. Chile. © NICK ATHANAS, TROPICAL BIRDING TOURS

In Brazil, the mixed conifer forests share many species with the surrounding NEOTROPICAL SEMI-EVERGREEN FOREST, NEOTROPICAL CLOUD FOREST, and NEOTROPICAL LOWLAND RAINFOREST, so endemism is limited compared to the Chilean forests. Species typical of the Brazilian forests include Vinaceous-breasted Parrot, Striolated Tit-Spinetail, and Southern Brown Howler Monkey.

Endemism: Because these forests are so distinct and generally limited in range, birds have evolved in isolation in each of them. Black-throated Huet-huet, Chucao Tapaculo, and Ochre-flanked Tapaculo are among the endemics of the Chilean forests. Striolated Tit-Spinetail, Araucaria Tit-Spinetail, Black-capped Piprites, and Red-spectacled Parrot are all restricted-range species primarily found in this habitat in se. Brazil.

DISTRIBUTION: In Chile, these forests originally occurred from Concepción to the MAGELLANIC RAINFOREST of s. Chile, and from the coast to the steppes of the high Andes. Much of the Valdivian rainforests and the vast majority of the conifer forests have been logged. In se. South America, mixed conifer forests are found in se. Brazil and extreme ne. Argentina, from 1,640 ft. to 5,200 ft. (500–1,600m) elevation.

WHERE TO SEE: Nahuelbuta National Park, Chile.

NEOTROPICAL DESOLATE DESERT

IN A NUTSHELL: A desolate, cold desert often completely lacking vegetation. **Habitat Affinities:** EAST ASIAN COLD DESERT; AFROTROPICAL DESERT. **Species Overlap:** NEOTROPICAL SEMIDESERT SCRUB.

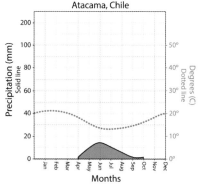

DESCRIPTION: This habitat includes hot deserts, cold deserts, and fog deserts. All very barren, they are treated together here. In n. Perú, the rainfall can be as low as 0.5 in. (12.7mm) annually, and near Antofagasta, Chile, decades can go by with no precipitation at all. In the northernmost part of this habitat, called the Sechura Desert, precipitation comes as winter rain, especially associated with El Niño years. Most of the vegetation near the coast is halophytic (salt-tolerant) grasses and small shrubs. Farther away from the coastline, the sand dunes are barren, but the rocky areas contain mesquite bushes and cacti that can grow to 30 ft. (9m). The southernmost section of this habitat is the Atacama Desert, which is the driest nonpolar place on the planet and best described as a moonscape. Most of this area is absolutely devoid of vascular plants; however, on the very steep western slope approaching the Pacific Ocean, fog forms between 1,000 and 3,000 ft. (300–900m), supporting limited vegetation called *loma* scrub, with columnar cacti, prostrate cacti, euphorbia shrubs, and small sclerophyllous shrubs growing in bands determined by fog thickness. At the southern edge of the Atacama, these bands merge into the MATORRAL SCLEROPHYLL FOREST AND SCRUB, which extends farther south. The most diverse section of Neotropical desolate desert is found from n. Perú to northernmost Chile. Here, winter fogs (*camanchaca* or *garúa*) form from the cold waters of the Pacific Ocean and come ashore at night (see sidebar 2.1). In exposed areas, or on the flatter areas with broad coastal plains below the thick fog zone, most of the dew evaporates in the first minutes of sunlight, but in regions with greater relief and protected gullies, moisture has enough time to filter into the ground and support plant life. In these *loma* scrub areas, cacti around 12 ft. (4m) tall, halophytic bushes around 3 ft. (1m) tall, and ground bromeliads can form thick stands, mimicking deserts of Baja California with much higher precipitation. Higher on these mountains, from 3,500 ft. (1,070m) to the mountaintops, there is slightly higher precipitation from occasional orographic rains (see sidebar 3.3), supporting columnar cactus thickets in protected gullies. In some areas with at least some fog and underlying groundwater, you can find groves of Tamarugo trees that resemble a casuarina forest in c. Australia. These trees grow with very few grasses and small sage-like bushes beneath them but otherwise seem incongruous in this harsh environment.

The Atacama Desert in n. Chile.
© NICK ATHANAS, TROPICAL BIRDING TOURS

WILDLIFE: The core of the Atacama Desert in n. Chile is the most desolate place on earth away from the poles; just Gray Gull and Grayish Miner use this habitat. North of the Atacama, the Peruvian–Chilean desert has many more species, with endemics like the Tamarugo Conebill (IS) in Tamarugo groves, Coastal Miner, and Seaside Cinclodes. Most of the birds of this desert habitat are found in the *lomas* in the *camanchaca* and *garúa* fog zones; these include such furnariids as Cactus Canastero (IS) and Thick-billed Miner.

Other typical birds are Peruvian Thick-knee, Short-tailed Field Tyrant, Dark-faced Ground-Tyrant, Slender-billed Finch, Raimondi's Yellow-Finch (IS), and Peruvian Pipit.

Compared to similar desert habitats in Namibia, Africa, few amphibians and surprisingly few reptiles live here. Mammal life is equally limited, consisting mainly of small rodents like Darwin's Leaf-eared Mouse. Few sizable carnivores can thrive here; the largest widespread carnivore is South American Gray Fox (aka Chilla or Patagonian Fox).

Peruvian Thick-knees inhabit this arid landscape.
© NICK ATHANAS, TROPICAL BIRDING TOURS

Mountain Viscacha occupies rocky outcrops within desolate desert (photo: Perú).
© JOSÉ ILLÁNES, TROPICAL BIRDING TOURS

Endemism: The deserts of Chile and Perú are rather contiguous, so the bird and plant species are not usually range-restricted. There is one area of endemism, the Perú–Chile Pacific slope, with restricted-range desert birds like Coastal and Thick-billed Miners and Slender-billed Finch.

DISTRIBUTION: The Neotropical desolate desert complex ranges along the coast from southernmost Ecuador, where it blends into NEOTROPICAL SEMIDESERT SCRUB, south to La Selena in c. Chile, where it then merges into MATORRAL SCLEROPHYLL FOREST AND SCRUB. To the east it is bounded by PUNA grasslands and PATAGONIAN STEPPE. Overall, it is rarely more than 60 mi. (100km) wide.

WHERE TO SEE: Paracas National Reserve, Perú; Atacama Desert, near Atacama City, along the n. Chile coast.

SIDEBAR 2.1   **THE EFFECTS OF WORLD OCEAN CURRENTS ON CLIMATE**

Continental climates are those that are not really modified by ocean currents. However, many of the world's climates are significantly modified by currents, making them very different from what they would be in the absence of these oceanic influences. This may also be said of the corresponding habitats found within these current-affected areas. Western Europe offers a prime example: Scotland and Norway are kept warmer than places farther east in Europe, like Poland and Belarus, because the North Atlantic Current brings warm water from the tropics; the waters of such currents warm the air and promote rain. Areas such as the southwest coasts of South America, Africa, and Australia, as well as the Mediterranean Sea coast and the southwest coast of North America, are influenced by cold offshore waters. As a result, these areas all have Mediterranean climates, with little rainfall, hot dry summers, and scant winter precipitation.

World Ocean Currents and Modified Habitats

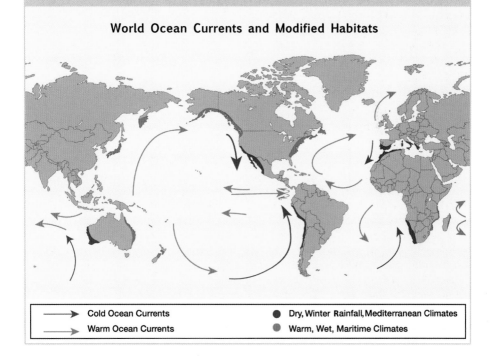

→ Cold Ocean Currents	● Dry, Winter Rainfall, Mediterranean Climates
→ Warm Ocean Currents	● Warm, Wet, Maritime Climates

GALÁPAGOS DESERT AND SCALESIA

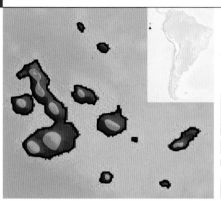

IN A NUTSHELL: A desert and an elfin forest in extreme proximity, separated from the rest of the region by hundreds of miles of ocean. **Habitat Affinities:** NEOTROPICAL THORNSCRUB. **Species Overlap:** None.

DESCRIPTION: The Galápagos archipelago straddles the equator 600 mi. (1,000km) west of the Ecuadorian coast. The archipelago covers 17,000 sq. mi. (44,000km²), with 3,000 sq. mi. (7,800km²) of land, plus a very small (5 mi./8km long) outlier, Cocos Island, 350 mi. (550km) southwest of Costa Rica.

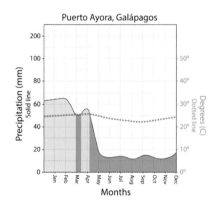

Puerto Ayora, Galápagos

The islands are a result of hot-spot volcanism under the eastward-moving Nazca Plate and get younger westward (see sidebar 2.2), but all have similar types of geology and geomorphology. Some, such as Santiago Island, are formed from very flat volcanoes and have no significant highland area, but most larger islands are formed from more conical volcanoes and have some areas above 1,000 ft. (300m), where the islands catch orographic rainfall (see sidebar 3.3).

Low-lying areas of these islands are covered with extensive desert scrublands, and because the islands are fairly flat, with volcanoes rising very gradually, this habitat makes up the vast majority of the area. This desert community is characterized by many large cacti, including multiple species of prickly-pears and candelabra cacti that grow up to 12–24 ft. (4–8m) tall. Mixed with the cacti are trees such as small acacias and the deciduous Palo Santo, which grows over 12 ft. (4m) tall and has wood that smells of incense, and Palo Verde, which can grow in shrub or tree form in the Galápagos, to 20–40 ft. (6–12m). Underneath these trees are small deciduous shrubs such as Darwin Cotton, Galápagos Lantana, and the small Lava Cactus.

The cacti of these desert shrublands become rare above 650 ft. (200m) elevation, where acacias and Palo Verde mix with trees of the genus *Scalesia* to become forest. This transition zone occurs lower on the islands' southern, wetter faces than on their northern, drier faces. Above 1,000 ft. (300m), most of the vegetation is evergreen, the scalesia trees become dominant, and the forest takes on a very different structure, similar to NEOTROPICAL SEMI-EVERGREEN FOREST on the mainland.

Above: **Desert scrublands along the shoreline of Santa Fé, in the Galápagos Islands of Ecuador.** © IAIN CAMPBELL, TROPICAL BIRDING TOURS

Right: **Many animals, like these Española Mockingbirds, are incredibly tame on the Galápagos, which are largely free of natural ground predators.** © SAM WOODS, TROPICAL BIRDING TOURS

Scalesia trees can form almost uniform canopies over 30 ft. (10m) tall, though usually the trees are around 15–18 ft. (5–6m) tall. The few other canopy or understory trees include Cat's Claw and Milk Berry. All the trees in this forest are laden with epiphytic bromeliads. There is a thick understory of low evergreen shrubs, with many mosses and liverworts as ground cover.

WILDLIFE: The wildlife of the Galápagos is quite depauperate, as in many archipelagos, but in similarity with them, the island chain has a high degree of endemism, or species found nowhere else. The arid zone is also the primary home of reptiles like the endangered Galápagos Pink Land Iguana, all seven species of endemic lava lizards, and all four endemic snake species. The only native land mammals on these islands are the Hoary Bat and the Galápagos Rice Rat.

The avifauna is embodied by many fewer bird groups than elsewhere in South America, comprising principally seabirds, waterbirds, and shorebirds; the few land birds are predominantly finches and mockingbirds. Other groups, like cuckoos, flycatchers, warblers, raptors, owls, and doves, are typically represented by only one or two species in the islands. Although many seabirds, naturally, may be found foraging in pelagic or coastal habitats outside of those covered here, some species breed in microhabitats within the Galápagos desert and scalesia habitats. This includes Galápagos Petrel, which breeds high in the scalesia zone, while Magnificent and Great Frigatebirds

The islands have a conspicuous reptile fauna, including this Galápagos Land Iguana and giant tortoises.
© PABLO CERVANTES DAZA, TROPICAL BIRDING TOURS

and Red-footed Boobies breed just outside the coastal zone, where trees provide nesting sites. Higher up, where cacti have given way to forest, is the main region of the islands' namesake, the giant tortoises. "Darwin's" finches, among the most talked-about Galápagos land birds, are a distinctive group within the tanager family (Thraupidae). They currently number 17 species, all of which are endemic to the Galápagos, and all of which occur in Galápagos desert and scalesia habitats. However, they vary in their preferred microhabitats. Vegetarian and Woodpecker Finches and all three *Camarhynchus* tree-finches occur mainly in the highlands, above the arid zone. Conversely, all nine *Geospiza* finches (seven ground-finches and two cactus-finches) are found only, or are most abundant, within the arid zone, and the critically endangered Mangrove Finch is confined to that microhabitat. The two species of *Certhidea* warbler-finches differ: Green Warbler-Finch is considerably more abundant in the highlands, while Gray Warbler-Finch is found only in the arid zone. All four species of endemic mockingbirds also occur in the arid zone, and Galápagos Dove nests among *Opuntia* cacti there. Other endemic bird species are generalists, like the Galápagos Hawk, which can be found at all elevations, and the widespread and conspicuous resident Yellow Warbler. The elusive Galápagos Rail occurs and nests only in the uppermost areas (usually above 1,640 ft./500m) of the scalesia zone.

Endemism: Almost all of the land birds of the Galápagos are endemic. This includes all the finches, mockingbirds, Galápagos Flycatcher, and Galápagos Rail, the lattermost restricted to the scalesia zone of the highlands in particular. While many of the birds have a preference for either the highlands (e.g., Vegetarian and Woodpecker Finches and Sharp-beaked Ground-Finch) or the arid zone (e.g., cactus-finches, ground-finches, mockingbirds), few species are completely restricted to either of these markedly different zones. All the "highland" species mentioned above, with the exception of Galápagos Rail, are known to wander to lower elevations at least occasionally, perhaps representing little known seasonal movements. Just two species seem to be truly restricted to the arid zone: Large Ground-Finch and Gray Warbler-Finch. Some of the others in this category can be explained away by simple geography; Española and Floreana Mockingbirds, Genovesa Cactus-Finch, and Genovesa Ground-Finch all occur on low-lying islands with no highland component, so they are bound to low elevations by geography. The mockingbird species on other islands reach into the highlands too. The relatively few birds exclusive to these microhabitats underline the fact that on islands like the Galápagos, only a very small number of birds are unable to drift between microhabitats, a situation unlikely to be seen on the mainland, where greater numbers of species would provide competition and a more distinct avifauna from one microhabitat to the next (and would in turn define these habitats). Most of the reptiles are also endemic to the Galápagos, including the giant tortoises, land iguanas, and all of the lava lizards.

DISTRIBUTION: This account covers all terrestrial habitats on the Galápagos Islands.

WHERE TO SEE: The Galápagos Islands are best explored by boats that provide cruises around the islands. Most flights to the Galápagos land on Baltra Island.

SIDEBAR 2.2   HOT-SPOT VOLCANISM AND PLATE MOVEMENTS

Many of the world's island archipelagos and underwater seamounts are caused by hot-spot volcanism. In the earth's mantle are areas of extreme heating and melting of rock. Periodically (i.e., over millions of years), the molten rock moves to the surface as magma and forms volcanoes or lava flows. As the earth's plates (which form the lithosphere) move over the mantle, these volcanoes form islands atop seamounts in the ocean, and on land form volcanoes, for example in Yellowstone National Park, Wyoming (United States), or flood basalts, such as the Deccan traps of India or the Paraná flood basalts of Brazil and Uruguay. In the cross-section diagram, you can see that the age of the Galápagos Islands gets younger in a westerly direction; Isla Española, at 3.2 million years old (3.2 Ma), is very flat and eroded. Isla Isabela, in the west of the archipelago, is still forming and is the highest island in the chain. The Hawaiian Islands also formed this way, and in the globe diagram you can see that this hot spot has formed an island and seamount chain, which is 65 million years old, extending across much of the Pacific Ocean. What is fascinating about this chain is that about 43 million years ago, the chain changed direction. There are two main explanations for this: The first is that the Pacific Plate was moving north over the hot spot between 65 and 43 million years ago, and then it started moving to the west and continued to do so up to the present. The other explanation is that the hot spot itself moved south within the mantle before becoming stationary around 43 million years ago, and then the Pacific Plate moved west over it. Geology is just so cool!

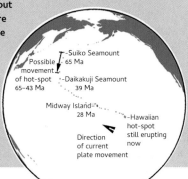

Hot-spot Volcanism Forming the Galápagos Islands

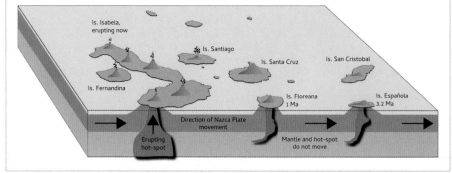

NEOTROPICAL THORNSCRUB

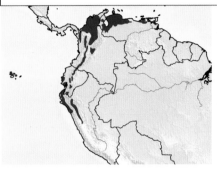

IN A NUTSHELL: A low, scrubby habitat made up of thorny bushes with small leaves and abundant cacti, found throughout much of the Caribbean and in dry inter-Andean valleys. **Habitat Affinities:** NEARCTIC MESQUITE BRUSHLAND AND THORNSCRUB; INDO-MALAYAN THORNSCRUB; AFROTROPICAL DRY THORN SAVANNA AND THORNSCRUB. **Species Overlap:** NEOTROPICAL SEMIDESERT SCRUB.

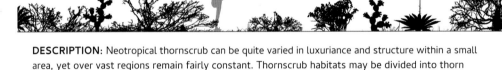

DESCRIPTION: Neotropical thornscrub can be quite varied in luxuriance and structure within a small area, yet over vast regions remain fairly constant. Thornscrub habitats may be divided into thorn savanna, thornscrub, thorn brush, and thorn thicket in regions like the Afrotropics and Palearctic,

Cactus ground cover under a thorny acacia canopy, near Quito, Ecuador.
© PABLO CERVANTES DAZA, TROPICAL BIRDING TOURS

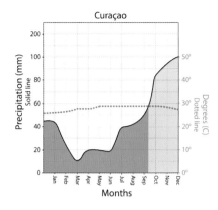

Curaçao

where all these forms have their own plant species assemblages. But in the Neotropics, the plants remain fairly uniform, and the bird assemblages of both thornscrub and thorn thickets are usually the same, so they are treated as a single habitat. Neotropical thornscrub habitat includes deciduous, sclerophyllous, and evergreen shrubs, with very small leaves, such as *Mimosa* species, and an understory and ground cover of cacti and other succulents.

Many of the deciduous shrubs flower during the dry season, when they don't have leaves, creating a profusion of flowers in an otherwise spartan scene. The very open canopy ranges from 10 ft. to 30 ft. (3–10m) in height, and beneath it are thick shrub and ground layers of succulent plants. The ground cover has a seemingly incongruous combination of cacti with bromeliads and orchids, and the abundance of Spanish Moss also seems out of place in this environment. Combined, these attributes of arid and wet environments together give the Neotropical thornscrub a distinctive feel. Walking through this habitat is very difficult, as the brush layer is thick and much of the canopy is at head height.

Absolute rainfall can be high in these areas, but it is periodic, leaving prolonged spells of drought. It is the erratic and intense nature of the wet and dry seasons that promotes thornscrub over other types of arid vegetation. Neotropical thornscrub forms over a wide variety of conditions that are unfavorable for more lush forest growth. Cyclical extreme droughts stop more semi-evergreen forests from colonizing these areas. Aridity and locally harsh growing conditions are caused by a variety of factors. In some coastal locations, localized aridity is caused by easily drained sandy soils. In some mountain areas, soils formed from recent volcanic ashfalls from andesitic volcanoes are very nutrient-deficient and retard forest development. In some other regions, such as inter-Andean valleys, the continent's two cordilleras have caused a rain shadow from both directions, so rainfall is simply too low to promote lusher vegetation.

WILDLIFE: Endemic mammals are not well represented in this habitat, which mostly hosts more widespread species like Ocelot, Margay, and White-nosed Coati.

Reptiles are prominent, including Guatemalan Beaded Lizard, Black Iguana, and multiple species of banded geckos.

Birds typical of this kind of environment in the inter-Andean valleys include White-tailed Shrike-Tyrant, Tufted Tit-Tyrant, Blue-and-yellow and Scrub Tanagers, Purple-collared Woodstar, Black-tailed Trainbearer, and Golden-rumped Euphonia. Species typical of

Scrub Tanager in thornscrub within a semiarid inter-Andean valley in Ecuador.
© IAIN CAMPBELL, TROPICAL BIRDING TOURS

low-lying thornscrub areas include Russet-crowned Motmot, Lesser Roadrunner, Banded Wren, Lesser Ground-Cuckoo, Varied Bunting, Crested Bobwhite, and Plain-capped Starthroat.

DISTRIBUTION: Thornscrub occurs on the Caribbean coast of Venezuela and Colombia, on many of the islands in the Caribbean, on the coast of the Yucatán Peninsula, along the Guatemalan Pacific coast, and in many of the equatorial inter-Andean valleys from e. Colombia to Perú. There is significant dispute about the nature of this habitat in the inter-Andean valleys such as those north of Quito, Ecuador. Many authors suggest that this is a modified environment akin to the MAQUIS of the Mediterranean basin (covered in the Palearctic chapter).

WHERE TO SEE: Jerusalem Reserve, north of Quito, Ecuador.

PUNA

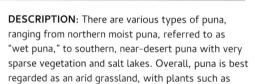

IN A NUTSHELL: Puna is a high-elevation, temperate, arid or semiarid grassland. **Habitat Affinities:** PALEARCTIC TEMPERATE DESERT STEPPE. **Species Overlap:** NEOTROPICAL SEMIDESERT SCRUB; PARAMO.

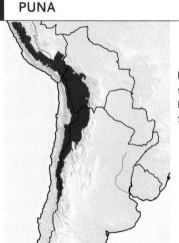

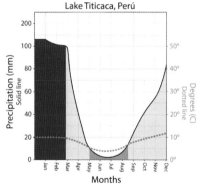

Lake Titicaca, Perú

DESCRIPTION: There are various types of puna, ranging from northern moist puna, referred to as "wet puna," to southern, near-desert puna with very sparse vegetation and salt lakes. Overall, puna is best regarded as an arid grassland, with plants such as feathergrasses adapted to much harsher conditions than those found in the lower PAMPAS grasslands to the east. In the north, this habitat occurs from above 13,000 ft. (4,000m) to over 16,400 ft. (5,000m). In southern regions it occurs from 9,800 ft. (3,000m) to 16,400 ft. (5,000m), though the MONTE habitat verges into the puna above 6,500 ft. (2,000m).

The average annual precipitation over most of the puna is 12–40 in. (300–1,000mm), though in the south it decreases to as little as 4 in. (100mm), which means that it is a true desert grassland. The landscape is called an *altiplano*, which basically means a high-elevation, large flat area, even though there is significant topography, such as towering volcanoes and expansive valleys, on the top of the Andes. In the north, the thick grasslands are similar to dry PARAMO, with moist grasses and small polylepis bushes and other shrubs growing in the protected areas, but cushion plants are much rarer and stop being a substantial part of the assemblage in s. Perú (except in the highest part of the landscape). Southward in the puna habitat, grasses become sparser and sage-like

Extensive high Andean puna in n. Argentina.
© ANDRÉS VÁSQUEZ, TROPICAL BIRDING TOURS

shrubs dominate the non-grass flora. Some better-watered areas are nourished by surrounding glacial melt from volcanoes, allowing the growth of a meadow-like, boggy habitat known as *bofedales*, which is quite different from the drier surrounding puna.

WILDLIFE: The high lakes of the puna region host a suite of waterbirds, including the three species of highland flamingo: Andean, Chilean, and James's. There are also species shared with the PARAMO such as Andean Condor, Andean Gull, Ornate Tinamou, and Andean Lapwing. The typical land birds of the puna include an assemblage of ground furnariids, such as Puna Canastero, Slender-billed Miner and other miners, earthcreepers, and cinclodes.

This is the zone of the camelids, and Vicuña and the much larger Guanaco both live in this habitat. Carnivores are represented by Puma, Andean Fox (aka Culpeo), and the highly endangered and rarely seen Andean Mountain Cat.

The enigmatic Diademed Sandpiper-Plover is found locally in bogs within high Andean puna. © ANDRÉS VÁSQUEZ, TROPICAL BIRDING TOURS

DISTRIBUTION: The puna occurs from c. Argentina north to n. Perú. North of Perú it merges with the PARAMO of the n. Andes. In the east there is a very sharp boundary with the narrow bands of ELFIN FOREST and the YUNGAS cloud forests of Bolivia and Argentina. There is a more gradual

Lesser Rheas live in the grasslands, low heaths, and bogs of the puna in n. Argentina.
© ANDRÉS VÁSQUEZ, TROPICAL BIRDING TOURS

mixing of habitats between puna and MONTE, along with a fusion of the respective bird assemblages. This transition zone is referred to as "pre-puna" habitat, which can also be regarded as a SEMIDESERT SCRUB. To the west in Perú and Chile, the boundary is, again, very sharp, here meeting with desert.

WHERE TO SEE: Lake Titicaca, Bolivia and Perú.

PATAGONIAN STEPPE

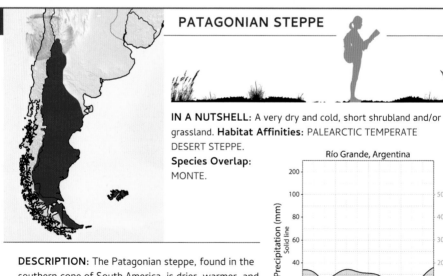

IN A NUTSHELL: A very dry and cold, short shrubland and/or grassland. **Habitat Affinities:** PALEARCTIC TEMPERATE DESERT STEPPE.
Species Overlap: MONTE.

Río Grande, Argentina

DESCRIPTION: The Patagonian steppe, found in the southern cone of South America, is drier, warmer, and has a longer growing season than the high-latitude steppes of the Palearctic. This is a mostly flat, arid habitat, with near-constant, bitterly cold, desiccating dry winds in winter and strong winds in summer.

Most of the precipitation occurs as limited snowfall in winter, which then thaws in spring. The aridity is caused by the region's location in the rain shadow of the Andes Mountains, which block prevailing moist winds from the Pacific Ocean (see sidebar 2.1). Based on the temperatures and precipitation alone, you might expect a lusher environment on the Patagonian steppe, but the desiccating winds cause it to be quite stark.

This habitat is the Neotropical equivalent of PALEARCTIC TEMPERATE DESERT STEPPE. It has harsh winters and cool summers. Some of the shrubs have reduced leaves and perform photosynthesis through their stems and thorns. Most of the shrubs with small leaves are deciduous. In the most exposed areas, the vegetation is limited to cushion plants or low hummocks, generally covering around 60–80% of the ground but as little as 30% in some areas. Shrub height is generally about 3 ft. (1m) but ranges from 1 ft. (0.3m) to 8 ft. (2.4m). The shrubs are intertwined and thorny, making walking in this habitat quite difficult.

WILDLIFE: The harsh climate and distinctive nature of the Patagonian steppe vegetation supports a particular suite of animals. Guanaco, a camelid, remains quite common, roaming between wild areas and pastures that it shares with domestic sheep. Other typical mammals include South American Gray (or Patagonian) Fox, Geoffroy's Cat, Humboldt's (or Patagonian) Hog-nosed Skunk, and Puma.

Right: **Herds of Guanacos are abundant in Patagonian steppe, where they are a major prey species for local Pumas.** © IAIN CAMPBELL, TROPICAL BIRDING TOURS

Below: **Rufous-chested Dotterel on Patagonian steppe in Tierra del Fuego, Chile.** © IAIN CAMPBELL, TROPICAL BIRDING TOURS

Patagonian birds are also quite distinctive. This is an area where ground-feeding birds dominate. Lesser Rhea and Elegant Crested-Tinamou are both iconic species of these steppes. Striated and Chimango Caracaras are two ground-feeding, scavenging falcons. The niches of larks and sparrows of the Northern Hemisphere are filled by furnariids, such as Patagonian Canastero, Plain-mantled Tit-Spinetail, and Band-tailed Earthcreeper. Shorebirds like Least Seedsnipe feed in the barer areas, while the Magellanic Plover, representing a monotypic family, occurs on saline lakes. Ruddy-headed Goose is a typical inhabitant of the numerous ponds in this habitat. Other typical species include Rusty-backed Monjita, Carbonated Sierra-Finch, Chocolate-vented Tyrant, and Cinnamon-bellied Ground-Tyrant.

DISTRIBUTION: The Patagonian steppe ranges from about 40°S in Argentina down to Tierra del Fuego. The habitat is also found in a small area of extreme se. Chile. In the north there is a gradual transition into the MONTE habitat, while in the south and southwest, there is a rather sharp boundary with the broadleaf *Nothofagus* and conifer forests of s. Chile and Argentina (MAGELLANIC RAINFOREST).

WHERE TO SEE: Pali-Aike National Park, Chile; n. Tierra del Fuego, Argentina.

NEOTROPICAL SEMIDESERT SCRUB

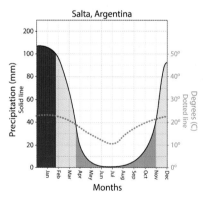

IN A NUTSHELL: A dry habitat consisting of cacti, euphorbias, and low shrubs. In n. Argentina this includes a subset with more grasses called "pre-puna." **Habitat Affinities:** PALEARCTIC SEMIDESERT THORNSCRUB. **Species Overlap:** MONTE; PUNA; NEOTROPICAL DESOLATE DESERT; PATAGONIAN STEPPE.

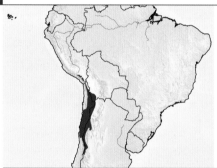

Salta, Argentina

DESCRIPTION: Between the semiarid habitats like MONTE, NEOTROPICAL THORNSCRUB, and PUNA, and the hyperarid DESOLATE DESERT of South America, is an area of semidesert scrub. In this habitat, the ground is nearly devoid of grasses, and what ground cover there is consists of succulents like Cardón Cactus and agave along with small xerophytic shrubs. In some places, such as the "pre-cordillera" area (see Distribution, below), the ground is covered by a low shrub layer made of succulents and perennial year-round flowering shrubs, and there is no taller scrub layer. Over most of the region this habitat is characterized by the large amount of bare ground between isolated shrubs or succulents. Sporadic rains trigger the rapid burst of ephemeral plants that spend most of their life cycle as seed, filling the bare ground between permanent bushes with these smaller plants. In the

Neotropical semidesert scrub in nw. Argentina.
© NICK ATHANAS, TROPICAL BIRDING TOURS

e. Andes, such as around Mendoza, Argentina, the pre-cordillera semidesert scrub differs from the monte in that the shrubs are more isolated and shorter, the grasses between the shrubs are replaced by succulents, and the whole landscape has a much more desolate appearance. Although the hillsides may be very sparse, with large rock faces seemingly devoid of vegetation, the rivers are fed from snowmelt in the high Andes, providing a source of water. The small areas on the edges of valleys in the semidesert can be quite lush compared to the surrounding terrain, looking much more like lower-elevation monte scrub. Walking around in this habitat is very easy, though this is generally a steep environment, and you have unlimited visibility at eye level.

The variation of semidesert scrub called pre-puna is a sparsely vegetated grassland with open meadows in the arid, high-elevation montane habitat of the high Andes at elevations of 6,500–11,500 ft. (2,000–3,500m). The most distinctive plant is Cardón Cactus; trees including the Churqui and shrubs like the wild tobacco Palán-Palán occur here as well. The climate in the high Andes is cold and dry, with precipitation in the form of snow and hail. The rolling valleys and highland meadows are a continuous landscape dominated by shrubs of varying densities.

WILDLIFE: Despite the rather different vegetation and landscape, there is extensive overlap in bird species between the semidesert scrub and the PATAGONIAN STEPPE. This habitat also has more typical birds such as Elegant Crested-Tinamou, Gray-breasted Seedsnipe, Patagonian Sierra-Finch, Buff-breasted Earthcreeper, Andean Tinamou, Scale-throated Earthcreeper, and Burrowing Owl.

Mammals here include Pampas Cat, Andean Fox (aka Culpeo), Mountain Viscacha, and Andean Hairy Armadillo. The little-known Andean Mountain Cat is also found in this habitat, though chances of seeing it are extremely remote.

Gray-breasted Seedsnipe and Patagonian Sierra-Finch share a rock in semidesert scrub in Chile. © IAIN CAMPBELL, TROPICAL BIRDING TOURS

DISTRIBUTION: The Neotropical semidesert scrub habitat is found in coastal areas, along the Pacific coast from s. Perú to mid-Chile, and on the east side of the Andes into Argentina and Bolivia, where it is referred to as "pre-puna" (grassier areas more common in n. Argentina) or "pre-cordillera" (areas richer in succulents).

WHERE TO SEE: Around the small town of Amaicha del Valle, near Tucumán, Argentina.

MONTE

IN A NUTSHELL: A dry scrubby desert habitat of the colder parts of South America. **Habitat Affinities:** PALEARCTIC SEMIDESERT THORNSCRUB. **Species Overlap:** CHACO SECO AND ESPINAL; PATAGONIAN STEPPE.

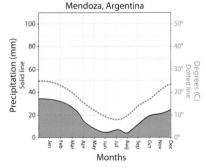

Mendoza, Argentina

DESCRIPTION: Monte has a unique combination of characteristics, which makes it hard to define as a habitat. It is not as dry as the hyperarid deserts to its west but is too thick to be a steppe shrubland and not thorny and high enough to be a thorn thicket, nor grassy enough to be a grassland savanna. Yet, it takes on characteristics of all of these habitats, with corresponding animal life (though no animals are restricted to this habitat alone), which makes it a fascinating habitat to visit. Most of the thorny shrubs in the monte are between waist and head

Monte landscape in Mendoza, Argentina.
© IAIN CAMPBELL, TROPICAL BIRDING TOURS

height, while occasional trees are up to 12 ft. (4m) tall. Most of the shrubs have narrow sclerophyllous leaves, and many photosynthesize through their stems and branches, giving them a green appearance, while the leaves generally have an olive color, similar to shrubs in the MAQUIS of the Mediterranean region (covered in the Palearctic chapter). Cacti are also common in this habitat, though never make up a significant portion of the vegetation. At the upper elevation zones of monte, *Rosa canina*, a wild rose from Europe, is rapidly invading more humid areas, replacing native plants in most accessible areas. The overall experience of being in this habitat is similar to that of the Sonoran Desert of North America, the KAROO region of South Africa, the scrublands of sw. Queensland in Australia, or the semidesert shrublands of the Palearctic. The bushes in the moister areas are thick, intertwined, and spiny, which makes walking through this habitat off-trail cumbersome.

WILDLIFE: Bird assemblages in the monte habitat have much in common with surrounding habitats, while a greater percentage of mammals, by contrast, are much more typical of this environment. Most herbivores are rodents, including Patagonian Mara and numerous species of cavy. Pink Fairy Armadillo is found only in this environment. Carnivores are represented by South American Gray Fox, Pampas Fox, and Lesser Grison.

Birds typical of the monte include Burrowing Parakeet, tinamous, Sandy Gallito, White-throated Cacholote, and an array of smaller seedeaters, including numerous finches and warbling-finches, such as Ringed Warbling-Finch.

DISTRIBUTION: Monte is confined to Argentina, where it forms an arc around the southern and western limits of the PAMPAS. The boundaries with surrounding habitats, which are based on gradual

South American Gray Fox is one of the carnivores found in the monte (photo: Chile).
© SAM WOODS, TROPICAL BIRDING TOUR

rainfall and temperature changes, are extremely diffuse. Monte blends into CHACO SECO in the north and PATAGONIAN STEPPE in the south. Where it merges with pampas in the northeast, the hills are dominated by monte, and the lower drained areas are covered by pampas. The boundary with the dry NEOTROPICAL SEMIDESERT SCRUB is sharper in the lower Andes (the pre-cordillera) at around 5,000 ft. (1,500m). The west edge of this zone is prime wine-growing land, and much of the monte around Mendoza, Argentina, has been turned into vineyards (of extremely good wineries).

WHERE TO SEE: East of Mendoza (around San Martín), Argentina; Valdés Peninsula, Argentina.

Burrowing Parakeet is a characteristic bird of monte habitat (photo: Argentina).
© NICK ATHANAS, TROPICAL BIRDING TOURS

NEOTROPICAL PINE-OAK WOODLAND

IN A NUTSHELL: This habitat is an extension of the MADREAN PINE-OAK WOODLAND habitat of the Nearctic and is treated comprehensively in that chapter. This habitat is classified as a temperate habitat in the Neotropics, whereas its sister is grouped in the conifer biome in the Nearctic.

WILDLIFE: The bird assemblage in this forest differs from that of the surrounding more humid forests by having fewer fruit-eating species like tanagers and cotingas and instead featuring species such as Pink-headed and Olive Warblers, Goldman's subspecies of Yellow-rumped Warbler, Ocellated Quail, Band-backed Wren, and the Guatemalan subspecies of Northern Flicker.

Mammals in this forest include Deppe's and Variegated Squirrels, Long-tailed Weasel, White-tailed Deer, and Derby's Woolly Opossum.

DISTRIBUTION: In the Neotropics, this dry conifer forest extends south from Mexico through c. Guatemala into Nicaragua, though it has been largely destroyed there.

WHERE TO SEE: Around Antigua and Lake Atitlán, Guatemala.

Above left: **Goldman's Warbler is a distinctive resident form (sometimes considered a full species) of Yellow-rumped Warbler that occurs in pine-oak woodlands in Central America.**
© DANIEL ALDANA SCHUMANN, TROPICAL BIRDING TOURS

Left: **Pine-oak woodland is a restricted habitat in Guatemala.**
© DANIEL ALDANA SCHUMANN, TROPICAL BIRDING TOURS

NEOTROPICAL LOWLAND RAINFOREST

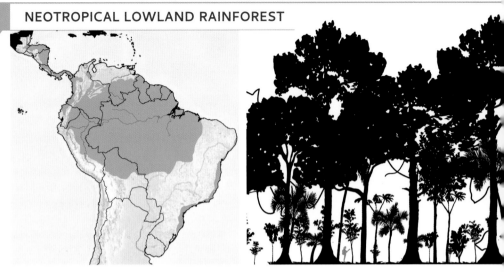

IN A NUTSHELL: One of the hottest, wettest, tallest, and thickest-canopied habitats in the world. **Habitat Affinities:** AFROTROPICAL LOWLAND RAINFOREST; AUSTRALASIAN TROPICAL LOWLAND RAINFOREST; INDO-MALAYAN TROPICAL LOWLAND RAINFOREST. **Species Overlap:** NEOTROPICAL SEMI-EVERGREEN FOREST; NEOTROPICAL CLOUD FOREST.

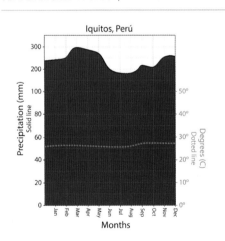

An observation tower in the Amazon rainforest of Ecuador allows dramatic views of the canopy of lowland rainforest. © ANDRÉS VÁSQUEZ, TROPICAL BIRDING TOURS

DESCRIPTION: A must-see for anyone with the vaguest interest in nature, this is the habitat that many people think of when they first imagine South America. Neotropical lowland rainforest is found in three large blocks: the massive Amazon rainforest, the Atlantic rainforest along the se. Brazilian coast, and the Chocó/Central American rainforests. There are variations within the forest type, but as a whole, the habitat and the bird assemblages in it remain fairly constant. However, as will be explained here, the experience of being in the lowland rainforest varies greatly depending on where you are in the canopy profile.

The uniformity of regular rainfall (in other words, the lack of extreme drought periods) and constant temperatures results in a surprisingly uniform leaf structure in canopy trees: medium-size leaves that appear similar to those of temperate trees from the laurel family. Conifers are notably rare in tropical lowland rainforests, as are deciduous trees. With no drought or temperature extremes, deciduous species had no reason to evolve, and nearly all the trees are broadleaf evergreens, most of which lose their old leaves and grow new ones throughout the year. The few trees that do drop their leaves at the same time do so at random times and are not triggered by drought or season, unlike many trees of NEOTROPICAL SEMI-EVERGREEN FOREST. The uniform nature of leaf size and texture in the canopy masks remarkable diversity, though the forest may superficially look no richer than a deciduous forest of the e. United States in summer.

Palms make up only a minor fraction of the rainforest canopy and are noted more for being easily recognizable than for their abundance. They tend to occur in thick groves in swampy areas and much less commonly as an understory tree. Although from a distance the forest may look uniform, underneath it is far from it. The understory of the rainforest is lush, with a variety of smaller trees, which mostly have oversize leaves with large drip tips. When canopy trees fall and create an opening in the forest, plants

Rufous Motmot occurs in lowland rainforests as well as hillier forests. © SAM WOODS, TROPICAL BIRDING TOURS

like cecropias rapidly fill the light gap. Over time, saplings from the canopy trees, lying in wait on the forest floor, grow through these pioneer cecropias and reestablish the closed forest canopy.

We tend to talk of visiting the lowland rainforest as a single entity, yet this habitat is so variable that you can spend many days in it and always concentrate on a new environment with a different suite of species. For example, when looking for animals in the richest areas on the western edge of the Amazon, in e. Ecuador, Perú, or s. Colombia, you will have different experiences and find different species in the understories and canopies of IGAPÓ FLOODED FOREST, palm forest, and lowland rainforest's dry terra firma.

Looking for animals on the trails of the terra firma, away from the thick forest edge and any light gaps, you will be surprised by just how cool and open it is under the thick, closed canopy and the multiple sub-canopies. Under the sub-canopy, which is almost always over 15 ft. (5m) high, you can see for 30 ft. (10m) at eye level in most directions. The style of birding here is walking to find territories of resident birds, such as manakins, while hoping to bump into sub-canopy feeding flocks of flycatchers, foliage-gleaners, and tanagers. The highlight of birding at ground level is when you come across an army ant swarm. Here, you have antbirds, such as White-plumed,

Golden-mantled Tamarin near the Napo River in the Amazonian lowlands of Ecuador.
© JOSÉ ILLÁNES, TROPICAL BIRDING TOURS

White-cheeked, Hairy-crested, and Lunulated Antbirds, as well as terrestrial birds feeding on the first wave of insects fleeing the marauding army ants. Woodcreepers, ant-tanagers, and a host of different types of flycatchers feast on these many escapees. Other ant swarm specialists perch on sapling stems or low vines a few feet above the ground, just out of danger from the ants but close enough to have a full buffet of small invertebrates.

Experiencing the birds and mammals of the rainforest canopy can be a frustrating venture, because although you may hear scores of bird species calling, catching a mere glimpse is difficult. Even from the roadside or clearings, most animals are not visible. Luckily, many of the Amazonian lodges, as well as reserves in Central America and nw. Ecuador, have canopy towers over 100 ft. (30m) tall, from which you can see into and above the canopy. Most of these are built around huge, emergent kapok trees; some are metal towers, and some even have walkways through the canopy. Up here, it is a different world, where you can see spider monkeys and howler monkeys moving around or hanging out at a fruiting tree, and marmosets and tamarins running back and forth. Most visits to the canopy should result in a sloth encounter, as these strange animals move (or more often don't move) up and down canopy trees. Birding in the canopy is spectacular, with flocks of fruit-eating tanagers mixing with cotingas, while toucans and parrots tend to move in single-species flocks that concentrate on fruiting trees. Well over 100 species of birds can be recorded from a single tower over the course of a day.

River islands within the vast Amazon system are consistently building and eroding, supporting a series of subclimax scrub. The taller trees are mainly cecropias, and there is thick grass underneath. Chasing animals here is tough and generally unpleasant because it can be blisteringly hot, and the birding is concentrated on some rare, limited-range specialists. Birding on these islands is something to do once all other microhabitats have been covered.

The palm groves in low-lying areas are always worth exploring, though they are almost impossible to walk through, meaning that you'll need to stand on the terra firma and look in. In these stands are highly specialized birds such as Point-tailed Palmcreeper (a kind of insect-eating furnariid that looks like a giant treecreeper) that are found only in this microhabitat. See also the IGAPÓ AND VÁRZEA FLOODED FOREST habitat account in this chapter.

WILDLIFE: Most bird families and subfamilies that typify lowland rainforest have at least some members that are also found in NEOTROPICAL CLOUD FOREST or NEOTROPICAL SEMI-EVERGREEN FOREST, but more species are found in this habitat. These include the hermits (a subfamily of hummingbirds), with very long and decurved bills; and fruit eaters, like smaller cotingas and purpletufts. There are many tanagers, including Opal-rumped and Opal-crowned Tanagers; toucans, including small species like Collared Aracari and large ones like White-throated and Channel-billed Toucans; hunters like nunbirds and puffbirds; and ground species like gnateaters and ant-following antbirds.

In the Neotropics, most primates have prehensile tails, are arboreal and very agile, and are predominantly fruit eaters. New World primates are found mostly in rainforest, whereas Old World primates use a wider variety of habitats and many species are fairly terrestrial.

Cream-colored Woodpecker within the canopy
of Amazon lowland rainforest in Ecuador.
© IAIN CAMPBELL, TROPICAL BIRDING TOURS

Herps are well represented in the rainforest, and frogs make up a much larger percentage of the assemblage than they do in higher CLOUD FOREST or in drier SEMI-EVERGREEN FOREST.

Endemism: Despite their massive area in South America and variety of microhabitats, continental lowland rainforests have few localized endemics. The endemic areas are concentrated along the western border of the Amazon at the edge of the Andes, such as along the Napo River valley in e. Ecuador. The Amazon River and its tributaries, such as the Napo, act as barriers between species—for example, we find such primate assemblages as Black-mantled Tamarin and Brown Woolly Monkey north of the river and their sister counterparts Golden-mantled Tamarin and Silvery Woolly Monkey south of the river. The Atlantic forests of coastal Brazil are centered on coastal mountain ranges and contain many restricted-range species. In Central America, areas of endemism include the Caribbean slope from Honduras to Panama, and the Darién lowlands of Panama. The richest endemic area is the Chocó region of nw. Ecuador and w. Colombia. Bounded by the Andes Mountains to the east, the Tumbesian dry forests to the south, and the Pacific Ocean to the west, this area has over 60 endemic species of birds.

DISTRIBUTION: The bulk of the lowland rainforests of the Neotropics are in the Brazilian Amazon, extending to e. Ecuador in the west, s. Venezuela in the north, and n. Bolivia in the south. In Ecuador, the Andes narrow to just 70 mi. (112km), so that the Amazon and Chocó rainforests are separated by thin bands of PARAMO, NEOTROPICAL CLOUD FOREST, and inter-Andean NEOTROPICAL THORNSCRUB. The Chocó/Central American lowland rainforest runs from w. Ecuador through w. Colombia to Central America and on to s. Mexico. Atlantic lowland rainforest exists in se. Brazil and extreme ne. Argentina. Some islands of the Caribbean also have remnants of lowland rainforest, including Jamaica, Hispaniola, Cuba, and Puerto Rico. Much of this habitat has already been destroyed in Central America and on the coastal plains of Ecuador and Colombia. The Atlantic forests have been decimated, and those that remain are almost all protected in official reserves. Amazon lowland rainforest is still largely intact but is being rapidly encroached upon from all sides. There is little protection in Perú, Colombia, or Ecuador, where large-scale African Oil Palm plantations have been developed. The forest is being cleared for soy and cattle in Brazil, with little control from the government. The situation is bleak.

WHERE TO SEE: Amazon lodges of the Napo River valley in Ecuador; lodges in Manu National Park, Perú; Rio Cristalino Lodge, Mato Grosso, Brazil.

NEOTROPICAL SEMI-EVERGREEN FOREST

IN A NUTSHELL: A slightly drier version of NEOTROPICAL LOWLAND RAINFOREST (and often mistakenly referred to as lowland rainforest in Central America), which differs in that some trees lose their leaves in drier periods of the year. **Habitat Affinities:** INDO-MALAYAN SEMI-EVERGREEN FOREST. **Species Overlap:** NEOTROPICAL LOWLAND RAINFOREST.

DESCRIPTION: Much of what is called rainforest in e. Brazil, Central America, and even parts of the Amazon is actually semi-evergreen forest. The confusion arises because for much of the year, Neotropical semi-evergreen forest appears very similar to NEOTROPICAL LOWLAND RAINFOREST, with tall trees that grow to around 130 ft. (40m), have large, laurel-shaped leaves, and are draped by numerous lianas. In the drier months, however, semi-evergreen forest looks very different, with numerous bare branches on those trees. This semi-evergreen nature is not evident in periods when rain remains steady throughout the year. Only 30% of the canopy trees are deciduous, losing their leaves in

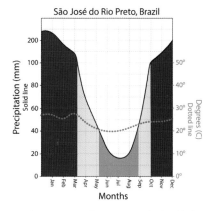

São José do Rio Preto, Brazil

the dry season. Another significant proportion of the canopy trees are semi-evergreen, losing their leaves only in the years when rainfall is reduced to a degree that stresses them.

Fewer of the trees in this forest have the massive buttress roots typical of lowland rainforest trees. The canopy of this forest is dominated by broadleaf species such as Sapodilla, Honduran Mahogany, Cuban Cedar, and Pink Poui. It is a little more open than lowland rainforest, with upper canopy coverage at around 80%, which allows for an overall two-canopy structure to the forest, including a distinct and complete middle canopy. Most of the middle-canopy tree species are microphyllous (having small, narrow leaves) rather than deciduous and are more similar to the species of the CAATINGA or CHACO SECO habitats than to lowland rainforest trees. These lower-level trees tend to flower in the dry season, and the combination of a colorful lower canopy and a more barren upper canopy makes this forest easier to identify at this time of the year. Epiphytes and mosses are not as numerous as in NEOTROPICAL LOWLAND RAINFOREST or NEOTROPICAL CLOUD FOREST, a fact that becomes obvious only when you are inside the forest looking up.

Right: **Semi-evergreen forest can form at the base of escarpments where the Cerrado savanna habitat grows above (photo: c. Brazil).** © NICK ATHANAS, TROPICAL BIRDING TOURS

Below: **Jaguar is the apex predator in this habitat (photo: Brazil).** © IAIN CAMPBELL, TROPICAL BIRDING TOURS

Undergrowth in the semi-evergreen forest is also diminished, due to the heavy lower canopy, so walking around in this forest feels similar to walking in lowland rainforest. Rainfall is comparable to that of the lowland rainforests, approximately 48–75 in. (1,200–1,900mm) annually, but the seasonality is far more pronounced, with a dry season of reduced rainfall of up to six months and a few exceedingly dry months with almost no rainfall. It is this intense dry season, and the excessive evapotranspiration rates associated with it, that produces this forest rather than lowland rainforest.

WILDLIFE: Although its vegetation is distinctly different, semi-evergreen forest is remarkably similar to lowland rainforest in terms of wildlife. Most mammal species freely use both when the two habitats are adjacent. Amphibian numbers are lower in semi-evergreen forests, while lizards become relatively more common.

Birdlife is similar in the two habitats, with some of the same families represented, but overall species numbers are lower in the semi-evergreen forests of Brazil than in the Amazon lowland rainforests at their western edge in Ecuador or Colombia. The Brazilian semi-evergreen forests are notable in their lack of ground-cuckoos and general lack of obligate antbirds, and in having far fewer barbets, puffbirds, and gnateaters relative to the lowland forests. Compared to the nearby CAATINGA deciduous forest, semi-evergreen forest has many more parrots, antbirds, and tanagers.

Endemism: Despite the large area this habitat originally covered, most of the species in this forest type are widespread. It contains few officially designated range-restricted species and therefore few EBAs (see "Endemic Bird Areas" in the Introduction). This is a prime example of where the parameters of restricted range—at less than 19,300 sq. mi. (50,000km^2)—hide the fact that these areas are centers of endemics.

DISTRIBUTION: Neotropical semi-evergreen forest is found in Mesoamerica from s. Mexico through the Caribbean slope of Central America to Panama. It also occurs through many of the Caribbean islands, where it is the predominant humid forest type. It is, or to be exact, *was*, the predominant forest type in the noncoastal parts of se. and e. Brazil between the Atlantic rainforest and the CERRADO region but has essentially been annihilated to the same degree as the Atlantic lowland rainforests. It also occurred between the Cerrado and the Amazon rainforest proper, but here also has been largely cleared, burned, and turned over to agriculture. The seasonal nature of the rainfall and intense dry seasons have resulted in baking and duricrust development (see sidebar 2.5) in these cleared lands, with little chance of regeneration of this forest once it's cleared. Even creating corridors between the small remaining reserves is very difficult. Significant patches remain within the Amazon, surrounded by lowland rainforest, and these remain better protected than the forests farther east.

WHERE TO SEE: Tikal, Guatemala.

NEOTROPICAL CLOUD FOREST

IN A NUTSHELL: Cloud forests, with large trees laden with mosses and bromeliads, and constant water dripping off the leaves, are the quintessential forests of the tropical Andes and the highlands of se. Brazil. **Habitat Affinities:** AFROTROPICAL MONTANE FOREST. **Species Overlap:** NEOTROPICAL SEMI-EVERGREEN FOREST; NEOTROPICAL LOWLAND RAINFOREST.

DESCRIPTION: Both sides of the equatorial Andes, between 3,300 ft. and 9,900 ft. (1,000–3,000m), receive precipitation not only as rainfall but also as a perpetual billow of cloud and fog cover that emanates humidity. This humidity, combined with the very steep nature of the slopes on which this habitat grows, produces frequent landslides and a constant recolonization by cloud forest. The elevation of the cloud forest lowers at latitudes away from the equatorial regions, and it is hundreds of meters lower in Central America than in the Andes.

The canopies of this evergreen forest are lower than those of the rainforest—trees rarely grow higher than 80 ft. (25m)—and branches extend much lower on the trunk than in trees of NEOTROPICAL LOWLAND RAINFOREST or NEOTROPICAL SEMI-EVERGREEN FOREST. Because of the constant landslides, the forest is disjointed, with trees of different ages and heights forming a noncontiguous canopy that lets far more light hit the understory than in lowland rainforest. Most of the trees are close relatives of species found in rainforest, though pioneer species like the extremely fast-growing cecropias make up a much higher proportion of the canopy than in more stable forests. Bromeliads are also a major component of the cloud forest canopy, and it is rare to see an old tree that is not enveloped by bromeliads and draping mosses.

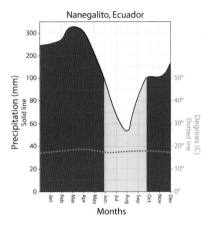

Nanegalito, Ecuador

Because of the open nature of the canopy, there is a strong and irregular sub-canopy as well as a very thick brush layer. In contrast with rainforest, the undergrowth in the cloud forest is extremely thick, made up of many broadleaf saplings and massive-leaved bushes such as *Gunnera* spp., whose leaves can approach 10 sq. ft. (1m²) in area. Heliconias (which resemble banana plants) are very common in light gaps, and in Ecuador above 6,600 ft. (2,000m). Chusquea bamboo forms

Cloud forest near Tandayapa Bird Lodge in Ecuador, home to extraordinary numbers of hummingbirds and tanagers, and many birds endemic to this distinctive Chocó bioregion.
© PABLO CERVANTES DAZA, TROPICAL BIRDING TOURS

dense thickets on the ground up to 10 ft. (3m) tall, and species of climbing bamboos reach up into the canopy. The cloud forest of se. Brazil also has copious bamboo growing in the understory, and in some places it becomes so dominant it is reminiscent of EAST ASIAN TEMPERATE BAMBOO FOREST. Due to the thick understory, it is nearly impossible to walk through cloud forest away from a trail without a machete to cut a path.

WILDLIFE: The habitat's great range of elevation and its growth in many isolated side ranges have produced huge biodiversity, resulting in many localized endemics. Frog and reptile diversity is very high below 4,000 ft. (1,200m), but above this elevation the forest gets cold, and species diversity drops dramatically. Very few venomous snakes make it above 5,000 ft. (1,500m), and frog diversity is reduced, with the species present often being localized endemics.

Mammals of the cloud forest include the flocculent Spectacled Bear, Mountain Tapir, and a newly discovered species of small carnivore, Olinguito, which resembles a Kinkajou and is related to raccoons. Monkeys occasionally move up into the cloud forest during warmer or drier spells but generally find the climate too cold.

Birdlife in the cloud forest is dominated by frugivorous families such as tanagers, barbets, and cotingas. Many of these canopy birds move in large mixed-species flocks, sometimes numbering hundreds of individuals. These flocks move as waves: First come the flock leader species, often mountain-tanagers or bush tanagers. They are followed by a whole suite of species that wander great distances; this group is joined for short periods by species that are more loyal to one territory but will participate in the feeding frenzy in their area. All this means that flocking may last for hours and go for miles with a changing mix of species. There are also understory flocks that may move with the canopy flocks or independently of them. These undergrowth flocks are dominated by spinetails, barbtails, leaftossers, antwrens, chat-tyrants, and thrushes. Hummingbird diversity reaches extreme levels in the cloud forests of the Andes. Brazil has 80 species of hummingbirds, while Ecuador boasts 140, despite being a fraction of Brazil's size. Indeed, one set of hummingbird feeders at Tandayapa Bird Lodge, in Ecuador, attracts a colorful pastiche of over 30 species. Tanagers also dominate the avifauna of the cloud forests of se. Brazil, but antbirds and antwrens are more prominent and diverse in these Atlantic forests than in cloud forests of the west. While the mid-elevation cloud forests on the west slope of the Andes have Zeledon's Antbird and Western Slaty Antshrike, the same type of forest in the Atlantic region of Brazil

Male Andean Cock-of-the-rocks gather at dawn for dramatic displays in the cloud forests of the Andes (photo: nw. Ecuador).
© PABLO CERVANTES DAZA, TROPICAL BIRDING TOURS

Right: **Tayra photographed in Andean cloud forest in the Tandayapa Valley of nw. Ecuador.** © PABLO CERVANTES DAZA, TROPICAL BIRDING TOURS

Below: **Booted Racket-tail at Tandayapa Bird Lodge in the cloud forests of nw. Ecuador.** © PABLO CERVANTES DAZA, TROPICAL BIRDING TOURS

has Bertoni's and White-bibbed Antbirds, White-bearded and Large-tailed Antshrikes, and Rufous-backed Antvireo.

Endemism: Because of the isolated distribution of many cloud forests and the habitat's occurrence on both sides of the Andes, it supports many endemic species. New research on avian vocalization is showing that many species found on both sides of the Andes may be multiple species. The degree of endemism on the spurs of the Andes and Central America are matched only by those in the mountains of New Guinea.

DISTRIBUTION: Cloud forests are found in Central America on mountain ranges from Mexico south to Panama. They become common on either side of the Andes in Colombia and continue down to the

Ecuadorian–Peruvian border on the west slope. On the Andes' east slope, this habitat continues south through Perú into Bolivia, where it is replaced by the YUNGAS forest. Cloud forests also occur along the Atlantic ranges of Brazil, above either NEOTROPICAL LOWLAND RAINFOREST or NEOTROPICAL SEMI-EVERGREEN FOREST.

WHERE TO SEE: Tandayapa Valley, Ecuador; Monteverde, Costa Rica; Manu Road, Perú.

YUNGAS

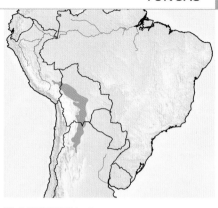

IN A NUTSHELL: A colder, drier sister habitat of the NEOTROPICAL CLOUD FOREST farther north, with a broken canopy and smaller-leaved plants in the undergrowth. **Habitat Affinities:** AFROTROPICAL MONTANE FOREST **Species Overlap:** NEOTROPICAL CLOUD FOREST.

DESCRIPTION: In the northern portion of its range in Bolivia and s. Perú, yungas forest is found between the high grasslands and ELFIN FOREST at higher elevations and the humid forests below. In the southern part of its range, in n. Argentina, it is wedged between the drier habitats of the Andes: the PUNA and desert scrub to the west, and the savanna of the Gran Chaco to the east. In the north it forms between 3,300 ft. and 10,000 ft. (1,000–3,000m) elevation in areas where the climate is hot and humid, the rain falls mainly in spring and summer, and northerly winds create high precipitation of over 100 in. (2,500mm) per year. In the south, where the climate becomes far more seasonal, with a long dry

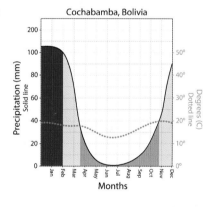

spell during the winter months and an annual rainfall of less than 50 in. (1,250mm), it is mainly found between 1,500 ft. and 9,000 ft. (450–2,750m). Within this large elevation range there are distinct bands, with the lower elevations being wetter and warmer, and the higher bands drier and colder and starting to resemble and blend into elfin forest above 8,000 ft. (2,400m).

The vast majority of the trees in the yungas canopy are evergreen and large-leaved, though in the southern part of the range, smaller, microphyllous and deciduous trees start to become more prominent. In the northern yungas forests, broadleaf trees like Nogales and figs overwhelmingly dominate, but at higher elevations and farther south, the forest contains many more conifers, such as *Podocarpus,* as well as alders, especially Andean Alder. In summer, the yungas looks much like NEOTROPICAL CLOUD FOREST, but in the winter months the semi-evergreen nature of the forest becomes more obvious, with branches devoid of leaves yet covered with mosses and bromeliads.

Above: **Yungas forest near Jujuy, Argentina, home to the rare Red-faced Guan and Yungas Pygmy-Owl.** © ANDRÉS VÁSQUEZ, TROPICAL BIRDING TOURS

Right: **Red-faced Guan within yungas forest in nw. Argentina; it is confined to this habitat.** © SAM WOODS, TROPICAL BIRDING TOURS

The canopies of yungas are lower and more disjointed than those of cloud forest, with canopy cover at around 70%, markedly different tree heights, and many more gaps. The canopy is usually around 45 ft. (15m) high, though it can be 60 ft. (18m) high with 100 ft. (30m) emergents in Bolivia. Some trees have gnarled and twisted shapes, with limbs branching off the trunk low on the tree and even growing downslope from the tree. With the many light gaps, the undergrowth of the forest is very thick but not nearly as luxuriant as that of cloud forest. Bamboo and *Gunnera* bushes, with their massive leaves, are still present, but they share the forest floor with terrestrial bromeliads and many tree ferns. Another notable difference from cloud forest is the increasing number of sclerophyllous shrubs in the undergrowth. The southern yungas forests even have cacti and euphorbias, two groups completely lacking from the cloud forests of the equatorial Andes.

WILDLIFE: This habitat is rich in birds, its avian communities similar to those of NEOTROPICAL CLOUD FOREST, though some bird groups, like the barbets and fruit eaters (a subset of cotingas), do not make it to Argentina. Hummingbirds such as the Red-tailed Comet and tanagers form a much greater portion of the bird assemblage than in surrounding habitats. Most of the endemic bird species are undergrowth and sub-canopy specialists, such as White-throated Antpitta, Zimmer's Tapaculo, White-browed Tapaculo, Yungas Tody-Tyrant, Yellow-striped Brushfinch, Rusty-browed Warbling-Finch, and Yungas Manakin, though there are many canopy species as well, including Green-cheeked Parakeet, Brown-capped Redstart, Tucuman Parrot, Red-faced Guan, Yungas Pygmy-Owl, Slaty Elaenia, and Rust-and-yellow Tanager.

Yungas contains fewer reptiles and amphibians than NEOTROPICAL LOWLAND RAINFOREST, though frogs, especially marsupial frogs, are a prominent feature in Argentinean yungas.

The mammals are mainly widespread species shared with other habitats, such as White-lipped and Collared Peccaries, Jaguar, Jaguarundi, and Margay.

Endemism: Yungas forest is found only on the east slope of the Andes and is contiguous, with fewer localized forest areas and therefore less endemism than NEOTROPICAL CLOUD FOREST. Three EBAs occur in this habitat: the Bolivian and Peruvian lower yungas, the Bolivia and Peruvian upper yungas, and the Argentinean and s. Bolivian yungas. These three zones have around 30 endemic birds in total.

DISTRIBUTION: The yungas habitat extends for 800 mi. (1,250km) along the east slope of the Andes, from c. Perú to n. Argentina around Tucumán. Remarkably, this band of habitat is never more than 60 mi. (100km) wide. It is at its most expansive in Bolivia, where it forms a constant barrier between the highlands of the Andes and the lowlands to the east. Much of the habitat in the lower zones has been cleared for timber and farming, such as sugarcane production, but much of the yungas remains protected by the steep terrain of the e. Andes, which makes timber extraction and farming difficult.

WHERE TO SEE: Río Los Sosa Valley, near Tucumán, Argentina; Yala Valley, near Jujuy, Argentina.

SIDEBAR 2.3 ▶ TREE-LINE AND KRUMMHOLZ FORESTS

The tree line is the elevation above which forests can no longer exist. Both the elevation of the tree line and the hardiest tree species that live at its limit are highly varied, but tree-line forests do all have the same basic form. Most trees in these environments develop a krummholz structure, a stunted, crooked form similar to trees that grow on windy coastlines. These hardy trees can withstand

POLYLEPIS   CONIFER

heavy cover of snow and, in areas with strong prevailing winds, will bend toward the lee side, while branches break off the windward side; they end up with a crown like a tipped right-angle triangle and are generally dwarfed compared to the same species in more protected areas. In the harsh conditions of both high elevation and high winds, as the habitat changes to alpine tundra, the trees take on a bonsai shape, growing in thickets with stunted, tangled limbs.

In all regions, the tree line is generally higher at lower latitudes and in areas away from coastlines, so it occurs at 12,000 ft. (3,650m) in the Himalayas, at 3,000 ft. (900m) in Scotland, at 9,000 ft. (2,750m) in the Tian Shan range in Kyrgyzstan, at 6,000 ft. (1,800m) in the Altai Mountains in Russia, and at 1,000 ft. (300m) in Kamchatka in Russia. In any one area, the tree line can vary dramatically, depending on predominant wind direction, aspect, soil type, and rain shadow effect. Air, drainage, sun exposure, avalanches, cold or wet soil, and rocky slopes all affect the growth of trees. Larch and pine are the two most common tree-line groups in the taiga (or boreal forest), except in Fennoscandia, where birch becomes an important tree-line group. In the Tian Shan, pine and juniper dominate this zone; in the Austrian Alps, Creeping Pine prevails; in the Pyrenees, Common Beech and Umbrella Pine become important; and in e. Russia and Japan, Dwarf Siberian Pine abounds. In the Andes of South America, the trees that grow at the highest elevations and take on this gnarled shape are *Polylepis* spp., which can grow well into the PARAMO in areas protected from wind.

ELFIN AND STUNTED CLOUD FOREST

IN A NUTSHELL: Stunted mixed forest near the tree line, above the humid forest and below the PARAMO.
Habitat Affinities: In protected areas, INDO-MALAYAN TROPICAL MONTANE RAINFOREST; in exposed areas, AFROTROPICAL FYNBOS and AUSTRALASIAN ALPINE HEATHLAND. **Species Overlap:** NEOTROPICAL CLOUD FOREST.

DESCRIPTION: Elfin forest exists near the tree line above most of the NEOTROPICAL CLOUD FORESTS and YUNGAS forests in the Neotropics, existing as a transition zone between these forests and the PARAMO or polylepis groves above. This is a stunted, gnarled version of the forest below it, with which it shares most canopy tree species. The trees are generally 5–30 ft. (1.5–10m) tall, form a single canopy, and often take on the krummholz form (see sidebar 2.3), bending away from the direction of the prevailing wind. The trees are species from the lower cloud forests or yungas as well as different, almost microphyllous trees, their smaller leaves presumably adapted to the strong winds. In this habitat, tree branches and most trunks are covered in mosses and epiphytes, which are much more common and make up a much higher proportion of the biomass of elfin forests than of other forest types. The undergrowth is extremely thick with ferns, large-leaved bushes, and massive-leaved *Gunnera* shrubs. There is a lot of bamboo, such as *Chusquea* spp., usually found as undergrowth but sometimes forming large independent stands that reach the canopy, as in Utuana in s. Ecuador. Bushes with conspicuous flowers make up a significant and often prominent proportion of the understory of this forest. In more exposed areas, there is no canopy cover at all, and sclerophyllous bushes can dominate in the thick brush, giving the area an appearance remarkably similar to that of Afrotropical FYNBOS or AUSTRALASIAN ALPINE HEATHLAND. Polylepis groves are common in the transition from elfin forest to paramo, and although it is a forest tree, polylepis grows in the more protected snow crannies in the paramo.

Elfin and stunted forests are generally found at high elevations, between 9,000 ft. and 13,000 ft. (2,750–4,000m), and above Neotropical cloud forest. Elfin forest can be found as low as 1,000 ft.

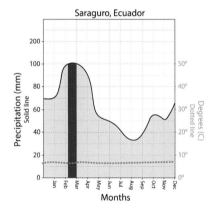

Saraguro, Ecuador

Elfin forest in s. Ecuador, showing extensive bamboo stands and stunted trees.
© IAIN CAMPBELL, TROPICAL BIRDING TOURS

(300m) on isolated hills where the soils are very poor, or in areas with high rainfall combined with poor drainage. Elfin forests can also be produced by chemical poisoning from concentrated metal deposits in the underlying soil. In essence, they are versions of the forest below them that are undergoing stress. They are extremely fire intolerant, and the same kinds of conditions, combined with fire, would form heathlands in more temperate parts of the world.

WILDLIFE: Elfin forests and stunted cloud forests have a very distinctive bird assemblage, and many species are found only in these forests. The bamboo stands have the Gray-headed Antbird, Rusty-breasted Antpitta, and Piura Hemispingus. Areas with thick bushes have the Crescent-faced

Polylepis trees at the boundary between elfin forest and paramo in n. Ecuador. These are among the highest-living trees in the world, having been recorded at 17,000 ft. (5,200m). A number of specialist birds occur in this habitat, such as Giant Conebill, Ash-breasted Tit-Tyrant, Royal Cinclodes, and White-cheeked Cotinga.
© SAM WOODS, TROPICAL BIRDING TOURS

Antpitta, Rufous-necked Foliage-gleaner, Masked Mountain-Tanager, Golden-crowned Tanager, and Jelski's Chat-Tyrant, along with the exceptionally cute Black-crested Tit-Tyrant. Hummingbirds make up a significant portion of the bird population of this habitat, including elfin forest specialists like the Purple-throated Sunangel, Rainbow-bearded Thornbill, Shining Sunbeam, Viridian Metaltail, and Neblina Metaltail.

Mammals are not well represented, and there are no primates in elfin forests or stunted cloud forests. Shrews and rodents make up most of the mammals.

Endemism: The EBA system (see "Endemic Bird Areas" in the Introduction) of overlaying polygons on a map to designate areas fails to capture how important these limited forests are for endemics, because this habitat is almost always incorporated into larger areas covering multiple habitats.

DISTRIBUTION: Elfin forests and stunted cloud forests are found above NEOTROPICAL CLOUD FOREST and YUNGAS forests from Central America southward, down either side of the Andes from Venezuela to Perú, and then down the Andes east slope to around Salta in n. Argentina. They also occur in limited areas above the Atlantic cloud forests of se. Brazil.

WHERE TO SEE: Bottom of Papallacta Pass, n. Ecuador; Cerro Toledo area of Podocarpus National Park, s. Ecuador; upper trail at Yanacocha Reserve, nc. Ecuador; Utuana Reserve (for stunted forest), s. Ecuador.

Rainbow-bearded Thornbill in elfin forest near timberline in s. Ecuador.
© SAM WOODS, TROPICAL BIRDING TOURS

MAGELLANIC RAINFOREST

IN A NUTSHELL: Evergreen and semi-deciduous forests that have moss-covered floors and very little undergrowth. **Habitat Affinities:** NEW ZEALAND BEECH FOREST. **Species Overlap:** VALDIVIAN RAINFOREST AND NEOTROPICAL MIXED CONIFER FOREST.

DESCRIPTION: From the outside, this forest appears notably uniform in tree species, made up of a few beech species. Many of these trees have a bent krummholz structure (see sidebar 2.3) caused by relentless winds, and plenty of others have been blown down. Because the cold climate of this rainforest results in a decelerated speed at which the dead trees decompose, the ground is littered with slowly rotting logs. The forest is dominated by an evergreen beech (*Nothofagus*, a Gondwanan relict genus) called Magellanic Coihue, but the forests closer to the drier PATAGONIAN STEPPE are dominated by Lenga, a shorter, deciduous beech. The transition between the Magellanic Coihues and the Lenga Beeches is very gradual, and this area is best described as a broadleaf semi-deciduous mix.

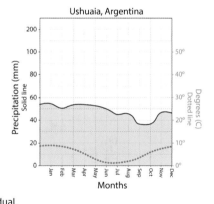

Ushuaia, Argentina

In Magellanic rainforest, the canopy is generally quite low, at around 30 ft. (10m), and the forest generally lacks a mid-story. Trunks and branches are swathed in mosses and lichens. The undergrowth of Magellanic Coihue is very sparse, with few bushes and many ferns on the ground. Lenga Beeches grow closer together and can form thicket-like stands with little undergrowth. Mosses and fungi completely blanket the forest floor, so there is almost no bare earth. Many areas have poorly drained soils and support bogs covered with sphagnum moss. This forest generally occurs in areas with high rainfall, little snow, high winds, and temperatures that do not go much below freezing. Although this climate is cold, freezes are rare due to the moderating winds of the s. Atlantic. On the leeward side of the forest, in the rain shadow of the low Andes, the winds gradually become colder and drier, and therefore much more desiccating. Here, the Magellanic rainforest becomes completely deciduous and dominated by Lenga Beech; then it transitions into what appears very similar to PALEARCTIC FOREST-STEPPE over a few miles before merging into the PATAGONIAN STEPPE. Magellanic rainforest extends to nearly 56°S, which is the austral equivalent of Churchill, Canada, or Copenhagen, Denmark.

WILDLIFE: Although this habitat is distinctive, it is quite extensive and homogeneous, and does not support any range-restricted species. Birdlife consists mainly of specialized species, such as Austral Pygmy-Owl, Magellanic Woodpecker, Rufous-tailed Hawk, White-throated Treerunner, and Thorn-tailed Rayadito. In the open areas between forest blocks are species such as the Chilean Swallow, Austral Thrush, and Patagonian Sierra-Finch.

Right: **Magellanic Woodpecker is one of the most highly sought-after birds in this habitat.**
© MIKE NEUBAUER

Below: **Beech forest in Tierra del Fuego, Argentina.**
© IAIN CAMPBELL, TROPICAL BIRDING TOURS

There are very few mammals in these forests; Fuegan Red Fox (the local race of the Andean Fox or Culpeo) and Puma are the main predators, feeding on Southern Pudu. Smaller mammals include rodents like the small Magellanic Tuco-tuco and the semiaquatic Nutria. As would be expected this far south, there are few reptiles or amphibians.

DISTRIBUTION: Magellanic rainforest occurs from Coyhaique in s. Chile all the way around Cape Horn in southernmost Tierra del Fuego. It is the main forest type around Ushuaia, covers the southernmost Andes mountains, and occurs on the northern slope where the Andes strikes east–west. It merges with the PATAGONIAN STEPPE.

WHERE TO SEE: Tierra del Fuego National Park, Argentina.

NEOTROPICAL DRY DECIDUOUS FOREST

IN A NUTSHELL: Dry tropical forest that appears lush in the wet season but stark in the dry season, when the canopy loses many of its leaves. **Habitat Affinities:** MALAGASY DRY DECIDUOUS FOREST; AUSTRALASIAN TROPICAL SEMI-DECIDUOUS FOREST. **Species Overlap:** NEOTROPICAL SEMI-EVERGREEN FOREST; CAATINGA; NEOTROPICAL THORNSCRUB.

DESCRIPTION: Dry deciduous forests of the Neotropics vary greatly throughout the year, parched in the dry season and lush during the wet season. This habitat is found across a wide range, from high-precipitation areas with enough annual rainfall to support rainforest to drier areas that would normally sustain open woodland. Dry deciduous forest has the same seasonal rainfall regimen as savanna but is less resistant to fire. The defining feature of this habitat is the intense wet season, which alternates with long periods of low rainfall, including a few months of near-total drought that is enough to kill most large, evergreen trees. In the wet season this habitat appears very similar to NEOTROPICAL SEMI-EVERGREEN FOREST or rainforest: water abounds, the canopy appears thick (though close inspection shows that the leaves are generally smaller than those in rainforest), and there is abundant, lush undergrowth. Visiting the forest during the dry season gives a very different impression: many of the plants have lost their leaves, the ground is covered with a thick layer of dry, brittle leaves, and the layer of brush at head height seems almost impenetrable.

The canopy of deciduous trees sits at a height of 40–65 ft. (12–20m) and is generally very open, with less than 50% cover during the wet season. The open nature of the canopy allows light to penetrate the forest, resulting in a thick lower mid-story of evergreen trees. When this habitat is more open during the dry season, you can see that most trees are stout, quite thick for their height, and not as straight as trees in wetter forests. They are usually angled and gnarled, with branches starting much closer to the base of the trunk than in rainforest. The trees have a variety of canopy forms, with many of the smaller-leaved trees having a flatter shape than trees of closed forests. One of the prominent features of this kind of forest, despite the wet season, is the abundance of baobab-like kapok trees (*Ceiba* spp.), which store water in their trunks. Kapoks dominate the forest and are usually higher than the surrounding trees as well as much broader. The canopy has fewer vines and epiphytes than rainforest, but because of the deciduous nature of this forest, they are often more conspicuous than their rainforest counterparts.

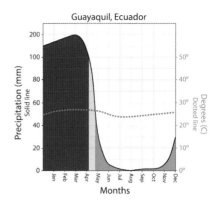

Undergrowth in tropical dry deciduous forest in s. Ecuador, which hosts White-headed Brushfinch and Elegant Crescentchest.
© SAM WOODS, TROPICAL BIRDING TOURS

The understory usually consists of small evergreen trees and shrubs, with a few younger deciduous canopy trees. Thornbushes, similar to those in thornscrub habitats, exist as ground-cover shrubs. Grasses and herbaceous plants are limited, distinguishing dry deciduous forest from woodland habitats such as CHACO SECO, and most dry deciduous forests lack cacti, which distinguishes them from similar habitats such as CAATINGA. Because of the thick undergrowth, walking through this habitat away from trails is generally difficult, especially in the wet season. In the dry season you can see the ground for 20–50 ft. (6–15m); in the wet season ground visibility is limited to 10–20 ft. (3–6m). At head height, visibility is limited year-round due to the thick understory.

WILDLIFE: Generally, the avifauna of dry deciduous forest has representatives of most Neotropical bird families, although fruit-eating groups such as toucans, barbets, and cotingas are not well represented. Furnariids such as woodcreepers, foliage-gleaners, and spinetails make up a larger proportion of the avian assemblage than in wetter areas. In the Tumbesian region of sw. Ecuador and n. Perú, the tropical dry forests have White-headed Brushfinch, Elegant Crescentchest, and

White-tailed Jay. In Costa Rica such forests are home to Turquoise-browed Motmot, Black-headed Trogon, Orange-fronted Parakeet, White-throated Magpie-Jay, and Spot-breasted Oriole. In Brazil, typical birds include Moustached Woodcreeper, Minas Gerais Tyrannulet, and Slender Antbird. Frogs have a difficult time in this environment, while reptiles are well represented.

Mammals are more obvious in the dry deciduous forest than in wetter tropical forests. A southern subspecies of White-tailed Deer inhabits this forest. The most obvious mammal of Ecuador's dry deciduous forest is the Guayaquil Squirrel, and both Brown-throated and Hoffman's Two-toed Sloth also occur here.

Endemism: Because this habitat is usually surrounded by wetter areas in the Neotropics, it has developed endemic bird species. In sw. Ecuador and nw. Perú, dry deciduous forest forms the core of the extremely important Tumbesian region, which has over 50 endemic birds. There are smaller areas of endemism along the Pacific coast of Costa Rica and Guatemala. Eastern Brazil also holds some dry deciduous forest, where Minas Gerais Tyrannulet is limited to a small area.

DISTRIBUTION: Dry deciduous forest has a variety of names and forms. It is called *baja caducifolia* (low deciduous) in Central America, where it occurs along the Pacific coast from Guatemala to Panama. It is extensive in sw. Ecuador, where it borders the rainforests of the Chocó region, and in nw. Perú, where it borders the deserts of the Pacific coast. It is patchily distributed along the Venezuelan coastline and on many islands in the Caribbean.

WHERE TO SEE: Jorupe Reserve, s. Ecuador; Santa Rosa National Park and La Ensenada, Costa Rica; Boa Nova, Bahia, Brazil.

Above: **Guayaquil Squirrel** occurs in this habitat in the Tumbesian region of s. Ecuador and n. Perú.
© NICK ATHANAS, TROPICAL BIRDING TOURS

Left: **White-tailed Jay is confined to dry deciduous forest in s. Ecuador and n. Perú.** © IAIN CAMPBELL, TROPICAL BIRDING TOURS

CAATINGA

IN A NUTSHELL: Caatinga is thornscrub on steroids: a dry forest of deciduous trees with thornscrub and cactus ground cover. **Habitat Affinities:** AFROTROPICAL MALAGASY SPINY FOREST. **Species Overlap:** CERRADO; NEOTROPICAL DRY DECIDUOUS FOREST.

DESCRIPTION: The trees of the caatinga habitat are deciduous, but many drop their leaves only in times of stress, rather than seasonally. They may remain fully leafed right through the dry season in a wet year, and likewise, they may not leaf out in an abnormally dry wet season—all of this contributing to an utterly bizarre forest type. Essentially, the main features that distinguish caatinga are the thorny (rather than evergreen) understory and the prominence of cacti. It is broadly similar to the thornscrubs

The caatinga, a habitat of ne. Brazil, becomes verdant in the wet season, as shown here.
© NICK ATHANAS, TROPICAL BIRDING TOURS

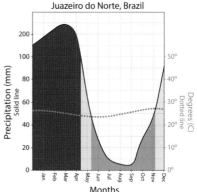

The rare Araripe Manakin is confined to a tiny area of lush caatinga in ne. Brazil.
© NICK ATHANAS, TROPICAL BIRDING TOURS

Juazeiro do Norte, Brazil

of Central America or the CHACO SECO farther south but is a much more diverse, full forest with many more plant species than those more uniform habitats.

Caatinga forests have a thorny mid-canopy (or understory) of acacia and mimosa as well as other thorn trees, but looming over this is a canopy of dry deciduous trees that can lose their leaves in the dry season but look ridiculously lush in the wet season, appearing incongruous with the understory. There are Brazilian Bottle Trees that resemble the baobabs of Western Australia, candelabra- and octopus-shaped cacti that resemble African euphorbias, and drought-tolerant palm trees.

The non-cactus xerophytic plants that do exist in this predominantly deciduous forest have small, leathery leaves more similar to the leaves of the eucalypts of Australia than the larger, more delicate, laurel-like leaves of the nearby NEOTROPICAL LOWLAND RAINFOREST. The canopy varies from a height of around 10 ft. (3m) in the drier zones to 25 ft. (8m) in the wetter areas, with the average canopy height at 15 ft. (5m). Regardless of canopy height, the plant assemblages remain fairly constant. The habitat is very dense, considering its dryness, with near-complete canopy cover and very little grass cover underneath it. Rainfall is variable throughout the region. In the north, this forest has the same weather systems as the nearby Amazon, although it is drier. The southern caatinga has its wet season in the Southern Hemisphere summer and has droughts in the middle of the year. In contrast with the adjoining CERRADO, the caatinga is not fire-tolerant and is easily destroyed by fire, which turns it into a desertlike scrub.

WILDLIFE: In bird assemblages, there is more overlap between the caatinga and the CERRADO than there is between the caatinga and Amazon broadleaf forests. Some species are strongly associated with caatinga. The most notable of the large birds of the caatinga, the Spix's Macaw, went extinct in the wild in 2000. Its close relative Indigo Macaw remains in small numbers. Another rare bird of this area is the Araripe Manakin, known from one small area of lusher than normal caatinga. There are many other birds endemic to the caatinga region and habitat, such as Great Xenops and Scarlet-throated Tanager, and most are under pressure due to habitat loss. Other

Many caatinga mammals, like this Six-banded Armadillo, are also found in Cerrado savanna (photo: Brazil).
© IAIN CAMPBELL, TROPICAL BIRDING TOURS

species include Cactus Parakeet (IS), Stripe-backed Antbird, Spotted Piculet, Tawny Piculet (IS), Silvery-cheeked Antshrike, Red-shouldered Spinetail (IS), and Pygmy Nightjar (IS).

There is extensive mammal overlap between caatinga and Cerrado, which is not surprising given that they are both open habitats with strong wet/dry climates. About 80% of mammal species are shared with the Cerrado, including most of the bat species. However, roughly half of the total mammal species of the caatinga are bats, contrasting with the Cerrado, where they account for only 30% of the animal assemblage, and with the NEOTROPICAL SEMI-EVERGREEN FOREST and rainforests, where they represent an even smaller percentage. Larger mammals tend to be more widespread and more catholic in habitat choice between open habitats, and some, such as the carnivores, occur in a broad range of open and closed habitats. Rodents are more restricted to caatinga than any other mammal group. Primates are poorly represented here, but Barbara Brown's Titi Monkey is a caatinga endemic primate.

DISTRIBUTION: The caatinga forests, centered on the ne. states of Ceará and Bahia, Brazil, are bounded to the west by Amazon SEMI-EVERGREEN FOREST and to the south by the CERRADO savanna. Most of the caatinga has been cleared for monoculture and cattle farming, and there are few remaining large blocks of forest. The habitat persists primarily as corridors along roads and in small isolated islands within farmland.

WHERE TO SEE: Chapada do Araripe, Ceará and Pernambuco, Brazil; Catimbau National Park, Pernambuco, Brazil.

CHACO SECO AND ESPINAL

IN A NUTSHELL: Chaco seco is an open woodland that resembles a typical savanna at a distance but has thicker ground cover. **Habitat Affinities:** AFROTROPICAL MOPANE SAVANNA; AUSTRALASIAN MULGA WOODLAND AND ACACIA SHRUBLAND. **Species Overlap:** MONTE; PAMPAS.

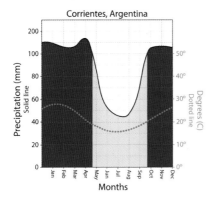

Corrientes, Argentina

DESCRIPTION: *Chaco* is a confusing term because it designates both a region and a habitat, which do not completely overlap. The Gran Chaco (or Chaco) region is a huge, arid plain where the rains average 16–48 in. (400–1,200mm) per year. It is centered on Paraguay but includes large areas of e. Bolivia, nw. Argentina, and a small area of sw. Brazil. The Gran Chaco includes palm groves (as does the Pantanal of the NEOTROPICAL FLOODED GRASSLAND AND WETLAND habitat), small amounts of moist evergreen gallery forest resembling NEOTROPICAL LOWLAND RAINFOREST, and shrubby thorn thickets.

The habitat covering most of the region is Chaco seco (dry Chaco), which is akin to a combination of an open savanna woodland, with a canopy about 30 ft. (10m) high, and an underlying, extremely thick, thorny shrub layer, 10–20 ft. (3–6m) in height. The undergrowth and ground cover are limited, though there is some grass. The trees and bushes include both xerophytic plants and sclerophyllous plants with small leathery leaves, along with broadleaf deciduous trees that lose their leaves in April–July during the dry winter. Most of the trees bloom at the start of the summer wet season, between October and December. Chaco is generally very difficult to walk through, and in areas without canopy trees, the xeromorphic scrub can become extremely thick and nearly impenetrable. Espinal can be regarded as a lower, less diverse version of the more widespread Chaco seco, as there is massive overlap of bird and other animal species. In regions where the habitat becomes scrubbier, espinal forms its own canopy, around 20–27 ft. (6–8m) high, and has significant grassy undergrowth.

WILDLIFE: There are many mammals in the Chaco seco, including the endangered Chacoan Peccary; numerous small cats, such as Little Spotted and Geoffroy's Cats; Giant Armadillo; Nine-banded, Six-banded, and Southern Three-banded Armadillos; Giant Anteater; Azara's Agouti; and Capybara.

Chaco birdlife consists of both grassland and woodland assemblages. Grassland birds include species from families typical of the PAMPAS region, such as the Greater Rhea and Crested Hornero.

Left: **The Gran Chaco region in Argentina.**
© IAIN CAMPBELL, TROPICAL BIRDING TOURS

Below: **Southern Tamandua feeds exclusively on termites and ants and is one of many mammals found in the Chaco seco habitat.**
© PABLO CERVANTES DAZA, TROPICAL BIRDING TOURS

The woodland birds include Nanday Parakeet, Spot-winged Falconet, Chaco Chachalaca, Turquoise-fronted Parrot, Scimitar-billed Woodcreeper, Black-bodied Woodpecker, Black-capped Warbling-Finch, Crested Gallito, and Many-colored Chaco Finch.

DISTRIBUTION: The Chaco seco is centered on c. and w. Paraguay, where it borders Bolivia's flooded grassland to the north and the CERRADO region to the northeast. To the west it is bordered by the YUNGAS cloud forests of the e. Andes. In the south it continues down to the Córdoba province in Argentina, where it merges into MONTE shrublands. Espinal woodland lies farther to the east and southeast of the Chaco seco. It is the dominant vegetation in Argentina's Corrientes province, to the west of Uruguay. It surrounds the PAMPAS grassland in e. Argentina and extends to the Atlantic Ocean in s. Buenos Aires province.

WHERE TO SEE: This is common vegetation north and west of Córdoba, Argentina.

Red-crested Cardinal in the Chaco seco of Argentina. © IAIN CAMPBELL, TROPICAL BIRDING TOURS

SIDEBAR 2.4   WHAT MAKES A HABITAT?

It is not only rainfall and temperature that define habitats. Habitats are also influenced by soil type, bedrock, wind direction, local topography, and perhaps most importantly, the intensity of dry seasons and the occurrence of fire. For example, in subtropical environments, both seasonal forests and open woodlands can receive the same amount of rainfall, but the seasonal forests cannot grow in areas with more frequent fires. Another example is in the temperate regions, where European deciduous forests and juniper forests may both exist in low-fire areas, but the juniper forests can better survive periods of drought.

The likely forest type given similar temperatures but with varying amount of exposure to fire, amount of precipitation, and periods of drought.

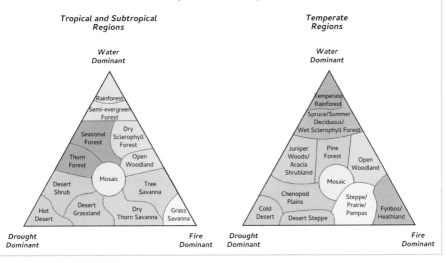

CERRADO

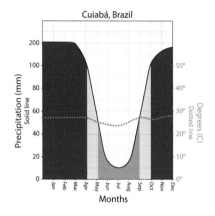

IN A NUTSHELL: A shrubby savanna habitat in which the bushes usually are not thorny, and most have sclerophyllous leaves. **Habitat Affinities:** No direct affinities; appears like a cross between AUSTRALASIAN LOWLAND HEATHLAND and AUSTRALIAN MULGA AND ACACIA SHRUBLAND. **Species Overlap:** CHACO SECO.

Cuiabá, Brazil

Precipitation (mm) Solid line / Degrees (C) Dotted line / Months

DESCRIPTION: Cerrado habitat has a range of expressions: It can be almost pure grassland, with a few trees, resembling CAMPO; this is often called *Cerrado limpio*. It can appear similar to AUSTRALASIAN LOWLAND HEATHLAND; this is *campo rupestre*. In some places it blends from thorn forest to woodland so thick that it resembles a thinned evergreen or semi-evergreen forest. In areas protected from fire, and along rivers, ribbons of NEOTROPICAL SEMI-EVERGREEN FOREST extend from the Amazon and the Atlantic forests and act as corridors for animals through the drier savannas of the Cerrado. Over most of its area, Cerrado is a dense savanna dominated by shrubs and small trees that are multi-stemmed or twisted and gnarled, with limbs branching off short, stout trunks. Canopy size varies markedly with rock types and local soils. Where the topsoils developed on completely inert, nutrient-deficient sandy soils, growth is retarded, resulting in a canopy as low as 9 ft. (3m); this habitat is sometimes called *campo Cerrado*. The ancient, nutrient-poor lateritic soils (see sidebar 2.5) that are extensive throughout c. Brazil support Cerrado with a canopy 12–18 ft. (4–5.5m) high, while the depressions just a few feet lower, with more nutrient-rich erosional or depositional soils, have Cerrado with a canopy around 20 ft. and up to 35 ft. (6–11m) tall. The characteristic of all Cerrado habitat is the very limited canopy cover, always less than 50%, creating an open environment for the smaller trees and shrubs in the understory. The shrub cover is between 40% and 80%, while thick grass tussocks around 2 ft. (0.6m) tall fill the voids between shrubs, so that there is almost no bare ground in this habitat.

Rainfall in the Cerrado is between 40 in. and 80 in. (1,000–2,000mm) annually. In other environments, this is enough moisture to allow the growth of evergreen forest, but that is not the case here, since the rainfall comes as a single summer wet season of six months. These areas must survive the other half of the year in conditions of near-total drought, which makes it impossible for evergreen forests to survive. Though there are some sclerophyllous bushes with thick leathery leaves, the trees and bushes are mostly semi-deciduous, setting Cerrado apart from campo or CHACO SECO habitats, which do not have deciduous trees. Cerrado is also prone to fire, so most

Left: **Cerrado supports a significant mammalian fauna, including Maned Wolf (photo: Brazil).** © KEITH BARNES, TROPICAL BIRDING TOURS

Below: **Cerrado habitat in c. Brazil.** © NICK ATHANAS, TROPICAL BIRDING TOURS

of the bushes and small trees are fire-tolerant, with very thick bark, in contrast to the trees of the CAATINGA. The thick nature of the understory and brush layer makes walking through Cerrado habitat difficult, though visibility is quite good at eye level.

WILDLIFE: Mammals are a major feature of the Cerrado savanna. Maned Wolf, Giant Armadillo, and Giant Anteater are present, and Jaguar and Ocelot still exist in decent numbers. Discounting the riparian forests, which can be viewed as outliers of rainforest and contain primates otherwise not seen in Cerrado, the mammal assemblage is most similar to that of the CHACO SECO savannas to the south.

Because Cerrado is a shrubby type of savanna, in contrast to the grasslands of the Pantanal (part of the NEOTROPICAL FLOODED GRASSLAND AND WETLAND habitat), CAMPO, and PAMPAS, its bird assemblages include not only strictly ground-dwelling birds, like Red-legged Seriema, tinamous, and ground doves, but also shrubland and woodland species like Horned Sungem, Curl-crested Jay, Black-throated Saltator, Collared Crescentchest, White-banded and White-rumped Tanagers, Plumbeous Seedeater, and flycatchers. Cerrado is surprisingly rich in parrots, such as Red-shouldered and Red-and-green Macaws and White-eyed and Yellow-chevroned Parakeets.

Endemism: Because of the Cerrado's large extent and its lack of mountains, most bird species are widespread. There is a minor center of endemism in the hills of the Minas Gerais state of Brazil, which holds Hyacinth Visorbearer and Gray-backed Tachuri.

Woodland species like Toco Toucan also occur in Cerrado (photo: Brazil).
© IAIN CAMPBELL, TROPICAL BIRDING TOURS

DISTRIBUTION: A huge Cerrado belt covers a massive area of c. Brazil south of the Amazon rainforests. It is centered in the state of Tocantins but reaches east into Minas Gerais and Bahia and down to Cuiabá, where it merges into the Pantanal (NEOTROPICAL FLOODED GRASSLAND AND WETLAND). Much of the NEOTROPICAL SEMI-EVERGREEN FOREST of Brazil has been destroyed and turned into a pseudo-Cerrado, while the majority of the original Cerrado itself has been destroyed by grazing. Cerrado habitats also exist in small patches within the Amazon rainforest, as well as extensively in the Rupununi savanna of s. Guyana and in se. Colombia. Some of the best-preserved examples of this habitat are found in Brazil's Chapada dos Guimarães National Park.

WHERE TO SEE: Serra do Cipó National Park, Minas Gerais, Brazil; Chapada dos Guimarães National Park, Mato Grosso, Brazil; Serra da Canastra National Park, Minas Gerais, Brazil.

SIDEBAR 2.5   DURICRUSTS AND DESERTIFICATION

Resistant elements such as iron, aluminum, silica, and calcium move around naturally in soils for thousands of years, causing gradual changes in soil type and vegetation type. But when vegetation over a soil that is balanced with a hot, shaded environment is cleared or overgrazed, these elements solidify, and the soils turn into desert wasteland, regardless of the natural precursor, be it desert, savanna, dry deciduous, or even semi-evergreen habitat.

In tropical and subtropical environments, the elements within the soil are constantly being carried by groundwater and deposited in other areas. In some environments such as warm, wet regions, very prolonged chemical weathering and seasonal movement of elements over thousands of years result in complex regolith (weathered material overlying fresh rock) profiles, including laterite formed from iron oxides (ferricrete) and aluminum oxides (bauxite or alcrete). In arid terrains, over very large timescales, natural silcretes and calcretes form as hardpan in natural soils. In both humid and arid terrains, the soils on top of these ancient surfaces are very nutrient-poor, and the vegetation on hillsides (where soils are younger and contain more nutrients) is usually lusher than on tops of plateaus.

In some cases, such movements can be much faster and form duricrusts—hard concretions of residual elements in soils. In more arid environments where chemical weathering is much less intense, silica and calcium get concentrated as silcrete and calcrete duricrusts at the bottom of slopes. In wetter environments, iron and aluminum moved downslope in groundwater crystallize on slopes and in unconsolidated colluvial regolith in valley bottoms. Duricrusts can be exposed and baked on valley sides, but usually they remain as concretions in soil profiles. However, when these materials are exposed to desiccation and heating through forest clearing and topsoil removal, materials that may have lasted for hundreds of years in colloidal form can harden and solidify into an extremely impermeable and durable duricrust cap. Once formed, either as ferricrete in savanna environments or as silcrete and calcrete in arid environments, these duricrust caps form a wasteland in which forests, savannas, and desert shrublands can no longer grow. The small areas on this diagram marked as A, B, C, and D can exist as massive desertified wastelands in areas that were formerly semi-evergreen forests in Brazil or dry thorn savanna in the Sahel of Africa. A represents a human-induced hardening of laterite and infusion of ferricrete formed from clearing and topsoil removal in savanna landscapes. B is a natural ferricrete formed at break in slope, and where vegetation is stunted. C is a human-induced ferricrete barren area formed when forest is cleared and topsoil is removed. D is human-induced silcrete/calcrete desertification formed from clearing and overgrazing in an arid environment.

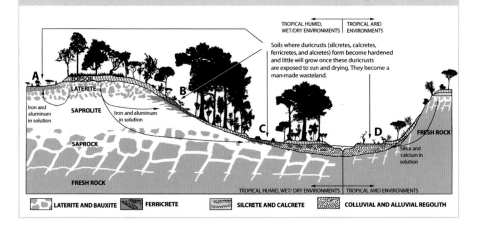

PAMPAS AND CAMPO

IN A NUTSHELL: A massive and contiguous grassland area called the campo in the north and the pampas in the temperate south. **Habitat Affinities:** AFROTROPICAL TROPICAL GRASSLAND. **Species Overlap:** NEOTROPICAL FLOODED GRASSLAND; CHACO SECO.

DESCRIPTION: The grassland landscape is extremely flat in the pampas (this habitat extends beyond the Argentinean region known as the Pampas) and consists of low rolling hills in the campo region. The general climate in the campo is subtropical, with warm weather and up to 60 in. (1,500mm) of annual rainfall, much of which falls in the austral winter (July–August). The pampas habitat is much colder and has as little as 16 in. (400mm) of yearly rainfall and a very dry austral summer (December–February).

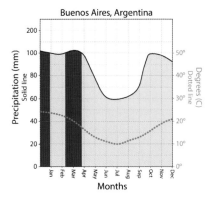

While grasses are, by far, the dominant vegetation type in both these habitats, the campo has more of a savanna feel to it, with forests occurring along rivers and in protected areas, such as breaks in the slope where the valley sides become flatter or near rock outcrops. There can be what appears to be an incongruous combination of North American Great Plains–type TALLGRASS PRAIRIE with stands of riparian forest like those that cut through the Pantanal (part of the NEOTROPICAL FLOODED GRASSLAND AND WETLAND habitat). The pampas is dominated by feathergrasses and tussocks of bunchgrasses, and has fewer and more isolated trees than the campo. As you move west and south, the grasses get shorter, and both shrubs and bare patches of earth become common. The boundary between the pampas and the MONTE to the south is gradual, with the relative abundance of grasses and shrubs changing, and this is reflected by the similar bird communities in the two habitats. In the west, many of the grass species of the pampas change to the same species found in the CHACO SECO AND ESPINAL, and thorn trees begin to dot the grassland. Eventually, these trees become close enough that the bird communities change to woodland rather than grassland groups, and the habitat can be regarded as Chaco seco and espinal. Walking through the pampas and campo is easy, with grass at waist height and few trees to obscure your vision.

OPPOSITE PAGE:
Top: **Pampas grassland west of Buenos Aires, Argentina.** © IAIN CAMPBELL, TROPICAL BIRDING TOURS

Bottom: **Crab-eating Fox is one of the mid-sized predators found in the campo (photo: Brazil).**
© PABLO CERVANTES DAZA, TROPICAL BIRDING TOURS

Long-winged Harrier is a characteristic raptor of South American grasslands, including pampas (photo: Argentina). © IAIN CAMPBELL, TROPICAL BIRDING TOURS

While some authors regard the pampas and campo as true natural climax communities, others suggest that the grasslands Europeans first saw when they arrived in South America constituted a human fire-induced environment. It is interesting to note that the bird assemblage of the pampas and campo is almost totally derived from species that are also found in one of the surrounding habitats (monte, Chaco seco, or the Pantanal).

WILDLIFE: Numerous mammals occur in the pampas, such as Pampas Cat, a range of skunks and armadillos, and the endangered Pampas Deer. Giant Anteater was characteristic of the campo grasslands but has been largely extirpated from this habitat.

The campo and pampas support grassland bird species, mostly passerines such as Spectacled Tyrant, Campo Miner, Pampas Meadowlark, sparrows, and wintering Bobolinks. Raptors such as White-tailed Kite, Roadside and White-tailed Hawks, and Long-winged and Cinereous Harriers patrol the sky. Large birds that hunt on the ground include Red-legged Seriema, Maguari Stork, Giant Wood-Rail, and Greater and Lesser Rheas.

DISTRIBUTION: The campo is found through s. Brazil, Uruguay, e. Paraguay, and n. Argentina, and the pampas mainly from Buenos Aires south. They blend into the MONTE in the south and the CHACO SECO in the west. Combined, these grasslands formerly covered a massive block of over 250,000 sq. mi. (650,000km²), the vast majority of which has been converted to agriculture. In the southern pampas, grazing is more important than monocultural farming, meaning that much more habitat remains in a seminatural state.

WHERE TO SEE: Pampas: Lihué Calel National Park, Argentina; the areas surrounding Buenos Aires, Argentina. Campo: Bañados del Este Biosphere Reserve, Uruguay.

MATORRAL SCLEROPHYLL FOREST AND SCRUB

IN A NUTSHELL: Woodlands and scrublands that are fire-prone, small-leaved, and dry for much of the year. **Habitat Affinities:** AFROTROPICAL FYNBOS; AUSTRALASIAN LOWLAND HEATHLAND; AUSTRALASIAN BRIGALOW WOODLAND. **Species Overlap:** NEOTROPICAL SEMIDESERT SCRUB.

DESCRIPTION: This habitat includes many sclerophyllous (thick-, leathery-leaved) shrubs and is similar to Australian BRIGALOW WOODLAND in structure and plant mixture, though the eucalypts of Australia are replaced here with plants of other families. Acacias occur, as in the Australian sclerophyllous forest. In matorral sclerophyll forest and scrub, the canopy is low and very irregular, with no sub-canopy. The trees are generally gnarled, with limbs branching off the trunk close to the ground, though some trees, like the Peumo and Boldo, can have more of an oak-like structure. The undergrowth in these forests is dominated by thornbushes and other small-leaved shrubs, which can make walking difficult. Sometimes

Matorral sclerophyll forest in La Campana National Park, c. Chile.
© NICK ATHANAS, TROPICAL BIRDING TOURS

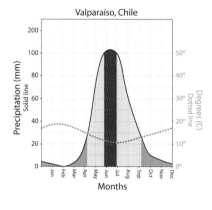

White-throated Tapaculo, pictured in matorral sclerophyll forest in c. Chile, is a specialist of this habitat. © FABRICE SCHMIDT, WINGS BIRDTOURS WORLDWIDE

the canopy is open, and the forest is reduced to a height that seems more shrublike. This version of the habitat is similar to the PACIFIC CHAPARRAL of California or the MAQUIS of the Mediterranean. Grasses grow between the trees and shrubs, and the landscape can sometimes resemble an urban park. The third main form of this habitat is a thornbush *espinal*, with spiny acacias, that resembles the CHACO SECO AND ESPINAL of Argentina. Here the trees can grow with enough space to allow luxurious grass growth during the wet season, giving the habitat a savanna-like appearance.

As with the other c. and s. Chilean habitats, matorral sclerophyll forest and scrub encompasses a broad grouping of similar forests. The coast of c. Chile has a Mediterranean climate, with hot and dry summers and cold and wet winters, with annual rainfall in the range of 12–48 in. (300–1,200mm). In various forms, this habitat occurs from the coast to about 6,000 ft. (1,800m) elevation in the Andes. Above 3,000 ft. (900m), it has less of a Mediterranean climate and taller trees.

WILDLIFE: As with other Chilean forests, although few animals exist in these habitats, the ones that are here are very localized. This is especially true of the tapaculos. In the tropics, these birds are mostly small, dull, and mouselike, but here they have radiated into a range of very attractive species, such as the Moustached Turca and White-throated Tapaculo. The furnariids are also well represented, with species like Crag Chilia (aka Crag Earthcreeper) and Dusky-tailed Canastero.

Two snakes and 23 lizards, the vast majority from the iguana family, are found in these matorral habitats. Mammals are limited, though there are some spectacular species, such as Elegant Fat-tailed Mouse-Opossum.

Endemism: Both the sclerophyllous scrubs and humid forests of Chile are contained within one EBA, Central Chile.

DISTRIBUTION: The Mediterranean sclerophyllous habitats of Chile occur from Copiapó, in nc. Chile along the coast, to around Santiago, then south in the central valley to Tempo in sc. Chile. Most of this habitat has been turned over to agriculture, and this is where most of Chile's wines are grown. The remnant habitats are often found as corridors along roadsides, and there are only a few large reserves.

WHERE TO SEE: La Campana National Park, Chile.

PARAMO

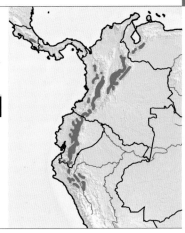

IN A NUTSHELL: A shrubby and grassy, tundra-like habitat of very high elevations. **Habitat Affinities:** AFROTROPICAL MONTANE GRASSLAND; EURASIAN ALPINE TUNDRA AND HIMALAYAN MONTANE DESERT; NEARCTIC ALPINE TUNDRA. **Species Overlap:** PUNA.

DESCRIPTION: The paramo is the Neotropical equivalent of alpine tundra, found only in equatorial areas at elevations generally above 10,000 ft. (3,000m), from Costa Rica to n. Perú, where the Marañón River separates this habitat from high Andean habitats farther south. Precipitation in the paramo is not high, with annual amounts in the range of 28–80 in. (700–2,000mm), but because the area is so high, cold, and covered in mist, overall moisture is abundant, making this a very humid habitat. Soils in the paramo are almost always acidic, nutrient-deficient, and sandy, with very little relationship to the underlying rock type. Essentially, because of the very cold climate, chemical weathering is so weak here that the soils form from organic material rather than underlying material. In some areas, pronounced dry seasons occur, and vegetation varies accordingly. Over most of the area, nighttime temperatures drop below freezing, but temperatures can rise to around 65–70°F (18–21°C) during the middle of the day.

El Angel, Ecuador

The paramo is a series of hills, mires, and marshes, with extremely irregular drainage patterns. In wetter areas, the landscape is dominated by cushion plants with rosette leaves, dwarf shrubs, mosses, and tussock grasses. The cushion plants are usually around 1 ft. (0.3m) high. Most shrubs are less than 1 ft. (0.3m) tall, though in protected areas they can grow to 3 ft. (1m). In slightly drier paramo, such as that found on Ecuador's Antisana, tussock grasses dominate, and there are fewer cushion plants. In a few locations, such as El Angel in n. Ecuador, and in Colombia and Venezuela, rosette-forming plants called frailejones grow to about 10 ft. (3m) in height and form prominent stands in an otherwise low habitat. In more protected areas, polylepis trees can grow as shrubs far above the tree line (treated here in the ELFIN AND STUNTED CLOUD FOREST habitat). There are ancient relationships between some paramo plant species and relatives that grow in the NEARCTIC ALPINE TUNDRA. However, over half the species in the Ecuadorian paramo have no close relatives outside this habitat.

The rosette-forming sunflowers known as frailejones are abundant in northern paramo.
© NICK ATHANAS, TROPICAL BIRDING TOURS

Paramo is a very fragile habitat that takes many decades to stabilize, is extremely slow growing, and has very long-lived vegetation. Once it has been overgrazed or cleared, leaving the landscape in an impoverished state, regeneration requires an incredibly long time.

Rufous-bellied Seedsnipe dwells among the cushion plants of the high Andean paramo. © PABLO CERVANTES DAZA, TROPICAL BIRDING TOURS

Spectacled Bears feed on puya plants in the paramo of the high Andes (photo: Ecuador).
© SAM WOODS, TROPICAL BIRDING TOURS

WILDLIFE: Mammals found in the paramo include the Andean Fox (aka Culpeo) and Andean Cottontail. Spectacled Bears regularly visit this habitat from lower-elevation NEOTROPICAL CLOUD FOREST and ELFIN FOREST. At this extreme elevation, reptiles and amphibians are rare.

Although this habitat is poor in birds, most of the species found in the paramo are endemic, specialized to survive in harsh tundra. Paramo birds include Andean Lapwing; Tawny Antpitta (IS); Jameson's Snipe (IS); furnariids such as canasteros, tit-spinetails, and cinclodes, like Buff-winged Cinclodes (IS); and hummingbirds including helmetcrests, Ecuadorian Hillstar (IS), Blue-mantled

Thornbill, and metaltails. In Central America, Volcano Junco and Volcano Hummingbird are prominent species.

DISTRIBUTION: Paramo is found at the highest elevations of Central America, from Costa Rica to Panama. It is found in the Andes from w. Venezuela throughout the highlands of Colombia and on both cordilleras through Ecuador south to n. Perú, where the Marañón River separates it from the PUNA farther south. Paramo generally occurs between 9,800 ft. and 15,500 ft. (3,000–4,800m) elevation.

WHERE TO SEE: Papallacta Pass, n. Ecuador (wet paramo); Antisana Ecological Reserve, Ecuador (dry paramo); Cerro de la Muerte, Costa Rica.

SIDEBAR 2.6   SUBDUCTION AND THE ANDES

The Andes is a very young mountain range: it started to form around 45 million years ago, making it younger than the first modern birds, and is still forming now. It is this mountain chain, running along the western edge of South America, that makes this continent such a megadiverse region. Its great elevational variations, with strikingly different microclimates, have created a vast array of niches, promoting the evolution of a very large number of species. The variation is so intense because, for much of the extraordinary length of this mountain chain, there are actually two parallel ranges, the Eastern and Western Cordilleras, with dry valleys between them. The reason for the two ranges is even more fascinating. As the Nazca Plate pushes toward the South American Plate, the Nazca Plate is being submerged, because its dark rocks, such as basalt, are denser than lighter continental rocks like granite, so they sink. As the ocean floor subducts (descends) under South America, the temperature in the earth's lithosphere becomes much hotter, and the ocean-floor sediments melt first because they are mainly made of quartz. This first series of melts comes to the surface as magma, forming the volcanoes of the Western Cordillera. The remaining ocean floor continues to sink until, at much hotter temperatures, the basalt melts and forms the volcanoes of the Eastern Cordillera. The Coast and Cascade Ranges of the Washington–Oregon coast also formed in this way.

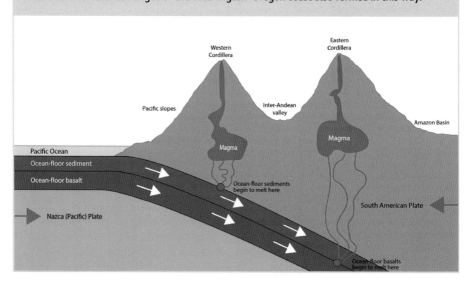

ANTARCTIC TUNDRA AND TUSSOCK GRASS

IN A NUTSHELL: The only vegetation on the Antarctic Peninsula and subantarctic islands. **Habitat Affinities:** NEARCTIC ROCKY TUNDRA; NEARCTIC BOGGY TUNDRA. **Species Overlap:** Very little with PATAGONIAN STEPPE.

King Penguins sometimes breed within the tussock grasses (photo: South Georgia Island).
© IAIN CAMPBELL, TROPICAL BIRDING TOURS

DESCRIPTION: On the Antarctic mainland, where conditions are very harsh in terms of climate and snow or ice cover, the tundra landscape is dominated by algae and cold-tolerant, tiny nonvascular plants such as lichens (over 200 species, with *Xanthoria, Caloplaca,* and *Usnea* the dominant genera), mosses (100 species, including local species like *Grimmia antarctici, Schistidium antarctici,* and *Sarconeurum glaciale),* liverworts, and fungi (over 600 species). There are just two flowering plants on the mainland of Antarctica, Antarctic Hairgrass and Antarctic Pearlwort. Tolerant of long periods when photosynthesis is not possible due to snow cover, which is usually complete in winter and can also extend into spring, these plants take advantage of short seasonal periods of growth during the Antarctic spring and summer. The constant snow cover through much of the year, interspersed by only brief periods when plant life can grow, limits the species in mainland Antarctica to shorter plants, preventing larger shrubs and trees from growing. The plant life is usually concentrated in the warmer, wetter areas of the mainland (predominantly on the west side of the Antarctic Peninsula) and also on the significantly warmer and wetter subantarctic islands, which have a much greater diversity of plant life supported by the more favorable climatic conditions. Other areas of the region are so dominated by snow cover or extreme exposure that plants cannot get a grip on the landscape, and bare rock persists.

The vegetation of the Antarctic Peninsula and the South Shetland Islands to its north is very similar, the latter differing in being more extensively vegetated, particularly with abundant moss growth. The subantarctic islands, such as South Georgia off South America and Macquarie off New Zealand, are similar to the Antarctic mainland in having high numbers of nonvascular species—lichens, liverworts, mosses, and fungi—but differ in their greater variety of vascular plants (around 25 species on South Georgia and 46 on Macquarie) and visually by the large swaths of tussock grasses, which are concentrated in the warmer coastal regions. The tussocks reach 6 ft. (2m) high and represent the tallest vegetation on both islands. Inland from the coasts, shrubs and mosses dominate the landscape. The greater variety of vascular plants on some subantarctic islands like Macquarie is explained by the relative lack of snow cover. On South Georgia Island, snow cover is complete in winter, while on Macquarie Island it typically occurs for only a few days each year, allowing for a longer growing season.

WILDLIFE: Much of the wildlife of the Antarctic region is naturally concentrated along the rocky coastlines, where the dominant bird and mammal populations are marine foragers. However, tundra and particularly tussock grasses provide valuable breeding habitat for a variety of birds in the region, including Wandering, Royal, Gray-headed, Black-browed, Light-mantled, and Sooty Albatrosses, which nest in the open among tall tussock grasses. Many seabirds of the region nest colonially in burrows on well-vegetated slopes, often in areas of abundant tussock grasses,

Southern Elephant Seals often rest among tussock grasses. © IAIN CAMPBELL, TROPICAL BIRDING TOURS

Snowy Sheathbills breed within the tussock grasses.
© IAIN CAMPBELL, TROPICAL BIRDING TOURS

including gadfly and other petrels, shearwaters, and prions, such as Soft-plumaged, White-chinned, Blue, and White-headed Petrels; Sooty Shearwater; and Antarctic Prion. On South Georgia Island, the world's most southerly songbird, South Georgia Pipit, lives and breeds in tussock grasses, although it can also forage in rocky areas outside of this habitat. On Macquarie Island, species like Pacific Black Duck and Mallard breed within the tussock grasses, along with seabirds such as Common Diving-Petrel (in burrows within the grass), while Great Skua nests in shorter, tundra vegetation. On Macquarie, pests have been a threat to populations of seabirds like Blue Petrel; there has been a massive program to eradicate rabbits in recent years to combat this. Overfishing of pelagic waters also remains a problem for many pelagic birds, as does being caught directly in fishing nets.

DISTRIBUTION: Antarctic tundra is widely distributed across the region as well as other subantarctic parts of the world, from the tip of the Cape of Good Hope, Africa, through parts of the Falkland Islands to South Georgia Island, and on to the continent of Antarctica and also to the southernmost subantarctic islands of New Zealand, including Macquarie Island. (Note though that the main habitat of the Falklands is PATAGONIAN STEPPE, which is also dominant on the tip of the South American mainland.) The Antarctic tussock grass element of this habitat is absent from the Antarctic mainland but is conspicuous along the warmer coastal regions of the subantarctic islands and provides vital breeding habitat for nesting birds.

WHERE TO SEE: South Georgia Island; Macquarie Island; Falkland Islands; South Shetland Islands; Antarctic Peninsula.

NEOTROPICAL FLOODED GRASSLAND AND WETLAND

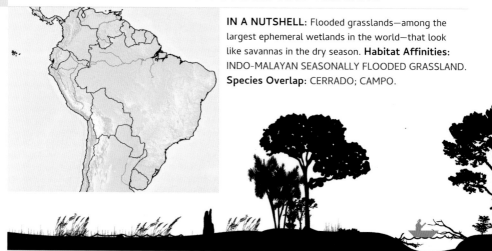

IN A NUTSHELL: Flooded grasslands—among the largest ephemeral wetlands in the world—that look like savannas in the dry season. **Habitat Affinities:** INDO-MALAYAN SEASONALLY FLOODED GRASSLAND. **Species Overlap:** CERRADO; CAMPO.

DESCRIPTION: Most of the flooded grasslands in the Neotropics, such as the Pantanal of Brazil, Paraguay, and Bolivia; the Llanos of Colombia and Venezuela; and the Churute area of Ecuador, are essentially just massive wetlands that share much in common with smaller wetlands around the region. The Pantanal and Llanos are regarded as some of the largest seasonal wetlands in the world. The landscapes change dramatically through the seasons, oscillating between massive flooded areas that are marshes in the wet season to dry grasslands that are essentially of same

The Pantanal spans three countries and extends over about 80,000 sq. mi. (200,000km²), so qualifies as the world's largest wetland (photo: Brazil). © PABLO CERVANTES DAZA, TROPICAL BIRDING TOURS

type as the CAMPO and PUNA grasslands in the dry season. Interspersed among these grasslands are permanent marshes with deep water, muddy boundaries, and thick stands of reeds, where birds and animals that were widespread during the wet season are concentrated to near mind-boggling numbers during the dry season. Within the grasslands are small patches of riverine forest that is usually NEOTROPICAL LOWLAND RAINFOREST or NEOTROPICAL SEMI-EVERGREEN FOREST, though the undergrowth is normally more open than in the true forest, and there are many more palms in these riverine patches. Other marshes occur throughout the forests and savannas of the Neotropics; a vast number of oxbow lakes in the Amazon Basin form permanent marshes surrounded by IGAPÓ FLOODED FOREST and Neotropical lowland rainforest.

Sunbittern displaying within Brazil's Pantanal, where it is an abundant species.
© PABLO CERVANTES DAZA, TROPICAL BIRDING TOURS

WILDLIFE: The marshes within these grasslands usually contain various widespread bird species that occur over most of the Neotropics, such as Horned Screamer, Wood Stork, Jabiru, Wattled Jacana, and a plethora of ducks and egrets and herons. The reeds around these wetlands contain specialist species such as Black-capped Donacobius and the highly localized Scarlet-headed Blackbird. Mammals of the rivers and marshes include Giant Otter and Capybara. The reptiles include Yacare Caiman, which can concentrate in astronomical numbers, plus Yellow Anaconda, Velvety Swamp Snake, and Paraguay Caiman Lizard.

The flooded grasslands of this habitat during the dry season become home to grassland birds like Red-legged Seriema, Greater Rhea, and Buff-necked Ibis. Lots of smaller birds enter the area, including seed-eating birds like doves, finches, tanagers, and cardinalids, and flycatchers like monjitas. Mammals of the grasslands include the Giant Anteater, Maned Wolf, Crab-eating Raccoon, Crab-eating Fox, and Giant Armadillo.

The riverine forest patches of this habitat usually contain the same kind of bird assemblages as the forest proper but in much lower diversity. Whether this is due to the island effect, where smaller

Giant Otters are often encountered along Neotropical rivers, like this one in the Pantanal of Brazil. © IAIN CAMPBELL, TROPICAL BIRDING TOURS

patches can hold less species variety, or because the forests are a simplified version of the forest proper is still disputed, but it is likely to be a combination of the two. A notable species of the Pantanal that requires forest patches to breed is the Hyacinth Macaw, the largest of all parrots and a spectacular sight. Mammals that concentrate in the riverine forests include Jaguar, Southern Tamandua, Black-and-gold Howler Monkey, Azara's Night Monkey, and Brazilian Porcupine.

DISTRIBUTION: Wetlands are found throughout the Neotropics. Significant flooded grasslands occur in n. Costa Rica and in w. Ecuador around Guayaquil. Massive flooded grasslands exist in the Pantanal, which covers about 80,000 sq. mi. (200,000km²), from Cuiabá in sw. Brazil to e. Bolivia and Paraguay. The Llanos covers much of c. Venezuela and e. Colombia. Other massive flooded grasslands include the Iberá wetlands of Corrientes, n. Argentina.

WHERE TO SEE: The Pantanal is best accessed from Cuiabá, in Mato Grosso, Brazil, where the Transpantaneira roadway runs through this spectacular wetland to the Paraguay River, at Porto Jofre.

Giant Anteater is one of the most conspicuous large mammals of Neotropical grasslands (photo: Brazil).
© PABLO CERVANTES DAZA, TROPICAL BIRDING TOURS

IGAPÓ AND VÁRZEA FLOODED FOREST

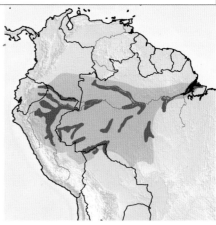

IN A NUTSHELL: Freshwater variations of mangrove forests that for months of the year are partially submerged by slow-flowing or still river waters. **Habitat Affinities:** INDO-MALAYAN FRESHWATER SWAMP FOREST. **Species Overlap:** NEOTROPICAL LOWLAND RAINFOREST.

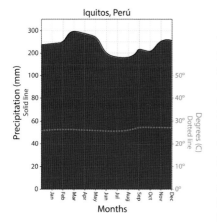

Iquitos, Perú

DESCRIPTION: Várzea forest and igapó forest are both flooded for part of the year. Várzea forests are dry more often than igapó forests and are flooded to depths over 10 ft. (3m) by water that contains very high sediment content, called "whitewater." In contrast, igapó forests are flooded for up to six months of the year, and their waters are more stationary, sediment-poor, and rich in tannins, which give the water a tea color, and are called "blackwater." Flooded forests are much lower in elevation than the surrounding NEOTROPICAL LOWLAND RAINFOREST but have many Moriche Palms growing in thick stands, which can be over 65 ft. (20m) tall, on slightly elevated and relatively dry areas. A small set of tree species dominates this forest, in contrast to the diverse canopy of lowland rainforest. The canopy is also much lower than that of the rainforest, at 85 ft. (26m), and this habitat is more open, with only 50% canopy cover. There is an understory of stunted broadleaf trees, which is not as thick as the rainforest understory, allowing more light to reach the forest floor. Given that amount of light, it is a surprise that the undergrowth is not thicker than it is, but this is due to the relatively constant flooding, which prevents most bushes from growing. The undergrowth that does exist consists of low palms and ferns. Even in the light gaps, grasses do not grow. Lianas are prominent and also tend to grow on the many trees that lean over in this forest. When you visit the flooded forests, you will be traveling in a canoe or walking on a raised boardwalk, such as those around many Amazonian lodges. The stilt-root systems of the trees of this forest limit visibility from a canoe, but from a boardwalk, visibility at head height is quite good, except near dense palm groves.

WILDLIFE: These flooded forests are nowhere near as rich in bird species as the surrounding terra firma NEOTROPICAL LOWLAND RAINFOREST, though they support a different set of birds, such as Zigzag Heron; Cocha Antshrike; Orange-crowned Manakin; Straight-billed, Striped, and Long-billed Woodcreepers; Plumbeous, Dot-backed, White-shouldered, and Silvered Antbirds; Citron-bellied

Tourists birding by canoe inside igapó forest in Amazonian Ecuador. © ANDRÉS VÁSQUEZ, TROPICAL BIRDING TOURS

A Brazilian Tapir and its calf traveling by river through flooded forest in Brazil. These tapirs are most abundant in this habitat. © NICK ATHANAS, TROPICAL BIRDING TOURS

Attila; Gray-headed Tanager; and White-chinned Jacamar. As we would expect from this watery habitat, kingfishers are common, including American Pygmy and Green-and-rufous Kingfishers.

Few terrestrial mammals can survive in this habitat, though tapirs are more common in this kind of forest than in the surrounding Neotropical lowland rainforest. Monkeys and other arboreal mammals abound but are usually taking advantage of seasonal fruit and also live in the surrounding terra firma forest. As would be expected, fish are seasonally common; they are hunted by Pink River Dolphin and Giant Otter. Amazonian Manatee feeds on plants in the várzea forests during flooding. Jaguars and Ocelot hunt in the seasons when the water levels are reduced.

DISTRIBUTION: Igapó and várzea forests are found throughout the Amazon Basin, with the difference in distribution being a function of the type of river (whitewater or blackwater) draining into the forest and the length of time it is inundated. Várzea covers a massive area of the Amazon Basin. One of the best places to see igapó is along the Napo River of Ecuador, where lodges have large stands of this forest around them, with numerous boardwalks, canoe channels, and towers overlooking the igapó canopy. Similar habitats exist in limited extent along the mouth of the Amazon, on the Pacific coast in nw. Ecuador and sw. Colombia, and in Central America.

WHERE TO SEE: Lodges along the Napo River, e. Ecuador.

Above: **Hoatzin is a noisy and conspicuous bird found at the borders of this habitat and local wetlands (photo: Amazon, Ecuador).** © PABLO CERVANTES DAZA, TROPICAL BIRDING TOURS

Right: **When waters recede in the dry season, predators like Ocelot hunt in this habitat (photo: Brazil).** © ANDRÉS VÁSQUEZ, TROPICAL BIRDING TOURS

NEOTROPICAL MANGROVE

IN A NUTSHELL: Tidal, salt and brackish, periodically flooded forests that form in coastal areas where wave action is limited. They often line the edges of beaches. **Habitat Affinities:** AUSTRALASIAN MANGROVE. **Species Overlap:** NEOTROPICAL LOWLAND RAINFOREST.

DESCRIPTION: Mangrove habitat is treated more thoroughly in the AUSTRALASIAN MANGROVE account. The mangroves found in the Neotropics, although as extensive in area as they are in other parts of the world, are nowhere near as diverse as they are in Asia and Australasia, with far fewer tree species. Of the 70 mangrove species in the world, only 11 occur in the Neotropics, where Red Mangrove (*Rhizophora mangle*) is the most common and widespread. However, Neotropical mangroves have a much closer association with surrounding humid forest species, and there is often a wide assortment of non-mangrove species such as ferns and palms in slightly elevated areas of mangrove swamps. In some areas adjacent to rainforest, such as in far nw. Ecuador, clumps of detritus can get caught at the tops of mangrove spider roots, and these form a mat on which plants of many species from surrounding rainforest take hold. They rarely grow too large, as the episodic saltwater flooding kills them.

Mangroves near the Pacific Ocean in Costa Rica.
© IAIN CAMPBELL, TROPICAL BIRDING TOURS

American Pygmy Kingfisher often hunts from sheltered, low mangrove branches and roots, just above the water.
© IAIN CAMPBELL, TROPICAL BIRDING TOURS

WILDLIFE: Most mammals that live in the Neotropical mangroves are visitors from surrounding drier habitats. Agoutis, Key Deer (a subspecies of White-tailed Deer), and Brazilian Tapir visit at low tide, though are not typical of mangroves. Carnivores include Crab-eating Raccoon, Crab-eating Fox, and Jaguar.

The Neotropical mangroves have far fewer specialist bird species than the mangroves of Asia or Australasia, with a scant number of endemics. These forests do support most of the Neotropical waders and shorebirds, as well as many of the species from surrounding NEOTROPICAL LOWLAND RAINFOREST. Species found only in the mangroves include Mangrove Hummingbird, Mangrove Vireo, Mangrove Cuckoo, Sooty-capped Puffbird, and Little Wood-Rail.

DISTRIBUTION: Mangroves are a diversified habitat found from coastal mudflats and offshore islands to far inland along large tidal rivers. They occur all the way around the Caribbean coast of Central and South America, and down the Atlantic coast to s. Brazil; the world's largest mangrove swamp is centered on São Luís in n. Brazil. Along the Pacific coast, mangroves occur from Honduras to Ecuador; a very accessible mangrove swamp is in Muisne, Ecuador.

WHERE TO SEE: Manglares Churute Ecological Reserve, Ecuador; Muisne, Ecuador; Tárcoles River, Costa Rica.

NEOTROPICAL TIDAL MUDFLAT

IN A NUTSHELL: The muddy and silty tidal areas between mangroves and salt marsh and the ocean.
Habitat Affinities: AUSTRALASIAN TIDAL MUDFLAT. **Species Overlap:** NEOTROPICAL SANDY BEACH.

DESCRIPTION: Tidal mudflats and beaches are widespread throughout Central and South America. These are flat, silt- and mud-dominated, unvegetated, open coastal areas within the intertidal zone that undergo daily flooding and exposure as a result of the action of the tides. Tidal mudflats occur in calm coastal areas of low energy, such as bays, inlets, and estuaries, and are composed of fine sediments, like clay, silt, sand, gravel, and marine detritus, which have been deposited by rivers or the tides. Such areas are typically gently sloping and shallow and are home to a good diversity and abundance of invertebrates that provide an important food source for coastal birds. The flats and beaches are often closely associated with mangroves, which frequently occur alongside them. This habitat occurs between the seas and inland coastal habitats and so is also an important barrier to coastal erosion, as well as providing key sites for feeding and roosting coastal birds. While this open habitat can provide some of the most spectacular concentrations of birds, their presence is governed by the state of the tide.

WILDLIFE: Most animals use mudflats only when the tide is low and the mud exposed; when tides are high, most of the birds may relocate to another area to sit out the period when their food is submerged and out of reach. However, when the flats become exposed, thousands of birds can be sprawled across them, making for an incredible sight. Some of the most conspicuous birds are migrant shorebirds, which distribute differently across the intertidal zone, due to variations in leg, body, and bill morphology among species. Godwits, Willet, Whimbrel, and dowitchers can often been seen clustered in deeper-water zones, which their long bills and relatively lengthy legs allow them to exploit, while shorter-billed, shorter-legged species (like small sandpipers) are prevented from foraging there and tend to forage closer to the beach on either open flats or within shallower tidal pools. Feeding strategies also limit some shorebirds to certain zones within the tideline. There are two great divisions of shorebirds, the plovers and the sandpipers. The latter search for their food by feel, using their bills to find prey by touch, and therefore can find invertebrates even when they are not visible. Plovers, however, typically find their prey by sight and therefore are restricted to feeding in mudflat areas with little or no water or on the adjacent beaches, where they are able to succeed with this foraging strategy.

Tidal flats attract an assortment of waterbirds, including terns, gulls, shorebirds, pelicans, egrets, and herons, like this Tricolored Heron. © IAIN CAMPBELL, TROPICAL BIRDING TOURS

These habitats are important for great numbers of wintering migrant birds, many of which travel long distances to and from their breeding and wintering grounds. Flats and beaches are also important stopover sites for feeding and roosting stopovers during the migrations. These sites are heavily threatened, exploited for coastal developments and frequently disturbed by human recreation. Nonetheless, there are a number of excellent examples of extensive flats in the Neotropics. In the Bay of Panama, beside Panama City, over a million shorebirds visit each year, and any one day in the boreal winter can involve hundreds of thousands, including Western and Semipalmated Sandpipers, Short-billed Dowitcher, Willet, and Black-bellied Plover. In e. Tierra del Fuego, Argentina, flats around Río Grande are significant wintering grounds for White-rumped Sandpiper and Hudsonian Godwit, which breed in the far north of North America. Other good examples of tidal flats can be found on the n. Pacific coast of Costa Rica and around Lima, Perú, both of which host large numbers of coastal birds, which vary in their presence and abundance based on the state of the tide. As well as shorebirds, many other coastal birds regularly utilize this habitat too, such as gulls, terns, herons, and frigatebirds.

DISTRIBUTION: Tidal mudflats are found all around the Neotropical coastline, though are much more common in the estuaries of large rivers.

WHERE TO SEE: Tárcoles River, Costa Rica; Salinas, Ecuador.

NEOTROPICAL ROCKY COASTLINE AND SANDY BEACH

IN A NUTSHELL: The beaches and headlands around the coast. **Habitat Affinities:** AUSTRALASIAN ROCKY COASTLINE AND SANDY BEACH. **Species Overlap:** NEOTROPICAL TIDAL MUDFLAT.

DESCRIPTION: Rocky coastline—the exposed rock at the continent's edges—is widespread throughout the Neotropics, though more extensive on the Pacific (west) side of the region. Its existence is dictated by numerous factors, including the local rock type, the age of the coastline, and the extent of local weathering. Rocky shorelines typically occur in areas where rock types resistant to erosion prevail, such as igneous granite and basalt or metamorphic quartzite and gneiss. In areas of clay stones, shales, and limestones, which tend not to form cliffs, sandy beaches form, the sand accumulating in the depressions where the rocks have eroded.

Birds can often be tame around island breeding colonies, like these Magellanic Penguins and geese in the Falklands. © SUE POST AND MICHAEL JEFFORDS

WILDLIFE: Rocky shorelines and cliffs are often broken up by sandy beaches and TIDAL MUDFLATS that are important foraging areas for coastal birds such as shorebirds. Intertidal rocky areas with a wave-cut platform or cliffs, where seaweed accumulates, barnacles grow, and rock pools form, are preferred feeding places for such shorebirds as Surfbird and oystercatchers. Many other species feed and roost in the area above the high-tide line, especially on sea stacks and islets, both of which are better protected from predators than onshore areas. These safer areas are also critical nesting sites for many seabirds, such as cormorants, tropicbirds, boobies, and petrels. In the Galápagos, Marine Iguanas and such birds as Flightless Cormorant, Galápagos Penguin, and Lava Gull use this habitat, while at Pucusana, Perú, the avifauna includes Inca Tern, Guanay Cormorant, Belcher's Gull, Peruvian Booby, Peruvian Pelican, and Humboldt Penguin. Rocky headlands can also be good areas to search for pelagic birds passing by offshore during migration, like albatrosses, shearwaters, boobies, terns, and gulls. As coastlines are often subjected to human disturbance by way of coastal developments and leisure activities, the most utilized nesting areas are often on offshore islands and islets.

Left: Inca Terns breed on cliffs and rocky islands along the coast of South America (photo: near Lima, Perú). © JOSÉ ILLÁNES, TROPICAL BIRDING TOURS

Below: A cluster of Marine Iguanas in the Galápagos Islands of Ecuador. The only marine lizard in the world, this species is capable of diving down to 65 ft. (20m) and staying underwater for more than 40 minutes! © SAM WOODS, TROPICAL BIRDING TOURS

DISTRIBUTION: Rocky coastline and beach habitat exists around the entire coastline of the Neotropics. Good and very accessible examples are found in the Galápagos Islands of Ecuador; at Pucusana, south of Lima, Perú; and on Chiloé Island, Chile, but this habitat also extends all the way along the Pacific coastline to Tierra del Fuego at the tip of South America.

WHERE TO SEE: Galápagos Islands, Ecuador; coastline near Lima, Perú; Valdés Peninsula, Argentina.

NEOTROPICAL AND ANTARCTIC PELAGIC WATERS

IN A NUTSHELL: The oceanic waters surrounding the Neotropics, south to the Southern Ocean. **Habitat Affinities:** AUSTRALASIAN PELAGIC WATERS. **Species Overlap:** NEOTROPICAL ROCKY COASTLINE AND SANDY BEACH.

DESCRIPTION: *Pelagic* refers to the open waters offshore. Many birds that occur in this oceanic habitat breed or feed in remote areas and are often difficult to see from land; a pelagic boat trip is often required to see them. The most productive pelagic regions in the Neotropics and Antarctica occur in areas of cold waters, such as those influenced by the Humboldt Current on the western side of South America and in the Southern Ocean between s. South America and Antarctica. Outside these main areas, pelagic birding is relatively underexplored but rather limited.

Red-billed Tropicbird
off the Galápagos
Islands of Ecuador.
It breeds on the cliffs
there but forages in
pelagic waters. © IAIN
CAMPBELL, TROPICAL
BIRDING TOURS

WILDLIFE: The Humboldt (or Peruvian) Current extends from the Galápagos, at its northern end, down to s. Chile. This current, which brings large upwellings of cold water to the surface, is rich in phytoplankton and provides habitat for the world's largest fisheries. The abundant marine life makes it one of the richest marine ecosystems on earth, and a diverse assemblage of seabirds has evolved to exploit this. The fish provide food for birds like penguins and boobies, while phytoplanktonic fauna provide food for sea mammals like whales as well as for seabirds such as petrels. As an indication of the significance of this cold-water upwelling, around a third of all the world's seabirds are found there—some breeders, but others migrant species visiting from the Southern Ocean much farther south. The best way to experience this firsthand is to take a boat cruise to the Galápagos of Ecuador or a day trip out of ports in the Lima area of Perú or n. or c. Chile.

In the Galápagos, locally breeding seabirds like Waved Albatross, Galápagos Shearwater, Galápagos Petrel, and Elliot's and Wedge-rumped Storm-Petrels may be seen cruising between islands, along with mammals like Bryde's and Sperm Whales and Common Bottlenose Dolphin. Pelagic trips out of n. Chile (Arica or Iquique) or c. Chile build up massive lists of seabirds, recording Humboldt Current specialties like Inca Tern; Humboldt Penguin; Northern Royal, Buller's, and Salvin's Albatrosses; Juan

Snow Petrel inhabits Antarctic pelagic waters. © LISLE GWYNN, TROPICAL BIRDING TOURS

Fernandez, Masatierra, and Westland Petrels; Peruvian Booby; Peruvian Diving-Petrel; and Red-legged and Guanay Cormorants. The more northerly trips offer better chances at Markham's and Ringed Storm-Petrels. Pelagic trips out of Perú (Lima and Paracas) are also very diverse, recording more than 50 species over the years, including Swallow-tailed Gull; Peruvian and Inca Terns; Waved, Salvin's, and Buller's Albatrosses; Ringed and Elliot's Storm-Petrels; Southern Giant-Petrel; and Peruvian Diving-Petrel.

While species like Black-browed Albatross and giant-petrels may be seen from the s. South American mainland (e.g., Tierra del Fuego), for the highest diversity of seabirds the Southern Ocean is best explored via cruises combining the Falkland Islands, South Georgia Island, and Antarctica. Moving south from mainland South America, cruises traverse the Drake Passage and in doing so move from one distinct pelagic area to another as they cross the Antarctic Convergence. In this 20–30 mi. (30–50km) stretch of ocean, colder Antarctic waters from the south meet warmer subantarctic waters from the north, and the cooler waters submerge beneath the warmer. This is a rich feeding area and also marks an invisible boundary for some bird species and mammals that prefer to stay south of this zone. Species like Light-mantled and Sooty Albatrosses; Antarctic Shag; South Polar Skua; Snow Petrel; Black-bellied Storm-Petrel; Blue and Antarctic Petrels; and Slender-billed and Antarctic Prions are all likely to be seen, sometimes in abundance, in the colder waters south of the convergence. Nine of the world's penguins also breed south of this zone, and sea mammals like Minke, Blue, and Humpback Whales also become more abundant. Examples of species from the northern side, which decrease in abundance as you head south of the convergence, include Gray-headed Albatross; Chilean Skua; Soft-plumaged and White-chinned Petrels; Common Diving-Petrel; and Great and Sooty Shearwaters. The crossings between islands, such as the Falklands and South Georgia, each important breeding sites for seabirds, are also productive pelagic waters for species like Royal and Wandering Albatrosses, Gray-backed Storm-Petrel, Cape Petrel, giant-petrels, and jaegers.

In the Humboldt Current destinations, pelagic trips can be productive year-round, although albatross diversity is said to best off Perú from June to August. Antarctic cruises usually run only between November and March, as many of the birds nest then but at other times of year wander vast distances away. The biggest threats to these ecosystems are overfishing, pollution of marine environments, and birds being directly caught and damaged by fishing nets, which has been particularly problematic among albatrosses.

DISTRIBUTION: Pelagic waters surround all of the Neotropics, but they are richer in wildlife in the more temperate parts of the region and especially so where the continental slope is close to shore.

WHERE TO SEE: The best pelagic boat trips are those that cruise around the Galápagos Islands, Ecuador; and boats out of Lima, Perú, and Antarctica cruises, Santiago, Chile.

NEOTROPICAL CROPLAND

IN A NUTSHELL: Areas of formerly natural habitat used primarily for the production of human-planted crops. **Habitat Affinities:** None. **Species Overlap:** Some overlap with surrounding natural habitats.

There are extensive stands of African Oil Palm in Colombia and Ecuador.
© DAVID FRIED

DESCRIPTION: Most large-scale crop farming throughout the Neotropics, as elsewhere in the world, reduces diversity through degradation or clearance of natural habitat. Significant crops within the Neotropics include African Oil Palm, soybeans, bananas, cacao, and coffee. Of these, the latter two can thrive in partially shaded environments and so can be farmed while retaining more of the original diversity than the other crops mentioned.

As a general rule, the most intensely farmed areas result in the most extreme losses of diversity. Studies of cacao plantations in Costa Rica have shown that if run as shade-grown plantations close to blocks of natural forest, they can actually enhance the diversity of an area, if managed with this in mind by selectively keeping certain native forest trees. The same results can be seen with shade-grown coffee plantations too. In Antioquia, Colombia, for example, such plantations were particularly productive for wintering boreal migrant songbirds like Blackburnian, Mourning, Canada, and Cerulean Warblers.

African Oil Palm monocultures, soybean production, and banana plantations generally lead to wholescale clearance of the original vegetation, which is then replaced by habitats occupied by only a small number of widespread, tolerant bird species. While most oil palm production is centered in Malaysia and Indonesia, a significant shift toward Latin America is occurring for two principal reasons: global pressure on those se. Asian nations led to tightening of regulations on oil palm plantations, and a lack of available land in those heavily populated countries. At present, the major area affected in the Neotropics is Colombia, which is the fourth-largest producer of palm oil in the world. However, Ecuador, Honduras, and Brazil are also within the top 10 producers, while Costa Rica and Perú lie within the top 20, and global demand is projected to double by 2050. The increase of oil palm plantations almost always occurs by clearance of native habitat. Plantations have even been documented to have encroached into national parks in Honduras, and in Ecuador they have decimated forest areas of the Chocó EBA (home to more than 60 endemic species), which now has less than 5% of native habitat remaining. Overwhelmingly, the habitats that have been cleared for oil palm are lowland forests. While most studies on bird populations and oil palm plantations have been conducted in se. Asia, a Brazil-based study mirrored their findings that a 60% loss in bird species diversity can result from changing land use to oil palm, in this case from native NEOTROPICAL LOWLAND RAINFOREST in the Amazon. Some of the bird species that have been recorded within oil palm plantations in Ecuador include Tropical Kingbird; Rusty-margined Flycatcher; Pale-vented Pigeon; and Blue-gray, Flame-rumped, and Palm Tanagers. If a grassy layer is permitted to grow underneath the palms, seedeaters such as Variable and Yellow-bellied Seedeaters and Dull-colored Grassquit can occur too.

Soybean production in the region has meant large-scale clearance of native forests and savannas, particularly in the CHACO SECO, CERRADO, and Amazon habitats. The main countries producing large amounts of soybeans in the Neotropics are Brazil, Argentina, Paraguay, and Bolivia, which are all among the top 10 world producers, and Uruguay is not far behind them. The resultant monoculture is extremely low in diversity of birds, with only persistent, highly tolerant species surviving in this harsh habitat. Bananas should be mentioned too: large-scale production occurs in Ecuador, Brazil, Guatemala, Costa Rica, and Colombia, which are all within the top 15 world producers of this crop. As with the other types of large-scale crop farming mentioned here, this has typically come at the cost of the native habitats that are cleared for this use, which results in a dramatic drop in bird diversity. This has mostly affected tropical lowland forest habitats, but marshes have been drained and lands poisoned from overuse of chemicals too, which has also affected nearby rivers. In banana plantations in Ecuador, some of the bird species that have been recorded include Flame-rumped and Palm Tanagers; Tropical Kingbird; Rusty-margined and Social Flycatchers; and Pale-vented Pigeon.

DISTRIBUTION: LOWLAND RAINFORESTS and CLOUD FORESTS of the tropical areas of Central and South America have been converted to coffee, banana, and African Oil Palm production. Much of the NEOTROPICAL SEMI-EVERGREEN FOREST of Brazil has been converted to soybean cultivation. The PAMPAS and MONTE of Argentina have been turned over to wheat production.

WHERE TO SEE: Except in the wildest areas of the Amazon or in the most inhospitable deserts, croplands are ubiquitous.

Blackburnian Warbler winters in the Neotropics, often utilizing coffee plantations in Central and South America. © SAM WOODS, TROPICAL BIRDING TOURS

NEOTROPICAL GRAZING LAND

IN A NUTSHELL: The areas cleared for cattle or sheep farming. **Habitat Affinities:** AFROTROPICAL GRAZING LAND. **Species Overlap:** Often shares species with surrounding native savanna habitats.

Above: **Cattle Tyrant, like this one in the Pantanal of Brazil, often forages around livestock.** © ANDRÉS VÁSQUEZ, TROPICAL BIRDING TOURS

Below: **Capybaras can be found in the wetter parts of ranchlands in South America.** © IAIN CAMPBELL, TROPICAL BIRDING TOURS

DESCRIPTION: Many habitats have been converted to grazing lands across the Neotropics. Most notable among these are the species-rich PAMPAS and CAMPO habitats of se. South America. Vast areas of these grasslands have been converted to cattle grazing for meat production and, to a lesser extent, sheep grazing for wool production. Many millions of heads of cattle graze across this region, and this trend has generally been on the increase, but even where it is declining, the same land is often being converted to large-scale soybean crop production, which has even greater detrimental effects on local wildlife populations. Conversion of these areas has led to significant declines in native birds and other fauna that are strongly reliant on natural, unmodified grassland, little of which now exists after centuries of modification for human use.

Nearly half of Argentina's threatened birds (25) are grassland species, with Saffron-cowled Blackbird, Strange-tailed Tyrant, and Spectacled Tyrant among the most affected. In addition to these South American species, migratory birds from North America like Bobolink and Upland Sandpiper use these habitats during the boreal winter. Pampas Deer and Guanaco are some of the mammals

that have been adversely affected as well. On the PATAGONIAN STEPPE of mainland South America, birds like Chocolate-vented Tyrant and Ruddy-headed Goose have declined with the increase in grazing, the latter not helped by advice the Argentinean government offered in the 1960s, which promoted the goose as a pest species and falsely claimed it was harmful to grazing lands.

It should be noted that grazing lands not only threaten grassland areas but have also led to clearance of large swaths of forests and other habitats too, causing catastrophic declines in diversity. While research tends to focus on the largest of these operations, the multitude of smaller-scale grazing areas has had an impact no less significant, as economic factors often lead to lack of new technologies and poor land management, leading to lower production rates and encouraging further clearance of land to increase yield. Studies in Belize in Central America have shown that clearance of Neotropical lowland forest led to a decline in bird species of more than 50%, obliterating both the resident bird community and the migratory species that were found in the area. Grazing lands generally support a low diversity of bird species (relative to the native vegetation that was cleared), which are typically widespread species tolerant of many open habitats, such as Smooth-billed Ani, Tropical Kingbird, Southern Rough-winged Swallow, Olive-crowned Yellowthroat, and Variable Seedeater. Where open-country species have benefited from land clearance, it has been found that their populations only truly boom if the land is not too intensively managed (e.g., not heavily overgrazed, or chemical use is not extensive). However, it should be noted that careful management of grazing properties, with retention of certain plant species, can reduce the effects on bird populations or even be beneficial to them in specific cases. The Hyacinth Macaw has profited from species-specific management practices of cattle ranches in the Pantanal of Brazil (NEOTROPICAL FLOODED GRASSLAND AND WETLAND), where numbers have increased in recent years since the institution of altered practices that prevented the clearance of macaw food resources and roosting sites.

DISTRIBUTION: Most of the cattle-grazing areas are in Brazil and Argentina where CERRADO or CHACO SECO AND ESPINAL habitat existed, or in c. Brazil where NEOTROPICAL SEMI-EVERGREEN FOREST has been cleared. The sheep farming is concentrated in the PAMPAS and CAMPO, MONTE, and PATAGONIAN STEPPE of Argentina.

The world's largest parrot, Hyacinth Macaw, often visits cattle ranches in the Brazilian Pantanal.
© ANDRÉS VÁSQUEZ, TROPICAL BIRDING TOURS

HABITATS OF INDO-MALAYSIA
(SOUTHEAST ASIA AND INDIA)

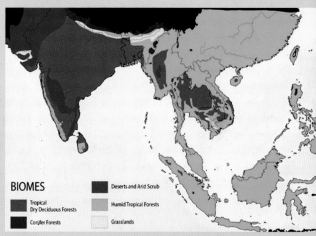

BIOMES

Tropical Dry Deciduous Forests

Conifer Forests

Deserts and Arid Scrub

Humid Tropical Forests

Grasslands

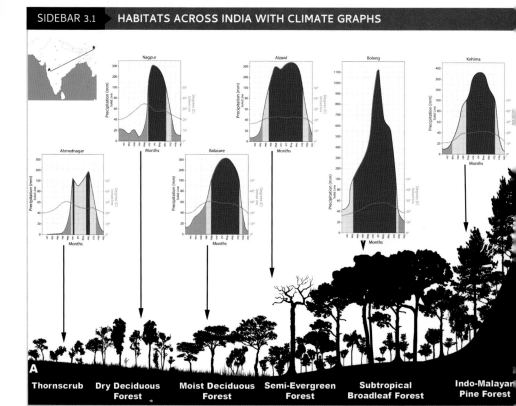

Thornscrub   Dry Deciduous   Moist Deciduous   Semi-Evergreen   Subtropical   Indo-Malayan
Forest   Forest   Forest   Broadleaf Forest   Pine Forest

INDO-MALAYAN PINE FOREST

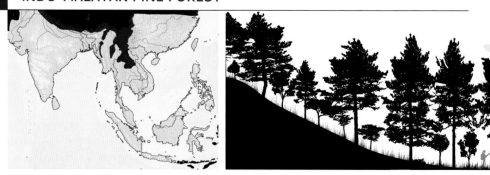

IN A NUTSHELL: Forest dominated by conifers, often at higher elevations and usually with some grass cover. **Habitat Affinities:** EURASIAN MONTANE CONIFER FOREST. **Species Overlap:** EURASIAN MONTANE CONIFER FOREST; INDO-MALAYAN TROPICAL MONTANE RAINFOREST.

DESCRIPTION: Coniferous forests are not what you might expect to find in the tropics and subtropics, but examples exist in the Indo-Malayan region. They often occur in areas with less rainfall than other types of forest require, such as mountain slopes in rain shadow. Pines can be

Pine forest in the Arakan Mountains of Myanmar.
© CHARLEY HESSE, TROPICAL BIRDING TOURS

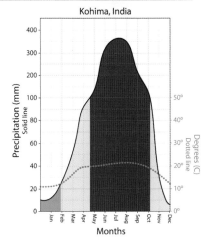

Kohima, India

pioneering species, and after disturbances, either natural or human-caused, such as typhoons, landslides, or fires, they may become the dominant vegetation in some areas. Pine forests usually occur at higher elevations than broadleaf montane habitats. Some pine forests have a parkland structure of grass and dispersed trees, which is maintained by regular fires that prevent broadleaf trees from establishing but don't affect the fire-resistant pines.

In Sumatra and some parts of the Philippines, Sumatran Pine dominates. In Luzon, Philippines, nw. Thailand, and the Arakan Mountains of Myanmar and ne. India, it is Benguet Pine; in the Himalayas, Chir Pine; and on the Dalat Plateau in Vietnam, Dalat Pine.

At the lower end of the elevation range, coniferous trees occur with broadleaf trees, creating a mixed forest, and many montane species of animals inhabit both coniferous and broadleaf forests. Higher up, pines mix with rhododendrons, the flowers of which add a splash of color. Some forests have an understory of bushes and bamboo that add to the structural diversity of the forest and are inhabited by numerous skulking birds such as laughingthrushes. The mountainous locations of these forests and the beautiful tree shapes create very picturesque scenery.

WILDLIFE: Indo-Malayan pine forests contain lower biodiversity than neighboring broadleaf forests. Mammals are thin on the ground, but Asiatic Black Bear, serows, and Sambar are still present in some areas. Smaller mammals such as squirrels, flying squirrels, and other rodents are more common but still not obvious.

Birdlife is much more conspicuous. Not many bird species are restricted to pine forests in the Indo-Malayan region, but some species show a preference for it. Birds present include: in Himalayan subtropical pine forests, Cheer Pheasant, Western Tragopan, and Elliot's Laughingthrush; in the pine forests of ne. India and Myanmar, Black-bibbed Tit, Burmese subspecies of Black-browed Tit (aka Burmese Bushtit), Bar-tailed Treecreeper, and White-browed Nuthatch; in the Luzon pine forests, Luzon Scops-Owl, Gray-capped Shrike, and Green-backed Whistler; in the pine forests of Taiwan, Eurasian Nutcracker, White-whiskered Laughingthrush, Flamecrest, and Gray-headed Bullfinch; in the pine forests of nw. Thailand, Hume's Pheasant, Gray Treepie, Giant Nuthatch, Hume's Treecreeper, and Ultramarine Flycatcher

Red-and-white Flying Squirrel inhabits pine forest in the mountains of Taiwan. © KEITH BARNES, TROPICAL BIRDING TOURS

(IS); and in the pine forests of s. Vietnam, Vietnamese Greenfinch, Vietnamese Cutia, Slender-billed Oriole (IS), and an endemic subspecies of Red Crossbill.

Endemism: Areas of endemism within this zone occur in Taiwan, the Philippines, nw. Thailand, the Arakan Mountains of Myanmar, the Dalat Plateau in s. Vietnam, and the Himalayas. Some coniferous trees themselves have restricted ranges, such as Dalat Pine in Vietnam and Coffin Tree in Taiwan.

DISTRIBUTION: Indo-Malayan pine forests are distributed in many highland areas of the region. They are found in n. Sumatra, around Lake Toba and along the Barisan mountain range, south to Mt. Kerinci and Mt. Talang in c. Sumatra. In the Philippines, they are found within the central cordillera of nw. Luzon on Mts. Polis, Data, Puguis, and Pulag, and some occur on Mindoro island. In Taiwan, they are common at mid- to upper elevations, although the higher forests are closely related to EURASIAN MONTANE CONIFER FOREST (covered in the Palearctic section). In Vietnam, they are found on the Dalat Plateau plus other mountains of c. Vietnam and adjacent Laos. In nw.

Vietnamese Cutia is found in the pine forests of Vietnam and Laos, especially where there is a broadleaf component. © KEN BEHRENS, TROPICAL BIRDING TOURS

Thailand, pine forests are found in several mountainous areas, including Doi Chiang Dao, Doi Ang Khang, and Doi Lang. In ne. India and Myanmar, they occur along the Arakan range from Patkai in the north down to the Chin Hills and Nat Ma Taung (Mt. Victoria) in the south. The largest area of subtropical pine forests ranges from Pakistan, along the lower stretches of the Himalayas, to Nepal and Bhutan. Higher up in the Himalayas and beyond are large stretches of Eurasian montane conifer forest.

WHERE TO SEE: Nainital, Uttarakhand, India; Nat Ma Taung National Park (Mt. Victoria), Myanmar; Doi Ang Khang, Thailand; Kerinci Seblat National Park, Sumatra, Indonesia; Dalat Plateau, s. Vietnam; Mt. Polis, Luzon, Philippines.

INDO-MALAYAN THORNSCRUB

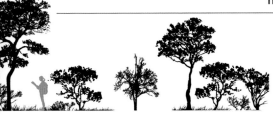

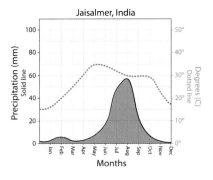

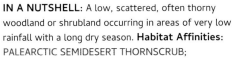

IN A NUTSHELL: A low, scattered, often thorny woodland or shrubland occurring in areas of very low rainfall with a long dry season. **Habitat Affinities:** PALEARCTIC SEMIDESERT THORNSCRUB; NEOTROPICAL CHACO SECO AND ESPINAL. **Species Overlap:** PALEARCTIC SEMIDESERT THORNSCRUB; INDO-MALAYAN DRY DECIDUOUS FOREST; PALEARCTIC HOT DESERT.

DESCRIPTION: Indo-Malayan thornscrub is an arid habitat with low, scattered bushes and other thorny plants. The ground between these plants can be bare or have some sparse grass cover. With its open nature, the habitat is fairly easy to pass through, although, as the name suggests, it can be thorny. Also, some areas have an uneven or rocky substrate that can be tricky to walk on. In this region, thornscrub occurs on the Indian subcontinent, where it receives less than 30 in. (750mm) of rain per year, and in Myanmar, where it can get up to 35 in. (900mm). Most of this falls in the short monsoon season. Average humidity is less than 50%, and mean temperatures are 77–86°F (25–30°C).

Trees are typically 20–30 ft. (6–9m) tall, with short trunks and low, branching crowns, and are widely scattered. They have long roots that penetrate deep into the soil to find water. Their leaves tend to be small, an adaptation to prevent water loss. Some common trees in nw. Indian thornscrub include White-bark Acacia, Sponge Tree, and Tooth Brush Tree; in the south, acacias and euphorbias are very prominent, and other common tree species include Neem, Golden Shower, and Indian Wild Date. In the drier areas of the south, Umbrella Thorn is dominant. Spiny, xerophytic shrubs (plants requiring little water) are also present. Some grasses also grow in the rainy season, including species in the genera *Heteropogon*, *Eremopogon*, *Chrysopogon*, and *Dactyloctenium*. One particularly distinctive plant from Myanmar's thornscrub is the cactus-like Antique Spurge.

Some people believe that thornscrub is a degraded state of tropical dry deciduous forest. This is probably the case in some areas, but lower rainfall and a set of unique species show it to be a valid natural habitat. Indo-Malayan thornscrub habitat is threatened by cutting of trees for firewood, and overgrazing.

WILDLIFE: Chinkara, Blackbuck, and Four-horned Antelope are found in thornscrub but not restricted to it, and Leopard and Caracal persist in some areas. A wide variety of smaller mammals also can be found, including numerous rodents and bats. Reptiles inhabit thornscrub, including Scaly Gecko, Reticulate Leaf-toed Gecko, Leschenault's Snake-eye, and Nagarjun Sagar Racer.

Thornscrub alongside the temples of Bagan in Myanmar.
© KEN BEHRENS, TROPICAL BIRDING TOURS

Despite its dry nature, this habitat has good avian diversity, and either of the two areas can hold 350–400 species. It can be surprisingly productive for birding and, with ample light, low vegetation, and lack of leaves, good for bird photography. However, with the lack of shade, the heat in the middle of the day can be intolerable. White-naped Tit (IS), Sind Sparrow, and the western subspecies of Rufous-vented Grass Babbler are mostly restricted to this habitat, although in different areas of it—the sparrow and the grass babbler to the north and west, the tit to the east and south. The critically endangered Jerdon's Courser, the endangered Lesser Florican, and the critically endangered Great Indian Bustard are found very locally in Deccan (s. Indian) thornscrub. Myanmar thornscrub holds Hooded Treepie (IS) and Burmese Bushlark.

Endemism: The nw. Indian thornscrub holds the Sind Sparrow and Rufous-vented Grass Babbler, which are mostly restricted to this habitat. Farther south, the Deccan thornscrub contains the restricted-range Jerdon's Courser and also a near-endemic mammal, Indian Roundleaf Bat. White-naped Tit is restricted to this habitat but is found in both the Deccan and northwestern areas. The Irrawaddy plains endemic area, which is completely within Myanmar, has the range-restricted Hooded Treepie (IS), Jerdon's Minivet (IS), and White-throated Babbler (IS).

DISTRIBUTION: Indo-Malayan thornscrub surrounds the Thar Desert, between India and Pakistan, which we consider the border between the Indo-Malayan and Palearctic regions. Thornscrub also covers the drier parts of the Deccan Plateau, on the inland side of the Western Ghats mountain range, across the states of Tamil Nadu, Andhra Pradesh, Karnataka, and Maharashtra, stretching as

far as Sri Lanka. On the drier side, thornscrub grades into the Thar Desert, and on the moister side, it grades into INDO-MALAYAN DRY DECIDUOUS FOREST, although the transition is sharper where it meets the Western Ghats. An isolated patch of thornscrub occurs on the Irrawaddy plains of Myanmar, surrounding the ancient city of Bagan.

WHERE TO SEE: Sultanpur National Park, Gurugram, Haryana, India; Bagan Archaeological Zone, Myanmar.

Left: **Brahminy Starling occupies this habitat in India.**
© IAIN CAMPBELL, TROPICAL BIRDING TOURS

Below: **Blackbucks feeding within thornscrub in India.**
© VIKRAM PODKAR

INDO-MALAYAN TROPICAL LOWLAND RAINFOREST

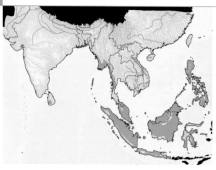

IN A NUTSHELL: Lush forest with a high canopy and complex structure that remains moist and humid throughout most or all of the year. **Habitat Affinities:** AFROTROPICAL LOWLAND RAINFOREST; NEOTROPICAL LOWLAND RAINFOREST. **Species Overlap:** INDO-MALAYAN SEMI-EVERGREEN FOREST.

DESCRIPTION: Indo-Malayan tropical lowland rainforests are thought to be some of the oldest in the world and have some of the highest biodiversity. One thing that distinguishes the Asian rainforests from other similar forests around the world is the dominance of the Dipterocarpaceae family of trees. Tropical rainforests tend to be stratified, and the canopy can be split into three layers. The emergent layer in Indo-Malayan rainforests is typically taller than in African or Neotropical rainforests. The emergents are usually dipterocarps, which often reach 200–230 ft. (60–70m) in height. The next level is a closed canopy at about 100–135 ft. (30–41m) that is again dominated by dipterocarps but may also contain figs, laurels, mahoganies, and legumes. The mid-story can be at around 30–60 ft. (9–18m); below it are shrub and ground-cover layers.

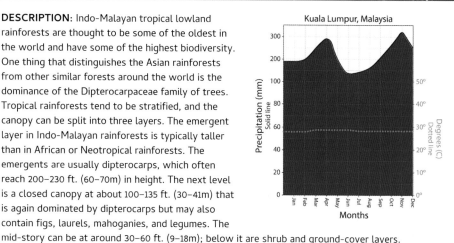

Kuala Lumpur, Malaysia

Tropical rainforests in this region have minimum monthly temperatures of 64°F (18°C) throughout the year. Temperatures average around 87°F (31°C) during the day and 72°F (22°C) at night. Humidity is 70–80% year-round. Rainfall is high and is distributed throughout the year. Some authors suggest a forest must receive a minimum of 100 in. (2,500mm) of rain per year to be classified as rainforest. It must also have a positive water balance, in which precipitation exceeds evaporation. Tropical lowland rainforest in the Indo-Malayan region is affected by the southwest monsoon from May to September and the northeast monsoon from November to March. Some areas are affected by one more than the other. Rainfall in the inter-monsoonal months is lighter, although there are no significant dry periods.

High rainfall and rampant plant life often result in nutrient-deficient soils below the topsoils, due to the leaching of nutrients and plant uptake. Most of the recycling of nutrients takes place in the rapidly decomposing layer above the soil and is undertaken by insects, bacteria, and fungi. For this reason, the roots of tropical rainforest plants tend to be shallow or even above the ground. Plants may have specialized structures like buttress roots, which add support and increase carbon dioxide intake, and leaves with drip tips, which aid in the drainage of rainfall. Emergent trees often have small leaves to reduce water loss to evaporation, whereas leaves from the lower canopy and shrub layers tend to be large in order to capture as much light as possible. The understory is usually dark, with only 1% of the light that hits the canopy reaching the ground.

Despite the abundance of life, tropical lowland rainforests are some of the most challenging places to find wildlife, and one of the things that strikes people on a first visit is how little they see. These are not easy places to explore. The canopy is so high that its inhabitants are virtually out of view, while those of the undergrowth are adept skulkers. Walking through the forest is not easy, with its many hooks and spines and tangles (not to mention the leeches and the danger of getting lost). Despite these obstacles, wildlife-viewing and especially birding can be very rewarding. Some rainforest sites have canopy towers or walkways, which are a tremendous help in exploring the canopy levels. Some have a network of trails and good habitat along roads and tracks. A combination of all these modes of access is needed to find a good selection of wildlife. A knowledge of vocalizations is essential for tracking down birds in this habitat, perhaps more than in any other.

The area covered by Indo-Malayan tropical lowland rainforests is rapidly decreasing due to logging and clearing for agriculture, and these are now among the most threatened ecosystems.

WILDLIFE: Indo-Malayan tropical lowland rainforests exhibit high biodiversity and are home to half of all the living animal and plant species on the planet, with millions of species still undescribed. Tree

Tropical lowland rainforest in Malaysian Borneo.
© CHARLEY HESSE, TROPICAL BIRDING TOURS

diversity is amongst the highest of any forest in the world, with as many as 300 species per 2.5 acres (1 ha). Many flowering plants are present, including the amazing *Rafflesia* species, which grow on terrestrial vines and produce the world's largest flowers. Several groups of animals of these forests are adapted to gliding, including flying squirrels, colugos, frogs, lizards, and even snakes.

Mammal groups represented include pangolins, monkeys, squirrels, tapirs, elephants, and rodents, plus many predators, including Leopard and in some areas Tiger. There is also a high diversity of bats. Some examples of tropical lowland rainforest mammals include: Purple-faced Langur and Grizzled Giant Squirrel at Sinharaja Forest in Sri Lanka; Malayan Tapir at Taman Negara in Peninsular Malaysia; Asian Elephant, Leopard Cat, and Bornean Orangutan at Danum Valley Conservation Area in Sabah, Malaysia; and Philippine Tarsier and Philippine Flying Lemur at Raja Sikatuna National Park on Bohol island in the Philippines.

Bird diversity is extremely high among several typical groups including pittas, broadbills, and hornbills. Some examples of typical tropical lowland rainforest birds include: Serendib Scops-Owl, Yellow-fronted Barbet, and Ashy-headed Laughingthrush in Sri Lanka; Crested Partridge,

Above: **Gursky's Spectral Tarsier is a nocturnal primate that is seen regularly in the lowland rainforest at Tangkoko in n. Sulawesi (Indonesia), where regular roosting hollows are known.**
© SAM WOODS, TROPICAL BIRDING TOURS

Below: **Black-and-yellow Broadbill is a vocal and fairly common resident of tropical lowland rainforests.**
© KEN BEHRENS, TROPICAL BIRDING TOURS

Malaysian Rail-babbler (IS), Garnet Pitta, Black-capped Babbler (IS), and Crested Shrikejay (IS) in Peninsular Malaysia; Palawan Peacock-Pheasant, Rufous Coucal, and Northern Sooty-Woodpecker in the Philippines; and Crested Fireback (IS), Rhinoceros Hornbill, and Blue-headed Pitta in Sabah, Malaysia.

Endemism: Tropical lowland rainforests contain many areas of endemism in the Indo-Malayan region. One of the most significant areas of bird endemism in the region is Sundaland, which comprises the Malay Peninsula, Sumatra, Borneo, and Java, plus many smaller islands. Other endemic areas may be found in the Andaman and Nicobar Islands; the Annamese lowlands of n. Vietnam; Sulawesi; and, in the Philippines, Cebu, Luzon, Mindanao, the Eastern Visayas, Mindoro, Negros and neighboring Panay, Palawan, and the Sulu Archipelago.

DISTRIBUTION: In the Indo-Malayan region, tropical lowland rainforests are distributed from the southwest corner of Sri Lanka to the Andaman and Nicobar Islands, Sumatra (including Enggano and the Mentawi Islands), Java, Bali, Christmas Island, Cocos (Keeling) Islands, and Borneo. Continental areas include Peninsular Malaysia and relicts in n. Vietnam and coastal areas of Myanmar. These rainforests are also found in the Philippines, from Luzon through Mindoro, the Visayas, Mindanao, Palawan, and the Sulu Archipelago, as well as many other smaller islands.

WHERE TO SEE: Sinharaja Forest Reserve, Sri Lanka; Taman Negara National Park, Peninsular Malaysia; Danum Valley Conservation Area, Sabah, Malaysia; Subic Bay, Luzon, Philippines.

SIDEBAR 3.2   ISLAND ARC COLLISIONS: BUILDING ARCHIPELAGOS

When two oceanic plates collide, one of the plates will be subducted under the other. This is especially evident where the Pacific Plate crashes into the Asian Plate, forming archipelagos like Japan and the Philippines. The example illustrated below shows the Australian Plate moving north into the Asian Plate. As the oceanic basalts and sea-floor sediments get subducted under the Asian Plate, they melt and rise under the Asian Plate to form a string of volcanic islands, here creating the Sundaic Islands. As long as the Australian Plate continues to move north, islands of the Greater Sundas will continue to form. These islands, a long distance from the continental landmasses, and often very steep and high, create many niches for animals to evolve into different species and therefore have extremely high levels of endemism. Thus, plate tectonics determines habitat types and animal distribution.

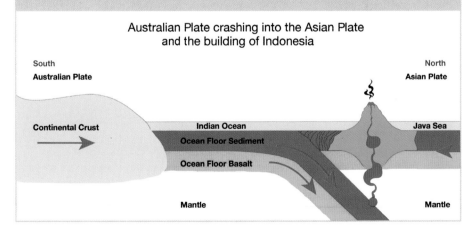

Australian Plate crashing into the Asian Plate and the building of Indonesia

INDO-MALAYAN SEMI-EVERGREEN FOREST

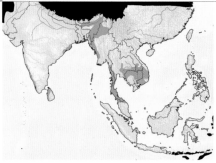

IN A NUTSHELL: Tall, multilayered forest growing in areas of more seasonal and lower annual rainfall than tropical lowland rainforest.
Habitat Affinities: NEOTROPICAL SEMI-EVERGREEN FOREST. **Species Overlap:** INDO-MALAYAN TROPICAL LOWLAND RAINFOREST; INDO-MALAYAN MOIST DECIDUOUS FOREST.

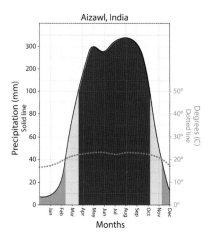

DESCRIPTION: Semi-evergreen forests have a mixture of evergreen and deciduous trees, and although a significant proportion of canopy trees can lose their leaves at the height of the dry season, the forest as a whole may appear green all year long. This forest type is transitional between INDO-MALAYAN TROPICAL LOWLAND RAINFOREST and INDO-MALAYAN MOIST DECIDUOUS FOREST. Annual rainfall is in the range of 80–100 in. (2,000–2,500mm), but there is a marked dry season (lacking in tropical lowland rainforest). The canopy is closed, and the forest structure is more stratified and the trees usually taller than in deciduous forests, with some reaching almost 200 ft. (60m). Trees usually have buttressed trunks and host abundant epiphytes.

The species composition of these forests varies geographically. In the semi-evergreen forests on the east coast of India, Sal is dominant. In the Brahmaputra Valley, evergreen trees include myrtles (*Syzygium* spp.) and laurels (*Cinnamomum* spp.). Forests in e. India, Myanmar, peninsular Thailand, and Indochina (Laos, Cambodia, Vietnam), have several dipterocarp species including Resin Tree. Another feature of many semi-evergreen forests is the presence of bamboo patches, which are the preferred habitat of several bird species and also popular with Asian Elephant.

Semi-evergreen forests are fantastic places for birding (although very slow at times during the middle of the day), and by exploring edge, canopy, and forest interior, you can rack up large day lists.

Semi-evergreen forest in Kaeng Krachan National Park, Thailand. © CHARLEY HESSE, TROPICAL BIRDING TOURS

Birds are numerous and easy to see, and often a good fruiting tree can be a treasure trove, attracting large numbers of species. The forests often have a developed understory and are not particularly easy to walk through, but luckily many reserves have networks of trails allowing easy access to the forest. Exploring the forest interior is less productive, but with effort, many of the more secretive species can be found.

WILDLIFE: Semi-evergreen forests in India still contain Tiger, Asian Elephant, and Indian Rhinoceros. Protected areas of Thailand's semi-evergreen forests, such as those in Kaeng Krachan and Khao Yai National Parks, still hold impressive collections of mammals, including Asian Elephant, Leopard, Dhole, Gaur, Sambar, White-handed Gibbon, and Stump-tailed and Southern Pig-tailed Macaques.

These forests contain high avian diversity, holding 400–500 species, and are great places for birding. Some species showing an affinity for semi-evergreen forest are White-throated Bulbul; Abbott's Babbler; Pale-chinned Blue

Leopard Cat occurs in a variety of habitats, including semi-evergreen forest (photo: Borneo).
© SAM WOODS, TROPICAL BIRDING

Great Hornbill occurs in se. Asian semi-evergreen forests. © KEN BEHRENS, TROPICAL BIRDING TOURS

Flycatcher (IS); several pheasants and partridges; Great, Wreathed, Brown, and Rusty-cheeked Hornbills; Silver-breasted, Banded, and Black-and-red Broadbills; Blue and Eared Pittas; plus many woodpeckers and numerous other species. Until recently, the critically endangered Gurney's Pitta inhabited this habitat in far s. Thailand, but it is now found only across the border in Myanmar. Semi-evergreen forests in e. Cambodia and s. Vietnam, such as those in Cat Tien National Park, contain rare species such as Black-shanked Douc Langur and Siamese Fireback (IS).

   Endemism: Each separate area within the range of Indo-Malayan semi-evergreen forest holds a quite distinct community, although some areas show slightly more overlap with neighboring habitats than others. One of the most important areas of endemism in the region is the s. Vietnamese semi-evergreen forests, which hold the range-restricted Orange-necked Partridge and Germain's Peacock-Pheasant (IS). Semi-evergreen forests on the Malabar Coast of sw. India hold several Western Ghats endemics, although none of those is restricted to this habitat, and the same goes for Sri Lankan endemics found in expanses of this habitat. Forests between s. Thailand and s. Myanmar contain the range-restricted and critically endangered Gurney's Pitta.

DISTRIBUTION: Semi-evergreen forests are distributed along the Malabar Coast of India, locally in Sri Lanka, in the Brahmaputra Valley of Assam (ne. India), the lower slopes of the e. Himalayas, Odisha (e. India), far e. India and adjacent Myanmar, the Andaman Islands, and the Tenasserim Range of Thailand and adjacent Myanmar. There are also semi-evergreen forests on the Indochinese Peninsula from c. Thailand across s. Laos and Cambodia to Vietnam.

   In the Brahmaputra Valley, most of the original forests have been cleared, leaving only a few remnant patches. The original forest type of the Malabar Coast is thought to have been tropical lowland rainforest, but most has been replaced by or interspersed with native Teak and become semi-evergreen in nature. There is a north–south rainfall gradient in peninsular Thailand, with more rain falling in the south, where semi-evergreen forests grade into INDO-MALAYAN TROPICAL LOWLAND RAINFOREST. On the drier slopes, they grade into INDO-MALAYAN MOIST DECIDUOUS FOREST.

WHERE TO SEE: Manas National Park, Assam, India; Kaeng Krachan National Park, Thailand; Khao Yai National Park, Thailand; Cat Tien National Park, Vietnam.

INDO-MALAYAN TROPICAL MONTANE RAINFOREST

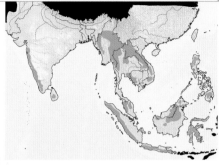

IN A NUTSHELL: Medium- to low-stature forest with abundant epiphytes and a developed understory, growing at upper elevations with high rainfall. **Habitat Affinities:** AFROTROPICAL MONTANE FOREST; NEOTROPICAL CLOUD FOREST. **Species Overlap:** INDO-MALAYAN PINE FOREST; INDO-MALAYAN SUBTROPICAL BROADLEAF FOREST.

DESCRIPTION: Montane rainforests are a lush, moist habitat found at higher elevations than INDO-MALAYAN TROPICAL LOWLAND RAINFOREST, with relatively smaller trees and many epiphytes. They generally occur above 3,300 ft. (1,000m), although this lower limit can vary with geography or topography, and on up to the tree line, which is usually above 9,800 ft. (3,000m) in the Indo-Malayan region. The nature of these forests changes with elevation. Trees can be around 30–65 ft. (10–20m) tall, becoming smaller higher up, and have slenderer trunks and smaller leaves than lowland trees and lack buttresses. There is a developed understory with ferns, palms, herbs, and ground orchids. Epiphytic mosses, lichens, and liverworts are prevalent, and they increase with elevation. Annual rainfall in montane rainforests generally exceeds 80 in. (2,000mm); they have a short dry season and hence are evergreen in nature. The dipterocarps that dominate many Indo-Malayan lowland forests are replaced by trees of other families, such as oaks (Fagaceae), laurels (Lauraceae), and myrtles (Myrtaceae). Lower montane forests, at 3,300–6,600 ft. (1,000–2,000m), are often dominated by evergreen oaks (*Quercus* spp.), stone oaks (Lithofagus spp.), and chinquapins, although other types such as cinnamon trees, podocarps, and kauris can also be prominent. Higher up, broadleaf trees mix with conifers, and higher still, rhododendrons may be common.

One subtype of tropical montane forest is cloud forest, which gains its moisture from clouds and fog. Cloud forests often exhibit an abundance of mosses covering both ground and vegetation.

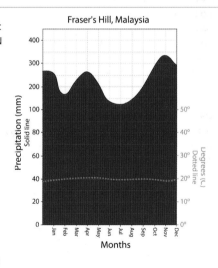

Fraser's Hill, Malaysia

The highest montane rainforests are often elfin forests, which are loaded with epiphytes and, with their stunted trees and gnarled trunks, have an otherworldly atmosphere. Montane rainforests are generally fascinating places to visit and host lots of unique wildlife. Due to high elevations and steep slopes, they can be physically challenging places to explore, although some sites have roads providing easy access.

As lowland forests are more often targeted for logging, montane forests tend to be better conserved. They also have escaped many of the devastating fires that have ravaged lowland forests in the past few decades. Some montane forests are, however, locally threatened by clearance, especially for grazing domestic animals, and hunting threatens wildlife.

WILDLIFE: In general, montane rainforest is not as diverse as INDO-MALAYAN TROPICAL LOWLAND RAINFOREST, although regional endemism is higher. There are also not as many larger mammals, although some primates are present, and squirrels are especially abundant. Due to wide-scale destruction of lowland habitats, some larger species of mammals are now confined mainly to montane rainforests. Some examples of montane rainforest mammal species include: Nilgiri Langur in the Western Ghats in India; Red Slender Loris in Sri Lanka; Chinese Serow and Chinese Goral in nw. Thailand and Myanmar; Phayre's Leaf Monkey, Assam Macaque, and Lar Gibbon in Sumatra and Peninsular Malaysia; Hose's Palm Civet, Masked Palm Civet, and Whitehead's Pygmy Squirrel in Borneo; and Mindanao Flying Squirrel in the Philippines.

Left: **Some species of rafflesias found within this habitat on the island of Borneo produce the largest flowers in the world. This particular species (*Rafflesia keithii*) can grow to over 3 ft. (1m) in diameter.** © KEN BEHRENS, TROPICAL BIRDING TOURS

Below: **Tropical montane rainforest on the slopes of Mt. Kinabalu, Malaysian Borneo.** © SAM WOODS, TROPICAL BIRDING TOURS

Montane rainforests are especially rich in birdlife, and the many isolated areas give rise to a profusion of interesting species. Certain families are particularly well represented, such as pheasants and partridges, pigeons, owls, laughingthrushes, flycatchers, and sunbirds. Some examples of montane rainforest bird species of the region include: White-throated Fantail (IS), Gray-throated Babbler (IS), Eyebrowed Wren-Babbler (IS), Snowy-browed Flycatcher (IS), and Mountain Tailorbird (IS), which are found over much of the region; Nilgiri Wood-Pigeon, Nilgiri Laughingthrush, and Nilgiri Sholakili in the Western Ghats in India; Dull-blue Flycatcher, Sri Lanka Bush Warbler, and Yellow-eared Bulbul in Sri Lanka; Rufous-throated Partridge, Scarlet-faced Liocichla, and Green Cochoa in nw. Thailand; Vietnamese Cutia, Dalat Shrike-Babbler, and Gray-crowned

Bornean Green-Magpie within the montane rainforest on Mt. Kinabalu in n. Borneo.
© KEN BEHRENS, TROPICAL BIRDING TOURS

Crocias in the Dalat Plateau of s. Vietnam; Chestnut-headed Partridge and Cambodian Laughingthrush in the Cardamom Mountains of Cambodia; Mountain Peacock-Pheasant in Peninsular Malaysia; Bronze-tailed Peacock-Pheasant, Sumatran Ground-Cuckoo, and Schneider's Pitta in Sumatra; Pink-headed Fruit-Dove, Javan Cochoa, and Blood-breasted Flowerpecker in Java; the trio of Whitehead's Trogon, Whitehead's Broadbill, and Whitehead's Spiderhunter in Borneo; and Scale-feathered Malkoha and Apo Myna in the Philippines.

Endemism: Montane rainforests of the Indo-Malayan region coincide with many areas of endemism, in the Western Ghats mountain range of India; the montane zone of Sri Lanka; some remnant patches in Hainan, China; the Dalat Plateau in s. Vietnam; the mountains of Sumatra and Peninsular Malaysia; Borneo; Java; Bali; and several separate areas on different mountainous Philippine islands, including Luzon, Mindoro, Palawan, Negros, Panay, and Mindanao.

DISTRIBUTION: Montane rainforests occur at higher elevations throughout the tropical parts of the region, including: the Western Ghats mountain range of India, the central ranges of Sri Lanka, the Arakan Mountains in w. Myanmar, the northern parts of the Tenasserim Range between Myanmar and Thailand, the Annamite Range in Laos and Vietnam, the Dalat Plateau of s. Vietnam, the Cardamom Mountains of sw. Cambodia, the mountains of Peninsular Malaysia, the Barisan mountain range of Sumatra, the mountainous areas of Java and Bali, the mountainous central region of Borneo, and the mountainous areas of the Philippines. Their range does not include the Himalayas, because in terms of the vegetation and animal life, the montane forests there have more in common with the lowland subtropical forests to the east than with the tropical montane forests to the south. Note that the moist broadleaf forests of lower elevations of the Himalayas are treated as a subset of the INDO-MALAYAN SUBTROPICAL BROADLEAF FOREST and are treated in that habitat account.

WHERE TO SEE: Ooty, Tamil Nadu, India; Horton Plains National Park, Sri Lanka; Nat Ma Taung National Park (Mt. Victoria), Myanmar; Doi Inthanon National Park, Thailand; Fraser's Hill, Peninsular Malaysia; Mt. Ijen, Java, Indonesia; Mt. Kinabalu, Sabah, Malaysia; Mt. Kitanglad, Mindanao, Philippines; Dalat Plateau, Vietnam.

SIDEBAR 3.3   WHY ARE THERE CLOUDS IN THE CLOUD FOREST?

There are many different phenomena that produce precipitation and therefore have a profound effect on local habitats. One of the most significant is the orographic effect. This occurs when moist air is pushed upward by a mountain, forcing it to precipitate its moisture in the form of rain or fog. This effect is common in warm coastal areas, where evaporation from the ocean seeds the atmosphere with moisture. The orographic effect waters the NEOTROPICAL CLOUD FOREST of Central and South America and the INDIAN OCEAN RAINFOREST along the eastern escarpment of Madagascar. A related effect is the rain shadow: areas of unusual aridity on the leeward side of the mountain from the prevailing, moist winds. Once the air has dropped its moisture, only dry air flows to this side, creating the rain shadow. A classic example of this is seen in the Atacama Desert of w. South America (see NEOTROPICAL DESOLATE DESERT), which is the driest nonpolar area on earth due to its position in the rain shadow of the towering Andes Mountains.

INDO-MALAYAN SUBTROPICAL BROADLEAF FOREST

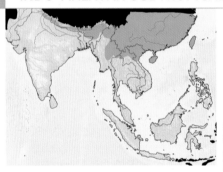

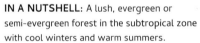

IN A NUTSHELL: A lush, evergreen or semi-evergreen forest in the subtropical zone with cool winters and warm summers.
Habitat Affinities: EAST ASIAN TEMPERATE DECIDUOUS FOREST. **Species Overlap:** EAST ASIAN TEMPERATE DECIDUOUS FOREST; INDO-MALAYAN TROPICAL MONTANE RAINFOREST.

DESCRIPTION: This is a lush, tall, often evergreen forest found in areas with a mild climate. It can have a fairly developed understory, but it is still possible to walk through with relative ease, although visibility is reduced. The pleasant climate and abundant greenery make it an agreeable habitat to explore at any time of year. The subtropics concerns the areas between the tropical and temperate zones, here from the Tropic of Cancer, at 23.5°N, to the start of the temperate region at 35°N. In the Indo-Malayan region, however, due to the effects of elevation, the corresponding subtropical climate doesn't follow these straight lines; this habitat exists in the Himalayas as a montane forest and farther north in e. China as a lowland forest. In a subtropical climate, the coldest month doesn't average below freezing, and the warmest is above 71.6°F (22°C).

Climatic conditions of subtropical broadleaf forests vary widely throughout this region, with mean annual rainfall in the range of 31–100 in. (800–2,500mm) and an average annual temperature

Left: **Subtropical broadleaf forest on the island of Taiwan.**
© KEITH BARNES, TROPICAL BIRDING TOURS

Below: **Red Panda is found within subtropical broadleaf forests in the Himalayas.** © ANDREW SPENCER

Boleng, India

Precipitation (mm)
Solid line

Degrees (C)
Dotted line

Months

range of 59–77°F (15–25°C). The northeast monsoon sweeps down from the Pacific during the boreal winter bringing rainfall. The coastal areas receive the most rain, and the effects are gradually reduced moving inland. This creates an east–west moisture gradient. The composition of these forests changes accordingly, with evergreen species becoming mixed with deciduous elements. Conifers are also present in these forests, especially at higher elevations in mountain ranges such as the Himalayas. In the subtropical broadleaf forests of the Ryukyu Islands of Japan, Taiwan, and s. China, Japanese Blue Oak, chestnut (*Castanopsis hystrix*), and Chinese Cryptocarya are common. Moving through n. Indochina toward e. India, the forests take on more Himalayan endemic components, such as Needlewood Tree and Indian Chestnut. Continuing to the drier west, species such as Sal and the beautiful Mountain Ebony are present. It could be argued that the Himalayan forests look like INDO-MALAYAN

TROPICAL MONTANE RAINFOREST, and while they appear similar in structure and topography, the Himalayan forests are more closely related to other subtropical broadleaf forests and grow at much lower elevations than the forests to the tropical south.

Swinhoe's Pheasant inhabits subtropical broadleaf forest in Taiwan. © IAIN CAMPBELL, TROPICAL BIRDING TOURS

WILDLIFE: Subtropical broadleaf forests of the Indo-Malayan region contain spectacular and varied wildlife. Some of the more widespread species found through many areas of subtropical forests include mammals such as Assamese Macaque (IS) and Spotted Linsang, and birds like Black-throated Tit (IS) and Red-billed Leiothrix (IS). The Ryukyu Islands of Japan have some really exciting species on various island groups: Amami Island in the north has an endemic woodcock, jay, thrush, and black rabbit; the main island of Okinawa has Okinawa Woodpecker and Okinawa Rail; the Yaeyama Islands have Iriomote Tit and Iriomote Wildcat; and the isolated Ogasawara (or Bonin) Islands have Bonin White-eye.

Taiwan's subtropical broadleaf forests are full of amazing endemics; standout species include Formosan Rock Macaque, Formosan Serow, Swinhoe's Pheasant, Taiwan Partridge, and Taiwan Blue-Magpie. On the adjacent mainland in se. China, some localized birds include Cabot's Tragopan, Elliot's Pheasant, and White-necklaced Partridge. In the streams here dwells the Chinese Giant Salamander, perhaps the world's largest amphibian. The forests of s. China and n. Vietnam hold the rare François' Langur. Off China's coast, the large island of Hainan holds three endemic birds: Hainan Leaf Warbler, Hainan Partridge, and Hainan Peacock-Pheasant. Some typical species from the forests of s. China and n. Indochina include Chestnut Bulbul (IS), Fork-tailed Sunbird (IS), Indochinese Yuhina (IS), and also the range-restricted Northern White-cheeked Gibbon. The forests of n. Myanmar have the range-restricted Gongshan Muntjac and Rusty-bellied Shortwing.

The Meghalaya forests of e. India mark the western extent of the range of many species, including Western Hoolock Gibbon. They also hold Rufous-necked Hornbills plus many interesting laughingthrushes and babblers. The subtropical broadleaf forests farther west in the Himalayas contain the range-restricted Golden Langur and Chestnut-breasted Partridge. Forests of the Himalayas change gradually with the east–west rainfall gradient, as do the distributions of species that live in them.

Endemism: Islands, by their isolated nature, are more likely to have endemics. The Ryukyu Islands and the outlying Ogasawara Islands contain several separate small centers of endemism. Ogasawara has suffered avian extinctions but still has one extant endemic, Bonin White-eye. Amami Island, at the north end of the Ryukyu archipelago, Okinawa in the center, and the Yaeyama Islands at the south end each hold several endemic taxa. Taiwan is a very important area of endemism, with almost 30 endemic birds, plus many mammals, reptiles, and amphibians. Many of these are found in the subtropical broadleaf forests. Hainan also holds several endemic bird species. Even in the continental parts of this habitat's range, we find areas of endemism. Several bird species are restricted to the se. Chinese mountains. In s. China and n. Indochina, animal communities change gradually, but some areas still contain pockets of range-restricted species.

DISTRIBUTION: In the Indo-Malayan region, subtropical broadleaf forests are distributed from the Ryukyu Islands in the very south of Japan (Yakushima in the north to Iriomote in the south) and the outlying Ogasawara Islands farther east to the large islands of Taiwan and Hainan, se. China below the Yangtze River, and across large areas of s. China. They are also found down to n. Vietnam and west through n. Indochina and Myanmar to the forests of Meghalaya on the Shillong Plateau in e. India, and finally along the base of the Himalayas to just north of New Delhi, India.

WHERE TO SEE: Kalaw, Myanmar; Wuyishan, se. China; Daxueshan National Forest Recreation Area, Taiwan; Kunigami, Okinawa, Japan; Sattal, Uttarakhand, India.

SIDEBAR 3.4

WHY THERE ARE NO VOLCANOES IN THE HIMALAYAS: THE CONTINENTAL-CONTINENTAL COLLISIONS

As explained in sidebars 2.6 and 3.2, when oceanic plates collide with each other or with continental plates, one of the plates will subduct under the other plate, melt, and form volcanoes. However, continental rocks are lighter than oceanic rock (strange but true), so they cannot be subducted into the earth's mantle; therefore, there is no melting subducting rock, and thus no volcanoes form. When continents crash into each other, huge mountains such as the Himalayas are formed instead, in a process called orogeny. These mountain ranges drastically affect weather patterns, creating rain shadows and very wet areas of orographic rain (sidebar 3.3). The great ranges in elevation and precipitation in the Himalayas have created six different habitats, and with those many endemic birds.

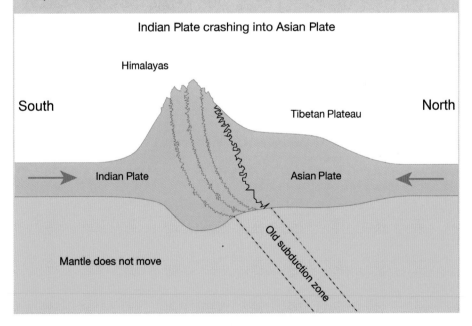

INDO-MALAYAN PEAT SWAMP FOREST

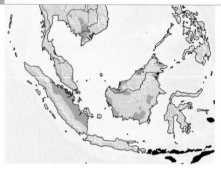

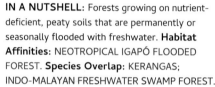

IN A NUTSHELL: Forests growing on nutrient-deficient, peaty soils that are permanently or seasonally flooded with freshwater. **Habitat Affinities:** NEOTROPICAL IGAPÓ FLOODED FOREST. **Species Overlap:** KERANGAS; INDO-MALAYAN FRESHWATER SWAMP FOREST.

DESCRIPTION: Peat swamp forest is a form of INDO-MALAYAN TROPICAL LOWLAND RAINFOREST growing on nutrient-poor, inundated peat soils. It is often found between INDO-MALAYAN MANGROVE FOREST and tropical lowland rainforest and can be formed when a river deposits organic debris behind the inland edge of a mangrove and it gets trapped behind the mangrove roots, or when soil is waterlogged and plant material is unable to fully decay; both these methods form peat. Peat deposits are at least 20 in. (50cm) thick but can go as deep as 65 ft. (20m). The soils do not drain after flooding and tend to be nutrient-deficient and acidic. Peat swamp forests often grow in concentric circles from the wettest part outward. Trees in the center tend to be small, those on the edges much taller. Many trees in flooded forests have stilt roots or air-breathing roots (pneumatophores). The dominant trees in Borneo's peat swamp forests are Miyapok, Ramin, and Alan Bunga. This habitat also has some of the same tree species as tropical lowland rainforest, although it lacks many of that forest's trees. There are often Strangler Figs around the edges of peat forests, and various palms and pandanus species in the understory. Interestingly, peat swamp forests (like KERANGAS) have an abundance of trees with hollow trunks. It has been suggested that these have evolved to encourage bats to roost inside and thereby contribute valuable nutrients through their feces.

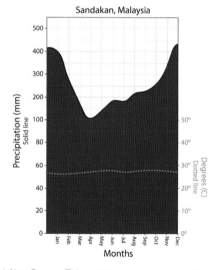

The waterlogged nature of this forest means that it is difficult to get through on foot. Some sites have boardwalks that allow access; otherwise, the best way to experience this habitat is by

Above: **Peat swamp forest in Kalimantan, Indonesian Borneo.**
© RUANDA AGUNG SUGARDIMAN/AUSAID

Right: **Peat swamp forest is an important habitat for the endangered Storm's Stork (photo: Borneo).** © JAMES EATON, BIRDTOUR ASIA

boat. Peat swamp forests were previously considered to be less threatened than INDO-MALAYAN FRESHWATER SWAMP FORESTS because of their low nutrient levels, which make them less suitable for conversion to agriculture. However, in recent years, unusually dry conditions have meant that extensive fires were able to sweep through many of these forests, destroying the peat deposits permanently. It has been suggested that both droughts precipitated by El Niño and climate change may have played a role in this. Many areas have also been cleared for African Oil Palm plantations. It has been found that peat swamp forests, unlike Indo-Malayan tropical lowland rainforest, are very difficult to reforest. Possibly more than any other habitat, peat swamp forests store carbon dioxide. When they burn, massive carbon emissions result.

WILDLIFE: Peat swamp forests are not species-rich, but they do still hold some of the region's threatened mammals, including Asian Elephant, Sumatran Rhinoceros, Malayan Tapir, Bornean and

Muller's Gibbon lives in the canopy above peat swamp.
© KEN BEHRENS, TROPICAL BIRDING TOURS

Sumatran Orangutans, and Clouded Leopard. Residents of the Bornean peat swamp forests include Proboscis Monkey and Otter Civet, both of which may show some preference for the habitat. Despite most mammals being less abundant in this habitat, some Bornean peat forests have elevated densities of primates. One such site, Sabangau National Park, has important populations of the scarce Red Leaf Monkey, Bornean White-bearded Gibbon, and Bornean Orangutan, plus the more common Proboscis Monkey, Silvered Leaf Monkey, Crab-eating Macaque, and Southern Pig-tailed Macaque. This is likely because the trees here produce fruit year-round, but reduced hunting due to the difficulty in accessing the habitat may also play a part. Peat swamp forest in Peninsular Malaysia has been found to provide important roosting sites for flying foxes. Reptiles found in blackwater peat swamp rivers include Saltwater Crocodile, False Gharial, and Asian Water Monitor.

Although less diverse than adjacent INDO-MALAYAN TROPICAL LOWLAND RAINFOREST, peat swamp forests hold many of the same species of birds. In a recent study, 223 bird species were counted in peat swamp forests of se. Pahang in Peninsular Malaysia. These forests are particularly rich in hornbill species and appeared to be a local stronghold of the endangered Wrinkled Hornbill. On Borneo, 218 species of birds have been found in Tanjung Puting National Park, many of them in this habitat. Some birds, such as Crestless Fireback, Black Partridge, Orange-backed Woodpecker, Red-crowned Barbet, Bornean Bristlehead, Gray-breasted Babbler (IS), Gray-chested Jungle-Flycatcher, and Hook-billed Bulbul (IS), show a preference for this habitat together with KERANGAS. Birds such as Cinnamon-headed Green-Pigeon, Black-bellied Malkoha, and Javan White-eye favor this habitat together with INDO-MALAYAN MANGROVE FOREST. Other species of conservation importance that are present in Borneo's peat swamp forests are Storm's Stork, Wallace's Hawk-Eagle, and Large Green-Pigeon.

DISTRIBUTION: Peat swamp forests are found in a few small areas of Peninsular Malaysia, along the east coast of Sumatra, patchily on Borneo, in the Tonlé Sap basin of Cambodia, and in some areas of the Mekong Basin of Vietnam. The largest area of peat swamp forest in Peninsular Malaysia is in se. Pahang (on the east side of the peninsula) and occurs within several small forest reserves. About 90% of both the Tonlé Sap and Mekong peat swamp forests have been destroyed, and only small ones remain in Vietnam, protected by two small national parks: U Minh Thuong and U Minh Ha. On Borneo, 80% of peat swamp forests in Sabah and Sarawak have been lost, 20% in Brunei, and 70% in Kalimantan (which still has the largest remaining area). Well over half of the peat swamp forests in e. Sumatra have been cleared in recent years, with the largest remaining areas to be found in Riau province.

WHERE TO SEE: Tanjung Puting National Park, Kalimantan, Indonesia; Klias Peat Swamp Forest, Sabah, Malaysia.

KERANGAS

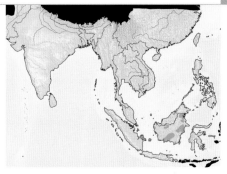

Banjarmasin, Indonesia

IN A NUTSHELL: Forest with low, densely packed trees growing on nutrient-poor, sandy soils. **Habitat Affinities:** None. **Species Overlap:** INDO-MALAYAN PEAT SWAMP FOREST; INDO-MALAYAN TROPICAL LOWLAND RAINFOREST.

DESCRIPTION: Kerangas, or Sundaland heath forest, is a bizarre and almost Tolkienesque habitat that shares some common features with mossy, upper INDO-MALAYAN TROPICAL MONTANE RAINFOREST. Kerangas is found on ridges, raised beaches, or plateaus, and is often adjacent to INDO-MALAYAN PEAT SWAMP FOREST. Kerangas forests have well-drained, acidic, white-sand soils, which lack nutrients (especially nitrogen) and are very poor for agriculture—an Indonesian name, *kerangas* means "the land where rice will not grow." Some tree species (e.g., *Myrmecodia* spp.) use symbiotic relationships with ants to gain nutrients, while others (e.g., *Gymnostoma nobile*) use nitrogen-fixing bacteria in their root nodules. Kerangas forests usually have a low, open canopy, and the trees are densely packed, with a thick understory. The trees are not buttressed, there are few vines, and the leaves of plants tend to be leathery. This habitat includes dipterocarps, though different species than INDO-MALAYAN TROPICAL LOWLAND RAINFOREST. There are also casuarinas, some palms, conifers, and small-leaved plants from the myrtle family.

Different types of kerangas can occur, from those containing large trees reaching 80–100 ft. (25–30m) in height to pole forests with many densely packed thin trees, 17–40 ft. (5–12m) tall, to more open woodland. The soils have a layer of humus or peat, which is lost if the forests are cleared, making them especially fragile. The ground in these heath forests can be waterlogged, and may be covered with brown peaty water from the leached soil. Permanently waterlogged kerangas is known as *kerapah* forest. Where the sand is not exposed, the ground layer can be covered in mosses, epiphytic plants, and carnivorous plants. Due to the nitrogen-poor conditions of the soil, kerangas forests are home to many carnivorous plants, including pitcher plants, sundews, and bladderworts.

Above: **Kerangas habitat, with pitcher plants visible on the right-hand side, in Borneo.** © KEITH BARNES, TROPICAL BIRDING TOURS

Above right: **The scarce Bornean Bristlehead represents a monotypic bird family restricted to the island of Borneo, where it reaches high densities within peat swamp forest.** © JAMES EATON, BIRDTOUR ASIA

Many of the trees of kerangas have an Australasian origin, including casuarinas (e.g., *Gymnostoma nobile*), conifers of the podocarp family (e.g., *Dacrydium* and *Podocarpus* spp.), and Sarawak Kauri. The myrtle (e.g., *Syzygium* spp.) and dipterocarp (e.g., *Shorea* spp.) families are also prominent. Kerangas share many of the same tree species as swamp forests, including several dipterocarps. Interestingly, kerangas (like INDO-MALAYAN PEAT SWAMP FOREST) have an abundance of trees with hollow trunks. It has been suggested that these have evolved to encourage bats to roost inside and thereby contribute valuable nutrients through their feces. The ground layer of vegetation is often very thick, and this combined with the tightly packed trees makes it difficult to walk around easily. Kerangas are vulnerable to logging and forest fires, and vast tracts in Borneo were lost in huge fires in 1982–83 and again in 1997–98.

WILDLIFE: As kerangas is a less-productive environment, its animal diversity is reduced, and it has fewer than half the number of reptile and amphibian species as surrounding forests. Mammal diversity is also lower, although the endemic Bornean Orangutan and Hose's Langur do make use of the habitat.

Some people have described kerangas as "virtually birdless," although, interestingly, one very special species appears to be common at least locally, the Bornean Bristlehead, a Bornean endemic in a monotypic family. Bonaparte's Nightjar, Brown-backed and Scarlet-breasted Flowerpeckers, Gray-breasted Babbler, and Hook-billed Bulbul (IS) also seem to prefer this habitat along with INDO-MALAYAN PEAT SWAMP FOREST. Red-crowned Barbet, although found in other habitats, seems particularly common in kerangas.

Endemism: Although kerangas contains some Bornean endemics, endemism within the habitat type is low. However, some plant species are shared endemics between kerangas and peat swamp forest, for example the pitcher plant *Nephentes bicalcarata*.

DISTRIBUTION: Kerangas forests are found on Borneo and the nearby islands of Belitung and Bangka to the west. They occur patchily within this zone and only in areas with the physical attributes that allow their formation. Around the edges, they grade into surrounding forest types.

WHERE TO SEE: Bako National Park, Sarawak, Malaysia.

INDO-MALAYAN LIMESTONE FOREST

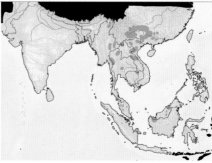

IN A NUTSHELL: Forests that are usually moist deciduous or semi-evergreen, and often stunted, which grow on or between limestone formations. **Habitat Affinities:** AFROTROPICAL INSELBERGS, KOPPIES, AND CLIFFS. **Species Overlap:** INDO-MALAYAN SEMI-EVERGREEN FOREST; INDO-MALAYAN TROPICAL LOWLAND RAINFOREST.

DESCRIPTION: Limestone forest grows on areas of limestone outcrop known as karst, a type of formation created when acidic rain irregularly degrades the exposed rock surfaces. In some areas, the soil is limited in quantity and low in mineral nutrients, leading to small, stunted trees and shrubs. The forests can be moist deciduous to semi-evergreen in nature, and the trees are often lower in stature than the rocky pinnacles that penetrate the canopy, giving the habitat a unique and picturesque appearance. Limestone forests often show a high level of plant diversity—in botanical studies of this habitat in sw. China, over 4,000 plant species were found, of which about a third were endemic or showed a preference for this habitat. These plants are known as calciphiles or limestone flora. In Malaysia, limestone forests contain 14% of the region's flora, including 130 species found only in this habitat. These include two endemic palm genera (*Liberbaileya* and *Maxburretia*) and many orchids. Some, such as the White Slipper Orchid, are in danger of extinction. Bryophytes, such as mosses and liverworts, are a noticeable feature of this habitat and can cover as much as 15–20% of the surface.

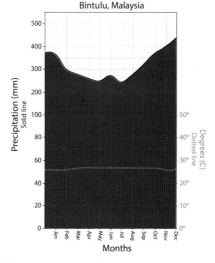

The physical nature of the dense limestone outcrops makes much of this habitat inaccessible to humans, and this offers some protection to the wildlife. While the inaccessible terrain makes entering the habitat difficult, wildlife enthusiasts with spotting scopes can enjoy views of some of the rare species of the limestone forest from viewpoints or the roadside along the edge of the habitat. Limestone forests are vulnerable to quarrying, and once the limestone features have been removed, they can never be regained and the habitat is lost. These forests are also threatened by burning and their animals by hunting.

WILDLIFE: The Limestone Wren-Babbler (IS) is found in the limestone forests of se. Asia from e. Myanmar across c. and n. Thailand, Laos, and Vietnam, and even into s. China. The limestone forest of Ban Nahin in Laos is an extremely exciting location biologically, home to some amazing recent discoveries, including Laotian Rock Rat (described in 2005), Limestone Leaf Warbler (IS) (2010), Bare-faced Bulbul (2009), and Sooty Babbler (IS) (rediscovered after 74 years in 1994 in adjacent Vietnam and then subsequently found here). Limestone Leaf Warbler's scientific species name *calciatilis* means "dwelling on limestone." Streaked Wren-Babbler and the recently discovered Nonggang Babbler also show a preference for limestone forests and have been found to nest in the rocky cavities, although nest predation is quite high. Even species not restricted to limestone forest breed readily in it. Studies in s. China showed that Fairy and Blue-rumped Pittas laid more eggs when nesting in this habitat.

Apart from the limestone specialist species, these forests also contain a selection of the surrounding habitats' wildlife. Despite being found in areas with high hunting pressure, limestone forests can retain populations of mammals due to the inaccessibility of the terrain. At Ban Nahin, Laos, Laotian Langur, Chinese Serow, and Rhesus Macaque are found, and there are unconfirmed reports of other primate species, bears, and wildcats. Limestone forests in s. China and n. Vietnam hold rare primates such as François',

Above: **Delacour's Langur lives on steep limestone cliffs within limestone forest.** © KEN BEHRENS, TROPICAL BIRDING TOURS

Below: **Bar-bellied Pitta within limestone forest in n. Vietnam.** © KEN BEHRENS, TROPICAL BIRDING TOURS

Delacour's and Hatinh Langurs. Caves within the limestone formations support important bat populations, and raptors such as Oriental Hobby and Peregrine Falcon are active at dawn and dusk to feed on them. Studies of bats at Borneo's Bau limestone area found 23 species in four families, almost a quarter of the island's total. Limestone caves on Borneo also provide nest sites for four species of swiftlets, the nests of some of which are traditionally harvested for use in bird's nest soup.

Endemism: The Bornean areas of limestone forest form an endemic area for several species of limestone orchids. In Bornean Malaysia, there are endemic genera of palms restricted to limestone forest. Limestone forests in s. China have been found to have many endemic plants. The Khammouan limestone area of Laos and adjacent Vietnam contains several range-restricted birds.

DISTRIBUTION: Limestone forests are found in several areas of se. Asia, including se. Myanmar; w., c., and n. Thailand; n. and c. Laos and Vietnam; s. China; Peninsular Malaysia and also some areas of Sabah and Sarawak in Malaysian Borneo. Some well-known sites include the Bau limestone area and Gunung Mulu National Park in Sarawak, Malaysia; Banggi Island off the northern tip of Sabah, Malaysia; Phrang Na-Ke Bang National Park and Cuc Phuong National Park in Vietnam; and Ban Nahin in Laos.

WHERE TO SEE: Ban Nahin, Laos; Gunung Mulu National Park, Sarawak, Malaysia.

INDO-MALAYAN MOIST DECIDUOUS FOREST

IN A NUTSHELL: A medium-stature, seasonal forest with sparse grass cover occurring in areas of moderate but seasonal rainfall. **Habitat Affinities:** NEOTROPICAL CERRADO. **Species Overlap:** INDO-MALAYAN DRY DECIDUOUS FOREST; INDO-MALAYAN SEMI-EVERGREEN FOREST.

DESCRIPTION: Moist deciduous forests are one of the dominant habitats of the Indo-Malayan region and home to much of its spectacular megafauna. They are closed stands with at least 80% tree cover. Trees are about 65–85 ft. (20–25m) tall, and there is usually sparse grass cover. These forests receive higher annual rainfall than INDO-MALAYAN DRY DECIDUOUS FOREST, most of which is from the southwest monsoon, usually 60–80 in. (1,500–2,000mm) per year but as low as 40 in. (1,000mm) in some areas. Rainfall is highly seasonal, and there is an extended dry season (usually of four to five months), during which many trees drop their leaves and the forest's general appearance is bare, especially in extreme years. The habitat is interspersed with riparian forest along watercourses where the trees retain their leaves all year long. Many trees are heavily buttressed, and the undergrowth is fairly thick. The forests can also contain bamboo, often in secondary areas. The tree species composition varies geographically: Teak is the dominant species in peninsular India and Thailand, while Sal is dominant in e. India and the Ganges River valley. Other important species include Indian Laurel, Kadam, and Burma Padauk. This is a fairly easy habitat to walk through, other than some dense thickets of bamboo.

As moist deciduous forests are transitional between Indo-Malayan dry deciduous forest and SEMI-EVERGREEN FOREST (or TROPICAL MONTANE RAINFOREST in the Western Ghats), they can contain species from both. Moist deciduous forests still cover large areas of the Indo-Malayan region, but many areas have been cleared for timber and/or agriculture, especially where they occur on fertile alluvial plains such as the Indo-Gangetic Plain.

WILDLIFE: Large areas of moist deciduous forest that once covered much of the Indo-Gangetic Plain in India have been cleared but formerly held much of the country's megafauna, including Indian Rhinoceros, Asian Elephant, Gaur, and Tiger. Corbett National Park retains this forest and protects populations of these animals. Other mammals here include Chital, Indian Hog Deer, and Rhesus Macaque. At the base of the Western Ghats, Mudumalai National Park has one of the largest remaining populations of Asian Elephant, and important populations of Tiger, Leopard, Dhole, and Sloth Bear. More commonly seen species are Wild Boar, Gray Langur, and Bonnet Macaque.

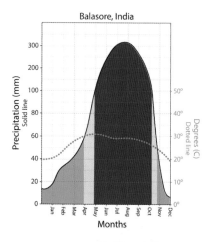

Balasore, India

Moist deciduous forest in Corbett National Park, India.
© IAIN CAMPBELL, TROPICAL BIRDING TOURS

Leopards reach their highest density in Asia within moist deciduous forest habitat in Sri Lanka.
© IAIN CAMPBELL, TROPICAL BIRDING TOURS

Asian Elephant in moist
deciduous forest in India.
© KEN BEHRENS, TROPICAL
BIRDING TOURS

Although large areas of
moist deciduous forest
remain in se. Asia, much of
the original megafauna has
been extirpated, though
smaller mammals such as
Red Muntjac and Golden
Jackal remain. At Alas
Purwo National Park in e.
Java, the huge Banteng (a
species of wild cow) can still
be seen.

Moist deciduous forests,
even though they are
transitional habitats
between DRY DECIDUOUS
FOREST and SEMI-
EVERGREEN FOREST, are
rich in birdlife. Several bird
species are restricted mainly
to them, while others are shared with one or the other of these abutting forest types. Some typical
species of moist deciduous forests include: Changeable Hawk-Eagle, Spot-bellied Eagle-Owl, and
Fulvous-breasted Woodpecker at Corbett National Park in India; Gray-headed Parakeet, Golden-
fronted Leafbird (IS), and Blue-throated Flycatcher at Ka Zun Ma in Myanmar (between Bagan and
Nat Ma Taung); Blue-throated Barbet, Red-breasted Parakeet (IS), and Collared Falconet in
Thailand; and Bali Myna and Black-winged Starling at Bali Barat National Park in Indonesia.

Endemism: Moist deciduous forests in the Western Ghats, Sri Lanka, the Andaman and Nicobar
Islands, Java, and Bali form respective endemic areas, although many of the endemic species are
shared with neighboring habitats.

DISTRIBUTION: Moist deciduous forests are widely distributed in the Indo-Malayan region. In Sri
Lanka they occur east and north of the hills. In India, they are found in a fairly narrow band around
the base of the Western Ghats, on the Indo-Gangetic Plain (although much of this fertile alluvial
plain has been cleared for agriculture), in the Eastern Ghats to Bangladesh, and on the Andaman and
Nicobar Islands. They cover large areas stretching from Myanmar to Thailand and across Indochina.
They occur locally on the island of Hainan, in w. Philippines, and on the lower slopes of e. Java and
Bali. Moist deciduous forest represents a transition from INDO-MALAYAN SEMI-EVERGREEN FOREST
on the wetter side to INDO-MALAYAN DRY DECIDUOUS FOREST on the drier side. On the steeper
slopes of the Western Ghats, moist deciduous forest transitions quickly to INDO-MALAYAN
TROPICAL MONTANE RAINFOREST.

WHERE TO SEE: Mudumalai National Park, Tamil Nadu, India; Corbett National Park, Uttarakhand,
India; Ka Zun Ma, Myanmar.

INDO-MALAYAN DRY DECIDUOUS FOREST

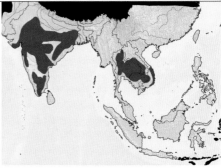

IN A NUTSHELL: A medium-stature, seasonal forest with a sparse grass cover occurring in areas of low and highly seasonal rainfall. **Habitat Affinities:** NEOTROPICAL DRY DECIDUOUS FOREST; MALAGASY DRY DECIDUOUS FOREST. **Species Overlap:** INDO-MALAYAN THORNSCRUB; INDO-MALAYAN MOIST DECIDUOUS FOREST.

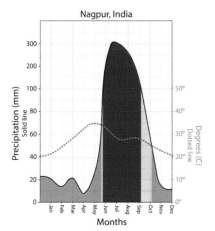

Nagpur, India

DESCRIPTION: Dry deciduous forests are found widely across the Indian subcontinent and se. Asia. These forests consist of trees that shed most of their leaves during the dry season. Those with a grassy understory can be easy to walk through, although they can also have some dense and spiky thickets that one must skirt around. For birding, winter and early spring are the best times to visit, with cooler temperatures and leafless trees that make it easier to spot birds. As in most habitats, the best time for birding is early morning, and by midmorning activity drops sharply and the forest can seem very quiet. Mixed flocks are an obvious feature of this habitat, especially in winter. In the late afternoon, the habitat is bathed in a warm light that can make for wonderful photography.

Indo-Malayan dry deciduous forests grow in areas of less rainfall than INDO-MALAYAN MOIST DECIDUOUS FOREST, with only 35–60 in. (900mm–1,500mm) annually. Most of the rainfall is concentrated in the rainy season, and the long dry season lasts several months. This habitat represents a transition between moist deciduous forests on the wetter side and INDO-MALAYAN THORNSCRUB on the drier side. The forests have a closed but uneven canopy reaching 50–80 ft. (15–25m) high, and a developed understory and undergrowth. Enough light reaches the ground to permit the growth of grass and climbing plants. The tree species composition and dominance in these forests vary geographically. In Indian dry deciduous forests, some important tree species include Teak, Axlewood, Indian Rosewood, Ceylon Satinwood, Anjan, and Bitter Albizia. In Indochina, dominant trees include Dark Red Meranti, Balau, *Dipterocarpus obtusifolius*, and Gurjun Tree. Some tree species present in dry deciduous forests are also found in moister habitats, where they may not lose their leaves to the same extent. Large tracts of this forest have been cleared for agricultural purposes, and they also suffer from overgrazing and fire.

Above: **Dry deciduous forest in n. Thailand.** © CHARLEY HESSE, TROPICAL BIRDING TOURS

Below: **Bengal Tiger occurs within much of India's protected dry deciduous forest habitat.** © IAIN CAMPBELL, TROPICAL BIRDING TOURS

WILDLIFE: India's dry deciduous forests harbor much of the subcontinent's impressive megafauna, including the only remaining Asiatic Lions, in the Kathiarbar-Gir forests; Asian Elephants, in the Chota Nagpur forests; and the magnificent Bengal Tiger, still present through many of these areas. Smaller carnivores such as Leopard, Dhole, and Sloth Bear also occur. A range of herbivores is present, including Gaur, Nilgai, Sambar, Chital, muntjac, Four-horned Antelope, Chinkara, and Blackbuck. In se. Asia, the large mammal fauna has mainly been extirpated, and the animals remaining, such as Eld's Deer, are endangered. Smaller mammals such as muntjacs are more likely to be seen here.

Some dry deciduous forest birds are found across the region, such as Yellow-footed Green-Pigeon, Small Minivet, White-browed Fantail, Rufous Treepie, Common Woodshrike, and Purple Sunbird, whereas others are more localized. Birds of the Indian dry deciduous forest include Sirkeer Malkoha, Indian Gray Hornbill, Black-rumped Flameback, White-bellied Minivet (IS), Indian Nuthatch, Indian Robin, and Yellow-throated Bush Sparrow. Dry deciduous forests in Indochina are home to Chinese Francolin, Black-headed Woodpecker (IS), Yellow-crowned Woodpecker (IS), Common Flameback, Blossom-headed Parakeet, White-rumped Falcon (IS), Indochinese Cuckooshrike, Indochinese Bushlark, Brown Prinia, and Burmese Nuthatch. A frequent bird of the edge of the forest in India is the Indian Roller. Dry forests of the Kulen Promtep Wildlife Sanctuary in Cambodia hold the last populations of the critically endangered Giant and White-shouldered Ibises. They make use of waterholes, known locally as *trapaing*, that are dotted throughout the forest.

DISTRIBUTION: Dry deciduous forests are widely distributed throughout India from the s. and c. Deccan Plateau through the Narmada Valley on the lee side of the Western Ghats to the Kathiarbar-Gir forests of the northwest and east to the Chota Nagpur Plateau. In se. Asia, they occur patchily, from the Irrawaddy dry forests of Myanmar, through n. Thailand to Cambodia and s. Laos, and a few remaining forests in s. Vietnam.

WHERE TO SEE: Gir National Park, Gujarat, India; Bandhavgarh National Park, Madhya Pradesh, India; Doi Inthanon National Park, Thailand; Tmatboey, Cambodia.

Dry deciduous forest in India is home to a good diversity of large mammals, including Gaur *(right)*, **Asian Elephant, Leopard,** and **Bengal Tiger.** © IAIN CAMPBELL, TROPICAL BIRDING TOURS

INDO-MALAYAN SEASONALLY FLOODED GRASSLAND

IN A NUTSHELL: Grasslands flooded seasonally by rivers and monsoon rains. **Habitat Affinities:** NEOTROPICAL FLOODED GRASSLAND AND WETLAND. **Species Overlap:** INDO-MALAYAN FRESHWATER WETLAND; PADDY FIELDS.

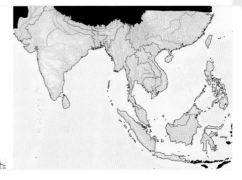

Sunset over the grasslands of Kaziranga National Park, India. © IAIN CAMPBELL, TROPICAL BIRDING TOURS

DESCRIPTION: This habitat is flooded with highly seasonal rainfall and inundation from rivers, and often contains plants adapted to such conditions. It can be important habitat for large numbers of migratory waterbirds. Some of these grasslands are found in a mosaic with forest, which increases the biodiversity of the area. Seasonally flooded grassland is relatively limited in the Indo-Malayan region. The Rann of Kutch seasonal salt marsh of Gujarat state in w. India and se. Pakistan is filled after the southwest monsoon rains from July to September.

It is flooded to a depth of 20 in. (50cm) or deeper, although some isolated areas of higher ground (known as *beyt*) serve as refuges for the area's wildlife. In the long dry season, the Rann of Kutch becomes a very dry and desolate place. Some common grass species here include *Apluda aristata*, *Dichanthium annulatum*, and Blue Panic Grass.

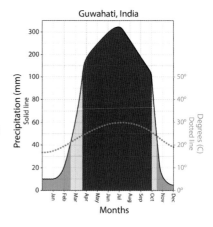

Tonlé Sap in Cambodia is se. Asia's largest lake; its surface area fluctuates between 965 sq. mi. (2,500km²) at the end of the dry season in late April to 6,200 sq. mi. (16,000 km²) at the end of September. It fills with the southwest monsoon rains, as does the Mekong River, with which it is connected. When the water recedes, it deposits sediment, giving nutrients to the exposed land. This area is very important for agriculture but also has some natural grasslands. Flooded grasslands at the base of the Himalayas in India and Nepal are the tallest in the world—indicative of the wet conditions and nutrient-rich soils. Their suitability for agriculture has meant that many areas of natural grassland have been converted. The annual monsoon floods deposit silt, and plants quickly recolonize these areas after the retreat of the water. Just a few days of flooding provide sufficient nutrients for the whole year. The grass species composition changes on the different terraces of the floodplain depending on moisture conditions. Very long grass occurs along the river's edge, while short grazing lawns

Seasonally flooded grasslands in India are home to a good variety of large mammals, like this Nilgai.
© IAIN CAMPBELL, TROPICAL BIRDING TOURS

White-throated Kingfisher is an abundant, widespread, and familiar species within this habitat on the Indian subcontinent. © IAIN CAMPBELL, TROPICAL BIRDING TOURS

are set farther back. Some important grass species include Baruwa and Kans Grass. There is also a moisture gradient from east to west throughout the range of these grasslands.

Seasonally flooded grasslands are best visited in the dry months during winter and early spring, when the habitat is the most accessible. During the flooded season, some of these reserves are closed. At the end of the dry season, animals are concentrated around remaining water, and the cooler temperatures are more pleasant for wildlife-viewing. The habitat is easy to walk through, although the presence of potentially dangerous mammals in some grasslands makes it more sensible to take a safari in a jeep.

WILDLIFE: In the Rann of Kutch, the mammal fauna numbers around 50 species, including Asiatic Wild Ass, Chinkara, Nilgai, and even Indian Wolf. Bird species number over 200 and include Lesser Florican, Macqueen's Bustard, Demoiselle Crane, and Lesser Flamingo, plus many other migratory birds in winter.

The Tonlé Sap floodplain is an important area for the critically endangered Bengal Florican; however, many natural grasslands in the area have been converted to agriculture in the last few years, and the number of remaining floricans has dropped.

Many of Asia's large mammals are found in the Terai-Duar grasslands at the base of the Himalayas, including: Asian Elephant, Indian Rhinoceros, Gaur, and Asian Buffalo. The productivity of these grasslands makes them very suitable for these large herbivores. Some other interesting small mammals are present, like Hispid Hare and Pygmy Hog, as well as some specialized grassland birds, including Bengal Florican, Black-breasted Parrotbill (IS), Marsh Babbler (IS), and Jerdon's Babbler (IS).

DISTRIBUTION: Rann of Kutch seasonal salt marsh is found in Gujarat state of nw. India and the Sind region of se. Pakistan. It is fed by the Luni River, which drains the Aravalli hills. Flooded grasslands also occur on the floodplain of Tonlé Sap in Cambodia. The Terai-Duar savanna and grassland forms a narrow belt at the base of the Himalayas only about 15 mi. (25km) wide. Essentially an extension of the Indo-Gangetic Plain, it ranges across the Indian states of Uttar Pradesh, Bihar, and West

Bengal to Assam, plus neighboring low-lying areas of Nepal. This habitat can also be found locally in small patches elsewhere.

WHERE TO SEE: Little Rann of Kutch, Gujarat, India; Chitwan National Park, Nepal; Kaziranga National Park, Assam, India; Stoung-Chikreng Bengal Florican Conservation Area, Cambodia.

SIDEBAR 3.5 ▸ WHAT IS A MONSOON?

The level of moisture is one of the most important factors in determining the habitat in any given place. There are many different phenomena that produce rainfall, including orographic rainfall on mountains (sidebar 3.3). But one of the most important contributors across the globe is the monsoons. These occur in areas with large landmasses that warm up more quickly during the summer (whether austral or boreal) than adjacent areas of ocean. As the land warms, the air begins to rise, which creates an area of low pressure and serves to pull in moist air from the cooler ocean. When this moist air arrives over land, it is drawn upward by the same dynamic, precipitating its moisture in the form of heavy rainfall. This phenomenon is powerful and pervasive enough to reach not just coastal areas but deep into continental landmasses. The term *monsoon* actually refers not just to the summer rains but also to the reverse phenomenon in winter, when a warmer ocean pulls air away from land and keeps it dry. Monsoons are a typical weather pattern of much of the tropics and subtropics, including the majority of sub-Saharan Africa, India, much of se. and e. Asia, Australia, se. South America, and w. Mexico and the sw. United States. Most areas with a monsoonal climate support savanna or dry deciduous forest habitats, which have evolved to survive sharply different wet and dry seasons and the latter's associated fires. Some monsoonal areas that manage to support forest have a secondary source of moisture to help them through the dry portion of the monsoonal cycle.

May *(top)* **and August** *(above)* **in the Western Ghats of sw. India.** © ARNE HÜCKELHEIM, WIKIMEDIA COMMONS, HTTPS://EN.WIKIPEDIA.ORG/WIKI/MONSOON#/MEDIA/FILE:MATHERANPANORAMAPOINTDRYSEASON.JPG, HTTPS://EN.WIKIPEDIA.ORG/WIKI/MONSOON#/MEDIA/FILE:MATHERANPANORAMAPOINTMONSOON.JPG

INDO-MALAYAN MONTANE GRASSLAND

IN A NUTSHELL: Grass-dominated habitat, found primarily above the tree line at high elevations but also locally down to middle elevations. **Habitat Affinities:** AFROTROPICAL MONTANE GRASSLAND. **Species Overlap:** INDO-MALAYAN SEASONALLY FLOODED GRASSLAND.

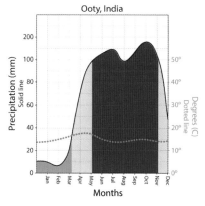

Ooty, India

Grasslands in Horton Plains National Park on the island of Sri Lanka, where Hill Swallows often hawk overhead.
© KEN BEHRENS, TROPICAL BIRDING TOURS

DESCRIPTION: Montane grasslands occur above the tree line, where temperature, precipitation, or soil conditions prevent tree growth; the elevation of the tree line varies greatly. This upper montane habitat stretches from the tree line up to where the soil cover disappears. There are very few naturally occurring montane grasslands in the Indo-Malayan region. One such example is the alpine meadows of Mt. Kinabalu on Borneo, found mainly at 8,500–10,500 ft. (2,600–3,200m) but also higher up in places that retain soil. At such elevations, temperatures are usually 43–57°F (6–14°C) but can drop to as low as

32°F (0°C) at night. The plant growth here consists mainly of shrubs, mosses, liverworts, ferns, and grasses as well as lichens. Even in the tropics, the open landscapes and fresh mountain air of these montane grasslands have a familiar feel for visitors from more northerly climes. The spectacular mountain scenery, cool temperatures, and ease of viewing wildlife make these locations very pleasant places to visit.

Another example is the shola grasslands of India. These are areas of stunted montane forest found in a matrix with rolling grasslands in the southern hills. They are found across the states of Karnataka, Kerala, and Tamil Nadu, and usually occur above 6,600 ft. (2,000m), although in some places as low as 5,250 ft. (1,600m). Some believe that these grasslands may have been created and maintained by early pastoralists, but pollen analysis has shown that grasslands existed here tens of thousands of years ago. The shola montane grasslands have frost and fire-resistant grass species including *Chrysopogon zeylanicus*, *Cymbopogon flexuosus*, and *Arundinella ciliata*. Montane grasslands of Sri Lanka are dominated by *Arundinella villosa* and *Chrysopogon zeylanicus*. These are maintained by fire, and there are also conflicting views about how they were originally formed. The tree line in Taiwan is at about 11,800 ft. (3,600m), and the highest mountain, Yushan, is 12,966 ft. (3,952 m) tall. Montane grassland cloaks this upper zone on the highest peaks, although it also reaches farther down locally.

WILDLIFE: Mt. Kinabalu holds an incredible diversity of plants, numbering more than 4,500 species, and has one of the highest concentrations of endemic plant species in the world. It is particularly rich in orchid species and also famous for pitcher plants (*Nephentes* spp.), some of which reach the alpine meadows.

Bird species that live in this habitat include Friendly Bush Warbler and Island Thrush. Nilgiri Tahr (IS) is found in the shola grasslands of India. Nilgiri Pipit (IS), Hill Swallow, Pied Bushchat, and Indian Blackbird also occur in this habitat. Montane grasslands in Horton Plains National Park of Sri Lanka have much species overlap with the shola grasslands (although Nilgiri Pipit is not present). A sizable population of Sambars feed on the Horton Plains grasslands.

Hehuanshan is an easily accessible montane grassland site in Taiwan. Some common species of birds include the endemics Taiwan Bush Warbler, Taiwan Rosefinch, and a subspecies of the Alpine Accentor. Taiwan Alpine Skink can also be found on rocky areas in the grassland.

Endemism: Nilgiri Tahr is restricted to the shola grasslands of s. India. Taiwan Bush Warbler is restricted mostly to the montane grasslands of Taiwan but also occurs in upper montane forest. The alpine meadows of Mt. Kinabalu hold several endemic plants and snails.

Nilgiri Tahr is restricted to the shola grasslands of s. India.
© KEN BEHRENS, TROPICAL BIRDING

DISTRIBUTION: Alpine meadows are found on Mt. Kinabalu, in the state of Sabah in Malaysian Borneo. Shola grassland mosaics are found in the hilly areas of Karnataka, Kerala, and Tamil Nadu in s. India. Some montane grasslands also occur in the mountains of s. Sri Lanka and the highest mountains in Taiwan, as well as locally on isolated peaks that reach above the tree line in other parts of the region, such as Mt. Apo and Mt. Pulag in the Philippines, Nat Ma Taung (Mt. Victoria) in Myanmar, and Mt. Kerinci in Sumatra.

WHERE TO SEE: Eravikulam National Park, Kerala, India; Horton Plains National Park, Sri Lanka; Kinabalu National Park, Sabah, Malaysia; Hehuanshan, Taiwan.

INDO-MALAYAN FRESHWATER SWAMP FOREST

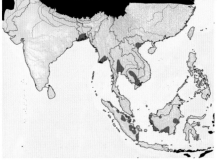

IN A NUTSHELL: Forests that are permanently or seasonally flooded with fresh water and lack peat deposits. **Habitat Affinities:** AFROTROPICAL SWAMP FOREST. **Species Overlap:** INDO-MALAYAN PEAT SWAMP FOREST; INDO-MALAYAN MANGROVE FOREST.

DESCRIPTION: Freshwater swamp forests usually occur where rivers pass through low-lying, flat, alluvial plains before reaching INDO-MALAYAN MANGROVE FOREST. These forests lack the deep peat layer of INDO-MALAYAN PEAT SWAMP FOREST. Due to the anaerobic conditions of the flooded soils, many species have developed pneumatophores, specialized structures to supply oxygen to the trees. Freshwater swamp forests are under great threat and have been cleared in many areas, due to the soils suitability for agriculture. In undisturbed habitats, the nutrient-rich soils produce taller trees and more diverse and productive forests than peat swamp forests. Tree size varies geographically: in Borneo, average tree height is about 115 ft. (35m), but at Tonlé Sap in Cambodia, where the forest is more stunted, trees reach just 24–50 ft. (7–15m).Tree species composition also differs in each area, although the presence of pandanus trees seems to be a common feature. At Ratargul, the last remaining freshwater swamp forest in Bangladesh, Koroch Tree is dominant. At Tonlé Sap in Cambodia, the main species are Freshwater Mangrove and *Diospyros cambodiana*. There is a very rare form of freshwater swamp forest in sw. India dominated by *Myristica* trees, although many other species are also present. Indonesian freshwater swamp forests are diverse but can be dominated by Weeping Paperbark in Sumatra and *Mallotus* species in s. Borneo.

The depth of water and amount of time trees are flooded also vary. At Ratargul, the forest can be flooded to a depth of 20–30 ft. (6–9m) in the rainy season, and forests in Tonlé Sap can be flooded

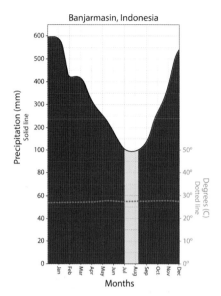

Banjarmasin, Indonesia

for six to eight months of the year. Naturally, freshwater swamp forest is best accessed by boat. As the water recedes toward the end of the dry season, it can become shallow, and the fish are more concentrated. This can sometimes lead to feeding frenzies of waterbirds, which provide wonderful photographic opportunities.

WILDLIFE: There is little endemism within freshwater swamp forests, although the widely spread patches do hold very different assemblages. In Indonesia, many species of mammals from neighboring INDO-MALAYAN TROPICAL LOWLAND RAINFOREST also inhabit freshwater swamp forests. At Way Kambas National Park in Sumatra, Asian Elephant, Sumatran Rhinoceros, Malayan Tapir, and Sumatran Tiger persist in small numbers. At Sungai Wain, a protected forest in Kalimantan, Indonesian Borneo, some regularly seen species include Bornean Orangutan, Red Leaf Monkey, and Pen-tailed Tree Shrew; Sun Bears are also present in the reserve and may forage in this habitat. Tonlé Sap in Cambodia, while not home to many mammal species, may still hold the critically endangered Siamese Crocodile.

Freshwater swamp forests can be havens for waterbirds and other species. At Prek Toal Bird Sanctuary in Tonlé Sap, Spot-billed Pelican, Greater and Lesser Adjutants and other storks, cormorants, and herons have large nesting colonies in the treetops. In Indonesia, the near-mythical Bornean Ground-Cuckoo seems to prefer alluvial forests, including freshwater swamp forests or at least the borders of them. The endangered White-winged Duck (IS) also exhibits a preference for this habitat

Above: **Bornean Orangutan occurs in high densities within Borneo's freshwater swamp forests.** © SAM WOODS, TROPICAL BIRDING TOURS

Right: **Freshwater swamp forest at Ratargul, ne. Bangladesh.** © M MAHFUZUL ISLAM RAHAT, LICENSED UNDER CC BY-SA 4.0

in the Sumatran part of its range, including Way Kambas National Park. Other typical species of the freshwater swamp forest habitat are Cinnamon-headed Green-Pigeon, Gray-headed Fish-Eagle, Stork-billed Kingfisher, Black-and-red Broadbill, and White-chested Babbler.

DISTRIBUTION: The main Indo-Malayan freshwater swamp forests, all of which are highly endangered, occur along the eastern alluvial plain of Sumatra, in s. Borneo, and around Tonlé Sap in Cambodia. This habitat previously existed along the Red River in n. Vietnam, the Irrawaddy River of Myanmar, the Chao Phraya River of Thailand, and behind the Sundarbans mangroves between Bangladesh and West Bengal in India, but very little remains at these sites. The Ratargul swamp forest on the Gowain River in ne. Bangladesh is one of the last remaining examples in the country. The rare form of *Myristica*-dominated freshwater swamp forest persists in remnant patches in Karnataka and Kerala states in sw. India.

WHERE TO SEE: Prek Toal Bird Sanctuary, Cambodia; Way Kambas National Park in Sumatra, Indonesia; Sungai Wain Protection Forest, Kalimantan, Indonesia.

INDO-MALAYAN FRESHWATER WETLAND

IN A NUTSHELL: Various aquatic habitats, including rivers, lakes, and marshes. **Habitat Affinities:** NEOTROPICAL FLOODED GRASSLAND AND WETLAND. **Species Overlap:** INDO-MALAYAN SEASONALLY FLOODED GRASSLAND; PADDY FIELDS.

DESCRIPTION: Freshwater wetland is a very broad category comprising many distinct habitat types. Some Indo-Malayan freshwater aquatic habitats are covered separately, such as PADDY FIELDS, SEASONALLY FLOODED GRASSLAND, FRESHWATER SWAMP FOREST, and PEAT SWAMP FOREST. Here we cover the remaining freshwater habitats, including rivers, lakes, marshes, and swamps. In

Man-made bunds are a widespread form of wetland in India.
© IAIN CAMPBELL, TROPICAL BIRDING TOURS

most habitats, water is the limiting factor for the abundance of life. The more available, the lusher the vegetation and the more animal life it supports. Wetlands have a preponderance of water, at least seasonally. This creates some very productive environments, often with large concentrations of wildlife. Wetlands are also very important for human populations, providing drinking water for people and domestic animals; food, either in the form of fish, forage for domestic animals, or water for irrigating crops; and shelter, either as building materials or accommodation for houseboats.

Rivers vary greatly in nature from crystal-clear, fast-flowing mountain streams to broad, slow-moving, sediment-filled rivers on plains. They are dynamic environments where the water changes in volume with the seasons, and they even change course through the years. Wildlife must be able to track these changes, and most species found on rivers have fairly broad ranges. Humans have sought to control these natural fluctuations by damming rivers and concreting their banks, which has had severe effects on wildlife. Natural marshes that once occurred widely on the plains of Asia have suffered greatly, and most have been converted to rice cultivation to feed the populace.

WILDLIFE: Wetlands are among the easiest places to observe, enjoy, and photograph wildlife. Birds are especially numerous and diverse, and there are species to fill every wetland niche available, including rushing streams, exposed rocks, open water, underwater, muddy riverbanks, river islands, lake edges, stony riverbanks or islands, and reed beds. Some common groups of waterbirds include ducks, herons, shorebirds, rails, and kingfishers. Wetlands can be very important wintering sites for migratory birds. Mammals, while not as numerous, include several species of otters and even a few freshwater cetaceans. Even species not tied closely to wetlands utilize them for drinking and bathing. Freshwater wetlands are of course vitally important for fish and also for amphibians that require water to complete their breeding cycles.

On rivers at the base of the Himalayas, such as at Nameri National Park in ne. India, species like Ruddy Shelduck, River Tern, and the unique Ibisbill can be seen. On the Mekong River in Chiang Saen, n. Thailand, are River Lapwing, Small Pratincole, and Gray-throated Martin. Much farther downstream, in Cambodia, one can see the recently described Mekong Wagtail and endangered Irrawaddy River Dolphin. The National Chambal Sanctuary in nc. India protects the critically endangered Gharial and Red-crowned Roofed Turtle. The Ganges River Dolphin, another endangered cetacean, is also found here, along with interesting birds like Indian Skimmer, Great Thick-knee, and Pallas's Gull.

Largest among the important lakes of the Indo-Malayan region is the huge Tonlé Sap in Cambodia, where the Prek Toal Bird Sanctuary contains many water channels providing habitat for Greater and Lesser Adjutants, Spot-billed Pelican, and Painted and Milky Storks. Thailand's largest lake, Bueng Boraphet, is famous for the White-eyed River Martin, which was discovered here in 1968 but last seen in 1980 and now presumed extinct. This huge lake also held the critically endangered Baer's Pochard, also heading toward extinction. Large numbers of ducks still occur here, including Lesser Whistling-Duck, Cotton Pygmy-Goose, and some Ferruginous Ducks. Gray-headed Swamphen and

The weird, narrow-snouted Gharial, here pictured with Eurasian Spoonbills, occurs along rivers in India. © IAIN CAMPBELL, TROPICAL BIRDING TOURS

Pheasant-tailed Jacana prefers well-vegetated wetlands, such as this human-made tank in Sri Lanka. © SAM WOODS, TROPICAL BIRDING TOURS

Bronze-winged and Pheasant-tailed Jacanas patrol the floating vegetation.

Among well-established, human-made wetlands, the dam at Ang Trapeng Tmor in Cambodia supports birds like Sarus Crane, Black-necked Stork, and Watercock. The most famous freshwater wetland site in the whole region is probably Bharatpur, a mecca for birders, hosting such species as Bar-headed Goose, Knob-billed Duck, Great White and Dalmatian Pelicans, Black-headed Ibis, Eurasian Spoonbill, Eurasian Marsh-Harrier, and Blyth's and Clamorous Reed Warblers.

DISTRIBUTION: Some important rivers in the region include the Indus in Pakistan, the Ganges and the Brahmaputra in India, the Irrawaddy in Myanmar, the Mekong in Indochina, and the Yangtze of China, which forms the northern border of the Indo-Malayan realm. All these rivers have their sources in mountains, and the ecosystems change in nature along their lengths. Large lakes include the enormous Tonlé Sap in Cambodia, whose volume and surface area expand and contract with the seasons, and Bueng Boraphet in c. Thailand. Some dams create large reservoirs that also support wildlife, especially well-established ones like Ang Trapeng Tmor in Cambodia. Small ponds and canals can be found anywhere in the region and almost always contain some sort of aquatic wildlife.

WHERE TO SEE: Bueng Boraphet, Thailand; Keoladeo Ghana National Park (Bharatpur), Rajasthan, India; Prek Toal Bird Sanctuary, Cambodia; National Chambal Sanctuary, Rajasthan, Madhya Pradesh, and Uttar Pradesh, India.

INDO-MALAYAN MANGROVE FOREST

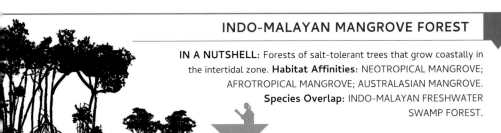

IN A NUTSHELL: Forests of salt-tolerant trees that grow coastally in the intertidal zone. **Habitat Affinities:** NEOTROPICAL MANGROVE; AFROTROPICAL MANGROVE; AUSTRALASIAN MANGROVE. **Species Overlap:** INDO-MALAYAN FRESHWATER SWAMP FOREST.

DESCRIPTION: Areas of transition from freshwater to marine systems, mangroves are found in salt or brackish tidal waters of tropical or subtropical regions. Due to their wide geographic range, they vary greatly in average temperature and rainfall, although this habitat is not reliant on rainfall for its water. Mangrove trees have stilt roots or specialized roots called pneumatophores, which grow up from the anaerobic mud to provide oxygen to the trees. They also have pores in the bark (lenticels) that allow absorption of air. Mangrove forests have a single canopy layer, and species often form pure stands. Those of the Indo-Malayan region are among the most diverse in the world; the dominant mangrove genera include *Avicennia*, *Rhizophora*, and *Brugueira*. The individual species' distributions depend on water level, salinity, and tidal regime. Red Mangroves (*Rhizophora mangle*)

Above: **Mangrove forest at Laem Pak Bia, in n. Thailand.**
© CHARLEY HESSE, TROPICAL BIRDING TOURS

Right: **Mangrove Pitta, a mangrove habitat specialist, in s. Thailand.**
© KEN BEHRENS, TROPICAL BIRDING TOURS

live in the most inundated areas and prop themselves out of the water on stilt roots, whereas Black Mangroves (*Avicennia germinans*) live on higher ground with more exposed mud.

Mangroves are vital for many species of fish and crustaceans that spend their juvenile stages among the roots. They are very important for people, as they protect coastlines from strong seas. Many mangroves in the region are under threat from clearing for firewood, timber, and alternative land use including agriculture and aquaculture. A number have been cleared for shrimp farms, for example. Mangroves have recently been found to sequester carbon dioxide from the air more efficiently than most other forests, and in some places, reforestation projects are underway. Covered in water or thick mud and having big tangles of roots, mangrove forest is not always easy to explore. Some reserves have boardwalks, which are an excellent resource and allow easy access. The best way to explore this habitat, however, is by boat.

WILDLIFE: Mangroves are not as diverse as other terrestrial systems but are home to some specialized wildlife. Some of the most easily seen inhabitants of the Indo-Malayan mangroves include mudskippers (family Oxudercidae), fish that can survive both in and out of the water; mangrove crabs, which dig burrows in the mud; and Asian Water Monitors, which are excellent swimmers and can exceed 6 ft. (2m) long. The crab larvae form a major food source for juvenile fish in the mangroves. Long-tailed Macaques (aka Crab-eating Macaques) are commonly seen in many mangrove areas in this region. The Sundarbans mangroves of e. India and Bangladesh are the only ones in the world to hold Tigers and are also home to the Saltwater Crocodile.

The mangroves of Myanmar, coastal Thailand, and Peninsular Malaysia have the large Brown-winged Kingfisher and the spectacular Mangrove Pitta, plus other species including Golden-bellied

Gerygone (IS), Copper-throated Sunbird, Mangrove Blue Flycatcher (IS), and Mangrove Whistler. The mangroves of the Indochinese Peninsula are home to the endangered Storm's Stork, and some mangroves in India support the rare Spot-billed Pelican. Mangroves often serve as rookeries and host nesting and roosting colonies of many waterbirds, including herons, storks, and cormorants. They are also important sites for birds on migration.

Endemism: Mangrove tree species tend to be widely distributed. Some mangrove bird species can be limited to certain areas. For example, Javan mangroves, which contain the range-restricted Sunda Coucal and Javan White-eye, form a small area of endemism.

DISTRIBUTION: Mangroves are found in many coastal areas of the region: coastal Pakistan and Gujarat, India; around the coast of Sri

Proboscis Monkey is confined to the island of Borneo, where it reaches its highest densities in mangrove forest and adjoining swamp forest.
© IAIN CAMPBELL, TROPICAL BIRDING TOURS

Lanka; along the east coast of India, including the Sundarbans mangroves of e. India and Bangladesh (the world's largest mangrove system); along the Myanmar coast south to Peninsular Malaysia; around the Gulf of Thailand; at the mouth of the Red River in n. Vietnam; s. China and Hainan Island; around Borneo and the east coast of Sumatra; along the north coast of Java; in the Philippines; along the west coast of Taiwan; and on the Ryukyu Islands of Japan. The most northerly natural mangroves in the world occur on the Japanese island of Tanegashima, north of the Ryukyu chain, at a latitude of 30°N.

WHERE TO SEE: Ao Phang Nga National Park, Thailand.

INDO-MALAYAN TIDAL MUDFLAT AND SALT PAN

IN A NUTSHELL: Mudflats are areas of nutrient-rich mud found in the intertidal zone of coastal areas. Salt pans are natural or human-made depressions filled with seawater and left to dry out to produce salt. **Habitat Affinities:** AFROTROPICAL TIDAL MUDFLAT AND SALT MARSH. **Species Overlap:** INDO-MALAYAN ROCKY COASTLINE AND SANDY BEACH; INDO-MALAYAN MANGROVE FOREST.

DESCRIPTION: Mudflats are a natural habitat, while salt pans can be natural or human-made. They are considered together here because they often hold the same set of species. Mudflats are found in the intertidal zone and are formed when sediment is brought down by a river and deposited in the river mouth or is deposited along the shore by tidal drift. The mudflat is exposed, usually twice a day, by tidal action. The nutrient-rich mud is full of invertebrates that serve as food for large numbers of shorebirds. Human-made salt pans are flat, square areas, surrounded by a low wall, or bund, and filled with seawater (usually by an electric pump and pipe), then left to evaporate. When dry, the salt is scraped off, taken away, and refined. When the water reaches a favorable depth and salt concentration, shorebirds flock to the pans to feed, finding in them a suitable substitute for the ever-decreasing mudflats. Mudflats are under pressure from land reclamation and coastal development, and now there is even a move away from traditional salt farms in favor of other forms of land use.

WILDLIFE: Indo-Malayan mudflats and salt pans are key sites for shorebirds. The eastern coastal side of the region forms part of the East Asia–Australasia Flyway, which is the main migration route for around 50 million migratory waterbirds that move between their wintering grounds in Australia and se. Asia to their breeding grounds in ne. Asia and even as far as Alaska. Dozens of species winter on the mudflats and salt pans of the Indo-Malayan region, including plovers, sandpipers, godwits, curlews, knots, dowitchers, and stints. Species of conservation importance include: Far Eastern Curlew, Great Knot, Nordmann's Greenshank, Asian Dowitcher, and last but not least, the unique Spoon-billed Sandpiper, which teeters on the brink of extinction. Other birds that use these habitats include gulls, terns, cormorants, herons, egrets (such as Chinese Egret), and kingfishers. Black-faced Spoonbill is another endangered species that feeds on mudflats; the majority of its population winters in Taiwan and in spring flies up to ne. China and Korea to breed.

Above: **These salt pans near Bangkok, Thailand, host thousands of wintering shorebirds and waterbirds, including globally threatened species like Chinese Egret, Great Knot, Far Eastern Curlew, Asian Dowitcher, and Spoon-billed Sandpiper.** © SAM WOODS, TROPICAL BIRDING TOURS

Left: **The endangered Nordmann's Greenshank and Great Knot mixing with more common shorebirds, like Terek Sandpiper, Ruddy Turnstone, and Black-bellied Plover.** © SIMON BUCKNELL

DISTRIBUTION: Mudflats are found locally in all coastal areas of the region. Some important areas include the Gulf of Thailand, the southeast coast of China, Hong Kong, and the west coast of Taiwan. Human-made salt pans, or salt farms, are also distributed throughout the region. Ecologically, the Gulf of Thailand is one of the most important areas, holding up to half a million shorebirds in winter.

WHERE TO SEE: Pak Thale, Thailand; Minjiang Estuary, China; Aogu, Taiwan.

INDO-MALAYAN ROCKY COASTLINE AND SANDY BEACH

IN A NUTSHELL: Coastal areas with poor rocky or sandy soil, usually lacking vegetation. **Habitat Affinities:** AFROTROPICAL ROCKY COASTLINE AND SANDY BEACH. **Species Overlap:** INDO-MALAYAN TIDAL MUDFLAT AND SALT PAN.

DESCRIPTION: Much of the shoreline in the Indo-Malayan region is rocky coastline or sandy beach. Sand comes from the weathering of rocks, shells, and coral, and is deposited by wave action. Sand beaches lack the nutritious sediment deposited on mudflats. Like INDO-MALAYAN TIDAL MUDFLATS, sandy shorelines are an intertidal habitat that is covered and exposed on a daily basis. The same goes for rocky shorelines, although these are more structurally diverse and may trap sea life in pools at low tide. Sand beaches are not a very rich habitat for wildlife but are nevertheless important for certain species of birds and as nesting sites for sea turtles. Sand spits and sandbars are extensions of sandy beaches, and their more isolated nature makes them popular roosting sites for seabirds like terns. Idyllic locations with palm-fringed, white-sand beaches are coveted by holiday makers, and this has led to wide-scale coastal development of this habitat in the region.

WILDLIFE: Despite their popularity with vacationers, sandy beaches are not a particularly productive habitat for wildlife—INDO-MALAYAN TIDAL MUDFLATS hold many more shorebirds. Nonetheless, certain species, such as Sanderling, Ruddy Turnstone, Malaysian Plover (IS), and the White-faced subspecies of Kentish Plover, do show a preference for sand beach. The distinctive Crab-Plover, although it feeds on mudflats, likes to roost on sand spits and sandbars. Sand spits also make popular roosts for Great Crested, Lesser Crested, Sandwich, Roseate, Black-naped, and Little Terns. These are often seen feeding along sandy shorelines, as is the huge White-bellied Sea-Eagle. Beach Thick-knee, as the name suggests, inhabits beaches, where it likes to feed on crabs. Sensitive to disturbance, it is now mainly restricted to remote beaches or small islands.

Malaysian Plover on a sandy beach in Thailand.
© CHARLEY HESSE, TROPICAL BIRDING TOURS

Rocky shorelines are favored by Pacific Reef-Heron (IS), which hunts for prey washed up on the rocks or trapped in the pools.

Sandy beaches throughout the region provide breeding habitat for Green, Loggerhead, Olive Ridley, Hawksbill, and Leatherback sea turtles. They are all threatened with extinction due to coastal development, hunting, and egg collecting.

DISTRIBUTION: Sandy shorelines are a common coastal habitat in the Indo-Malayan region. Long expanses of sandy coastline occur in India's Gujarat state; on the east coast of India; around Sri Lanka; on the east coasts of Peninsular Malaysia and Thailand; and in coastal Vietnam northward to coastal China. Many small OFFSHORE ISLANDS also have sandy shores; they are covered separately. Rocky coastlines and beaches are interspersed with INDO-MALAYAN MANGROVE FOREST, and TIDAL MUDFLAT and salt marsh in estuarine areas and gulfs.

WHERE TO SEE: Laem Pak Bia, Thailand; Bundala National Park, Sri Lanka.

INDO-MALAYAN PELAGIC WATERS

IN A NUTSHELL: The deep waters lying off the continental shelf in the Indian and Pacific Oceans. **Habitat Affinities:** AFROTROPICAL PELAGIC WATERS. **Species Overlap:** INDO-MALAYAN OFFSHORE ISLANDS.

DESCRIPTION: Landmasses of the Indo-Malayan region are surrounded by the Indian Ocean in the west and the Pacific Ocean in the east. The best areas for oceanic wildlife are beyond the continental shelf, where the shallow seabed drops steeply to deeper water. Here, upwellings of nutrient-rich water form the base of the food chain and allow abundant marine life to exist. Many areas of Indo-Malayan oceans have warm, shallow seas and hence little ocean life. In this region, dedicated boat trips rarely go offshore in search of pelagic birds, although the seas are crisscrossed by ferries serving the many islands. These usually are not very productive, but you can sometimes get lucky. Some cetaceans occur, and whale-watching trips are offered in some areas.

WILDLIFE: In se. Asia, some common seabirds, such as Brown Noddy; Sooty, Bridled, Roseate, Black-naped, and Great Crested Terns; Brown and Red-footed Boobies; and Lesser Frigatebird are found year-round across shallow and deep waters alike. These are joined by nonbreeding terns, jaegers, and some tubenoses, including Streaked and Wedge-tailed Shearwaters, Bulwer's Petrel, and Wilson's Storm-Petrel (IS). Some birds mainly restricted to deepwater areas beyond the continental shelf include White-tailed and Red-tailed Tropicbirds, Masked Booby, and a variety of tubenoses. Pelagic birding trips are rarely offered in se. Asia, but some off India see birds including Wilson's and Swinhoe's Storm-Petrels, Jouanin's Petrel, Flesh-footed Shearwater, Parasitic Jaeger, South Polar Skua, and Bridled Tern. More sea-watching takes place off Taiwan, and ferries linking outlying islands often record Bulwer's Petrel, Streaked and Wedge-tailed Shearwaters, and Brown Booby.

Some widespread marine mammals such as Humpback and Blue Whales and Spinner and Common Bottlenose Dolphins are found in the oceans of this region. Whale-watching trips depart from Mirissa on the south coast of Sri Lanka; November to April is the best time, when Blue Whales are regularly seen. Whale-watching is also possible off Okinawa, in the very south of Japan, where Humpback Whales are present from January to March.

DISTRIBUTION: The Indo-Malayan region is characterized by shallow seas in many areas, including both sides of the Malay Peninsula and east to Vietnam, then south to the area bordered by Sumatra, Java, and Borneo. These areas were historically joined during times of low sea level, and this strongly

A pelagic feeding frenzy, with birds and tuna hunting.
© LISLE GWYNN, TROPICAL BIRDING TOURS

influenced the biogeography of the region. West of Sumatra, south of Java, east of Borneo, and east of the Philippines, the continental shelf drops steeply, and there is deeper water close to the coast. The n. Indian Ocean is also fairly shallow on both sides of the subcontinent but gets deeper close to Chennai, in se. India, and on the south coast of Sri Lanka.

WHERE TO SEE: Whale-watching trip from Mirissa, Sri Lanka.

INDO-MALAYAN OFFSHORE ISLANDS

IN A NUTSHELL: Small offshore islands that provide refuge for oceanic wildlife and a small selection of specialist species. **Habitat Affinities:** AFROTROPICAL OFFSHORE ISLANDS. **Species Overlap:** INDO-MALAYAN TROPICAL LOWLAND RAINFOREST; INDO-MALAYAN SANDY BEACH.

DESCRIPTION: The Indo-Malayan region is dotted with islands of varying sizes. There are the huge islands of the Greater Sundas (Sumatra, Borneo, and Java) and the main islands of the Philippines (Luzon and Mindanao); and the smaller islands just off the continental mainland, which contain similar habitat and smaller subsets of the same wildlife as the mainland, with a few endemics thrown in. Finally, there are the numerous small offshore islands, many of which were formed by the breakdown of coral reefs, while others have volcanic origins. Although these small offshore islands look insignificant on a map, they are important habitat for some breeding seabirds and nesting sea turtles. Seabirds preferentially nest on them, as they were originally free of predators, although the introduction of non-native animals like rats and feral cats causes serious conservation challenges. Some of these small islands contain specialist birds that are found on other distant small islands but not on the adjacent mainland or larger islands. This is a strange

Nicobar Pigeon, despite being named for one Indian island group, is distributed across many offshore islands in Asia. © CHARLEY HESSE, TROPICAL BIRDING TOURS

phenomenon, and these species are known as "supertramps." Although supertramps are much more numerous in Australasia, there are several interesting examples in this region.

WILDLIFE: Christmas Island has many seabird colonies and hosts an impressive three species each of both boobies and frigatebirds, including the critically endangered Abbott's Booby and Christmas Island Frigatebird. It also has White-tailed and Red-tailed Tropicbirds and Brown Noddy. The beautiful White Tern breeds on the Cocos (Keeling) Islands. The Mu Ko Similan National Park, off the west coast of peninsular Thailand, is an easy place to see Nicobar Pigeon (IS) and Pied Imperial-Pigeon. Nesting colonies of seabirds occur on several islands around Borneo, including Pulau Lungisan (Mantanani Island), which has up to three species of frigatebirds and also hosts small island specialists like Tabon Scrubfowl, Metallic Pigeon, Gray Imperial-Pigeon, and Mantanani Scops-Owl. Other small islands around Palawan, Philippines, also host Gray Imperial-Pigeon and Mantanani Scops-Owl.

Green, Loggerhead, Olive Ridley, Hawksbill, and Leatherback sea turtles are found around the coasts of the Indo-Malayan region. Sandy offshore islands provide crucial breeding habitat for these species, all of which are endangered.

The Island Flying Fox roosts on many of these small offshore islands and flies across to larger bodies of land to feed at night.

Endemism: The remote Cocos (Keeling) Islands and Christmas Island each hold several endemic birds, but most of the smaller offshore islands contain species that are more widely distributed.

DISTRIBUTION: There are many thousands of small offshore islands in this region, and here we can name but a few. Starting with the real outliers: the Cocos (Keeling) Islands and Christmas Island in the Indian Ocean are both Australian external territories. In addition to the larger Andaman and Nicobar island groups, there are many small islands in the Andaman Sea off the coast of peninsular Myanmar, continuing south to the Surin and Similan Islands off peninsular Thailand. Farther south are several small islands that have been developed for tourism, like Ko Phi Phi, and on the other side of the peninsula are many more, including Ko Tao (a popular diving destination). Farther south, off the east coast of Peninsular Malaysia, are the Perhentian Islands. Borneo is surrounded by small offshore islands, some of the better known of which are Satang and Talang Islands off Sarawak; Manukan Island, the Mantanani Islands, and the Turtle Islands off Sabah; and the Derawan Islands off e. Kalimantan. There are several accessible small islands off Sumatra, Java, and Palawan and elsewhere in the Philippine island group.

WHERE TO SEE: Mu Ko Similan National Park, Thailand.

PADDY FIELDS AND OTHER CROPLAND

IN A NUTSHELL: A human-made habitat where rice is cultivated. **Habitat Affinities:** AFROTROPICAL CROPLAND. **Species Overlap:** INDO-MALAYAN FRESHWATER WETLAND; INDO-MALAYAN SEASONALLY FLOODED GRASSLAND.

Paddy fields in n. Thailand.
© SAM WOODS, TROPICAL BIRDING TOURS

DESCRIPTION: In tropical Asia, paddy fields are such an integral part of the landscape that they merit a separate account. The word "paddy" comes from *padi*, the Malay word for rice plant. Rice cultivation began when Wild Rice was first domesticated, probably around the Yangtze River in China, possibly more than 8,000 years ago; the oldest known paddy fields have been dated to at least 6,000 years ago. Paddy fields are now one of the main agricultural land uses, and rice forms the staple food for most countries in the Indo-Malayan region. Rice is a tropical crop and requires sufficient sunlight and warmth (temperatures above an average of 68°F/20°C) during the growing months. It also requires a level surface, sufficient rainfall (annual minimum 45 in./1,150mm), flooded conditions, suitable soils, and sufficient nutrients. Because of these requirements, the deltas and alluvial plains of Asia's great rivers have been the stronghold of rice cultivation. With development of irrigation techniques, paddy fields expanded into other flat areas and, with terracing, into the hills. Rice plants need to be covered by 4–6 in. (10–15cm) of water for about three-quarters of the growing season. They are usually flooded by rainfall and rivers during the monsoon, which dictates the growing season. Paddy fields have almost completely replaced the natural swamps and flooded forests that once occurred in this region. Many animal species have adapted well to living in rice paddies, although they are not a natural habitat. However, recent changes in rice cultivation, including breeding of new strains that require less water, plus the development of artificial fertilizers and pesticides, have seriously affected wildlife populations that live in and around paddy fields.

Other major monoculture crops of se. Asia include tea and African Oil Palm. The oil palm industry in particular has wreaked havoc on INDO-MALAYAN TROPICAL LOWLAND RAINFOREST and has nearly completely destroyed the lowland rainforests of Borneo and mainland Malaysia. Cropland on the Indian subcontinent is far more varied, with a patchwork of small mustard, millet, wheat, and corn farms, which allow for significant wildlife. Much of the nature observation in the plains of n. and nw. India takes place completely within farmed areas.

WILDLIFE: Paddy fields are an important habitat for birds, and two species even carry its name: Paddyfield Warbler and Paddyfield Pipit. The available habitat changes significantly with different stages of cultivation. During the nongrowing season, dry, bare fields are habitat for Paddyfield Pipit, Eastern Yellow Wagtail, and Oriental Skylark. When the fields are flooded, they provide a wonderful habitat for waterbirds, including many herons and egrets, Red-naped Ibis, Asian Openbill (IS), Gray-headed Lapwing (IS), and many other shorebird species. After the rice is planted and begins to grow, there is some cover, and more secretive birds like Ruddy-breasted Crake, Pin-tailed Snipe, and Greater Painted-Snipe move in. As the rice plants grow taller, some passerines use the habitat, such as Paddyfield Warbler and Golden-headed Cisticola. Finally, when the rice seeds develop, birds like Asian Golden Weaver, Chestnut Munia, and Yellow-breasted Bunting arrive to feed.

Rodents feast on the rice and can be common. Raptors, including Pied Harrier, Rufous-winged Buzzard, and Greater Spotted Eagle, feed on the ample prey. Paddy fields contain a lot more than just birds: they can be full of insects; fish come in when the fields are flooded and can breed here; frogs also use the habitat for breeding; and snakes come in to feed on these. A paddy field can function as an entire ecosystem.

Tea pluckers in Assam, ne. India.
© SAM WOODS, TROPICAL BIRDING TOURS

DISTRIBUTION: Paddy fields are distributed throughout the region but are most prevalent in large river deltas and alluvial plains, such as the Ganges River in India and Bangladesh, Indus River of Pakistan, Irrawaddy River in Myanmar, Chao Phraya River in Thailand, Mekong and Red Rivers in Vietnam, and Yangtze River in China. Other important areas include n. and e. India, Sri Lanka, e. Thailand, c. and s. China, Hainan, Taiwan, Luzon in the Philippines, and Java in Indonesia. Paddy fields are also present more patchily in the rest of the Philippines, Peninsular Malaysia, Sumatra, and Borneo. In fact, any flat area of tropical Asia with sufficient water can have paddy fields.

Sarus Cranes within Indian crop fields.
© VISHAL JADHAV

WHERE TO SEE: Chiang Dao, Thailand; Banaue, Luzon, Philippines.

INDO-MALAYAN CITIES AND VILLAGES

IN A NUTSHELL: Areas of human habitation. **Habitat Affinities:** AFROTROPICAL CITIES AND VILLAGES. **Species Overlap:** Surrounding habitats.

DESCRIPTION: The Indo-Malayan region has contained dense human settlements for millennia, and areas of the region now have some of the highest population densities in the world. With the rate of population growth, the area covered by human settlement will only increase. Cities, towns, and villages encompass a variety of land uses, from the concrete urban sprawl of mega cities to green town parks and leafy, suburban residential areas. How much nature lives alongside human populations depends very much on town planning. Some cities contain very little nature, while others that either have overgrown areas or have gone to the trouble of planting trees, rewilding water bodies, and creating green spaces have nature in abundance. Being surrounded by trees and other greenery not only helps filter out pollutants but is also beneficial for the psychological well-being of citizens. Trees around homes provide shade and shelter from wind; grassy lawns make nice places to let kids run around; and flowers are beautiful to look at and make people happy. People rarely go to the trouble of planting native species, but the structural diversity of this habitat still attracts a variety of birds and other animals.

Most cities nowadays have some urban parks, which are popular places for walking, jogging, picnics, or hanging out with friends and family. Water features are often part of such parks, along with grassy lawns and trees for shade. With a bit of careful planning, such places can become a real oasis for wildlife, and animals usually become tame. In dry regions, the availability of water and lush vegetation is greater than outside the urban area and can be a magnet for birds and other creatures. This is especially the case when birds are migrating and take advantage of water and greenery in stopover sites to shelter, rest, and refuel on their arduous journey. In a world with ever-increasing environmental problems, urban green areas provide some of the only exposure that most citizens get to nature. They can be used to educate and nurture an interest in wildlife for people who don't have the opportunity to visit true wild areas. Many birding clubs offer monthly guided walks in city parks to encourage people to take up the hobby.

WILDLIFE: Certain bird species have adapted well to city life in this region. Some of the commoner ones include House Crow, Eurasian Tree Sparrow (IS), House Sparrow, Common Myna (IS), and the introduced Rock Pigeon, all of which can be found in even the most sterile concrete urban environments. Areas with a little more greenery and a few trees will draw Zebra and Spotted Doves, and bulbuls (Red-vented in India and Yellow-vented in se. Asia), while flower plantings attract Olive-backed and Purple Sunbirds and Scarlet-backed Flowerpecker. The number of birds increases significantly in lush suburban gardens and leafy town parks. These areas hold birds like Coppersmith Barbet, Green Bee-eater, Asian Koel, Plaintive Cuckoo, and Greater Coucal.

Mammals also start to appear, such as striped squirrels and Variable Squirrel. Various swallows and swift species fly over, feeding on flying insects, and at dusk, bats emerge to take over the job. The presence of water features brings in birds such as egrets, kingfishers, and White-breasted Waterhen. Large buildings in cities make good hunting perches to scan for prey, and raptors such as Peregrine Falcon make a living picking off Rock Pigeons.

Urban habitats can also support plenty of reptiles. Most houses have many Common House Geckos, which feed on insects attracted to lights at night. The large Tokay Gecko gives away its presence with its extremely loud croak: *Tok-kay!* Rivers and ponds can harbor Asian Water Monitors. In rare instances, more exciting animals appear, such as the family of Smooth-coated Otters in Singapore Botanical Gardens and the Leopards that feed on feral pigs in some Indian cities.

Endangered Greater Adjutants feeding at a city dump in a decimated landscape in ne. India. © SAM WOODS, TROPICAL BIRDING TOURS

HABITATS OF THE AFROTROPICS
(SUB-SAHARAN AFRICA)

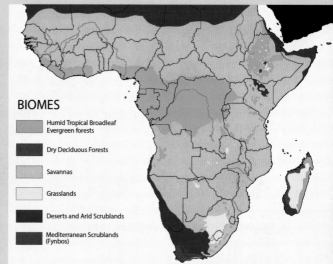

BIOMES

- Humid Tropical Broadleaf Evergreen forests
- Dry Deciduous Forests
- Savannas
- Grasslands
- Deserts and Arid Scrublands
- Mediterranean Scrublands (Fynbos)

SIDEBAR 4.1   HABITAT TRANSITION IN SENEGAL, AFRICA

Series of silhouettes illustrating the transition north to south in Senegal from desert thornscrub of the Sahel to moist mixed savanna to broadleaf Guinea savanna.

Sahel desert – desert thornscrub

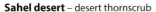

Northern Sahel – dry thorn savanna

Southern Sahel – dry thorn savanna/
                  tropical grasslands

Northern Guinea savanna – moist mixed savanna/tropical grasslands

Central Guinea savanna – moist mixed savanna/Guinea savanna/
                  isolated small pockets of
                  monsoon forest

AFROTROPICAL DESERT

IN A NUTSHELL: Arid, sparsely vegetated or unvegetated habitat that lacks consistent grass cover. **Habitat Affinities:** PALEARCTIC HOT DESERT; PALEARCTIC HOT SHRUB DESERT; NEOTROPICAL DESOLATE DESERT. **Species Overlap:** AFROTROPICAL DRY THORN SAVANNA; PALEARCTIC HOT SHRUB DESERT; KAROO; AFROTROPICAL TROPICAL GRASSLAND.

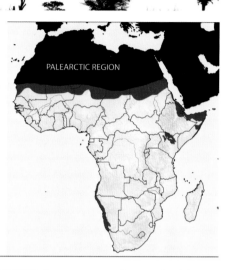

PALEARCTIC REGION

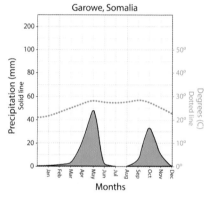

Garowe, Somalia

Gravel plains and sand dunes are classic components of deserts, such as the Namib of sw. Africa (photo: Sossusvlei, Namibia).
© KEN BEHRENS, TROPICAL BIRDING TOURS

DESCRIPTION: Deserts are very dry areas in which most of the ground is bare most of the time, even in areas with some vegetation. Despite being a fairly dry continent overall, Africa has only two sub-Saharan areas of desert, both relatively small: the Namib in the southwest and the deserts of the Horn of Africa and Kenya in the northeast. (The Sahara and Arabian Peninsula deserts and surrounding

shrublands are covered in the HOT DESERT and HOT SHRUB DESERT habitat accounts in the Palearctic chapter.) Desert habitat is where geology and geography can be enjoyed in their purest form, unobscured by concealing vegetation. The wide variety of landforms range from gravel plains to vast fields of sand dunes to rocky, eroded mountains with plateaus and valleys. All are open to exploration, on foot or by vehicle in areas with roads. The Namib, over 50 million years old, is thought to be the world's oldest desert. This has resulted in a flora and fauna that are highly adapted to the harsh environment. Despite its superficial bareness, this is one of the world's richest deserts.

Some areas are virtually free of vegetation, though a close examination will usually reveal the presence of lichens. The Namib supports remarkable *Wanderflechten*, or wandering lichens, which are unattached and blow in the wind, accumulating in shallow depressions. After exceptionally heavy rains, grasses will occasionally spring up, producing a crop of seeds to be blown away by the wind, an important component of desert ecology. Slightly moister areas, including those at the base of sand dunes or along drainage lines, allow the growth of sparse, shrubby vegetation. The fringes of the Namib and large areas of the northeastern desert are shrub desert, which is similar to KAROO, though far less diverse and with sparser shrub growth. One classic plant of the Namib is the spiny Nara Melon, which lacks leaves and instead photosynthesizes with its stems and thorns. The most famous plant of the Namib, and indeed, one of the strangest plants on earth, is the Welwitschia, a gymnosperm that comprises its own plant family. It has two leathery leaves, which emerge from a short trunk and grow continuously. The oldest specimens are thought to be over 2,000 years old. Trees grow very locally in areas with groundwater, such as valley bottoms, or at the desert fringe. Typical groups include *Commiphora* spp., euphorbias, acacias, desert dates, and yellowthorns.

Deserts receive less than 10 in. (250mm) of rain annually, and many areas are much drier; parts of the Namib receive as little as 0.2 in. (5mm), while the Eritrean coast receives less than 4 in. (100mm). Along the coast of the Namib, most of the moisture arrives in the form of dense fog that pushes inland from the Benguela Current–cooled Atlantic. The southern part of the Namib receives rainfall during the austral winter, and the northern part during the austral summer. Northeast Africa has two annual periods during which sparse rains can fall. The northeastern deserts and the inland part of the Namib are generally hot, though they cool off at night. Contrastingly, the coastal part of the Namib, buffered by cold water offshore, has a moderate climate without drastic daily temperature fluctuations. The Namib is remarkably well protected by large national parks and a huge diamond-mining concession. The conservation situation in the northeast is much worse, mainly due to long-term political instability. One problem across Africa's deserts is off-road driving, which can have a long-lasting effect on this fragile habitat.

WILDLIFE: Larger wildlife naturally tends to be thin on the ground in desert due to the scarcity of food. Nonetheless, the sub-Saharan deserts are home to a fair number of mammals and birds and are an excellent habitat for reptiles. Classic desert animals include gazelles, oryxes, foxes, bustards, sandgrouse, and larks. The southwestern and northeastern deserts share only a handful of species, such as Black-backed Jackal, Common Ostrich, and Double-banded Courser, remnants of times in the geological past when these two arid areas were connected.

The Namib has been less impacted by human activities and still supports good numbers of large mammals. Gemsbok and Springbok, the two most desert-adapted big mammals, are fairly common, especially at the desert fringes. Hartmann's Mountain Zebra and Angolan Giraffe are found in the adjacent habitats along the Namib Escarpment, and will sometimes descend into the Namib Desert. The major predators are Brown and Spotted Hyenas, Bat-eared Fox, Cheetah, and Black-backed Jackal. Lion, Black Rhinoceros, and African Bush Elephant were formerly widespread but now are localized, found mainly in remote parts of nw. Namibia. There are a couple of birds endemic to the

Dune Lark is endemic to the vast dune fields of the Namib Desert.
© KEN BEHRENS, TROPICAL BIRDING TOURS

Namib: Rüppell's Bustard, Gray's Lark (IS), and Dune Lark (IS). Other typical birds include Burchell's Courser, Namaqua Sandgrouse, Stark's Lark, and Tractrac Chat.

The reptile assemblage of the Namib is exceptional, comprising dozens of species, many of which are endemic. Several species have adapted to living on shifting sand dunes, such as Peringuey's Adder and Shovel-snouted and Wedge-snouted Lizards. All these species are capable of "swimming" into the sand to hide. The adder, which uses a side-winding motion to cross soft sand, drinks droplets of fog that condense on its body. Other reptiles include legless lizards, desert lizards (*Meroles*), sand lizards (*Pedioplanis*), plated lizards (*Gerrhosauridae*), Namib and Karoo Web-footed Geckos, barking geckos (*Ptenopus*), Namib day geckos (*Rhoptropus*), and Namaqua Chameleon.

The northeastern desert is less contiguous and homogeneous than the Namib, and many species are found in only a small portion of this region. This area has a strong affinity with the PALEARCTIC HOT DESERT of the Arabian Peninsula, and many species are found on both sides of the Red Sea and Gulf of Aden. The mammal population has been greatly reduced by hunting and overgrazing, and most of the larger species are now regionally extinct. Some fascinating mammals do remain locally, including Hamadryas Baboon; Speke's, Dorcas, and Soemmerring's Gazelles; Beira; Pale and Rüppell's Foxes; Leopard; and Striped Hyena. Deserts in Ethiopia and Eritrea support the world's last few African Wild Asses, the wild progenitor of the domestic donkey.

Birds of the Horn of Africa deserts include Heuglin's and Arabian Bustards; Chestnut-bellied, Spotted, and Lichtenstein's Sandgrouse; Desert (IS), Masked, Somali Long-billed, and Crested Larks; Greater and Lesser Hoopoe-Larks; Black-crowned and Chestnut-headed Sparrow-Larks; Somali Crow; and Somali Fiscal. The northeast is rich in reptiles, including wall lizards (*Mesalina*), Eritrean Rock Agama, Ocellated Spinytail, and Northeast African Carpet Viper.

Endemism: The Namib has a rich set of endemic species of plants, birds, and reptiles. The Horn of Africa has a few bird species, though most of them are not pure desert species but use both AFROTROPICAL DRY THORN SAVANNA and desert habitats.

DISTRIBUTION: The Namib Desert forms a narrow strip along the coast of Namibia and s. Angola in s. Africa. In the northeast, the Horn of Africa is rimmed with desert, which is closely allied with the deserts of the Arabian Peninsula (see PALEARCTIC HOT DESERT and PALEARCTIC HOT SHRUB DESERT). There is also an isolated desert in the driest parts of n. Kenya. All of Africa's deserts are bordered by AFROTROPICAL DRY THORN SAVANNA, and much of the northeastern desert occurs in a fine matrix alongside this habitat.

WHERE TO SEE: Namib-Naukluft National Park, Namibia.

KAROO

IN A NUTSHELL: Arid but highly diverse shrubland of sw. Africa with variable grass cover. **Habitat Affinities:** PALEARCTIC HOT SHRUB DESERT; NEOTROPICAL SEMIDESERT SCRUB; NEARCTIC SAGEBRUSH SHRUBLAND; AUSTRALASIAN CHENOPOD AND SAMPHIRE SHRUBLAND. **Species Overlap:** AFROTROPICAL DRY THORN SAVANNA; AFROTROPICAL DESERT; FYNBOS; AFROTROPICAL TROPICAL GRASSLAND.

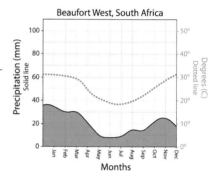

Beaufort West, South Africa

DESCRIPTION: The Karoo is a region of sw. South Africa, defined by a dry and open shrubland habitat that extends into Namibia. From a distance, the Karoo habitat looks like a typical desert or semidesert, and it often reminds visitors of the American Southwest. But a close look reveals an extraordinarily diverse and distinctive plant community unlike any other on earth, plus an associated community of wildlife. This is a dry and harsh environment, with huge fluctuations in temperature: summer maximums can exceed 95°F (35°C), while nights below freezing are common in the winter. In the interior, a single day can see a temperature fluctuation of 45°F (25°C). Rainfall is low throughout the Karoo, generally 4–20 in. (100–500mm) annually, though it averages only 3 in. (75mm) in the sw. Tanqua Karoo National Park, which lies in the rain shadow of the Cape Fold Mountains.

The dominant plants of the Karoo are short shrubs, less than 3 ft. (1m) tall. The most important shrub families are asters (Asteraceae), stonecrops (Crassulaceae), dogbanes (Apocynaceae), and ice plants (Aizoaceae). There is a sharp divide within the habitat between the smaller "succulent Karoo" region to the west and the more extensive Nama Karoo region to the east. This divide is based on the timing of the rainfall, which occurs in the austral winter in the succulent Karoo and during the austral summer in the Nama Karoo. This has had profound consequences for the respective plant communities. The succulent Karoo is remarkably diverse in plants, supporting at least 5,000 species, including a third of the world's succulents, of which 40% are endemic just to this portion of the Karoo. The Nama Karoo has fewer succulents and is less diverse botanically. It serves as a transition between the FYNBOS along the southwestern corner of Africa and the AFROTROPICAL DRY THORN SAVANNA of the Kalahari to the north. The prominence of grass varies across the Karoo. It is scarce in the succulent Karoo and becomes increasingly common in the Nama, especially to the north and east in the area of transition to the arid savanna of the Kalahari and the AFROTROPICAL MONTANE GRASSLAND of the Highveld. Larger bushes and trees are largely confined to drainage lines. Some of the common species include Wild Tamarisk, Karee, and

Above: **Rugged and open Karoo landscape, dominated by shrubs.** *Inset:* **A succulent plant typical of the succulent Karoo, South Africa.** BOTH IMAGES © KEN BEHRENS, TROPICAL BIRDING TOURS

A grassy variation of Karoo at the edge of the Kalahari, South Africa.
© KEITH BARNES, TROPICAL BIRDING TOURS

Sweet Thorn. These bands of riparian brush allow more typically savanna-dwelling species to survive in the Karoo. Despite the habitat's aridity, it does have wetlands, including many artificial farm ponds, which attract a wide variety of waterbirds.

Although the Karoo may look like pristine habitat to a casual observer, it has been profoundly impacted by humans. The vast majority of this habitat is on private farms, where sheep and other animals are grazed. Competition from domestic animals and persecution of predators have decimated the region's large mammal populations. Overgrazing also has the insidious effect of slowly replacing the full range of indigenous plants, which would have been browsed by wild animals, with a poor subset of plants that are toxic or unpalatable to domestic animals. Some moister parts of the Karoo, such as the Little Karoo, which borders the FYNBOS habitat (south of the Karoo), have been largely converted to agriculture. Another major problem is the introduction of invasive exotic plants, which can further exacerbate the environmental changes wrought by livestock.

WILDLIFE: In terms of big mammals, the Karoo fauna is a shadow of what it once was. The region used to host massive migrations of antelopes, mainly Springboks, a natural spectacle that matched or exceeded the wildebeest migration of the Serengeti. This habitat also supported nearly a full set of Africa's charismatic big mammals, but many of these species have either been completely eliminated or are now confined to small fenced-in reserves. This area was once the realm of the now-extinct Quagga, a distinctive and beautiful animal that may have been a subspecies of Common Zebra. The Karoo was the summer range of the Black Wildebeest, which was reduced to extinction in the wild before being resurrected by captive breeding. The formerly widespread Cape Mountain Zebra and Gemsbok are now largely confined to a handful of conservation areas. Although much has been lost, there are still big mammals in the Karoo. Despite heavy persecution, predators like Caracal, Black-backed Jackal, and Brown Hyena still remain. The Karoo is also excellent habitat for smaller predators, such as Bat-eared and Cape Foxes, Black-footed Cat, Aardwolf, Cape Gray and Yellow Mongooses, and Meerkat. Most of these are best sought on night drives, which can count among Africa's most exciting and may also turn up an Aardvark or a Cape Porcupine. Although the vast migratory herds are gone, there are still plenty of Springbok, plus Chacma Baboon, Cape and Scrub Hares, Steenbok, Common Duiker, and Red Hartebeest.

The Karoo also still holds some of Africa's charismatic big birds, including Common Ostrich, Secretarybird, and Martial Eagle. It's an essential destination for a birder, holding a rich set of endemic and near-endemic species, some of which are named for the region. Karoo specialists include Karoo Bustard (IS), Karoo Lark, Black-eared Sparrow-Lark, Rufous-eared Warbler, Kopje Warbler, Yellow-rumped Eremomela (IS), Sickle-winged and Karoo Chats, and Black-headed Canary (IS). More widespread birds that are at home in the Karoo include Black Harrier, Jackal Buzzard, Greater Kestrel, Burchell's and Double-banded Coursers, Namaqua Sandgrouse, Bokmakierie, Karoo Prinia, Layard's Warbler, Karoo Scrub-Robin, and Lark-like Bunting. Rivers and streams that penetrate the dry Karoo provide habitat for the critically endangered Riverine Rabbit and the much more common Namaqua Warbler. Bushes and trees along streams and other drainage lines support Pied Barbet, Pririt Batis, and Cape Crombec, species more typical of the AFROTROPICAL DRY THORN SAVANNA.

Above: **Kopje Warbler is restricted to a special microhabitat within the Karoo: rocky slopes, often with abundant aloes.** © KEN BEHRENS, TROPICAL BIRDING TOURS

Left: **Rufous-eared Warbler is a Karoo specialist preferring open plains.** © KEN BEHRENS, TROPICAL BIRDING TOURS

The Karoo has a diverse set of reptiles, many found only here; the succulent Karoo alone has dozens of endemic reptiles. Some of the groups with Karoo endemics include padlopers (minuscule tortoises); adders; sandveld (*Nucras*), sand (*Pedioplanis*), girdled (*Cordylus*), and crag (*Pseudocordylus*) lizards; and thick-toed geckos (*Pachydactylus*). Most of the endemic reptiles are highly localized to some small portion of the Karoo. And while some reptiles are conspicuous, the majority will be found only by those who carefully search.

With their differing rainfall patterns, there is a profound botanical division between the Nama Karoo and the succulent Karoo. But this difference in plants doesn't translate to a major division for birds and larger mammals (this is the main reason these two components of the Karoo are not split out as separate habitats in this book). A few bird species are shared between the FYNBOS and the succulent Karoo but not the Nama Karoo. There are many highly localized endemic reptile species, and the Nama Karoo and succulent Karoo have very different assemblages of reptiles.

DISTRIBUTION: Karoo habitat occurs mainly in South Africa, though it extends well into Namibia. According to some habitat classification schemes, Karoo is found all the way into Angola, though this isn't well substantiated by the presence of Karoo wildlife. Much of the succulent Karoo is coastal and at lower elevations, while the majority of the Nama Karoo lies above 3,000 ft. (900m) on the vast Southern African Plateau. Karoo is replaced by FYNBOS to the south and southwest and on mountain ranges.

WHERE TO SEE: Karoo National Park, South Africa.

MALAGASY SPINY FOREST

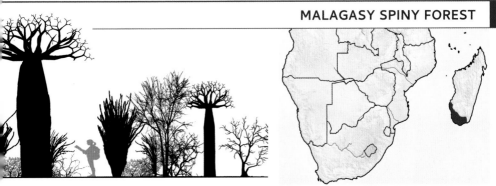

IN A NUTSHELL: Dry forest or thicket on Madagascar that includes mostly deciduous trees and in which many plants are spiny. **Habitat Affinities:** NEOTROPICAL CAATINGA; NEARCTIC COLUMNAR CACTUS DESERT. **Species Overlap:** MALAGASY DRY DECIDUOUS FOREST; INDIAN OCEAN RAINFOREST.

DESCRIPTION: This arid biome of sw. Madagascar is one of the Afrotropics' most visually distinctive; it's unlike any other place on earth and rather like something dreamed up for an episode of *Star Trek*. Some of the most striking trees in this habitat are fat-trunked, water-absorbing baobabs and pachypodiums, many-armed octopus trees, and a variety of succulent euphorbias. Other common trees include commiphoras, flame trees, and moringas. Succulent plants such as aloes and kalanchoes are prominent in the understory. There are clues suggesting that, prior to the ancient moistening of the climate and development of the rainforest, Madagascar was entirely arid, making this biome the island's oldest. This idea is borne out by its extraordinary diversity of endemic plants; 95% of the plants are endemic not just to Madagascar but to this habitat. Usually referred to as "spiny forest" (as it is here) and sometimes as "spiny desert," it's actually neither a true forest nor a true desert but rather a semiarid or sub-desert brushland or thicket. The exact species composition and the feel of spiny forest vary a lot across its range, partly depending on whether it is growing on limestone or sand. Some areas are thoroughly mixed, while some are dominated by octopus trees and others by euphorbias, and in some places, it can take the form of fairly generic thornscrub, especially where spiny forest transitions to MALAGASY DRY DECIDUOUS FOREST and where it has been degraded by humans.

Southwestern Madagascar lies in the rain shadow of the eastern mountains and is too far south to be much influenced by the intertropical convergence zone. As such, its climate is dry, with annual rainfall of 12–20 in. (300–500mm) and a dry season that can be as long as 10 months. During drought periods, some areas can receive little rainfall for more than a year. Although the spiny forest is less degraded than the other two Malagasy forest zones (DRY DECIDUOUS FOREST and INDIAN OCEAN RAINFOREST), it is under increasing pressure as the local human population increases and people struggle to simply survive in an impoverished region with a harsh climate.

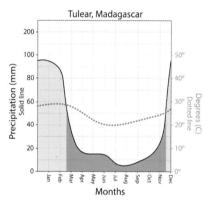

The odd and distinctive shapes of baobabs, octopus trees, and other trees give the spiny forest of sw. Madagascar a unique and otherworldly feeling.
© NICK ATHANAS, TROPICAL BIRDING TOURS

Verreaux's Sifaka lives in undisturbed spiny forest, where it leaps between the spiny trunks of octopus trees while somehow managing not to shred its hands or feet. © KEN BEHRENS, TROPICAL BIRDING TOURS

The roadrunner-like Long-tailed Ground-Roller is endemic to the spiny forest. It is a member of the Malagasy endemic ground-roller family, whose four other members are found in rainforest.
© KEN BEHRENS, TROPICAL BIRDING TOURS

WILDLIFE: This is the poorest of Madagascar's habitats for wildlife overall, though it holds a large number of local endemics. Verreaux's Sifaka and Ring-tailed Lemur are the classic big lemurs of spiny forest, though they have been eliminated from much of their former range. The other large mammals of the spiny forest are nocturnal species, namely Gray-brown (IS) and Gray Mouse Lemurs, White-footed and Petter's Sportive Lemurs, and Lesser Hedgehog Tenrec. Fossa was once widespread but is now seemingly rare, probably due to the elimination of the large lemurs upon which it preys. Two other members of the endemic Malagasy carnivorans family (Eupleridae) are found locally in spiny forest: Narrow-striped and Grandidier's Vontsiras.

The avifauna of the spiny forest consists mainly of widespread forest birds, plus a few species shared with the MALAGASY DRY DECIDUOUS FOREST. But there is a fascinating set of endemic birds, including Subdesert Mesite, Long-tailed Ground-Roller, Verreaux's and Running Couas, Littoral Rock Thrush, Subdesert Brush-Warbler (IS), Thamnornis, and Lafresnaye's Vanga (IS). Most are widespread in the southwest. The spiny forest is replete with reptiles, and the open nature of the habitat makes them easy to see. There are two endemic tortoises, Radiated Tortoise and Spider Tortoise, which were formerly common and widespread but are now considered critically endangered.

Several members of the endemic Malagasy iguanid family (Opluridae) are prominent, including Merrem's Madagascar Swift and Three-eyed Lizard. Other typical reptiles include Warty Chameleon, Three-lined and Four-lined Plated Lizards, Madagascar Keeled Plated Lizard, Gold-spotted Skink, Sakalava Madagascar Velvet Gecko, fish-scale geckos (*Geckolepis*), and Ocelot Gecko. Snake diversity is low, but Dumeril's Boa and Mahafaly Sand Snake are both present throughout.

Endemism: Although plant distribution is complex, with several centers of endemism within the spiny forest, the wildlife of this habitat is rather uniform throughout. Bizarrely, though, two of the habitat's most interesting birds, Long-tailed Ground-Roller and Subdesert Mesite, are found only in the northwest corner of the spiny forest, between the Mangoky and Fiheranana Rivers.

DISTRIBUTION: Malagasy spiny forest is confined to areas below 1,300 ft. (400m) in sw. Madagascar. Watercourses in this dry area support tall gallery forest that remains lush year-round. This riparian forest has a character reminiscent of MALAGASY DRY DECIDUOUS FOREST, though it is far less diverse. To the north, as moisture levels increase, spiny forest slowly blends into Malagasy dry deciduous forest. To the east, there is a remarkable, stark transition to INDIAN OCEAN RAINFOREST; some mountains have rainforest on their eastern slopes and spiny forest on their western slopes.

WHERE TO SEE: Parc Mosa, Reniala, and other small private reserves near Ifaty, Madagascar; Berenty Private Reserve, Amboasary, Madagascar.

DRAGON BLOOD SEMIDESERT

IN A NUTSHELL: Botanically unique semidesert habitat on Socotra and adjacent Indian Ocean islands, dominated in parts of the highlands by the umbrella-shaped Dragon Blood Tree.
Habitat Affinities: NEOTROPICAL THORNSCRUB; NEOTROPICAL SEMIDESERT SCRUB; NEARCTIC COLUMNAR CACTUS DESERT. **Species Overlap:** AFROTROPICAL DRY THORN SAVANNA AND THORNSCRUB; AFROTROPICAL DESERT.

DESCRIPTION: The large island of Socotra and smaller adjacent islands, including Abd al-Kuri, Samhah, and Darsah, lie in the Indian Ocean south of the Arabian Peninsula but closer to the Horn of Africa. The isolation and arid climate here have produced an unusual and distinctive semidesert flora unlike any other on earth. Although poor in wildlife, this habitat is given separate coverage here based on its botanical and visual distinctiveness. Dragon Blood semidesert is rugged, rocky, and open, and is sparsely vegetated with various plants that are adapted to a dry climate. Its namesake and most famous species is the markedly symmetrical, umbrella-shaped Dragon Blood Tree, one of the world's most bizarre and photogenic trees. This tree is restricted to the higher parts of the Socotran plateau. Another unusual species is the fat-trunked Cucumber Tree, the only member of the cucumber family that grows in tree form. Other characteristic trees of this semidesert habitat include Frankincense, Christ's Thorn Jujube, commiphoras, African Star-Chestnut, and *Euphorbia arbuscula*. Overall, there are around 250 species of plants, of which 30% are endemic, including 10 endemic genera. Much of the

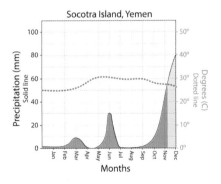

Socotra Island, Yemen

The Dragon Blood Tree *(below)* commands the semidesert heights of Socotra. Its blood-red resin *(right)* was used for magic and alchemy in ancient times and is still used locally for its medicinal properties. *Main photo:* © ROD WADDINGTON FROM KERGUNYAH, AUSTRALIA, WIKIMEDIA COMMONS, CC BY-SA 2.0, HTTPS://COMMONS.WIKIMEDIA.ORG/WIKI/CATEGORY:DRACAENA_ CINNABARI#/MEDIA/FILE:DRAGON'S_BLOOD_TREE,_SOCOTRA_IS_ (12473612124).JPG; *inset:* © MAŠA SINREIH IN VALENTINA VIVOD, WIKIMEDIA COMMONS, CC BY-SA 3.0, HTTPS://COMMONS.WIKIMEDIA.ORG/WIKI/ CATEGORY:DRACAENA_CINNABARI#/MEDIA/FILE:SANGUIS1.JPG

moisture in this arid environment comes from mist, though there is a rainy season from October to December. The wettest parts of the Socotra highlands receive as much as 40 in. (1,000mm) of precipitation annually, though most of this habitat is much drier, with some places receiving as little as 6 in. (150mm).

WILDLIFE: These islands have few indigenous mammals. Most of the terrestrial birds and reptiles are endemic. There are 10 endemic bird species: Socotra Buzzard, Socotra Scops-Owl, Socotra Warbler (IS), Socotra Cisticola, Socotra Starling, Socotra Sunbird (IS), Socotra Sparrow, Socotra Grosbeak, Socotra Bunting, and Abd al Kuri Sparrow. Additionally, there are many endemic bird subspecies. Reptiles show low diversity but a high degree of endemism; the majority of the 30-odd species are endemic. These include Socotra Chameleon, several rock geckos (*Pristurus*) and blind snakes (*Myriopholis*), and many leaf-toed geckos (*Hemidactylus*). Socotra is famous among spider-lovers as the home of one of the world's most beautiful tarantulas, the Socotra Island Blue Baboon.

DISTRIBUTION: Dragon Blood semidesert habitat is found throughout the archipelago, though biologically Socotra is by far the richest of the islands.

WHERE TO SEE: Rokeb di Firmihin, Socotra.

AFROTROPICAL LOWLAND RAINFOREST

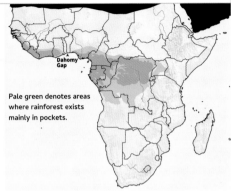

Pale green denotes areas where rainforest exists mainly in pockets.

Dahomy Gap

IN A NUTSHELL: Lush forest with a high canopy and complex structure that remains moist and humid throughout most or all of the year. **Habitat Affinities:** NEOTROPICAL LOWLAND RAINFOREST; AUSTRALASIAN TROPICAL LOWLAND RAINFOREST; INDO-MALAYAN TROPICAL LOWLAND RAINFOREST; INDIAN OCEAN RAINFOREST. **Species Overlap:** AFROTROPICAL SWAMP FOREST; AFROTROPICAL MONSOON FOREST; AFROTROPICAL MANGROVE; AFROTROPICAL MONTANE FOREST; GUINEA SAVANNA.

DESCRIPTION: Rainforest is renowned as a habitat of incredibly high biodiversity, and the rainforest of Africa is no exception. The continent holds one of the world's great tropical lowland rainforests, along with the Amazon of South America (NEOTROPICAL LOWLAND RAINFOREST) and the forests of se. Asia and New Guinea (AUSTRALASIAN TROPICAL LOWLAND RAINFOREST). As suggested by the name, this habitat will grow only in areas with high rainfall for most of the year. Typical annual rainfall is 63–80 in. (1,600–2,000mm), though locally it can be as low as 40 in. (1,000mm) or as high as 200 in. (5,000mm). Some areas remain drenched year-round, but most of the region's lowland rainforest experiences a short dry season between December and February. Some areas,

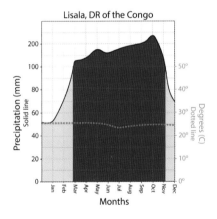

Lisala, DR of the Congo

such as much of the Congo Basin, have two short dry periods: December–January and June–July. Even the relatively dry periods still have some rainfall. Not surprisingly, humidity and temperature are high year-round. This complex habitat creates a variety of microclimates within itself. The canopy is subject to daily temperature fluctuations between lows of around 70°F (20°C) and highs of 85–90°F (30–33°C), whereas the sheltered and largely windless forest understory sees only minor temperature fluctuations of one to two degrees each day. Afrotropical lowland rainforest is found from sea level to around 5,000 ft. (1,500m) elevation in Uganda and w. Kenya. It is mostly found in very flat terrain, though there are some hilly areas, in particular the foothills of the Albertine Rift and West African mountains.

The rainforest has a high canopy that averages 100–130 ft. (30–40m), with some towering emergent trees as tall as 200 ft. (60m). It is tall forest but not nearly as tall on average as the rainforests of se. Asia; it is also very different and less diverse botanically. While Asian forests are usually dominated by dipterocarps, African rainforests tend to be dominated by members of the Caesalpinioideae subfamily of the enormous legume family, and dipterocarps are virtually absent. It is hard to generalize about the trees of Afrotropical lowland rainforest, as there is considerable diversity within a small area, and the common trees change across the vast range of this habitat. Limbali, an exception, can sometimes dominate large stretches of forest. Other typical lowland rainforest trees are Bokapi, Junglesop, Agba, African Walnut, Tallow Tree, Gabon Nut, African Teak, and African Mahogany, to name just a few. The transition from rainforest to savanna habitats is very gradual. Although Africa's lowland rainforest is technically an evergreen forest, much of it contains many deciduous trees, especially in the transitional areas. Although it would be possible to classify some of these forests as semi-evergreen or semi-deciduous, the gradual nature of the transition, and the fact that these slightly drier forests do not support a distinct set of wildlife species, argue against doing so for the purposes of this book.

The forest has a complex mid-story that includes smaller trees and many lianas. Epiphytes, especially orchids and ferns, are common. Undisturbed areas, where little light reaches the forest floor, can have a remarkably open understory and be easy to walk through. An example of this is in Uganda's Kibale National Park, where trackers of Chimpanzees often lead visitors off trails without much trouble. Disturbed areas quickly develop a virtually impenetrable understory. Tree-fall gaps are an important habitat in mature lowland rainforest. When a single, massive tree—connected by lianas to dozens of other trees—falls, it can create a considerable gap where sunshine suddenly penetrates to the floor. Fast-growing plants like members of the ginger family spring up and

Right: **Afrotropical lowland rainforest is the habitat of the Chimpanzee.** © KEN BEHRENS, TROPICAL BIRDING TOURS

Below: **Lush, complex, multilayered Afrotropical lowland rainforest in Kakum National Park, Ghana.** © KEITH BARNES, TROPICAL BIRDING TOURS

provide an important food source for mammals such as African Forest Elephant and Eastern and Western Lowland Gorillas. Africa's lowland rainforest zone shows stark contrast in terms of the degree of human impact. The forest from Nigeria west has been drastically impacted by the large human population, which can exceed hundreds of people per sq. mi. Larger mammals, even including squirrels and bats, have been wiped out of most of the region. This is driven by a booming market for bushmeat; in Nigeria, a large Grasscutter (aka Greater Cane Rat) can sell for the equivalent of two weeks' salary for an average person. Logging, forest clearing for agriculture and firewood, and even mining have all taken a toll. Meanwhile, the Congo Basin has vast stretches of nearly pristine forest, where the human population density can be 10 or fewer per sq. mi. (4/km²). But even the Congo Basin has been impacted by hunting for bushmeat and ivory, and larger mammals are now scarce along navigable waterways. Poverty and political instability have plagued much of the region and hamstrung most attempts at forest protection.

WILDLIFE: Very few mammals or birds are found across the entirety of the range of Afrotropical lowland rainforest, but there are many groups of animals that are found primarily or exclusively in this habitat. In conjunction with AFROTROPICAL SWAMP FOREST, this habitat has the continent's most distinctive assemblage of wildlife. There is little overlap between the wildlife of lowland rainforest/swamp forest and that of other habitats, save for some species shared with AFROTROPICAL MONSOON FOREST.

Lowland rainforest holds a bounty of remarkable mammals. Unfortunately, due to their shy nature, the prevalence of hunting, and the density of the habitat itself, most are very difficult to see. A successful rainforest walk might include a good look at one or two squirrels and fleeting glimpses of a troop of fleeing guenon monkeys. The major rivers of this zone, especially the Sassandra, Volta, Niger, Cross, and Congo, are serious barriers to mammals and profoundly affect their distributions, tightly delineating the edges of the ranges of many species. Classic lowland rainforest mammals are African Forest Elephant; African Brush-tailed Porcupine; pangolins, including the hefty Giant Pangolin; Red River Hog; Giant Forest Hog; African Forest Buffalo; and Bongo. The large predators are Leopard and African Golden Cat, while smaller predators include African Civet; oyans; many genets, including the Giant Genet; and kusimanses. The "holy grail" mammal of this habitat is Okapi, a beautiful, dark brown and white giraffe relative that seems virtually impossible to glimpse in the wild.

Afrotropical lowland rainforest is one of the world's most important areas for primate diversity. Its primate fauna includes four great apes, humans' closest relatives on the planet: Eastern Gorilla, Western Gorilla, Chimpanzee, and Bonobo. Chimpanzee and Bonobo show a classic pattern of ranges separated by a major river: Chimpanzee north of the Congo and Bonobo to the south. The great apes are joined by a wide range of monkeys: Mandrill and Drill; a diverse set of colobuses; mangabeys; a plethora of guenons, like the Red-tailed Monkey (IS south of

Great Blue Turaco is like a cartoon character come to life. © IAIN CAMPBELL, TROPICAL BIRDING TOURS

Right: **White-necked Picathartes (or Rockfowl) is one of two members of the genus *Picathartes*, both restricted to Afrotropical lowland rainforest.**
© KEITH BARNES, TROPICAL BIRDING TOURS

Below: **Tall cathedral-like trunks stretch upward, making wildlife observation a neck-craning activity (photo: w. Uganda).**
© IAIN CAMPBELL, TROPICAL BIRDING TOURS

Congo River); Potto; angwantibos; and galagos, such as the widespread Thomas's Galago (IS). Squirrels are also prominent. These include the otter-size Forest Giant Squirrel; sun squirrels, such as Red-legged Sun Squirrel (IS north of Congo River); African Pygmy Squirrel; Slender-tailed Squirrel; palm squirrels; at least nine species of rope squirrels; flying mice; and anomalures, such as Cameroon Scaly-Tail. The forest floor shelters a remarkable diversity of small antelopes, namely a diverse range of duikers and Royal and Dwarf Antelopes, plus a tiny deer, the Water Chevrotain.

Lowland rainforest is a very rich habitat for birds, with many centers of endemism. Birders wanting to come to grips with the majority of rainforest birds will have to make multiple trips, and most of the countries that host this habitat are challenging to access. While time spent in this forest will yield far more bird sightings than mammal sightings, birding here is hard work. Many species are scarce, and most are shy and unapproachable, hiding in the canopy or in the foliage away from trails. Some of the groups that reach peak diversity in rainforest are fishing-owls, spinetails, trogons, kingfishers, hornbills, barbets, honeyguides, woodpeckers, broadbills, cuckooshrikes, illadopsises, greenbuls, flycatcher-thrushes, forest robins, longbills, flycatchers, wattle-eyes, malimbes, and nigritas. Of all these groups, the most diverse is the greenbuls, a cryptic horde that gives headaches to birders trying to identify them.

Common and widespread species include Red-tailed Greenbul (IS) and Green Hylia (IS). Just a few more of the fascinating birds of the lowland rainforest are Long-tailed Hawk; Congo Serpent-Eagle; Latham's Francolin; Nkulengu Rail; Black-collared Lovebird; Great Blue Turaco; Maned Owl; Chocolate-backed Kingfisher; Black, Blue-headed, and Black-headed Bee-eaters; African Piculet;

Blue Cuckooshrike; two wattled-cuckooshrikes; and Tit-hylia. This habitat has wonderful hornbills, ranging from the tiny dwarf hornbills to the weird, long-tailed White-crested Hornbill and the massive Black-casqued and Yellow-casqued Hornbills. There are two *Picathartes* rockfowl, odd passerines with bare heads and long legs, that nest in forest caves. The avian equivalents of the near-mythical Okapi are Congo Peacock and Shelley's Eagle-Owl.

Reptiles are not easily seen, but that doesn't mean lowland rainforest lacks reptile diversity. Typical reptiles include forest geckos (*Cnemaspis*), tree snakes (*Thrasops*), Forest Vine Snake, snake-eaters (*Polemon*), Forest Cobra, Jameson's Mamba, night adders (*Causus*), Gaboon Viper, Rhinoceros Viper, and chameleons (*Chamaeleo*). Afrotropical rainforest is extremely rich in frogs, and this habitat holds a high proportion of the continent's 800-some amphibian species. It is also Africa's most diverse habitat for butterflies, which can be far more accessible than its mammals, birds, or reptiles.

Endemism: The biggest dividing expanse of savanna between lowland rainforest blocks is the Dahomey Gap between the Congo Basin and the Upper Guinea forests of West Africa, and there are many sister species on either side of this gap. Within the Upper Guinea region are two minor centers of endemism, which for mammals are divided by the Sassandra River. In the Congo Basin, many species are found only on the river's east or west side, either in the w. Cameroon–Gabon region or the parts of e. Democratic Republic of the Congo and Uganda that lie below the Albertine Rift mountains. The ranges of many local mammals are delineated by major rivers, especially the largest, the Congo. The volcanic islands of São Tomé, Príncipe, and Annobón, which lie in the Gulf of Guinea, are rich in endemics, including more than 20 endemic birds.

DISTRIBUTION: The heart of the African rainforest is the Congo Basin. From there it extends in a relatively narrow band into the Upper Guinea forests of West Africa; these are isolated by the dry Dahomey Gap, which covers much of e. Ghana, Togo, and Benin. The gap is dominated by the more open GUINEA SAVANNA habitat, with lowland rainforest restricted to rivers and other wet spots. To the north, lowland rainforest quickly gives way to MONSOON FOREST and GUINEA SAVANNA and again is found locally only around rivers. To the south, the transition to MIOMBO WOODLAND is more gradual, and rainforest, TROPICAL GRASSLAND, miombo, and other savanna habitats form a complex matrix. East Africa holds the easternmost outposts of the rainforest; the farthest east of all is Kakamega Forest in w. Kenya.

WHERE TO SEE: Kibale National Park, Uganda; Kakum National Park, Ghana; Loango National Park, Gabon.

SIDEBAR 4.2	**REFUGIA WITHIN EXPANDING AND CONTRACTING AFRICAN RAINFOREST**

During wet periods over the past several million years, the central and western AFROTROPICAL LOWLAND RAINFOREST would have spread far to the south, into the modern MIOMBO WOODLAND zone, and been connected to the humid forest (now AFROTROPICAL MONSOON FOREST) of the Indian Ocean coast. In drier periods, savanna habitats would have expanded, and rainforest would have shrunk down into two small refugia in the Upper Guinea area, one in Cameroon and one on the east side of the Congo Basin, below the Albertine Rift mountains. This dynamic of cooling and drying resulted in a complex pattern of speciation across the rainforest; despite its somewhat monolithic modern appearance, this habitat holds a great deal of regional variation, including several regional centers of endemism.

AFROTROPICAL MONSOON FOREST

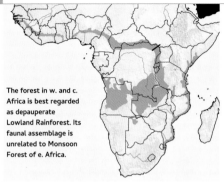

The forest in w. and c. Africa is best regarded as depauperate Lowland Rainforest. Its faunal assemblage is unrelated to Monsoon Forest of e. Africa.

IN A NUTSHELL: Lush, closed-canopy forest characterized by leaf litter on the forest floor and abundant lianas and palms in the mid-story. It is drier and more seasonal than AFROTROPICAL LOWLAND RAINFOREST and occurs mainly along watercourses and the coast. **Habitat Affinities:** NEOTROPICAL SEMI-EVERGREEN FOREST; AUSTRALASIAN MONSOON VINE FOREST; INDO-MALAYAN SEMI-EVERGREEN FOREST. **Species Overlap:** AFROTROPICAL MONTANE FOREST; AFROTROPICAL LOWLAND RAINFOREST; AFROTROPICAL SWAMP FOREST; MIOMBO WOODLAND.

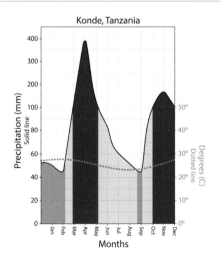

Konde, Tanzania

Precipitation (mm) Solid line

Degrees (C) Dotted line

Months

DESCRIPTION: To the north, east, and south of the vast Congo Basin rainforest is another type of humid forest called monsoon forest. Although it is also lush and has a closed canopy, this habitat has several significant differences from true AFROTROPICAL LOWLAND RAINFOREST. Most significantly, for several months of the year, little or no rain falls in monsoon forest, resulting in a level of stress to the plants that is rare in a true rainforest. This is manifested in the accumulation of leaf litter on the forest floor, giving a monsoon forest during the dry season an atmosphere closer to that of a Northern Hemisphere deciduous broadleaf forest in autumn than a rainforest. Another general difference is that monsoon forest has more lianas (hanging vines) and palms in its mid-story. The understory is often quite open, allowing relatively easy exploration on foot.

Monsoon forest rarely occurs in vast blocks, unlike lowland rainforest. It is found patchily, both along the coast and along watercourses in the interior. The timing of the distinct wet and dry seasons in monsoon forest varies around the continent. Along the n. Indian Ocean coast, near the equator, there are two rainy seasons: a long one from April to June and a short one from November to December. Elsewhere, most rain falls during the austral summer, approximately November to April. Typical rainfall is in the range of 30–60 in. (800–1,500mm), depending on the year and the

Afrotropical monsoon forest, Lake Manyara, Tanzania. © KEN BEHRENS, TROPICAL BIRDING TOURS

location. In general, the seasons are somewhat unpredictable and variable, and drought is common.

Throughout its range, monsoon forest is patchy and local, and this is especially true along the Indian Ocean coast. Long ago, there may have been a nearly solid band of coastal forest, but it has slowly been fragmented, first by a gradual drying trend and more recently and significantly by human activity. There is very little forest left along the northern part of the coast. Most of the Indian Ocean coast is covered in a matrix of grassland and savanna with scattered remnant monsoon forest patches. The exact character of monsoon forest varies a lot throughout its range. Plant diversity is high, with many localized endemics. Given the vast and discontinuous range and general complexity, it's hard to generalize about this forest's constituent trees. Many forests include *Hymenaea verrucosa*, Natal Guarri, figs, Panga Panga, *Marquesia acuminata*, and Pod Mahogany. Typical trees of sand and dune forest are Coastal Red Milkwood and Lebombo Wattle.

In drier parts of the MIOMBO WOODLAND zone, monsoon forest occurs along watercourses in very thin bands that may be only a few trees wide and can be walked through in mere moments. Farther north in that zone are wider forests called *mushitu*, which can be 1,500 ft. (450m) or more across. In w. Zambia and just across the border into Angola are extensive stands of dry evergreen forests known as *mavunda*, which are dominated by *Cryptosepalum exfoliatum*. Some of the monsoon forest growing on infertile sands along the coast of Kenya is open and miombo-like and even includes some *Brachystegia*, miombo's signature tree genus. Elsewhere on the Kenyan coast, but in close proximity, occurs a very different sort of monsoon forest: dense evergreen stands of *Cynometra webberi*. Along the s. Indian Ocean coast, there are vast areas of sandy soil, including huge dunes that can be over 650 ft. (200m) tall, where several different types of sand forest and dune forest grow. Sand forests near the coast that are exposed to salt spray and wind are dense and stunted and may have a canopy as low as 16 ft. (5m). Even the lushest monsoon forest generally has a lower canopy than lowland rainforest, around 80 ft. (25m) maximum. Another type of monsoon forest, called groundwater forest, grows in parts of East Africa where rainfall is insufficient to support forest but where oddly out-of-place patches of lush forest are nurtured by abundant groundwater.

Much of the e. African coast is heavily populated and has been impacted by woodcutting and conversion to agriculture for thousands of years. Hunting has eliminated most of the big mammals, even in miombo areas where human population densities are low. Few formally protected areas include monsoon forest habitat, and most of those that exist are poorly policed.

WILDLIFE: Monsoon forest is a complex and exciting wildlife habitat. Very few species are found throughout the complete range of this habitat and its many distinctive variations. Most of the species that are found throughout are also shared with AFROTROPICAL LOWLAND RAINFOREST. Typical monsoon forest mammals include Blue Monkey, Red-bellied Coast Squirrel, Suni, Natal Red Duiker, several species of galagos and sun squirrels, Bushy-tailed Mongoose, Bushpig, Bushbuck, and Blue Duiker. A few genera have different representatives in the north and south, such as Yellow Baboon and Ochre Bush Squirrel in the north and Chacma Baboon and Smith's Bush Squirrel in the south. The n. Indian Ocean forests hold three awesome species of giant sengi, or elephant shrew, as well as the Sokoke Dog Mongoose and the fur-cape-wearing black-and-white Angola Colobus. Two beautiful red colobuses are localized endemics in the north: Tana River Red Colobus along its namesake river and Zanzibar Red Colobus on its namesake island. Pemba and Mafia Islands host Africa's only flying foxes: an endemic species on Pemba and a species shared with the Comoros and Seychelles on Mafia.

Monsoon forest is one of Africa's more fascinating birding habitats. This is not only because of its own richness but because it tends to occur in a complex matrix alongside Afrotropical lowland rainforest, AFROTROPICAL MOIST MIXED SAVANNA, MIOMBO WOODLAND, and AFROTROPICAL FRESHWATER WETLAND and/or AFROTROPICAL MANGROVE, attracting a huge variety of birds. Very few birds are found throughout the entirety of the monsoon forest discussed here, but some widespread and typical birds include Fasciated Snake-Eagle, Green Malkoha, Crowned Hornbill, Livingstone's Turaco, White-eared Barbet, Common Square-tailed Drongo, Gray-olive and Sombre Greenbuls, Eastern Nicator, Black-backed Puffback, Gorgeous Bushshrike (a subspecies of Four-colored Bushshrike), Chestnut-fronted Helmetshrike, Black-bellied Starling, Olive Sunbird, Forest Weaver, Peters's Twinspot, Plain-backed Sunbird, and Black-tailed Waxbill.

Many species are shared with the Congo Basin lowland rainforest; just a few exciting examples are Palm-nut Vulture, Crested Guineafowl, Pel's Fishing-Owl, Narina Trogon, and African Broadbill. For visiting birders, some of the most appealing birds are localized endemics. The northern coastal zone is especially rich, boasting the likes of Fischer's Turaco, Sokoke Scops-Owl, Sokoke Pipit, Yellow Flycatcher, and Clarke's Weaver. The southern endemics include Rudd's Apalis, Livingstone's Flycatcher, Woodward's Batis, and Neergaard's Sunbird. Monsoon forest is an important habitat for the enigmatic and migratory African Pitta. The monsoon forests within the miombo zone have a different mix of species, a higher proportion of which are shared with the lowland rainforest. Some of the typical birds there are Schalow's and Ross's Turacos, White-tailed Blue Flycatcher, and Cabanis's Greenbul.

Golden-rumped Elephant Shrew is a weird and wonderful inhabitant of Afrotropical monsoon forest (photo: Arabuko Sokoke, Kenya). © PABLO CERVANTES, TROPICAL BIRDING TOURS

Monsoon forest is a prime habitat for Narina Trogon.
© KEN BEHRENS, TROPICAL BIRDING TOURS

Although reptiles are not especially conspicuous, this habitat does hold many species, including dozens of endemics. Typical reptiles include dwarf geckos (*Lygodactylus*), Tropical Girdled Lizard, Flap-necked Chameleon, several pygmy chameleons (*Rhampholeon*), blind snakes, Southeastern Green Snake, Cross-barred Tree Snake, and Savanna Vine Snake. Monsoon forest shares some of Africa's large and venomous snakes, such as Forest Cobra and Gaboon Viper, with the central and West African lowland rainforests.

Endemism: In monsoon forest there are many localized endemic birds, and some endemic smaller mammals, with some restricted to each of the three major zones: the miombo zone, the n. Indian Ocean coast, and the s. Indian Ocean coast. Within the n. Indian Ocean zone, s. Kenya and n. Tanzania are especially rich in localized endemics. The islands of Pemba, Zanzibar, and Mafia, off the coast of Tanzania, have a few endemics each.

DISTRIBUTION: There are two major areas of monsoon forest. One is along the Indian Ocean coast from s. Somalia south to e. South Africa. The other is inland, in the moister portions of the MIOMBO WOODLAND zone, where monsoon forest acts as a transitional habitat between miombo broadleaf savanna and AFROTROPICAL LOWLAND RAINFOREST. Throughout much of the miombo zone, monsoon forest is found locally along watercourses. The two major areas of monsoon forest are tenuously connected along the rivers that run through Tanzania and Mozambique. These connections are of great biogeographic importance, as they periodically would have connected the Indian Ocean coast monsoon forest with the greater Congo Basin, allowing for the exchange of species. The borders between monsoon forest and adjacent AFROTROPICAL MONTANE FOREST or Afrotropical lowland rainforest are not well defined. The forest changes character slowly with the transition to less seasonality and more rainfall (in the case of lowland rainforest) or higher elevation and a different plant community (in the case of montane forest). In the places where they abut, monsoon forest gives way to montane forest at around 2,600 ft. (800m) elevation. Yet, monsoon forest occurs at much higher elevations across much of the Central African Plateau, mostly at elevations between 3,300 ft. and 5,000 ft. (1,000–1,500m). Some of the forests in n. Tanzania, Kenya, and Ethiopia could be classified as monsoon forest but in this book are considered montane forest because they have wildlife more typical of that habitat. A third area that could be considered monsoon forest is a ribbon along the most northerly portions of the Congo Basin and Upper Guinea rainforest. These forests have a drier character than typical rainforest but don't have a set of wildlife that is distinctly different from that of lowland rainforest, but rather just a depauperate animal assemblage compared with the forest to the south.

WHERE TO SEE: Arabuko Sokoke National Park, Kenya; St. Lucia, South Africa; Lake Manyara National Park, Tanzania; Okavango Delta, Botswana.

AFROTROPICAL SWAMP FOREST

IN A NUTSHELL: Lowland rainforest that is completely flooded for at least part of the year, supporting a similar though subtly different set of wildlife than nonflooded forest and a very different set of plants. **Habitat Affinities:** NEOTROPICAL IGAPÓ AND VÁRZEA FLOODED FOREST; INDO-MALAYAN FRESHWATER SWAMP FOREST. **Species Overlap:** AFROTROPICAL LOWLAND RAINFOREST; AFROTROPICAL MANGROVE; AFROTROPICAL MONTANE FOREST.

Pale green denotes pockets within lowland rainforest or savanna.

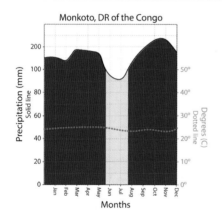

Monkoto, DR of the Congo

DESCRIPTION: The Afrotropical equivalent of the NEOTROPICAL IGAPÓ AND VÁRZEA FLOODED FOREST of South America, this habitat consists of forest that is completely flooded for at least part of the year. Although it looks very different when flooded and has a profoundly different botanical composition from normal AFROTROPICAL LOWLAND RAINFOREST, it is not drastically different in terms of its general structure and larger wildlife. The canopy of swamp forest is lower than that of rainforest, averaging 65–80 ft. (20–25m), with emergent trees up to 130 ft. (40m) tall. The soils are gleysols, typical of flooded environments. The diverse set of trees includes Boarwood, *Mitragyna*, *Uapaca*, and *Pseudospondias*, to name just a few.

The understory is usually dense and often includes raffia palms and other rattans, and African Oil Palm, here in its natural range. Accessing this habitat is even harder than accessing typical lowland rainforest and is usually possible only by boating along the waterways. The defining characteristic of this habitat is flooding, which happens on a seasonal basis, inundating the forest to a depth of up to 3 ft. (1m), then slowly drying out. The main flooding in the Niger Delta is August through December, while in the vast Congo Basin there are two major flood periods: December–January and May.

This is one of the most poorly known and least frequently visited of Africa's habitats. Despite its inaccessibility, this habitat has many conservation challenges. The Niger Delta is surrounded by heavily populated areas and is under great pressure for timber, fish, and bushmeat. The valuable Abura tree has already been virtually wiped out of the delta. The Congo Basin is much larger and has a much lower population, but African Forest Elephant and African Forest Buffalo have still been wiped out of most areas adjacent to navigable waterways.

WILDLIFE: This habitat supports a similar set of wildlife to AFROTROPICAL LOWLAND RAINFOREST, though it is preferred over lowland rainforest by a number of species, mainly primates. Unfortunately, wildlife-watching here is tough. Swamp forest is found mostly in countries and areas that are difficult to access, and the mammals and larger birds in this heavily hunted part of the world tend to be very shy. The critically endangered Niger Delta Red Colobus is endemic to its namesake delta. Pygmy Hippopotamus is associated with swamp forest, though sadly it has been eliminated from the eastern portion of its range, including the Niger Delta. Widespread West African species that are fond of this habitat include African Forest Elephant, Allen's Swamp Monkey, De Brazza's and Mona Monkeys, Forest Giant Squirrel, Thomas's Rope Squirrel, Alexander's Kusimanse, Long-tailed Pangolin, Red River Hog, and Water Chevrotain. The Congo River is a major biogeographic barrier for mammals; among species that favor swamp forest, Gray-cheeked Mangabey occurs exclusively on the north side, while Golden-bellied and Black-crested Mangabeys and Dryas Monkey live exclusively on the south side.

The swamp forest bird community has not been well studied; in general, it is a subset of the birds of Afrotropical lowland rainforest. No species are restricted to swamp forest, though it is prime habitat for White-crested Bittern, Spot-breasted Ibis, Hartlaub's Duck, African Finfoot, White-spotted Flufftail, Vermiculated and Rufous Fishing-Owls, Swamp Palm Bulbul (IS), Cassin's Flycatcher, Congo Sunbird, and Orange Weaver.

DISTRIBUTION: This is the main habitat in the central part of the Congo River basin and on the Niger River delta. Away from those vast areas, swamp forest is present only locally within normal AFROTROPICAL LOWLAND RAINFOREST. It also can be found very locally in the rainforest buffer zone and even within AFROTROPICAL MONSOON FOREST habitat (not mapped).

WHERE TO SEE: Semuliki National Park, Uganda.

Above: **Hartlaub's Duck is a specialist of Afrotropical swamp forest (photo: Ghana).** © KEN BEHRENS, TROPICAL BIRDING TOURS

Left: **The striking Palm-nut Vulture is attracted to *Raphia*, *Elaeis*, and *Phoenix* palms, which occur at high density in Afrotropical swamp forest (photo: Ghana).** © KEN BEHRENS, TROPICAL BIRDING TOURS

SIDEBAR 4.3   AFRICA'S GREAT RIFT VALLEY

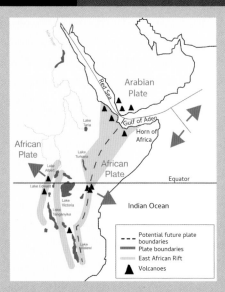

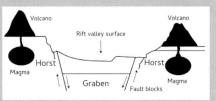

The Great Rift Valley is a remarkable geological feature, where you can stand in what may become a new sea between a big chunk of East Africa and the rest of the continent. It is actually part of a three-branched system, with sister rifts under the Red Sea and the Gulf of Aden, yet it did not fill with ocean water, unlike the other two arms of the system. As the Nubian (the main African) Plate and the new Somalian (East African) Plate get ripped apart, the land surface breaks along fault lines, and the area between these faults drops down. The area that drops down is called a graben, and the areas that are lifted up are called horsts. To make the whole scenario even more dramatic, areas of thinned crust on either side of the rift valley are prone to volcanism, giving us Mt. Kilimanjaro and many other spectacular peaks. These mountains, in turn, receive a lot more rain than the surrounding flatlands. So this rifting is the reason you can be in a montane forest on Tanzania's Mt. Meru while looking down at semidesert thornscrub and alkaline lakes in the bottom of the Great Rift Valley.

A painting of the Great Rift Valley in Kenya. © IAIN CAMPBELL, TROPICAL BIRDING TOURS

AFROTROPICAL MONTANE FOREST

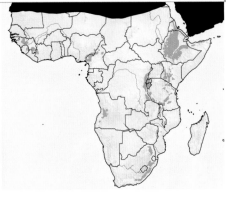

IN A NUTSHELL: Forest growing at moderate and high elevations and hosting a distinctive assemblage of wildlife. Varies from moderately dry to extremely wet. **Habitat Affinities:** NEOTROPICAL CLOUD FOREST; NEARCTIC CLOUD FOREST; INDO-MALAYAN TROPICAL MONTANE RAINFOREST; AUSTRALASIAN SUBTROPICAL AND MONTANE RAINFOREST; INDIAN OCEAN RAINFOREST. **Species Overlap:** AFROTROPICAL MONSOON FOREST; AFROTROPICAL LOWLAND RAINFOREST; AFROTROPICAL SWAMP FOREST.

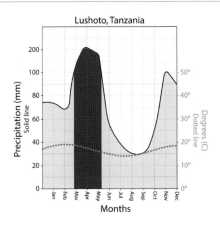

Lushoto, Tanzania

DESCRIPTION: Mountains tend to give rise to moist microclimates in which forest grows. This is because of the orographic effect: mountains force air upward, cooling it down and precipitating its moisture (sidebar 3.3). Even within very arid climates, such as that of n. Kenya, isolated mountains still produce enough moisture to foster forest. Although Afrotropical montane forest can look superficially similar to AFROTROPICAL LOWLAND RAINFOREST, the cooler high-elevation climate produces a subtly different sort of forest with a distinctive set of wildlife. One key difference is that in montane forest, much of the moisture comes in the form of mist, which gives rise to abundant mosses and epiphytes (and gives similar habitat in South America the name NEOTROPICAL CLOUD FOREST). Compared with superficially homogeneous lowland rainforest, montane forest is quite varied. Its classic form is similar to that of lowland rainforest: a lush and complex closed-canopy forest composed of broadleaf trees. Subtle differences include a lower canopy, more mosses, epiphytes, and lianas, and a different mix of plants. But this form of montane forest tends to grow only in the moistest areas, usually on the side of a mountain that receives the most moisture. Drier slopes have more open, semi-deciduous forest at lower elevations and coniferous *Podocarpus* and/or cedar-dominated forest at higher elevations. Most of the montane forest in the Arabian Peninsula and Somalia, and much of it in Ethiopia and Eritrea, is relatively dry and dominated by

Above: **Afrotropical montane forest in w. Uganda will feel familiar to naturalists who have visited South American cloud forests.**
© IAIN CAMPBELL,
TROPICAL BIRDING TOURS

Right: **Co-author Ken Behrens inspecting Afrotropical montane forest near Arusha, Tanzania.** © IAIN CAMPBELL,
TROPICAL BIRDING TOURS

cedar (*Juniperus*). The driest rain-shadow-affected slopes may completely lack forest but instead have savanna running nearly all the way to the mountaintop. In some cases, transitional woodlands of highland acacias like Flat-topped Acacia and Lahai Acacia grow in such areas. Even within the moistest montane forest there is considerable variation. Montane forest often grows patchily within a natural matrix of lush scrub and AFROTROPICAL MONTANE GRASSLAND. One common aspect of most of its permutations is a thick understory that makes it difficult to walk through except on existing roads and trails. There is oftentimes a band of African Alpine Bamboo just above the forest zone and below the heath zone, and bamboo is commonly mixed into the forest as well.

The climate that supports montane forest is generally warm during the day and cool at night. Temperatures below freezing sometimes occur in the southern part of the continent. Rainfall varies widely, and much of the moisture arrives in the form of mist rather than rain. The driest montane forest receives around 20 in. (500mm) of rainfall annually; this seems to be the approximate lower limit for supporting this habitat. The wettest forests on Mt. Cameroon and Mt. Bioko can receive 400 in., or 33 ft. (10,000mm), of rain in a year, making them among the wettest places on earth. Montane forests tend to receive some moisture throughout the year, but most have at least one major rainy season, and some mountains in East Africa have two distinct rainy seasons each year. It's also hard to generalize about the elevation at which montane forest is found. In most places with lowland forest, the transition to montane forest occurs at around 2,600 ft. (800m), though it can sometimes be as low as 1,600 ft. (500m). In Uganda, the transition occurs at around 5,000 ft. (1,500m), while in the cool climate of s. South Africa, montane forest is found all the way to sea level. The highest montane forests are found at around 11,500 ft. (3,500m), though these are exceptional, and the transition from forest to AFROTROPICAL MONTANE HEATH or Afrotropical montane grassland is much lower on most mountains.

Due to the vast and widely scattered range and complexity of Afrotropical montane forest, its tree species vary. On one level, the forest seems uniform, as most of the trees are superficially similar, aside from the few coniferous species and highland acacias. This stands in contrast to more arid environments, especially AFROTROPICAL MOIST MIXED SAVANNA, which has an astounding diversity of very obviously different-looking trees. But the superficial uniformity of montane forest hides great diversity, and a remarkable range of plants can be found in a small area, especially in the moister and more extensive montane forests near the center of the continent. The Eastern Arc, Albertine Rift, and Cameroon mountains each have around 1,000 endemic plants. With all that said, there are some classic montane forest trees with wide ranges across the continent. These include olives, Cape Beech, crotons, Cape Ash, *Nuxia*, hollies, African Cherry, *Ocotea*, East African Mahogany, and Waterberry.

The already scattered and isolated nature of this habitat makes it especially susceptible to human pressures. Foremost among these is the conversion of forest into farmland, as it grows in prime places for agriculture. Conflict between humans and animals can become a major issue in places where plantations and villages run right up to the edge of a national park. Another major threat, especially in areas with abundant natural grassland, is perennial human-started fires, which gradually shrink the forest, eventually leaving it confined to narrow gorges and valleys.

WILDLIFE: While montane forest is not as diverse as lowland forest or savanna, it does have fairly high diversity along with many localized endemics. It is a challenging habitat

L'Hoest's Monkey is a fairly common species in the Albertine Rift forests of w. Uganda.
© IAIN CAMPBELL, TROPICAL BIRDING TOURS

Mountain Gorilla lives in Afrotropical montane forest restricted along portions of the Albertine Rift.

SIDEBAR 4.4   ALBERTINE RIFT: THE HUB OF AFRICA'S MONTANE FOREST

The Albertine Rift is found along the eastern border of the Democratic Republic of the Congo and the western borders of Uganda, Rwanda, and Burundi. Historically, the mountain islands (sidebar 4.5) of East Africa provided important connections between the AFROTROPICAL MONSOON FOREST along the east coast and the rainforest of the Congo Basin. The Albertine Rift is the hub at the center of the continent, connecting all of Africa's humid forests. As such, it was colonized from similar habitats in multiple directions and as a result ended up having the continent's most diverse AFROTROPICAL MONTANE FOREST. The AFROTROPICAL LOWLAND RAINFOREST immediately west of the Albertine Rift mountains is also some of the continent's most diverse.

Sunbirds are usually prominent in Afrotropical montane forest. The gorgeous Northern Double-collared Sunbird is a common bird of forests and gardens in some highland areas of e. and c. Africa. © IAIN CAMPBELL, TROPICAL BIRDING TOURS

for wildlife-viewing—though not as tough as AFROTROPICAL LOWLAND RAINFOREST. The many localized endemic mammals of Africa's montane forest are mostly small and inconspicuous: mice, rats, and shrews. The patches of forest apparently weren't large or stable enough to allow for the evolution of many large endemic mammals. Some of the smaller mammals that specialize in this habitat are Eastern Tree Hyrax, L'Hoest's and Preuss's Monkeys, Carruther's Mountain Squirrel, the four subspecies of Tanganyika Mountain Squirrel, and Jackson's Mongoose. The Eastern Arc Mountains are home to some rare monkeys, of which the strangest is the Kipunji. Described only in 2005, this arboreal brown monkey has similarities to both baboons and colobuses. Most of the large mammals of montane forest are widespread species that use a broad range of habitats. These include Angola and Guereza Colobuses, African Bush Elephant, Leopard, Lion, Slender Mongoose, African Buffalo, Blue Duiker, and Bushbuck. Quite a few species are shared with either the AFROTROPICAL MONSOON FOREST to the east, the Afrotropical lowland rainforest to the west, or both. Examples of these are African Forest Elephant, Giant Forest Hog, and Harvey's and Weyns's Duikers. Montane forest is the most important component in the mix of habitats used by Mountain Gorilla.

For birders, montane forest is one of the continent's most exciting habitats due to its many localized endemics. For keen birders, while the sweeping savannas might inspire their first couple of trips to Africa, it is the forests that bring them back over and over. Some of the diverse and widespread groups of birds in montane forest include pigeons, turacos, greenbuls, robins, apalises, and sunbirds. Classic species of this habitat are Crowned Eagle, Rameron Pigeon, Bar-tailed Trogon, Silvery-cheeked and Black-and-white-casqued Hornbills, Western and Moustached Tinkerbirds, Gray-chested Babbler, Gray Cuckooshrike (IS), White-starred Robin, Pink-footed Puffback, Sharpe's Starling, and Brown-capped Weaver. The Albertine Rift holds the most diverse montane forest overall, and among the most interesting of its dozens of endemics are Grauer's Broadbill, Red-collared Mountain-Babbler, and Neumann's and Grauer's Warblers.

The next most endemic-rich forests are those in the ancient, nonvolcanic Eastern Arc Mountains, which run from se. Kenya all the way through Tanzania into n. Malawi. The most remarkable bird here is the Udzungwa Partridge, which is genetically similar to Asian partridges and was described only in the 1990s. Other Eastern Arc endemic birds are Usambara Eagle-Owl, Dapple-throat, Spot-throat, African and Long-billed Tailorbirds, and Uluguru Bushshrike. The mountains of n. Tanzania and Kenya are much newer, mostly of volcanic origin, and have a mix of the species of adjacent montane blocks. Such a mix of species is also found in the mountains of far s. Tanzania and Malawi.

The Ethiopian Highlands have many endemic species, though only a few, such as Yellow-fronted Parrot and Ethiopian Black-headed Oriole, are primarily forest-dwelling. In the southern part of the continent, there are two further minor centers of avian endemism: the highlands of e. Zimbabwe and w. Mozambique and the scattered montane forests of South Africa. The Angolan escarpment has some forest endemics such as Swierstra's Francolin and Red-crested Turaco. West Africa has one major area of mountains, along the west side of Cameroon. These hold the richest montane forests outside of East Africa, with endemics that include Mount Cameroon Spurfowl, Bannerman's Turaco, White-throated Mountain-Babbler, Green Longtail, and White-tailed Warbler. In far w. Africa, in the Upper Guinea region, are some low mountains with relict forests, but these mainly support species typical of Afrotropical lowland rainforest.

Montane forest holds many endemic reptiles. Some of the prominent groups include forest geckos (*Cnemaspis*), dwarf geckos (*Lygodactylus*), forest lizards (*Adolfus*), and bush vipers (*Atheris*). Around half of the world's chameleons are endemic to Madagascar, but most of the rest are found in African montane forest. The Eastern Arc Mountains are the richest, with 10 endemic chameleons, ranging from the hefty Usambara Two-horned Chameleon to the diminutive Uluguru Pygmy Chameleon. Nearly all the major mountain regions have at least one endemic chameleon. The abundant moisture in montane forest makes this a good habitat for amphibians. The Cameroon mountains have about 40 endemics; the Albertine Rift, 32; and the Eastern Arc, 25. Some of the diverse groups include reed frogs (*Hyperolius*), long-fingered frogs (*Cardioglossa*), puddle frogs (*Phrynobatrachus*), screeching frogs (*Arthroleptis*), and tree frogs (*Leptopelis*).

DISTRIBUTION: "Sky islands" of montane forest dot the east side of Africa from Eritrea all the way to s. South Africa (see sidebar 4.5). Outliers exist on the sw. Arabian Peninsula, in Cameroon and Upper Guinea, and in Angola. The majority of Africa's mountains are associated with the Great Rift Valley, whose formation forced up blocks of mountains and volcanoes (sidebar 4.3).

WHERE TO SEE: Nyungwe National Park, Rwanda; Arusha National Park, Tanzania; Wilderness National Park, South Africa.

SIDEBAR 4.5   AFRICAN SKY ISLANDS

Biogeographically, AFROTROPICAL MONTANE FOREST is the continent's most complex and fascinating wildlife habitat. This is because it exists as a vast archipelago within a sea of other habitats. While there is a certain continuity to this habitat, and some widespread species of plants and animals, each montane forest island is slightly different, and there are several major areas of endemism, as clusters of mountains that were frequently connected historically developed distinct sets of species. Over the geological past, as some mountains eroded and others were created by volcanism, and as the climate alternated between hot and dry and cool and wet, islands of humid montane forest were connected and then sundered, over and over. This dynamic has produced a patchwork of biological diversity that is one of the world's most complex, perhaps rivaled only by the actual archipelagos of se. Asia and Australasia.

INDIAN OCEAN RAINFOREST

IN A NUTSHELL: Lush broadleaf forest with a complex structure and a canopy of variable height that remains moist and humid throughout most or all of the year. Found on Madagascar, and the Mascarene, Comoros, and Seychelles archipelagoes. **Habitat Affinities:** NEOTROPICAL CLOUD FOREST; NEOTROPICAL LOWLAND RAINFOREST; INDO-MALAYAN TROPICAL MONTANE RAINFOREST; INDO-MALAYAN TROPICAL LOWLAND RAINFOREST; AUSTRALASIAN TROPICAL LOWLAND RAINFOREST. **Species Overlap:** MALAGASY DRY DECIDUOUS FOREST; MALAGASY SPINY FOREST.

Pale green denotes historic distribution.

Mauritius

DESCRIPTION: The overall character of Indian Ocean rainforest is superficially similar to that of other tropical rainforests, such as those on the African mainland (AFROTROPICAL LOWLAND RAINFOREST). But a close look reveals a high diversity of plants, the vast majority of which are endemic, along with an incredibly rich set of endemic wildlife. Within the Indian Ocean region, the rainforests of Madagascar are by far the most diverse, while the other Indian Ocean islands have vastly reduced sets of plants and animals. Because of its rainforests, Madagascar has almost continental-level diversity in some groups and fully deserves the moniker "the eighth continent."

Rainforest is found from sea level up to the tree line, which occurs at 6,500–9,000 ft. (2,000–2,750m) on the Indian Ocean islands. Within this broad elevational range, the forest varies significantly. At

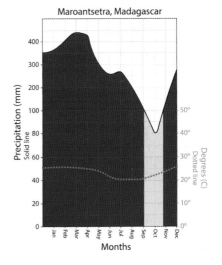

low elevations and along valleys, it can be tall, up to 130 ft. (40m), and lush, with abundant pandanus trees. At high elevations and on exposed sites such as ridgelines, the forest is more stunted, as short as 10 ft. (3m). Higher-elevation forest has the characteristics of montane or cloud forest, with abundant tree ferns, epiphytes, and mosses. The driest lowland areas receive as little as 30 in. (800mm) of rain annually, though 80 in. (2,000mm) is typical, and the wettest parts of the highlands can receive around 400 in. (10,000mm) of rain in a year. Some rain falls in every month of the year, but there is a pronounced cooler, drier season from April to October and a warmer, wetter season from November to March. This pattern is created by the intertropical convergence zone, which moves south during the austral summer. Some drier semi-deciduous forest was formerly found on some of the islands, especially Mauritius and Rodrigues in the Mascarenes, but this forest has been virtually obliterated by humans. The sandy island of Aldabra in the Seychelles is covered in scrubby habitat that is drier than the true rainforest on other islands. This range of forest types is considered together in this book because there are only minor differences in their assemblages of larger wildlife.

The rainforest on Madagascar holds thousands of plants, of which more than 80% are endemic. The other islands have much smaller plant assemblages, though endemism remains high. Some widespread tree genera on Madagascar are *Symphonia, Ficus, Tambourissa, Albizia, Diospyros, Ocotea, Dracaena, Dombeya,* and *Olea.* Typical families include coffee (*Rubiaceae*), ebony (*Ebenaceae*), myrtle (*Myrtaceae*), and laurel (*Lauraceae*). Some trees characteristic of higher-elevation forest are species of *Schefflera* and *Weinmannia.* There are many endemic palms, one of which is the spectacular Sea Coconut, a towering tree endemic to the Seychelles that boasts the largest seed in the plant kingdom. Scrubby secondary forest, known locally as *savoka,* is common on Madagascar and is the default habitat on all the other islands. It typically contains widespread exotic plants like Strawberry Guava, Common Lantana, gingers, and brambles. On Madagascar, vast areas of savoka are now dominated by the indigenous Traveler's Palm. Eastern Madagascar has been subject to vast deforestation. Lower-elevation forests away from the west side of the Masoala Peninsula have been decimated; most of the remaining rainforest is at middle and higher elevations. A look at satellite imagery shows that the continuity of the rainforest belt is being lost. In several places, there is no longer any intact rainforest remaining between the top and bottom of the east slope. This is of grave concern for the future integrity of Madagascar's biodiversity. There are several national parks that protect large tracts of rainforest, though their boundaries are poorly enforced.

As concerning as the conservation situation is on Madagascar, it is much worse on the other Indian Ocean islands. Due to hunting, the introduction of exotic mammals, and the destruction of forest, the Mascarene Islands have lost much of their wildlife. This included the charismatic Dodo

Classic Indian Ocean rainforest is found at Marojejy National Park, Madagascar.
© KEN BEHRENS, TROPICAL BIRDING TOURS

of Mauritius. (What other place on earth is more famous for an extinct animal it once hosted?) There have been fewer extinctions on the Comoros and Seychelles, but many of the endemic species there hang on by a thread. Well-organized conservation efforts are afoot in the Seychelles and the Mascarenes, but the same cannot be said for the poor and heavily populated Comoros.

WILDLIFE: Any discussion of Indian Ocean rainforest habitat will revolve around Madagascar, as it is far more diverse than the other islands and has been the main engine of evolution in the region. The other island groups have a much-reduced subset of Malagasy wildlife. Rainforest is by far Madagascar's richest habitat for most groups. Its marquee mammals are the lemurs, a whole endemic radiation of the primate order. The largest lemur is the Indri, which gives a shockingly loud song that is simultaneously reminiscent of both whales and wolves. There are also three beautiful species of sifakas and Black-and-white Ruffed and Red Ruffed Lemurs. These big lemurs have a diverse supporting cast of smaller species: mouse lemurs, dwarf lemurs, bamboo lemurs, brown lemurs, and woolly lemurs. Like all Malagasy forest, rainforest supports the rarely seen Aye-aye, a bizarre lemur with huge eyes, bushy fur, and an elongate, skeletal middle finger that makes up its own family.

Much of Madagascar's mammal diversity is within the endemic tenrec family. Its members include the large Tailless Tenrec, several hedgehog-like tenrecs, the stream-dwelling, otter-like Web-footed Tenrec, and a diverse range of smaller shrew tenrecs. Unfortunately, Tailless Tenrec has been introduced to most of the other Indian Ocean islands and has become an invasive species there. Madagascar has an endemic subfamily of rodents that includes the odd tree-dwelling Madagascar rats and a range of tuft-tailed rats. The Malagasy carnivorans (Eupleridae) are another endemic family, most of whose members are rainforest-dwelling. The largest is the lemur-hunting Fossa, which looks like a long and lean arboreal Puma. Smaller and more mongoose-like carnivorans are the Fanaloka, Ring-tailed Vontsira, and Eastern Falanouc. Madagascar has a rich assemblage of bats. The other Indian Ocean islands are virtually devoid of indigenous mammals save for bats, the most notable of which are several large species of flying foxes.

Rainforest supports more Malagasy endemic birds than any other habitat on the island, including the majority of the members of the island's endemic families. There are four spectacular ground-rollers: Pitta-like, Scaly, Short-legged, and Rufous-headed. Rainforest also supports one of the island's three mesites, three of four asities, and most members of the endemic Malagasy warbler family. Other typical rainforest birds include Madagascar Ibis, Madagascar Flufftail, Madagascar Blue-Pigeon (IS),

The Indri is the largest living lemur (there were much larger lemurs, but those are all now extinct). It is found exclusively in rainforest, where its wailing call is one of Madagascar's signature natural sounds.

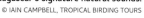
© IAIN CAMPBELL, TROPICAL BIRDING TOURS

Greater and Lesser Vasa Parrots, Blue and Red-fronted Couas, Madagascar Pygmy-Kingfisher, Cuckoo-roller, Madagascar Paradise-Flycatcher, and Nelicourvi Weaver. Rainforest supports a bounty of vangas, which run the gamut from the small and chickadee-like Red-tailed Vanga to the hefty and huge-billed Helmet Vanga. This is the main habitat for two near-mythical raptors: Madagascar Serpent-Eagle and Red Owl.

Surprisingly, the avifauna throughout the rainforest remains largely uniform, although diversity peaks at around 2,600 ft. (800m). In terms of birds, the most distinct subdivision is the higher-elevation forest, at approximately 3,300 ft. (1,000m) and higher. Several species are restricted to this zone, including Rufous-headed Ground-Roller and Yellow-bellied Sunbird-Asity. A handful of species, including Helmet and Bernier's Vangas, seem to be restricted to the northern half of Madagascar's rainforest belt, though for unknown reasons.

Scaly Ground-Roller is just one of four fabulous ground-rollers that are found exclusively in Malagasy Indian Ocean rainforest. This endemic family is a major target for visiting birders.
© KEN BEHRENS, TROPICAL BIRDING TOURS

Each of the other groups of Indian Ocean islands has its own assemblage of rainforest endemic birds. The Comoros have the richest set, more than 20 endemics, and the Seychelles the poorest, around a dozen. Most of these are closely related minor variants on Malagasy bird species, though there are exceptions, such as Echo Parakeet on Mauritius, Comoro Pigeon on the Comoros, and the odd Grand Comoro Flycatcher on Grande Comore. The Seychelles support the Seychelles Blue-Pigeon (IS), and the Comoros have the Comoro Blue-Pigeon (IS). Remote Aldabra has a flightless subspecies of White-throated Rail. Sadly, the most charismatic and distinctive birds of the Mascarene Islands, the likes of Dodo, Rodrigues Solitaire, Mascarene Parrot, and Réunion Starling, are long extinct.

Madagascar is fabulously rich in reptiles, boasting over 400 species, compared to around 800 for the whole continent of Africa. More than half of these are endemic to the rainforest. Madagascar's most famous reptiles are chameleons, and around half of the world's chameleons are endemic to the island, mostly to the rainforest. These include the world's two largest species, Parson's and Oustalet's Chameleons; a range of midsize species; and the dwarf chameleons (*Brookesia* and *Palleon*), some of which are among the world's smallest vertebrates. Another rich group in the rainforest is the colorful, diurnal day geckos (*Phelsuma*). The leaf-tailed geckos (*Uroplatus*) are wonderfully cryptic and include the huge Common Leaf-tailed Gecko, the bamboo-mimicking Lined Leaf-tailed Gecko, and the red-eyed Satanic Leaf-tailed Gecko. Other Malagasy rainforest lizards include plated lizards (*Zonosaurus*), Afro-Malagasy skinks (*Trachylepis*), and ground geckos (*Paroedura*). The diverse range of snakes includes the bizarre leaf-nosed snakes (*Langaha*) and a small group of biogeographically mysterious boas (*Sanzinia* and *Acrantophis*).

Rainforest on the other islands is far poorer in reptiles, which number no more than a few dozen species on each archipelago, mostly members of groups that are far more diverse in Madagascar, such as day geckos and skinks. One major exception is the Aldabra Giant Tortoise of the Seychelles, which has narrowly avoided extinction, making this the only other place in the world, besides the Galápagos, where you can see one of the formerly widespread giant tortoises. The average weight of a male tortoise is 550 lb. (250kg). Another exception is the Seychelles Tiger Chameleon, which belongs to a monotypic genus quite distinct from the chameleons of Madagascar.

Over 300 frogs have been described in Madagascar, and DNA barcoding suggests there may be around 500 species. The vast majority of these are endemic to the rainforest. Just a few of the noteworthy rainforest frogs are the Starry-night Reed Frog, Marbled Rain Frog, Madagascar fringed frogs (*Spinomantis*), a large array of bright-eyed frogs (*Boophis*), and the mantellas, which resemble the poison dart frogs of the Neotropics. Along with the smallest dwarf chameleons, the tiny stump-toed frogs (*Stumpffia*) count among the world's smallest vertebrates. The other Indian Ocean islands are very poor in frogs.

Madagascar is also rich in fascinating invertebrates, including a diverse array of butterflies, the huge Comet Moth, and the bizarre Giraffe-necked Weevil.

Endemism: On Madagascar, there are many mammals, reptiles, and amphibians that are restricted to a small portion of the rainforest. The most significant endemic areas for reptiles and amphibians are the n. Tsaratanana Massif and the Sambirano rainforest (sidebar 4.6). Other nodes of endemism for herps and mammals are the southeast of the island and the isolated Amber Mountain in the north. Most birds are found throughout the Madagascar rainforest, though a few species are restricted to the northern half of the rainforest belt. The flora and fauna of the Comoros, Seychelles, and Mascarenes are quite distinct, with many endemics on each island group.

DISTRIBUTION: Rainforest extends contiguously along most of the length of e. Madagascar, from Tsaratanana in the north to near Fort Dauphin in the south, and also locally in the far north and in isolated fragments in the central highlands. Though generally restricted to a narrow zone, the rainforest belt broadens in the north, nearly reaching the west coast near the island of Nosy Be. Rainforest formerly blanketed the other Indian Ocean islands but is now restricted to small inaccessible or formally protected areas on the Comoros, Seychelles, and Mauritius. Only Réunion, with its rugged highlands, has managed to hold on to around 40% of its forest cover.

WHERE TO SEE: Andasibe-Mantadia National Park, Madagascar; Ranomafana National Park, Madagascar; Black River Gorges National Park, Mauritius.

SIDEBAR 4.6   SAMBIRANO RAINFOREST

Most of Madagascar's rainforest is in a narrow belt, but the biome broadens in the north of the island. This westward extension of the rainforest zone, known as the Sambirano rainforest, is produced by the moist winds of the intertropical convergence zone, which lack a significant mountain obstacle north of the Tsaratanana Massif and wrap around that massif during the wet season, bringing moisture all the way to the west coast and the interior. This complex Sambirano area is rich in localized reptiles and amphibians and has a fascinating mixture of eastern rainforest and western dry forest bird species.

MALAGASY DRY DECIDUOUS FOREST

IN A NUTSHELL: Tall forest with a closed canopy that is dominated by deciduous trees and has a long dry season. **Habitat Affinities:** NEOTROPICAL DRY DECIDUOUS FOREST; INDO-MALAYAN DRY DECIDUOUS FOREST. **Species Overlap:** INDIAN OCEAN RAINFOREST; MALAGASY SPINY FOREST.

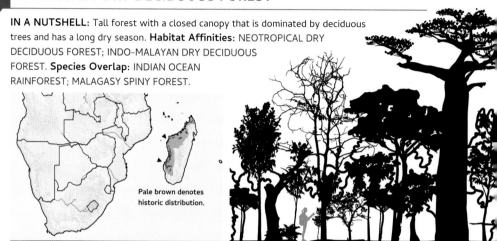

Pale brown denotes historic distribution.

DESCRIPTION: This is one of the world's richest dry forest habitats and the only such habitat in the Afrotropics. Like all the natural habitats of Madagascar, it is incredibly rich in fascinating, distinctive, and mostly approachable wildlife. Dry deciduous forest is fairly tall, usually reaching 30–50 ft. (10–15m), and has a closed canopy. In some areas on rich soil it can reach 80 ft. (25m). Away from watercourses, most of the trees are deciduous, losing their leaves during the dry season. The abundant leaf litter and open character of this habitat in the dry season almost make it resemble a temperate broadleaf deciduous forest in autumn. But at the height of the wet season, it feels almost like rainforest, dense and green. As with most forested habitats, there is little or no grass in the understory. This habitat lies in the rain shadow of Madagascar's

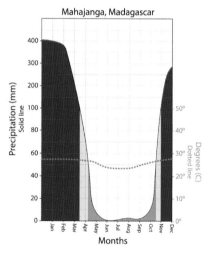

Mahajanga, Madagascar

eastern escarpment. It's much drier than the rainforest but wetter than the MALAGASY SPINY FOREST, as more moisture manages to curl around the northern end of the island and reach the northwest. The annual rainfall is generally 40–60 in. (1,000–1,500mm), though it can be as low as 20 in. (500mm) in the area of transition to spiny forest and as high as 80 in. (2,000mm) in the Sambirano region of the north, where this habitat blends with Sambirano rainforest (sidebar 4.6). Most of the year is dry, and the vast majority of the rain falls in a well-defined rainy season between November and March.

The plants of this habitat show a high degree of endemism, even greater than that of INDIAN OCEAN RAINFOREST. There are multiple species of baobabs, including the towering, columnar Grandidier's Baobab, one of the world's most spectacular trees. Other typical trees include ebonies, rosewoods, *Dupuya madagascariensis*, *Cedrelopsis*, pandanus trees, flame trees, and figs. In some areas, this habitat takes the form of thick scrub or thicket. Such places are likely to hold *Albizia*, commiphoras, acacias, and *Grewia*. A distinctive and beautiful community of plants forms on porous limestone, such as the limestone karst known as *tsingy*. These include rosewoods, *Adenia*, flame trees, euphorbias, and other succulents. Along rivers grows riparian forest of Tamarind, figs, and Mango.

This habitat has been decimated by human activity. About 97% of the original forest has already been destroyed, making it the Afrotropics' most severely threatened forest biome. Due mainly to fires but also to clearing for agriculture and wood, the vast majority of the west is now impoverished grassland or savanna in which Mango and fire-resistant palms survive. It is well on its way to following the central highlands down the path of near-complete deforestation.

WILDLIFE: Dry deciduous forest holds nearly as many mammals as the eastern rainforest (INDIAN OCEAN RAINFOREST), and they are much easier to see in this open habitat, which is a lemur-watching paradise. The mix of lemurs varies locally, as most of the mouse lemurs, sportive lemurs, and sifakas are restricted to only a portion of this habitat. The standout mammals are the big, diurnal sifakas. The six that use this habitat are, from south to north, Verreaux's, Von der Decken's, Crowned, Coquerel's, Perrier's, and Golden-crowned. Other diurnal lemurs include Western Lesser Bamboo Lemur; Common, Sanford's, and Red-fronted Brown Lemurs; and Crowned and Mongoose Lemurs. There are even more nocturnal lemurs: a bunch of species of mouse lemurs, including Madame Berthe's Mouse Lemur, the world's smallest primate; Fat-tailed Dwarf Lemur; Coquerel's Giant Mouse Lemur; Pale Fork-marked Lemur; a range of localized sportive lemurs; and Western and Cleese's Woolly Lemurs. This forest is also home to the Malagasy Giant Jumping Rat and is the best habitat for seeing the Fossa, a mongoose-cum-Puma and the island's largest carnivore.

There is not a large number of highly localized bird species in western dry forest, unlike mammals and reptiles. The majority of birds here are widespread forest species that are shared with the rainforest and/or MALAGASY SPINY FOREST, such as Madagascar Green-Pigeon; Giant Coua; Long-billed Bernieria; Blue, White-headed, Hook-billed, Sickle-billed, and Rufous Vangas; and Sakalava Weaver. Among the few birds found almost exclusively in western dry forest are White-breasted Mesite, Tsingy Wood-Rail, Coquerel's Coua (IS), Schlegel's Asity, Appert's Tetraka, and Van Dam's

Right: **Malagasy dry deciduous forest, Zombitse, Madagascar.**
© KEN BEHRENS, TROPICAL BIRDING TOURS

Below: **Blue Vanga lives in both Malagasy dry deciduous forest and Indian Ocean rainforest in Madagascar.** © PABLO CERVANTES, TROPICAL BIRDING TOUR

Malagasy dry deciduous forest is a paradise for a range of lemurs including the iconic Ring-tailed Lemur. © KEN BEHRENS, TROPICAL BIRDING TOURS

Madagascar's largest carnivorous mammal is the Fossa. Although it's also found in the rainforest, it's most readily seen in Malagasy dry deciduous forest. © LISLE GWYNN, TROPICAL BIRDING TOURS

Vanga. Most of these are quite local, even within the biome; this pattern may have been exacerbated by severe forest fragmentation, though forest destruction doesn't fully account for their scarcity and localized distribution.

Dry deciduous forest makes excellent habitat for reptiles, especially during the rainy season. There are many chameleons, including a bunch of dwarf chameleons (*Brookesia*), Rhinoceros Chameleon, and Labord's Chameleon. The latter is notable as the world's only "annual" reptile: its entire population dies off during the dry season, to be replaced by the hatching of eggs during the next rainy season. Lizards of western dry forest include Cuvier's Madagascar Swift, Western Plated Lizard, Western Skink, two Madagascar velvet geckos, several ground geckos, and three leaf-tailed geckos. Snakes are common and frequently sighted. Some of the most widespread include Western Madagascar Tree Boa, Madagascar Ground Boa, Madagascar and Blond Hognose Snakes, and Four-lined Snake. The western dry forest has high frog diversity. A few widespread species and groups are the reed frogs (*Heterixalus*), Madagascar Bullfrog, Western Bright-eyed Frog, bridge frogs (*Gephyromantis*), Madagascar frogs (*Mantidactylus*), and Betsileo Mantella.

Endemism: Major limestone massifs, especially the Tsingy de Bemaraha, the tsingy of Ankarana, and the Montagne des Français, support some highly localized endemics, mostly reptiles and amphibians but a couple of mammals and birds as well. Examples are the Tsingy Wood-Rail and Western Nesomys of Bemaraha, and the Tsingy Plated Lizard of Ankarana and Montagne des Français. Quite a few species are restricted to the southern portion of the western dry forest, as around Kirindy Forest. These include the Malagasy Giant Jumping Rat and the Narrow-striped Vontsira, shared with the Malagasy spiny forest.

DISTRIBUTION: Dry forest is the original habitat in the northern two-thirds of w. Madagascar and well into the southern interior, around the headwaters of the Mangoky River. It grades into rainforest in the Sambirano region of the north, around the Tsaratanana Massif. North of Tsaratanana, there is additional western dry forest, which locally reaches the east coast in the dry northern microclimate. Some of this northern forest is so dry that it actually resembles southwestern MALAGASY SPINY FOREST, though without any of that biome's unique wildlife. Wetlands in this zone have a character similar to FRESHWATER WETLAND in Africa and are Madagascar's richest. They support several endemics, such as Madagascar Jacana and Madagascar Fish-Eagle. The Mascarene Islands were historically dominated by humid forest but also had dry forest with a character similar to Malagasy dry deciduous forest. Sadly, this habitat is now virtually extinct.

WHERE TO SEE: Ankarafantsika National Park, Madagascar; Zombitse-Vohibasia National Park, Madagascar.

GUINEA SAVANNA

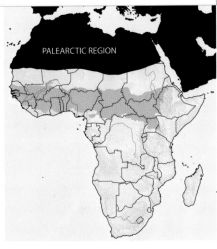

PALEARCTIC REGION

IN A NUTSHELL: Northern African savanna habitat that is dominated by broadleaf trees, especially Doka.
Habitat Affinities: AUSTRALASIAN BRIGALOW AND *CALLITRIS* WOODLANDS; NEARCTIC OAK SAVANNA; NEOTROPICAL CERRADO. **Species Overlap:** AFROTROPICAL DRY THORN SAVANNA AND THORNSCRUB; AFROTROPICAL MOIST MIXED SAVANNA; AFROTROPICAL LOWLAND RAINFOREST.

DESCRIPTION: Guinea savanna is moderately tall, broadleaf woodland with abundant grass on the ground and distinct wet and dry seasons. The canopy ranges from 16 to 60 ft. (5–18m), and is generally open, with less than 50% tree cover. This allows abundant light to reach the ground, permitting the growth of a thick cover of mostly perennial grasses, which persists year-round where not burned or grazed. This contrasts with the AFROTROPICAL DRY THORN SAVANNA AND THORNSCRUB of the Sahel to the north, where most grasses are annual, and grass cover can largely disappear during the dry season. Guinea savanna is park-like and easy to walk through during the dry season, when much of the ground cover has dried up and burned or been eaten by herbivores. During the wet season, the abundance of tall grass makes it hard to penetrate. For half of the year, Guinea savanna sees little rainfall, while heavy rains fall during the boreal summer. The typical annual precipitation range is 24–63 in. (600–1,600mm). Fire has always played an important role, burning the

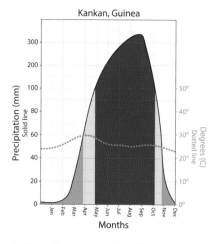

Kankan, Guinea

grasses and stimulating new growth, especially at the beginning of the wet season when the understory is still dry and thunderstorms bring lightning strikes. Today most fires are started by humans, resulting in a much higher frequency of burning and gradually opening up the habitat.

The vast majority of trees are broadleaf species that lose some or all of their leaves during the dry season. The most important of these is Doka, which dominates vast stretches of this habitat. Indeed, the Guinea savanna is sometimes known as Doka due to the importance of this tree. In some areas, Doka is absent or less important than other tree species, which can include Khaya, African Mahogany, Wild Seringa, African Locust Bean, and Shea. Bushwillows, and/or terminalias dominate some areas, especially in the northeastern and southeastern extensions of Guinea savanna, and in the north where it grades into Afrotropical dry thorn savanna. Guinea savanna is the northern equivalent of the

Recently burned classic Guinea savanna in Mole National Park, Ghana.
© KEITH BARNES, TROPICAL BIRDING TOURS

MIOMBO WOODLAND of e. and s. Africa. Both are open woodlands with grassy ground cover. Although miombo also has Doka, its most dominant and important trees are *Brachystegia* species, commonly known as miombo. The miombo habitat is also generally taller, with a more complex structure and more diverse plant composition. In terms of wildlife, there is only a modest degree of overlap between Guinea savanna and miombo. In many areas, smaller drainage lines support open AFROTROPICAL TROPICAL GRASSLAND that is seasonally flooded. Larger drainage lines and rivers give rise to deeper perennial wetlands, thickets, MONSOON FOREST, and riparian AFROTROPICAL LOWLAND RAINFOREST. These forests allow some rainforest species to penetrate deep into this savanna zone.

Some parts of this vast belt are remote and thinly populated by humans, such as ne. Guinea and sw. Mali. Others, such as c. and n. Nigeria, are heavily populated. Unfortunately, hunting is widespread, even in remote areas, and has wiped out most of the big mammals. Other major threats include tree clearing, frequent burning, and overgrazing. There are few national parks, and most of those that exist are poorly protected.

WILDLIFE: This habitat historically supported a wide variety of large mammals, though it always had lower diversity than the savanna habitats of East Africa. Unfortunately, the populations of many big mammals have been decimated by hunting and habitat modification. This is the stronghold of Kob (IS) and Giant Eland, though the latter is now restricted mostly to tiny pockets in Central African Republic, Cameroon, and Senegal. It is also one of the strongholds of Roan Antelope, which remains in good numbers, perhaps due to its wary nature. Widespread mammals include Olive Baboon; Patas, Green, and Tantalus Monkeys; Gambian Sun Squirrel; Common Warthog; Bushbuck; Common Duiker; Oribi; and Waterbuck. Black Rhinoceros and White Rhinoceros were once common, but Black Rhino hangs on the brink of local extinction, and the Northern subspecies of the White Rhino is now extinct in the wild. African Buffalo, Lion, Leopard, Cheetah, and African Wild Dog persist, though in ever-shrinking enclaves. This is excellent habitat for African Bush Elephant, though it has also been widely extirpated.

The Guinea savanna has excellent bird diversity, though it is inferior to that of the more topographically and climatically complex savannas of East Africa. There are a few species virtually restricted to this habitat, though most birds are shared with the more arid AFROTROPICAL DRY THORN SAVANNA to the north, especially the relatively moister southern portions. Typical Guinea savanna birds include Blue-bellied Roller (IS), Standard-winged Nightjar, Red-throated Bee-eater (IS), Yellow-fronted Tinkerbird, Vieillot's

Kob is a locally common and widespread Guinea savanna species in protected areas.
© KEN BEHRENS, TROPICAL BIRDING TOURS

Standard-winged Nightjar, whose wings are anything but "standard."
© KEN BEHRENS, TROPICAL BIRDING TOURS

Barbet, Fine-spotted Woodpecker, White-breasted Cuckooshrike, Piapiac, African Spotted Creeper, Yellow-billed Shrike, Western Violet-backed Sunbird, and Purple, Bronze-tailed (IS), and Lesser Blue-eared Starlings. The MONSOON FORESTS that penetrate the Guinea savanna generally have an impoverished subset of the birds of true AFROTROPICAL LOWLAND RAINFOREST. But there are a few birds that prefer this habitat, including Violet Turaco, Double-toothed Barbet, Thick-billed Cuckoo, Oriole Warbler, Red-shouldered Cuckooshrike, and White-crowned Robin-Chat. The Guinea savanna is an important wintering and migratory habitat for large numbers of European-breeding birds, including raptors, shrikes, and warblers.

Conspicuous reptiles include Common Agama (*Agama agama*), which remains abundant even in farmland and villages, as well as, locally, Nile Crocodile.

DISTRIBUTION: Guinea savanna runs in a broad belt across n. Africa from Senegal to w. Ethiopia and n. Uganda. This broadleaf-dominated habitat is the transition from the Congo Basin and Upper Guinea MONSOON FOREST to the south and the AFROTROPICAL DRY THORN SAVANNA of the

Sahel to the north, with some areas such as s. Senegal and far n. Ghana having AFROTROPICAL MOIST MIXED SAVANNA wedged between the Guinea savanna and the Sahel. The northern boundary of Guinea savanna is not well defined; broadleaf trees gradually give way to more open and thorny savanna vegetation, and perennial grasses give way to annual ones. Guinea savanna–like vegetation penetrates deep into the Sahel along watercourses. Guinea savanna, or something resembling it, is also spreading south into the former rainforest zone as the original rainforest is cut and burned by humans. Heavily degraded rainforest, despite having largely different plant species, quickly takes on the character and wildlife of Guinea savanna. Guinea savanna comes to within about 250 mi. (400km) of the nearest MIOMBO WOODLAND in the area west of Lake Victoria; despite their structural similarity, these habitats share relatively few species.

WHERE TO SEE: Mole National Park, Ghana; Murchison Falls National Park, Uganda.

Red-throated Bee-eater is often associated with riverbanks in Guinea savanna. © KEN BEHRENS, TROPICAL BIRDING TOURS

MIOMBO WOODLAND

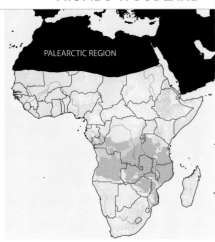

PALEARCTIC REGION

IN A NUTSHELL: Central African savanna habitat that is dominated by broadleaf trees, especially miombo trees (*Brachystegia* spp.). **Habitat Affinities:** AUSTRALASIAN BRIGALOW AND *CALLITRIS* WOODLANDS; AUSTRALASIAN TEMPERATE EUCALYPT WOODLAND; NEARCTIC OAK SAVANNA. **Species Overlap:** GUSU WOODLAND; MOPANE SAVANNA; AFROTROPICAL MOIST MIXED SAVANNA; GUINEA SAVANNA; AFROTROPICAL DRY THORN SAVANNA AND THORNSCRUB.

DESCRIPTION: This is a fairly tall broadleaf woodland with abundant grass on the ground and distinct wet and dry seasons. The canopy is typically at around 40 ft. (12m), though stunted miombo can be as short as 12 ft. (4m), and lush patches as tall as 60 ft. (18m). The soils are poor, heavily leached, and well drained, resulting in a low availability of nutrients compared with other savanna types. Miombo is generally open, with less than 50% canopy cover, allowing abundant light to reach the ground and permitting the growth of thick grass. It is park-like and easy to walk through during the dry season, and there is abundant leaf litter. During the wet season, the lush grass and flooded drainage lines make it harder to penetrate. Half of the year, miombo sees little rainfall, while heavy rains fall during the austral summer. The typical annual precipitation range is 24–60 in. (600–1,500mm). Just before the onset of the rains, most of the trees display reddish new foliage, the "miombo flush," which gives this woodland a strangely autumnal feeling, albeit during the austral spring.

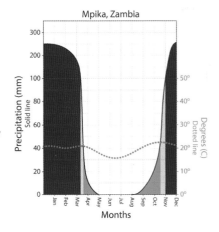

Mpika, Zambia

The vast majority of trees are broadleaf species, which lose some or all of their leaves during the dry season. The most important of these are the namesake miombo trees, as well as Munondo, Muchesa, and Doka; these trees dominate the miombo, but their precise mix varies locally. All tend

Above: **Miombo woodland in Dzalanyama, Malawi.**
© KEITH BARNES, TROPICAL BIRDING TOURS

Right: ***Brachystegia boehmii*** **showing the
hanging pinnately compound leaves typical
of the dominant trees in miombo habitat.**
© HTTPS://COMMONS.WIKIMEDIA.ORG/W/INDEX.PHP?
SEARCH=BRACHYSTEGIA+BOEHMII&TITLE=SPECIAL:
MEDIASEARCH&GO=GO&TYPE=IMAGE

to have a classic lollipop shape, with a
rounded ball of vegetation emerging from a
branchless trunk. They are all members of
the legume family and share a characteristic
leaf type: pinnately compound leaves without a terminal leaflet. The leaves vary by species,
bearing a few large oval leaflets to many small, fine leaflets. The prevalence of this leaf type gives
this forest a distinctive look, recognizable even at a distance. Trees with pinnately compound
leaves occur in other habitat types, especially AFROTROPICAL MOIST MIXED SAVANNA, but except
in GUSU WOODLAND, they rarely predominate to the degree they do in miombo woodland.
Miombo is the southern equivalent of the GUINEA SAVANNA of West Africa. Both are open
woodlands with grassy ground cover. Although Doka is important in both habitats, miombo
woodland is characterized by the additional presence of miombo, Munondo, and Muchesa trees.
Miombo woodland is also generally taller, with a more complex structure and more diverse plant
composition. In terms of wildlife, there is only a modest degree of overlap between the two. The
smaller drainage lines in miombo woodland support open, seasonally flooded grasslands, known as
dambo. Larger drainage lines and rivers give rise to deeper perennial wetlands, thickets, and
AFROTROPICAL MONSOON FOREST.

Fire has always played an important role in miombo woodland, burning the grass and stimulating
new growth, especially at the beginning of the wet season. Today, more frequent fires are started
by humans, gradually opening up the habitat. Another profound influence on this habitat is

termites, which occur in vast numbers, their towering mounds dotting the landscape. Poor soil, prevalence of the tsetse fly, and recent civil wars have resulted in low human populations in the miombo realm in much of Angola, Tanzania, and Mozambique. Other areas, such as Malawi and Zimbabwe, are heavily populated. Hunting is widespread, even in remote areas. Other major threats to miombo habitat include tree clearing, frequent burning, and overgrazing.

WILDLIFE: Due to its poor soil and the relatively low availability of nutrients, miombo has low diversity and density of wildlife compared with other African savanna and woodland habitats. Conditions are most favorable for grazing herbivores with a large body size, which are able to process large amounts of nutrient-poor feed during the dry season. Many mammals and birds also move around during the year, using seasonally flooded grasslands (AFROTROPICAL TROPICAL GRASSLAND), AFROTROPICAL MONSOON FOREST, and other habitats to get through the severe dry period that lasts for half the year. Despite its open nature, which is suggestive of other savanna environments in which you almost constantly encounter mammals and birds, miombo woodland is very different in that you can walk for long periods without seeing much wildlife. It feels uncannily quiet until you bump into a feeding flock of birds or enter a glade full of browsing Puku. This is especially true away from the busy austral spring season, when birds are breeding and active.

In general, miombo has a reduced subset of the savanna and grassland mammals typical of s. and e. Africa. It is the stronghold of the beautiful Sable Antelope, lanky Yellow Baboon, and shy Miombo Genet. African Bush Elephant is still widespread, though greatly reduced in population. Southern Tanzania's Selous Game Reserve has more elephants than anywhere else on the continent, though even here the population has been reduced by poaching from over 100,000 individuals to around 15,000. Selous is also the global stronghold of the endangered African Wild Dog. Other widespread mammals of miombo woodland include Vervet Monkey, Side-striped Jackal, Greater Kudu, Bushbuck, Common Duiker, Oribi, and Puku (mainly during the rainy season). Large

Miombo woodland is a global stronghold for the beautiful Sable Antelope.
© KEN BEHRENS, TROPICAL BIRDING TOURS

predators including Lion, Leopard, and Spotted Hyena are still fairly common in some areas. Due to the abundance of ants and termites, Aardvark and Ground Pangolin can be common in miombo woodland, though they are still hard to see.

Miombo woodland has excellent bird diversity, though inferior to that of the savannas to the east and south. A number of birds are restricted to this habitat, but few are found throughout, and many are generally uncommon. In terms of endemic birds, the richest area lies near the center of the biome, in Zambia, north of the Zambezi River. Miombo endemic birds include Anchieta's, Whyte's, and Miombo Barbets; Pale-billed Hornbill (IS); Miombo Rock-Thrush (IS); Eastern and Western Miombo Sunbirds (IS); Miombo Tit; Miombo Scrub-Robin; Red-capped Crombec; Böhm's Flycatcher; Anchieta's Sunbird; Lesser Blue-eared Starling (IS); Chestnut-backed Sparrow-Weaver; and Bar-winged Weaver. Many bird species are shared by miombo and other lush savanna and woodland habitats in the region, and some of these are more common in miombo. Typical non-endemic miombo birds include Bronze-winged Courser, Pennant-winged Nightjar, Racket-tailed Roller, Böhm's Bee-eater, White-breasted Cuckooshrike, African Golden Oriole, Rufous-bellied Tit, African Spotted Creeper, Arnot's Chat, Trilling Cisticola, Greencap Eremomela, Miombo and Stierling's Wren-Warblers, Yellow-bellied Hyliota, Retz's Helmetshrike, Western Violet-backed Sunbird, Orange-winged Pytilia, Broad-tailed Paradise-Whydah, and Cabanis's Bunting. The areas of Afrotropical monsoon forest that occur within miombo are species-poor but do support a small set of special birds.

Miombo does not have high reptile diversity, though it does have dozens of species of endemic snakes and lizards, most of which are inconspicuous. It also shelters some widespread and charismatic African reptiles like Flap-necked Chameleon, Savanna Monitor, African Rock Python, Black Mamba, Puff Adder, and Black-necked Agama.

DISTRIBUTION: This is the prevalent habitat in a vast swath of sc. Africa, from Angola across to nw. Tanzania and down to s. Mozambique. Most of this area is on the Central African Plateau, at elevations of 3,300–5,000 ft. (1,000–1,500m), though portions in Tanzania and Mozambique are much lower, down to 650 ft. (200m). Most of the terrain is mildly undulating, with isolated rocky outcrops in some areas. Miombo woodland is closely allied with two broadleaf woodland habitats that occur to its south and serve as the transition to the AFROTROPICAL DRY THORN SAVANNA of the Kalahari: MOPANE SAVANNA and GUSU WOODLAND. It also shares many species with the AFROTROPICAL MOIST MIXED SAVANNA to the south and east, though it has lower diversity. Throughout the miombo, there are patches of AFROTROPICAL MONSOON FOREST along drainage

lines and rivers; these forests support a mix of species from the se. African monsoon forests and the AFROTROPICAL LOWLAND RAINFOREST of the Congo Basin and provide an important link between these regions.

WHERE TO SEE: Kasanka National Park, Zambia; Selous Game Reserve, Tanzania; Dzalanyama Forest Reserve, Malawi.

Miombo Scrub-Robin is just one among a rich set of birds endemic to miombo woodland.
© KEN BEHRENS, TROPICAL BIRDING TOURS

GUSU WOODLAND

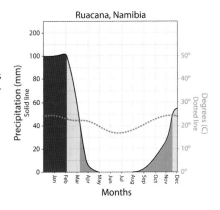

IN A NUTSHELL: South-central African savanna habitat that grows on Kalahari sands and is dominated by broadleaf trees, especially Zambezi Teak. **Habitat Affinities:** AUSTRALASIAN OPEN EUCALYPT SAVANNA; AUSTRALASIAN TETRODONTA WOODLAND SAVANNA; NEARCTIC OAK SAVANNA. **Species Overlap:** MIOMBO WOODLAND; MOPANE SAVANNA; AFROTROPICAL MOIST MIXED SAVANNA; AFROTROPICAL DRY THORN SAVANNA AND THORNSCRUB; GUINEA SAVANNA.

DESCRIPTION: Gusu is a fairly tall broadleaf woodland with abundant grass on the ground and distinct wet and dry seasons. It grows on deep Kalahari sands with little clay or silt. Trees that are able to penetrate the sand with their roots can access water that accumulates by lateral seepage. This water supply allows the growth of larger trees, as tall as 65 ft. (20m), than the local level of rainfall would normally allow. The dominant and characteristic tree is Zambezi Teak. This tree is similar in appearance to the dominant trees of MIOMBO WOODLAND, with a classic lollipop tree shape and drooping, pinnately compound leaves. The valuable, rich reddish-brown wood is often seen for sale along roadsides. Another important tree in gusu is African Teak, which can sometimes be the dominant tree, particularly in portions of sw. Zambia and Zimbabwe. Many other trees are common in gusu, including Marula, Large False-Mopane, monkey-oranges, Kalahari Podberry, Peeling Plane, Kalahari Clusterleaf, Wild Seringa, and Kalahari Bauhinia, which often form a mid-story of smaller trees under the open canopy. Trees other than teak tend to get the upper hand in heavily disturbed areas. Gusu woodland is easy to walk through, especially during the dry season.

Gusu has a similar feel and structure to miombo woodland, though it is generally more open. Miombo is more botanically diverse, with a mix of canopy species, in contrast to the dominance of Zambezi Teak in gusu. Half of the year, gusu woodland sees little rainfall, while heavy rains fall during the austral summer. Typical annual precipitation is 12–24 in. (300–600mm). As in miombo

Gusu woodland is characterized by the dominance of Zambezi Teak, growing on deep Kalahari sand. © ROGER CULOS, HTTPS://COMMONS. WIKIMEDIA.ORG/WIKI/ FILE:BAIKIAEA_PLURIJUGA_ ARBRE_MHNT.JPG

woodland, fire is important, and wildfires are common, especially at the onset of the rainy season. The smaller drainage lines support open grasslands, which are seasonally flooded. Larger drainage lines and rivers give rise to perennial wetlands, thickets, and AFROTROPICAL MONSOON FOREST. These are most prominent and important along the Kavango River, Okavango Delta, and Zambezi River. A characteristic tree of the monsoon forest within gusu woodland is Manketti. Due to its poor soil and lack of water, much of this habitat is sparsely populated by humans. Despite that, hunting is widespread. Fires are natural but have increased in frequency and severity due to human activity. Zambezi Teak is sensitive to fire and can be wiped out by frequent burning. Much of this woodland is also being overharvested for the valuable teak wood. Gusu is less resilient than other types of woodland; once land is cleared by a combination of cutting and fire, the hot sun destroys the organic matter in the soil, and the area tends to remain clear and never regenerate tall woodland.

WILDLIFE: Gusu woodland supports a similar suite of species to MOPANE SAVANNA—essentially, a much-reduced selection of the wildlife of the more diverse adjacent MIOMBO WOODLAND and AFROTROPICAL DRY THORN SAVANNA habitats. The dual influences of those habitats are evident in a list of some of the mammals of gusu woodland: Smith's Bush Squirrel, Rusty-spotted Genet, Roan and Sable Antelopes, Bushbuck, Greater Kudu, Steenbok, Puku, and African Buffalo. Primarily thanks to the existence of some well-protected areas, large predators persist,

The range of Bradfield's Hornbill largely coincides with that of gusu habitat. © KEN BEHRENS, TROPICAL BIRDING TOURS

Left: **Spotted Hyena is a generalist that occurs in gusu woodland and elsewhere.** © IAIN CAMPBELL, TROPICAL BIRDING TOURS

Below: **Pied Barbet occurs within gusu and other arid wooded African habitats.** © IAIN CAMPBELL, TROPICAL BIRDING TOURS

including Spotted Hyena, Leopard, Lion, and African Wild Dog. This habitat is used by large numbers of African Bush Elephants.

Although gusu has much lower bird diversity than miombo woodland or dry thorn savanna, it does have abundant birdlife, and usually feels more bird-rich than miombo. Bradfield's Hornbill (IS) is the sole avian gusu specialist, though it does also use adjacent habitats. Other typical birds of this habitat include Dark Chanting-Goshawk, Striped Kingfisher, Bennett's Woodpecker, Meyer's Parrot, Chinspot Batis, African Golden Oriole, Rufous-bellied Tit, Greencap Eremomela, Sharp-tailed Starling, Pale Flycatcher, and Yellow-fronted Canary.

Gusu supports many widespread snake and lizard species, including Bushveld Lizard, Kalahari Plated Lizard, and Ground Agama.

DISTRIBUTION: Gusu woodland occurs in a fairly narrow zone from s. Angola to w. Zimbabwe and Zambia. Gusu and MOPANE SAVANNA together form a belt across s. Africa and provide a transition from the moister MIOMBO WOODLAND to their north and the drier and thornier savanna types to their south and west.

WHERE TO SEE: Caprivi Strip, Namibia.

MOPANE SAVANNA

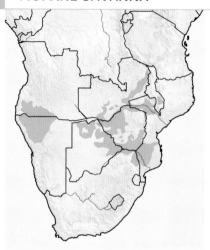

IN A NUTSHELL: Savanna habitat that is dominated by Mopane, a broadleaf tree with a diverse range of growth forms. **Habitat Affinities:** AUSTRALASIAN MALLEE WOODLAND AND SCRUBLAND; AUSTRALASIAN MULGA WOODLAND AND ACACIA SHRUBLAND; NEARCTIC OAK SAVANNA. **Species Overlap:** AFROTROPICAL MOIST MIXED SAVANNA; AFROTROPICAL DRY THORN SAVANNA AND THORNSCRUB; GUSU WOODLAND; MIOMBO WOODLAND.

DESCRIPTION: Any discussion of Mopane savanna should begin with its namesake dominant tree, easily recognized by its butterfly-shaped leaves, which smell like turpentine when crushed. The Mopane tree (*Colophospermum mopane*) has several odd characteristics: As a rule, broadleaf woodland grows on nutrient-poor, well-drained soil, but the Mopane is a broadleaf tree that grows on nutrient-rich, poorly drained soil. It is also unusual in its hugely variable growth forms, which are a response to local growing conditions and disturbance, especially from elephants. It most typically forms a broadleaf woodland similar to MIOMBO or GUSU, though with a lower canopy, at 24–33 ft. (7–10m). But it can also grow in forest-like stands known as "cathedral Mopane," with trees as tall as 80 ft. (25m). On the other end of the size spectrum, it can grow in scrubby stands that are 3–10 ft. (1–3m) tall and, exceptionally, even as low as 1 ft. (0.3m) tall in thickets interspersed with grass. Like many savanna trees, the Mopane is fire-resistant, allowing it to persist after fires sweep through and consume the understory grass. Although the Mopane tree tends to dominate, other trees are often mixed into this habitat, such as acacias, Wild Seringa, baobabs, African Blackwood, bushwillows (*Combretum*), Purple-pod Terminalia, and commiphoras. Like other woodland and savanna habitats, Mopane savanna grows in areas with highly seasonal rainfall; here the rain falls during the austral summer.

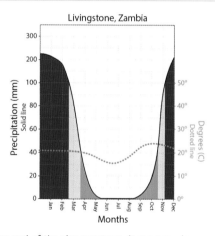

Livingstone, Zambia

WILDLIFE: Mopane savanna is generally not the most interesting habitat for wildlife; it does not support a distinctive or cohesive faunal assemblage but rather a much-reduced selection of the wildlife of adjacent habitats. On safari, you may find yourself quickly transiting through Mopane

Above: Tall "cathedral Mopane" can grow in uniform monospecific stands in the Luangwa Valley, Zambia. © HANS HILLEWAERT, WIKIMEDIA COMMONS/CC BY-SA 4.0

Right: Mopane can spread like a shrub and be quite bushy when it is young. © MUSÉUM DE TOULOUSE, CC BY-SA 3.0, HTTPS:// CREATIVECOMMONS.ORG/LICENSES/BY-SA/3.0, VIA WIKIMEDIA COMMONS, HTTPS://COMMONS.WIKIMEDIA. ORG/WIKI/FILE:COLOPHOSPERMUM_MOPANE_ARBRE_ MHNT.JPG

savanna to get to a more productive habitat. Vast regions of Mopane savanna have been set aside in well-protected reserves, but there are still problems, including poaching, especially of Black Rhinos. Most large mammals in the Angolan and n. Namibian Mopane savannas were killed during the Angolan civil war. On private land, farmers habitually kill predators like jackals, Cheetah, and Caracal.

Nonetheless, there is one mega-mammal that is very fond of Mopane savanna: the African Bush Elephant, which eats nearly every part of the Mopane tree and commonly knocks over full-size

Violet Woodhoopoe and Green Woodhoopoe meet and often hybridize in Namibia's Mopane woodlands.

African Wild Dogs are among the
mammals found in Mopane savanna.
© LISLE GWYNN, TROPICAL BIRDING TOURS.

African Wild Dogs are among the
mammals found in Mopane savanna.
© LISLE GWYNN, TROPICAL BIRDING TOURS.

trees to reach the leaves at the top.
Elephants make openings in the
woodland, which fill with grass,
providing fuel for bushfires, which play
a further role in keeping the habitat
open. There are concerns about the
impact of elephants on areas in which
their populations are out of balance
due to various human interventions.
Impala and Nyala also thrive in the
matrix of dense woodland and open
glades that is typical of Mopane
savanna. In Namibia's Etosha National Park, Mopane savanna is an important habitat for the localized
Black-faced subspecies of Impala. Many other mammals will use Mopane savanna, though it is not
a particularly preferred habitat, including Black-backed Jackal, Black Rhinoceros, Common Zebra,
Blue Wildebeest, Giraffe, Sharpe's Grysbok, and Greater Kudu. The western tracts of Mopane are
used by Springbok, Gemsbok, Hartmann's Mountain Zebra, and Brown Hyena. Many large reserves
include swaths of Mopane and support large cats, Spotted Hyena, and African Wild Dog.

The birds of Mopane savanna are heavily influenced by surrounding habitats. The western
Mopane habitat is used by some of the endemic birds associated with the Namib Escarpment. In
particular, Mopane along watercourses is a crucial habitat for the localized Violet Woodhoopoe and
Bare-cheeked Babbler. Upland western Mopane savanna can have White-tailed Shrike and Carp's
Tit. In the south-central part of its range, on the moderate elevations of the Central African
Plateau, Mopane savanna has a reduced subset of the birds of AFROTROPICAL DRY THORN
SAVANNA, such as Crimson-breasted Gonolek and Mariqua Flycatcher. Mopane savanna to the
north and east is more influenced by AFROTROPICAL MOIST MIXED SAVANNA and, to a lesser
degree, MIOMBO WOODLAND. Widespread savanna species occurring in Mopane habitat include
Pearl-spotted Owlet, Green Woodhoopoe, Greater Blue-eared and Meves's Starlings, and White
Helmetshrike. Miombo influence is shown by the local presence of Arnot's Chat, White-breasted
Cuckooshrike, and Miombo Wren-Warbler.

Mopane tree is the main food plant for the Mopane worm, the caterpillar form of a beautiful
species of emperor moth (*Gonimbrasia belina*). These caterpillars emerge in huge numbers in the
summer, providing an important food source for both wildlife and humans.

Mopane habitat has a subset of typical savanna reptiles, including conspicuous lizards like
Black-lined Plated Lizard and Tropical Spiny Agama. In the west, the diurnal Ovambo Tree Skink is
often found on Mopane trees.

Because Mopane savanna is strongly influenced by more diverse surrounding habitats, there are
major differences between the wildlife found in the western Mopane of Angola and Namibia, the
higher-elevation Mopane associated with the Kalahari Basin, and the lower-lying eastern and
northern Mopane of e. South Africa, Mozambique, Zimbabwe, and Zambia.

DISTRIBUTION: Mopane savanna is widespread though patchy from n. Botswana to e. Zambia and
south to e. South Africa. This is an important habitat in the Zambezi, Limpopo, and Luangwa River
valleys, and locally around the Okavango Delta. Mopane also occurs in a discrete area of n. Namibia

and s. Angola. Mopane can be found at both low elevations (such as the lowveld and river valleys of e. South Africa and Mozambique) and moderate ones (as on the Central African Plateau). Mopane savanna and GUSU WOODLAND together provide a transition from the moister MIOMBO WOODLAND to their north and the drier and thornier savanna types to their south and west. Mopane tends to occur in close conjunction with AFROTROPICAL DRY THORN SAVANNA and AFROTROPICAL MOIST MIXED SAVANNA, as it grows on a soil type that is also well suited to those habitats.

WHERE TO SEE: Etosha National Park, Namibia (central and western portions); South Luangwa National Park, Zambia; Kruger National Park, South Africa (northern portions).

AFROTROPICAL DRY THORN SAVANNA AND THORNSCRUB

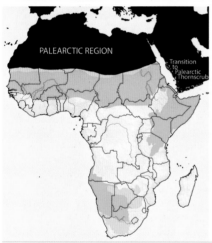

IN A NUTSHELL: Variably dry, open savanna that is usually dominated by spiny acacia and commiphora trees and has grass cover during the rainy season. **Habitat Affinities:** NEOTROPICAL THORNSCRUB; PALEARCTIC SEMIDESERT THORNSCRUB; PALEARCTIC HOT SHRUB DESERT; AUSTRALASIAN MULGA WOODLAND AND ACACIA SHRUBLAND; NEARCTIC CHIHUAHUAN DESERT SHRUBLAND. **Species Overlap:** AFROTROPICAL MOIST MIXED SAVANNA; MOPANE SAVANNA; GUINEA SAVANNA; GUSU WOODLAND; MIOMBO WOODLAND.

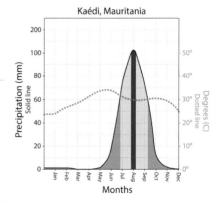

DESCRIPTION: This is a classic savanna environment in which there is variable cover from trees and bushes, and much or all of the ground is seasonally covered in grass. What makes this habitat distinctive is its generally dry and highly seasonal climate and the predominance of thorny and fine-leaved trees. During the long dry season, this habitat can seem superficially almost barren and desertlike, but during the rains, it bursts into fantastic productivity, an ephemerally lush and green world that bears little resemblance to the dry and dusty landscape of just a few weeks earlier. The grassland and woodland components of this habitat have fairly discrete sets of wildlife, and this account mainly treats the woodland aspect of this habitat; for more detail about the grassland aspect, see the AFROTROPICAL TROPICAL GRASSLAND account. Of course, separating the two is somewhat artificial and can never be done cleanly.

Even in very dry areas of dry thorn savanna, large acacia trees sometimes grow in areas with groundwater (photo: n. Tanzania).
© KEN BEHRENS, TROPICAL BIRDING TOURS

Dry thorn savanna is found in three distinct zones—in the Sahel of n. Africa, in the Somali-Masai zone in the northeast, and in the Kalahari and Namib Desert of the southwest—which has driven a fascinating pattern of colonization and endemic speciation among the zones (discussed further in the Distribution section). The exact structure of dry thorn savanna is variable. It can be an open landscape covered mainly in grassland, with scattered bushes. Or it can be woodland in which trees and bushes predominate, though not to the extent that they prohibit the growth of grass on the ground. In general, this habitat is open enough to make it easy to walk through, despite its thorny character. The height of the tallest trees is highly variable. In drier areas, this habitat may shrink to a semidesert shrubland in which the tallest woody plants are 3 ft. (1m) or less. In moister areas, and along watercourses or other areas with groundwater, there can be isolated tall trees up to 65 ft. (20m) tall. This is a generally dry and hot habitat in which the vast majority of the rain falls during short rainy seasons. The rains are produced by the intertropical convergence zone, which moves throughout the year. Rain falls during the boreal summer in the north, during the austral summer in the south, and in a more complicated pattern in parts of East Africa, with "long rains" between March and May, and "short rains" sometime between September and December. The rainfall is variable, typically 8–24 in. (200–600mm) annually, but it can fail entirely in some years. As such, this is an exacting habitat in which the animals and plants have had to adapt to harsh conditions. One such adaptation is local movements throughout the year, even among species that are considered residents. For example, many big mammals in the Serengeti feed on the open grasslands when they are green after the rains, then retreat to the dry thorn savanna when the grassy plains dry up. Fires have always been frequent, especially at the onset of the rainy season, but have been made more frequent by humans.

Unlike moister savanna habitats, dry thorn savanna can lose most or all of its grass during the dry season due to grazing, fire, and desiccation. In terms of its woody vegetation, this habitat is characterized by the dominance of spiny acacias (of multiple genera) and commiphoras. Acacias with the stereotypically African umbrella form can be prominent, such as Camel Thorn in the southwest and Umbrella Thorn elsewhere. Another distinctive acacia, found in the Somali-Masai zone, is Whistling Thorn, which can dominate large stretches of the landscape. Whistling Thorn has a mutualistic relationship with ants, which shelter in its swollen thorns, swarming out to protect the tree from herbivores in exchange. Other trees of this habitat include shepherd's trees, *Grewia*, Desert Date, jujubes, and mustardtrees. Succulent euphorbias and aloes are often common. Bushwillows and/or terminalias are prominent in some areas, especially where this habitat transitions to moister savanna and woodland. Tamarisk grows in areas with saline soils and fills

Acacias dominate Afrotropical dry thorn savanna in n. Tanzania.
© IAIN CAMPBELL, TROPICAL BIRDING TOURS

the same niche as casuarina in the dry parts of Australia. Most trees lose at least some of their leaves during the dry season, though the harshness of the environment and the small and fine nature of the leaves mean that there is rarely a significant accumulation of leaf litter. The most similar habitat covered in this book is AFROTROPICAL MOIST MIXED SAVANNA. These two mix with each other in South Africa, Botswana, and in a complicated patchwork across East Africa. Moist mixed savanna is characterized by a higher overall diversity of trees, a more complex and varied structure, and by the local prominence of broadleaf species. In most dry areas that are dominated by dry thorn savanna, Afrotropical moist mixed savanna (e. and s. Africa) or GUINEA SAVANNA (n. Africa) can still be found locally along watercourses.

In general, dry thorn savanna has low densities of humans. Despite this, their impact is profound, as they try to scrape out a living in a tough environment. Cutting trees for firewood and charcoal, grazing, and hunting are widespread. The southwestern and northeastern zones are in decent condition, with many protected areas and lots of big mammals. The Sahel, on the other hand, has been more heavily impacted by humans for much longer, and the vast majority of its big mammals have been wiped out. Much of the Sahel zone experiences political instability, meaning that people are struggling to survive, guns are widespread, and law enforcement is virtually nonexistent—a bad mixture when it comes to big mammal conservation. Another challenge facing this habitat, especially in the Sahel, is desertification, which is worsened by human activities such as attempts to cultivate very thin and dry soils during the brief rainy season.

WILDLIFE: In general, this is an excellent habitat for wildlife, harboring a high density and diversity of animals, which are easy to see due to its open nature. Prominent groups of mammals include oryxes, gazelles, and giraffe. Dry thorn savanna is an important habitat in many of Africa's greatest safari destinations, such Namibia's Etosha National Park and Tanzania's Serengeti National Park.

The birding in this habitat can be among the world's most electrifying, particularly in the Somali-Masai zone and especially after good rains. It can be absolutely pumping with birds and is delightful to walk through in places like s. Ethiopia, where there are few dangerous big mammals. Prominent groups of birds include bustards, sandgrouse, doves, scrub-robins, starlings, shrikes, warblers, sunbirds, and weavers. Many migrant warblers use this habitat, probably due to its structural similarity to the Mediterranean woodlands where they breed.

This habitat is also replete with conspicuous reptiles, including sand (multiple genera) and long-tailed (*Latastia*) lizards, agamas, skinks, geckos, and snakes, which you have a greater chance of bumping into here than in most other African habitats.

The only animals that occur throughout Africa's dry thorn savanna are widespread savanna species, which aren't restricted to this habitat, such as Honey Badger, Lion, Caracal, Spotted Hyena, Black Rhinoceros, Giraffe, Tawny Eagle, and Namaqua Dove.

This habitat's three discrete zones are central to any discussion of its wildlife. The most interesting wildlife is restricted to one or two of the three major zones. The Somali-Masai zone is by far the richest in terms of wildlife, as it lies between the Sahel and Kalahari, both of which historically acted as reservoirs of potential colonist species. Among the species shared with the Sahel are Striped Hyena, African Collared-Dove (IS), Blue-naped Mousebird, and Northern Crombec. Those also found in the Kalahari include Southern Springhare, Black-backed Jackal, Bat-eared Fox, and Kori Bustard. The Somali-Masai zone has some wonderful mammals that are virtually endemic, including Lesser Kudu (IS), Gerenuk (IS), Beisa Oryx, Dibatag, and the bizarre eusocial and virtually cold-blooded Naked Mole-Rat. Just a few of the many wonderful birds that are endemic or near-endemic are Buff-crested Bustard (IS), Golden-breasted Starling, Eastern Violet-backed Sunbird, Rosy-patched Bushshrike, and Somali Bunting. This zone has fantastic reptile diversity, and there are dozens of endemics.

The big mammals in the Sahel used to rival those in e. and s. Africa but have been decimated by hunting and other human pressures. This was a major habitat of the Scimitar-horned Oryx, which went extinct in the wild in the year 2000. Other Sahel mammals include Patas Monkey, Pale Fox, and Red-fronted (IS), Dama, and Dorcas Gazelles. Classic Sahel birds include Scissor-tailed Kite, Quail-plover, Yellow-breasted Barbet, and Black Scrub-Robin.

The Kalahari-Namib zone is the most distinct of the three, as its connection with the northern zones is more tenuous. Its wildlife has strong

Blue-naped Mousebird is most common in dry thornscrub and arid acacia savanna.

affinities with the AFROTROPICAL MOIST MIXED SAVANNA to its east and to a lesser degree to adjacent broadleaf woodlands (especially MOPANE SAVANNA). The Kalahari has a remarkable diversity of mammalian predators, from widespread species like Cheetah to localized ones like Brown Hyena, Black-footed Cat, Cape Fox, Aardwolf, and Meerkat. Classic Kalahari birds include Pale Chanting-Goshawk, Crimson-breasted Gonolek, Kalahari Scrub-Robin (IS), and Pririt Batis. A few birds, like Pygmy Falcon (IS), Black-faced Waxbill, and Mariqua Sunbird, are shared with the Somali-Masai zone.

Dry thorn savanna is a key component of the mix of habitats along the Namib Escarpment, which has a set of endemics including Black Mongoose, Hartlaub's Francolin, Rockrunner, White-tailed Shrike, and Namibian Rock Agama.

DISTRIBUTION: In general, dry thorn savanna lies between true desert environments and broadleaf woodland or moist savanna environments. This Afrotropical habitat comprises three discrete sectors: the Sahel of n. Africa, the Somali-Masai biome of the northeast (extending into the Arabian Peninsula), and the Kalahari and Namib of the southwest. The Sahel and Somali-Masai biomes are tenuously connected along the coast of the Red Sea, while the southwestern sector is completely disconnected from the other two blocks. In the geologic past, as Africa alternated between cool wet and warm dry periods, these zones would have been repeatedly isolated and then reconnected, resulting both in many species being shared, especially between the Sahel and Somali-Masai, and many species being restricted to only one block.

Gerenuk is a strange, extremely long-necked antelope that is restricted to Afrotropical dry thorn savanna in ne. Africa (photo: south of Yabello, Ethiopia).
© IAIN CAMPBELL, TROPICAL BIRDING TOURS

In areas like the south side of the Sahel and much of Tanzania, dry thorn savanna mixes into moister habitats and may be restricted to higher, drier areas. Along the northern verge of the Sahel, in the Namib Desert, the Karoo plateau, and parts of the Horn of Africa, dry thorn savanna extends only along drainage lines, and elsewhere gives way to drier habitats in which trees and bushes are lacking. These fringe areas generally lack grassy ground cover, though it can develop temporarily after good rains. Such desert-fringe habitats are far more widespread and important in the n. African deserts covered in the Palearctic chapter; for more information, see the sections on PALEARCTIC HOT SHRUB DESERT and PALEARCTIC SEMIDESERT THORNSCRUB.

WHERE TO SEE: Awash National Park, Ethiopia; Tsavo East and West National Parks, Kenya; Kgalagadi Transfrontier Park, South Africa and Botswana.

AFROTROPICAL MOIST MIXED SAVANNA

IN A NUTSHELL: Fairly moist and very diverse savanna that has a variety of both thorny, fine-leaved trees and broadleaf trees. **Habitat Affinities:** NEOTROPICAL CHACO SECO AND ESPINAL; NEOTROPICAL CERRADO; AUSTRALASIAN TEMPERATE EUCALYPT WOODLAND; AUSTRALASIAN OPEN EUCALYPT SAVANNA. **Species Overlap:** AFROTROPICAL DRY THORN SAVANNA AND THORNSCRUB; MOPANE SAVANNA; GUSU WOODLAND; GUINEA SAVANNA; MIOMBO WOODLAND.

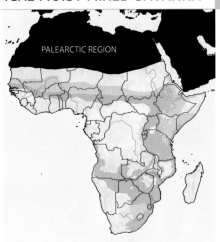

DESCRIPTION: This habitat, which offers some of Africa's most exciting wildlife-watching, may be the most complex of all those covered in this book. It's also one of the trickiest to define and is something of a catchall for several microhabitats that could be split out, except for the lack of space in this book and the fact that these microhabitats occur in such a fine and complex tapestry that they are almost impossible to tease apart. The name "moist mixed savanna" itself falls short, given that this habitat can be quite dry at times. It merges woodland/savanna habitats such as GUINEA SAVANNA and MIOMBO WOODLAND with the much drier thorn savanna. This is generally a warm to hot and seasonally dry habitat, in which the vast majority of the rain falls during intense

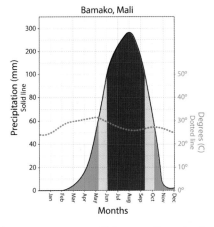

monsoonal rainy seasons, produced by the intertropical convergence zone, which moves throughout the year. In the south and west, the rains fall during the austral summer, while in much of East Africa, there are two rainy periods: one between March and June and one sometime between September and December. Typical annual rainfall is 16–40 in. (400–1,000mm). Fires have always been frequent, especially at the onset of the rainy season but have been made more frequent by humans.

Two characteristics of moist mixed savanna are complexity and diversity. Although it includes broadleaf components, broadleaf trees never dominate the landscape the way they do in miombo woodland, Guinea savanna, MOPANE SAVANNA, or GUSU WOODLAND. Likewise, although acacia and commiphora are important trees here, they don't completely dominate the way they do in typical AFROTROPICAL DRY THORN SAVANNA. The astounding array of different sorts of very distinctive trees is part of what makes this habitat a delight for a naturalist to explore, and the great diversity here is out in the open and easy to enjoy (not concealed, as in tropical forest). This is a major reason visitors flock to places like South Africa's Kruger National Park, where the abundant wildlife is relatively easy to see and the landscapes invite exploration. There is great structural complexity to this habitat. Like all savannas, it has a grassy ground cover for at least part of the year. And like dry thorn savanna, it can be very open and grassy locally, melting into pure grassland in places, such as the Serengeti. In some areas, this habitat occurs in thicket form, which is not truly savanna, as it can be so thick that there is no grassy ground cover. These thickets are included in this habitat because they have much the same bird and mammal species as the more open savanna. In areas with nutrient-poor soil, there can locally be a broadleaf-dominated moist mixed savanna that is a reminiscent of miombo woodland. Meanwhile, along lakes, rivers, and other watercourses, this habitat forms a lush riparian band with trees to 65 ft. (20m) or taller. Here it approaches or blends into AFROTROPICAL MONSOON FOREST, or even AFROTROPICAL LOWLAND RAINFOREST on the northwest side of Lake Victoria. Along watercourses, narrow bands of moist savanna penetrate deep into arid climates where the dominant habitat is Afrotropical dry thorn savanna. Due to its well-watered nature, both permanent and seasonal FRESHWATER WETLAND occur within this habitat, adding yet another dimension of diversity.

In terms of trees, various acacias are always prominent, especially in lower-lying areas with rich clay soils, as along rivers and other watercourses. These and other spiny trees and bushes such as commiphoras, jujubes, and Sicklebush are much more common than in broadleaf forests. One distinctive acacia is the yellow-barked Fever Tree, which often predominates in seasonally flooded areas, forming one of Africa's most distinctive and beautiful microhabitats. In areas with poor and well-drained sandy soil, as on the tops of hills, the savanna supports mainly broadleaf trees, including terminalias, bushwillows, and Marula. Moist mixed savanna is one of the key habitats for Africa's most famous tree, the mighty baobab. Other typical trees include *Albizia*, figs, Jackalberry, Sausage Tree, and

Baobabs and African Bush Elephants are two big and prominent members of the Afrotropical moist mixed savanna community (photo: Tarangire National Park, Tanzania). © KEN BEHRENS, TROPICAL BIRDING TOURS

Nyala Tree. Palms, including Wild Date Palm and *Hyphaene* spp., are more abundant and prominent in this habitat than in any other. Palms sometimes dominate a large stretch of country, forming yet another striking microhabitat under the moist mixed savanna umbrella.

Forest tends to turn into savanna-like habitat when it has been degraded by humans or natural climate change. Much of the moist mixed savanna habitat in e. and s. Africa may have been forest originally, especially on the Ethiopian Highlands, along the Indian Ocean coast, and around Lake Victoria. This habitat has probably never before been so common on the continent. But unlike human-created habitats in forest-dominated areas, which are remarkably devoid of wildlife, this anthropogenic savanna can be incredibly rich in wildlife, likely due to quick colonization from adjacent natural savannas. Many well-protected areas include moist mixed savanna, but this habitat certainly has its conservation challenges. Highly organized poaching for rhinos and elephants has reached catastrophic levels in recent years. In more populated areas, cultivation, grazing, hunting, and cutting for firewood and charcoal have eliminated most of the big mammals.

WILDLIFE: This locally very diverse habitat maintains a surprising cohesiveness across its wide range. Most of the species in e. South Africa are the same as those in c. Tanzania. There is a significant difference between the north and the east; and within the eastern moist mixed savanna, there is also a weak north–south division, with a few species found on only one side or the other. Diversity and complexity also characterize the animals that use this habitat. Very few wildlife species are confined to moist mixed savanna; rather, species are shared with adjacent habitats such as MIOMBO WOODLAND, GUINEA SAVANNA, MOPANE SAVANNA, GUSU WOODLAND, AFROTROPICAL DRY THORN SAVANNA, AFROTROPICAL MONSOON FOREST, and even AFROTROPICAL MONTANE FOREST. This

Above: **Southern Ground-Hornbill is one of the wonderful marquee birds of Afrotropical moist mixed savanna.** © IAIN CAMPBELL, TROPICAL BIRDING TOURS

Right: **One of the most-photographed birds in Africa, Lilac-breasted Roller is most common in moist mixed savanna habitat.** © IAIN CAMPBELL, TROPICAL BIRDING TOURS

is a sort of crossover or transitional habitat that serves to connect most of Africa's other habitats. The majority of Africa's birds and larger mammals are found in at least two habitats, and moist mixed savanna is situated perfectly, near the middle of the moisture-driven continuum of habitats, allowing it to provide refuge for a remarkably high proportion of the continent's species.

One of the best places on earth for big mammals, moist mixed savanna hosts widespread and charismatic beasts like African Bush Elephant, African Wild Dog, Spotted Hyena, Lion, Leopard, White and Black Rhinoceroses, Giraffe, and African Buffalo. A few mammals that are especially typical of this habitat, though not endemic to it, are Side-striped Jackal, Common Zebra, Greater Kudu, Nyala, Waterbuck, and Impala. The n. African savanna is far poorer in mammals due to the depravations of political instability and widespread hunting.

The bounty of different microhabitats makes moist mixed savanna a birding paradise, often harboring huge bird diversity within a small area. Indeed, this is the main habitat in Uganda's Queen Elizabeth National Park and South Africa's Kruger National Park, two of the national parks that have the longest bird lists in Africa. Moist mixed savanna is prime habitat for some of Africa's most beguiling birds, species even mammal-obsessed tourists can't miss, like Common Ostrich, Helmeted Guineafowl, Bateleur, ground-hornbills, oxpeckers, and the extremely well-photographed Lilac-breasted Roller. Other classic birds include Dark Chanting-Goshawk, Purple-crested Turaco, Crested Francolin (IS, except in n. Africa), Brown-headed Parrot, a variety of cuckoos, Speckled Mousebird, Green Woodhoopoe, Brown-hooded Kingfisher, Golden-tailed and Bearded Woodpeckers, Black Cuckooshrike, Black-headed Oriole, black-flycatchers, Bearded Scrub-Robin (IS, except in Ethiopia and n. Africa), Chinspot Batis, Black-backed Puffback, Southern and Tropical Boubous, Sulphur-breasted and Gray-headed Bushshrikes, Greater Blue-eared Starling, Scarlet-chested Sunbird, Spectacled and Village Weavers, and Green-winged Pytilia.

The southern areas hold Crested Barbet, White-throated Robin-Chat, and White-breasted Sunbird. The northeast has Silverbird and Variable Sunbird. There is a divide between the eastern and northern moist mixed savanna. The eastern savanna is more complex and diverse overall, with higher bird diversity, whereas the northern savanna has virtually no birds that aren't also found in dry thorn and/or Guinea savanna. The moist mixed savannas in the Ethiopian Highlands are the primary habitat of Black-winged Lovebird, Banded Barbet, and one of Africa's most enigmatic and sought-after birds, Prince Ruspoli's Turaco.

DISTRIBUTION: Moist mixed savanna is widespread in e. and s. Africa, from the Ethiopian Highlands and s. Arabian Peninsula to e. South Africa, and west to Angola. Due to its somewhat amorphous nature, the limits of moist mixed savanna are tricky to define, especially at the edge of the Kalahari in s. Africa and the edge of the GUINEA SAVANNA in n. Africa. It tends to be found in areas with more moisture than AFROTROPICAL DRY THORN SAVANNA, but less moisture than AFROTROPICAL MONSOON FOREST or AFROTROPICAL LOWLAND RAINFOREST, and more mixed soils and complicated topography than MIOMBO and GUSU broadleaf woodland. There is also a broad zone of this habitat all the way across n. Africa, between the dry thorn savanna of the Sahel and the broadleaf Guinea savanna, though the farther west in West Africa, the smaller the pockets of this habitat within the Guinea savanna–Sahel transition. This sector of moist mixed savanna is sometimes called "Sundaic savanna." The highlands of Ethiopia rise like an island of moister habitats in a sea of aridity. Although AFROTROPICAL MONTANE FOREST and grassland are the classic Ethiopian habitats, this area has also given rise to a montane form of moist mixed savanna. Humans probably played a role in converting highland forest into savanna.

WHERE TO SEE: Kruger National Park, South Africa; Queen Elizabeth National Park, Uganda; Tarangire National Park, Tanzania; Salalah, Oman.

INSELBERGS, KOPPIES, AND CLIFFS

IN A NUTSHELL: Exposed rocks that occur within other habitats and support a diverse and distinctive set of plants and wildlife. **Habitat Affinities:** INDO-MALAYAN LIMESTONE FOREST; AUSTRALASIAN SANDSTONE ESCARPMENTS. **Species Overlap:** Shares species with the surrounding habitats.

DESCRIPTION: Inselbergs are exposed rocks, usually isolated mountains that stand out of the surrounding terrain. *Koppie*, an Afrikaans word meaning "hill," is often used in the more specific sense of "small rocky hill," usually a pile of ancient granite boulders. These are just two catchily named examples of rock-dominated habitats, which can also include rocky canyons, eroded hillsides, limestone karst, cliffs, and lava fields. This sort of habitat can be found across Africa, especially in arid zones but locally even in rainforest. While visually striking, such rocks would normally qualify only as a microhabitat within other habitats if not for their possession of a rich and distinctive set of wildlife. Rocky habitat is almost analogous to wetlands: it usually occurs on a small scale within other habitats but must be sought out in order to find its suite of wildlife.

A wide variety of geological phenomena can produce exposed rock. In Africa, one of the most common is ancient granite sticking out of a more recently eroded plain. Sometimes these granite inselbergs are the compressed cores of ancient volcanoes. In some areas, sandstone or limestone massifs have been eroded down into complex series of canyons. The most remarkable example of this sort is the tsingy of Madagascar: huge blocks of limestone that have eroded into razor-sharp pinnacles. A less common type of exposed rock is the lava fields spewed by recently erupted volcanoes. Inselbergs in savanna environments often support a band of lusher savanna around their base, which is watered by the rain running off the rock face. Such microhabitats are important for some species, such as the Namib Escarpment endemic Herero Chat. Many rock-loving birds have adapted to human structures and are now found even in the concrete jungles of big cities.

The Simba Koppies of Serengeti National Park, Tanzania, are often used by Lions as a hunting outpost.
© IAIN CAMPBELL, TROPICAL BIRDING TOURS

Examples are White-collared Pigeon in Ethiopia and various red-winged starlings across the continent such as the Neumann's, Red-winged, and Bristle-crowned Starlings.

WILDLIFE: Across the Afrotropics, inselbergs, koppies, and cliffs form the exclusive habitat for the cliff-jumping Klipspringer (IS), as well as several species of hyraxes. In the Horn of Africa, rocky habitat is the home of Speke's Pectinator, an odd rodent, and Beira, an elegant small antelope. Along the Namib Escarpment, it holds the Black Mongoose and hyrax-like Dassie Rat. Leopards depend on cover, both to hunt and to avoid Lions and hyenas, and are very fond of rocky areas. They often prey on baboons, which range widely but also often find sanctuary in the same habitat. In Madagascar, the

Rock Hyrax is a classic species of rocky habitats. © IAIN CAMPBELL, TROPICAL BIRDING TOURS

Namibia's Spitzkoppe is a classic inselberg rising sharply from an arid savanna landscape. © KEN BEHRENS, TROPICAL BIRDING TOURS

Klipspringers are found exclusively in and around boulders and rocky hills (photo: Bale Mountains National Park, Ethiopia). © KEN BEHRENS, TROPICAL BIRDING TOURS

lush and inaccessible canyons within limestone tsingy and eroded sandstone massifs are important habitat for wildlife. Lemurs can sometimes even be seen jumping cautiously across the jagged tsingy.

Rocky habitat is the exclusive domain of several widespread birds, including Freckled Nightjar, Familiar Chat, White-winged and Mocking (IS) Cliff-Chats, Rock-loving Cisticola, rock-thrushes, and Cinnamon-breasted Bunting. Rocky mountains are patrolled by Verreaux's Eagles, which prey mainly on hyraxes. In order to escape their nemesis predator, hyraxes post lookouts and have evolved "sunshade" eyebrows to help them look into the bright midday sun. Along the Namib Escarpment, this habitat is the domain of two localized endemic birds: Hartlaub's Francolin and Rockrunner. In South Africa, it supports Ground Woodpecker, long-billed larks, and Kopje Warbler. In West Africa, there are three localized rock-loving species of firefinches, and on the Arabian Peninsula is the Yemen Serin. Cliffs and other rocky sites are crucial breeding habitat for many birds that feed primarily in other habitats. Classic examples are raptors, owls, swifts, ravens, and Rock Martin. Within AFROTROPICAL LOWLAND RAINFOREST, rocky caves are the breeding habitat of two of the continent's most bizarre birds, the White-necked and Gray-necked Picathartes (or Rockfowl), cave-dwelling, colonial-nesting passerines that look like no other birds and are notoriously difficult to find away from their nesting areas.

What wetlands are to birders, rocky habitats are to reptile enthusiasts: a habitat that covers a small portion of the landscape but is of disproportionate importance to the group of animals they seek. These habitats are crucial for many snakes and a high proportion of lizards. Examples include many skinks, girdled lizards (*Cordylus* and *Smaug*), flat lizards (*Platysaurus*), crag lizards (*Pseudocordylus*), agamas, flat geckos (*Afroedura*), leaf-toed geckos (*Hemidactylus*), dwarf geckos (*Lygodactylus*), thick-toed geckos (*Pachydactylus*), and Giant Plated Lizard. Many localized endemic reptiles in Madagascar are found exclusively on specific massifs or tracts of tsingy.

DISTRIBUTION: Inselbergs, koppies, cliffs, and other rocky habitats are most frequently encountered in more arid zones, where rocks are less likely to be veiled by thick vegetation. They are especially common in the ancient arid corridor that runs all the way from the Horn of Africa to the Namib Desert and the Karoo plateau in s. Africa. Occasional rocky hills and isolated boulders are found in c. and w. Africa, especially in the more arid Sahel zone but locally even in the rainforest. Some impressive granite inselbergs are found in Madagascar, especially on the deforested high plateau. There are a couple of vast sandstone massifs in the southwest of Madagascar, and three major areas of tsingy in the west and north.

WHERE TO SEE: Simba Koppies, Serengeti National Park, Tanzania; Spitzkoppe, Namibia; Tsingy de Bemaraha National Park, Madagascar.

AFROTROPICAL TROPICAL GRASSLAND

PALEARCTIC REGION

IN A NUTSHELL: Grass-dominated habitat with only occasional, scattered trees and bushes, found at low and middle elevations. **Habitat Affinities:** NEOTROPICAL PAMPAS AND CAMPO; AUSTRALASIAN TUSSOCK GRASSLAND; INDO-MALAYAN SEASONALLY FLOODED GRASSLAND; NEARCTIC DESERT GRASSLAND. **Species Overlap:** Afrotropical savannas (six different habitats); AFROTROPICAL MONTANE GRASSLAND.

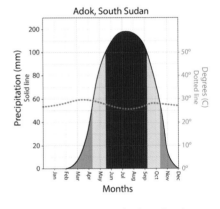

Adok, South Sudan

DESCRIPTION: Quite simply, grassland is habitat in which the dominant vegetation is grass and most of the wildlife is dependent on grass for its livelihood. Since savanna is a grassland with some trees, there is obviously broad overlap between grassland and the various savanna habitats. Anyone who has visited a national park in e. or s. Africa has seen that the interplay between tree-dominated and grass-dominated areas happens on a micro scale. One minute you're driving through a thicket-like area of acacia or MOPANE SAVANNA and then suddenly you emerge onto a grassy and treeless plain. Part of what's complicated about grassland is that this habitat can be produced by different phenomena. One factor is a climate in which nearly all the precipitation occurs in half, or less than half, of the year. During the dry half, the habitat dries out and becomes prone to burning, especially due to lightning strikes at the beginning of the wet season. These fires wipe out tree seedlings and maintain the grassland. A concurrent factor is the browsing of tree seedlings by herbivores, again preventing the growth of large trees. The other major influence that produces grassland is edaphic—related to soil types that don't allow the growth of trees. The most widespread example of such edaphic grassland is seasonally flooded areas with heavy clay soils, such as the dambo grasslands within the GUINEA SAVANNA and MIOMBO WOODLAND zones and the floodplains of the Okavango and Niger Deltas. Edaphic grasslands can also occur on hilltops or ridgelines with thin soils. Some

Opposite page: **East African grasslands** *(top to bottom):* **Lark Plains, Tanzania; Ngorongoro Crater, Tanzania; Masai Mara, Kenya; Ngorongoro Conservation Area, Tanzania; and Silale Swamp in Tarangire National Park, Tanzania.** © KEN BEHRENS, TROPICAL BIRDING TOURS

such grasslands occur even deep in the rainforest zone of c. Africa. The precise composition of grassland varies considerably, depending on its type and location on the continent. Classic species of flooded grassland include Antelope Grass, Wild Rice, Jaragua, and *Loudetia simplex*. Just a few of the many grasses common in drier grassland include Red Oat Grass, various *Andropogon* spp., Rhodes Grass, fountaingrasses, and lovegrasses.

Human disturbance also has a tendency to produce so-called derived grasslands in places that wouldn't typically have supported this habitat, such as the Western Cape of South Africa and the coast of East Africa. Cultivated fields of some crops have a grassland-like character and attract a few of the species of natural grassland. Remarkably, grasslands grow at both ends of the moisture gradient: into the Sahara and Namib Deserts after good rains and in seasonally flooded areas adjacent to permanent swamps in the wettest parts of the continent. Grassland is very much a boom-or-bust habitat. It grows with remarkable rapidity after good rains or recent flooding but can become stark and desertlike during the heart of the dry season. The burst of nutrients that can be produced by this habitat makes it very important for wildlife; it is at the heart of some of the greatest mammal concentrations and migrations left on earth.

The frequent presence of water and good forage for domestic animals tends to attract humans to grasslands. Large wild mammals have been mostly eliminated from the Niger Delta and Lake Chad floodplains by competition from domesticated animals and hunting. Massive water diversion schemes can have a drastic impact on floodplain environments. A dam on Zambia's Kafue River vastly reduced the seasonally flooded area and produced a 50% reduction in the local herd of Southern Lechwe, which formerly numbered over 100,000 individuals. Even worse results may follow from the still incomplete Jonglei Canal, which will divert water around the Sudd wetlands in South Sudan.

WILDLIFE: Grassland is of great importance to Africa's big mammals. Some species use this habitat throughout the year, while others migrate during the year between savanna and pure grassland habitats.

Grasslands are prime habitat for White Rhinoceros and their oxpeckers.
© KEITH BARNES, TROPICAL BIRDING TOURS

Secretarybird stalks snakes and other reptiles in long grass and grassy savanna.
© PABLO CERVANTES, TROPICAL BIRDING TOURS

Africa's grasslands support some of the greatest migratory mammal herds on earth. The best known of these is the Serengeti grassland of n. Tanzania, which hosts a circular annual migration of around 1.2 million Blue Wildebeests, 200,000 Common Zebras, and 400,000 Thomson's Gazelles. The White Nile floodplains of the Sudd in South Sudan support around 1 million Kobs and 30,000 Nile Lechwes. The Liuwa Plain of w. Zambia supports 30,000 migratory Blue Wildebeests and 3,000 Southern Lechwes.

For mammals, one major advantage of seasonally flooded grasslands is that they slowly expand and contract through the year, slowly opening up new areas of ideal forage. This habitat is so attractive that several mammal species have specifically adapted to it, namely Bohor and Southern Reedbucks, Puku, Southern and Nile Lechwes, and Common Tsessebe (IS). Many mammals found in savanna are fond of grassland. These include Cape Hare, Grévy's and Common Zebras, White Rhinoceros, African Buffalo, several species of gazelles, Blue Wildebeest, Red Hartebeest, and Oribi. Grassland is an important grazing habitat for Hippopotamus when it emerges from its aquatic habitation to feed at night. It is extensively used by predators such as Lion, Serval, African Wild Dog, and Cheetah. The soil types that tend to produce grassland also often attract termites, which in turn draw Aardvark, Aardwolf, and Ground Pangolin. Of course, grasslands are rodent paradise and support a wide range of mice and rats.

Grasslands that occur in conjunction with trees, in a savanna habitat, attract a huge variety of birds, most of which are covered in the accounts of specific savanna types. This account will focus exclusively on grassland species. The vast majority of African birds are adapted to use trees to at least some extent, and only a limited, specialized subset can be truly considered grassland birds. These include some francolins, quail, buttonquail, some bustards, several lapwings, some larks, most pipits and longclaws, several chats, and some widowbirds. Seasonally flooded grasslands attract a wide variety of wetland birds, both resident birds from adjacent permanent swamps and migratory ones that spend the boreal winter in Africa. A few species, such as Blue Quail, Black-rumped Buttonquail, and Streaky-breasted Flufftail, are highly specialized to seasonally flooded grassland. More general grassland species include Common Ostrich, Abdim's and White Storks, Black-chested Snake-Eagle, crowned-cranes, Temminck's Courser, Forbes's Plover, Yellow-throated Sandgrouse, Black Coucal, Flappet Lark (IS), Yellow-throated Longclaw (IS), African Pipit (IS), Banded Martin, Gray-rumped Swallow, Zitting and Desert Cisticolas, many migrant shrikes, wheatears, and queleas.

Pure grassland habitat is not particularly rich in reptiles. A few species, such as Grass Skink (*Trachylepis megalura*) and the grass lizards (*Chamaesaura*) are adapted to this habitat, and widespread species like African Rock Python will use it.

DISTRIBUTION: Grassland is found throughout most of the Afrotropics, though is naturally lacking in the KAROO and FYNBOS zones and the Namib Desert. However, even deep in the desert, in the occasional year with a good rain, grassland can quickly and temporarily spring up, producing a quick flush of nutrients, including windblown seeds, which are of great importance to this impoverished ecosystem. The continent's largest stands of pure grassland occur on the Niger, Sudd, and Okavango floodplains, on the Serengeti, and in parts of w. Tanzania and Zambia. Elsewhere, such as throughout the MIOMBO WOODLAND and GUINEA SAVANNA zones and along the southeast coast, grassland is a common habitat but one that occurs in a fine matrix alongside more treed habitats. Grassland is rare in the heart of the w. and c. African rainforest belt but becomes more common in the transition zones surrounding the Congo Basin and along the West African coast. At higher elevations, as in the mountains of Ethiopia and East Africa and the Highveld of South Africa, tropical grassland is replaced by AFROTROPICAL MONTANE GRASSLAND. The extent of grassland in Madagascar prior to the arrival of humans is disputed, but the general lack of grassland-adapted species seems to argue for this having been a rare natural habitat. Nonetheless, much of the island is now covered in derived grassland.

WHERE TO SEE: Serengeti National Park, Tanzania; Masai Mara National Park, Kenya; Awash National Park, Ethiopia; Kafue National Park, Zambia.

AFROTROPICAL MONTANE GRASSLAND

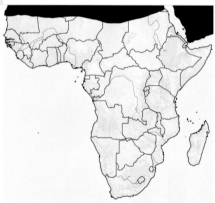

IN A NUTSHELL: Grass-dominated habitat found primarily at high elevations above the tree line but also locally down to middle elevations. **Habitat Affinities:** INDO-MALAYAN MONTANE GRASSLAND; NEOTROPICAL PARAMO; NEOTROPICAL PUNA; AUSTRALASIAN ALPINE HEATHLAND. **Species Overlap:** AFROTROPICAL MONTANE HEATH; AFROTROPICAL TROPICAL GRASSLAND; and Afrotropical savannas (six different habitats).

DESCRIPTION: While TROPICAL GRASSLAND is widespread across the Afrotropics, montane grasslands are much more localized, confined to mountainous areas. Although these grassy habitats can look superficially similar, they have significantly different assemblages of plants and animals. Classic African montane grasslands are found on the continent's highest mountains in a zone where taller plants simply won't grow due to the severity of the climate. On the highest portions of peaks such as Mt. Kilimanjaro in Tanzania and Mt. Stanley in the Rwenzori range on the Democratic Republic of the Congo–Uganda border, even grassland eventually gives out in the cold, desertlike climate, and little grows except lichens. Montane grassland also occurs at much lower elevations in areas that have poor soil or are prone to fire. It is difficult to know the extent of these middle-elevation grasslands before the advent of human-caused fires, but it's certain that they have greatly expanded in modern times at the expense of AFROTROPICAL MONTANE FOREST.

The climate in this habitat is harsh, with warm temperatures during the day and cold, often freezing temperatures at night. Some places have recorded a temperature fluctuation of 72°F (40°C) in

Afrotropical montane grasslands at the base of Sani Pass, Drakensberg, South Africa.

a single day. Montane grassland can develop across a broad range of rainfall levels, from the driest parts of the South African Highveld, which receive only 16 in. (400mm) of annual rain, to the wettest parts of the Ethiopian Highlands, which can receive 100 in. (2,500mm). The precise composition of the grasses that dominate this habitat varies considerably depending on elevation and location on the continent. Plants in the Ethiopian Highlands, including Junegrasses and Silver Hairgrass, are shared with Europe, evidence of the continents' close links during the ice ages. The most common grass in the Highveld of e. South Africa is Red Oat Grass. Other typical grasses include trident grasses, thatching grasses, fescues, *Loudetia simplex*, foxtails, and *Andropogon*. Wet areas often hold true sedges of the genus *Carex*. Herbaceous plants can be common, especially in moist and sheltered spots. A few examples are mouse-ear chickweeds, lady's mantles, felworts, Helichrysum, and buttercups. Giant lobelias, which grow massive flowers that can be up to 16 ft. (5m) tall, are among the most striking plants of this habitat. Several species are found in Ethiopia and East Africa. Another group of bizarre enormous flowering plants of East African montane grassland is the giant groundsels, which can have a thick, corky trunk and sometimes punctuate vast stretches of landscape.

Montane grasslands are among the habitats that are most frequently converted to cultivation. This has happened to the majority of the grasslands of the Ethiopian Highlands and the South African Highveld. The highest montane grasslands are protected by their severe climate, which allows for little human use save the grazing of livestock.

WILDLIFE: Montane grassland is rich in rodents even at high elevations, while most large mammals of this habitat are restricted to its occurrence at more moderate elevations. Typical mammals include baboons, Oribi, Mountain Reedbuck, and Common Eland. Lion, Leopard, and hyenas all sometimes use this habitat. Smaller predators include Black-backed Jackal and Serval. The Highveld once held

Gelada (aka Bleeding Heart Baboon) from the high grasslands of Ethiopia.
© KEN BEHRENS, TROPICAL BIRDING TOURS

Left: **Wattled Ibis is a specialty of montane grasslands in the Ethiopian Highlands.** © KEN BEHRENS, TROPICAL BIRDING TOURS

Below: **Barrow's White-bellied Bustard stalks the edges of Highveld montane grassland near Dirkiesdorp, e. South Africa.** © KEITH BARNES, TROPICAL BIRDING TOURS

a diverse set of big mammals, but most of them are now regionally extinct or confined to reserves. The Black Wildebeest was extinct in the wild but has now been reintroduced. Smaller mammals such as Cape Fox, Meerkat, and Gray Rhebok remain common. Montane grasslands in the Albertine Rift mountains are used by Mountain Gorilla.

The montane grasslands of West Africa and Angola are poor in birds, probably due to their isolation and perhaps to the scarcity of natural grasslands before frequent human-caused fires. The more contiguous archipelago of montane grassland from Eritrea to South Africa is somewhat richer in birds. Typical species include various harriers and other migrant raptors, Striped Flufftail, the rare migratory Montane Blue Swallow, Malachite and Red-tufted Sunbirds, and several species of cisticolas. The most interesting birds of this habitat are the more localized endemics.

Montane grassland is a poor habitat for reptiles, though a few species have adapted to it. These include Ornate Sandveld Lizard, Zimbabwe Girdled Lizard, Bayon's Skink, Grass Skink (*Trachylepis megalura*), and the grass lizards (*Chamaesaura*). The Highveld holds the remarkable Sungazer (*Smaug giganteus*), a big spiky lizard that is the largest member of the girdled lizard family.

Endemism: The scattered, archipelago-like nature of this habitat in Africa has led to many fascinating local endemics. The two major centers of endemism are the Ethiopian Highlands and the Highveld of South Africa. Ethiopia is home to Gelada, a large monkey that uses a complex vocabulary second only to that of *Homo sapiens*. It is also home to Ethiopian Wolf, the world's most endangered canid, Ethiopian Highland Hare, Walia Ibex, and Big-headed African Mole-Rat.

The region's wonderful lineup of endemic birds includes Wattled Ibis, Spot-breasted Lapwing, and Abyssinian Longclaw (IS), as well as Moorland Chat (IS), which is shared with Kenya and

n. Tanzania. South African Highveld is also rich in birds, including Blue Crane, Blue Bustard (IS), Buff-streaked Chat, Cape Grassbird, Botha's Lark, and Yellow-breasted Pipit. The mountains of Kenya and Tanzania have a few localized grassland species, like Jackson's Francolin and Sharpe's Longclaw. The mountains between the Ethiopian Highlands and the South African Highveld have some localized birds, and lots of localized endemic plants.

DISTRIBUTION: Higher-elevation montane grasslands generally occur above a zone of AFROTROPICAL MONTANE HEATH. Middle-elevation grasslands usually occur in a matrix alongside montane heath and AFROTROPICAL MONTANE FOREST. Near the equator, the tree line is around 11,500 ft. (3,500m), but equatorial montane grasslands can locally occur lower, down to around 6,000 ft. (1,800m). Away from the equator, montane grasslands are found at lower elevations; the Highveld grasslands of South Africa grow at 4,000–6,000 ft. (1,200–1,800m). The largest areas of natural montane grassland were originally found in the Ethiopian Highlands and South African Highveld. These two major blocks were loosely connected by an archipelago of mountains. Higher mountains such as Mt. Kilimanjaro, Mt. Kenya, and the Rwenzori and Virunga ranges would have always supported montane grassland. The grassland on many lower mountains, including those in far w. West Africa and the Eastern Arc Mountains, may have been very limited originally but have become much more widespread due to frequent human fires.

WHERE TO SEE: Wakkerstroom, South Africa; Bale Mountains National Park, Ethiopia.

FYNBOS

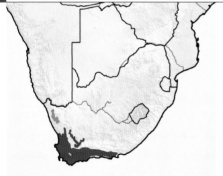

IN A NUTSHELL: A botanically incredibly diverse shrub habitat of w. South Africa that is dominated by ericas, proteas, restios, and geophytes. **Habitat Affinities:** AUSTRALASIAN LOWLAND HEATHLAND; PALEARCTIC MAQUIS; AFROTROPICAL MONTANE HEATH; NEARCTIC PACIFIC CHAPARRAL; NEOTROPICAL MATORRAL SCLEROPHYLL FOREST AND SCRUB. **Species Overlap:** KAROO; AFROTROPICAL MONTANE GRASSLAND; AFROTROPICAL MONTANE HEATH; AFROTROPICAL MONTANE FOREST; AFROTROPICAL MOIST MIXED SAVANNA.

DESCRIPTION: Fundamentally a shrubland growing on poor soil in a Mediterranean climate, this habitat is one of the Afrotropics' most extraordinary. Botanists have classified six floral kingdoms in the world, and this small corner of South Africa is one of them and by far the smallest. This treatment is merited by the monumental plant diversity and endemism, far and away the highest outside the tropics. There are around 9,000 plants in this floral kingdom, 69% of which are endemic. Compared with similar habitats growing in Mediterranean climates, fynbos is 1.7 times more diverse than sw. Australia, 2.2 times more diverse than the Mediterranean basin or California, and 3 times more diverse than Chile. Four groups of plants dominate fynbos: ericas, restios, proteas, and geophytes. These plants all show adaptations that allow them to grow in an environment where fire and long dry periods are common. Members of the erica family, commonly known as heaths, are shrubs with fine, leathery, evergreen leaves. Within this family, the single genus *Erica* has over 600 representatives in the fynbos. Restios, which make up their own family, are evergreen, rush-like plants that may look like grasses to the untrained eye. Proteas form a family of evergreen bushes that have cone-like flower heads that can be spectacular when blooming. Geophytes are plants with underground storage organs, in this case usually bulbs. Grasses and tall trees are both uncommon in fynbos, though they occur locally. The tallest shrubs are generally

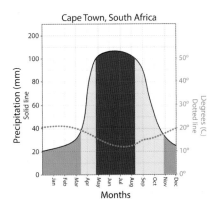

Cape Town, South Africa

Above: **Erica** *(left),* **restio** *(middle),* **and protea** *(right)* **plant families typify fynbos.** © KEN BEHRENS, TROPICAL BIRDING TOURS

Below: **Fynbos is a shrubland dominated by leathery-leaved plants (photo: Clarence Drive, near Rooi Els, South Africa).** © KEITH BARNES, TROPICAL BIRDING TOURS

Strandveld (fynbos growing on coastal dunes) at Koppie Alleen, De Hoop Nature Reserve, South Africa. © KEITH BARNES, TROPICAL BIRDING TOURS

3–10 ft. (1–3m) tall. Although fynbos looks inviting from a distance, the density of shrubs usually makes walking difficult, except in occasional grassy glades.

There are two seasons in the fynbos, a cooler austral winter, when the rain falls, and a warmer austral summer. The climate is fairly mild, especially along the coast, though snowfall can occur on mountaintops during the winter. Annual rainfall varies widely, from 10 in. to 80 in. (250–2,000mm), though a typical range is 12–30 in. (300–750mm). As suggested by its enormous plant diversity, the fynbos is far from uniform. There is tremendous heterogeneity and complexity to the local plant communities, driven by the wide variety of rainfall regimes, elevations, soil types, and slope aspects. The mix of plants varies a great deal from site to site. In one specific type of fynbos, called *renosterveld*, which grows in valleys and plains, asters dominate, while restios are completely absent and proteas rare.

Large areas of lowland fynbos have been cleared for agriculture and development, notably around the sprawling city of Cape Town. Less than 1% of the original renosterveld remains in its natural state. The montane fynbos has been less impacted by humans, though like the lowlands, it is subject to a more insidious threat in the form of introduced invasive plants. Most of the big mammals of this habitat were cleared out soon after Europeans arrived at the Cape. Areas that have been converted to treed suburbs and farmyard gardens can have a somewhat savanna-like character and attract species that historically wouldn't have been found in this zone.

WILDLIFE: Fynbos does not offer an abundance of easy nutrients to animals and therefore is a relatively wildlife-poor environment. Its mammal assemblage was further reduced by hunting after the arrival of Europeans. Large mammals like African Bush Elephant, Lion, Cheetah, and African Buffalo were locally extirpated, and the Bluebuck, a beautiful antelope closely related to Roan and Sable Antelopes, was hunted to extinction. The fynbos does still hold an assemblage of smaller mammals, including an impressive lineup of smaller antelopes. The Bontebok and Cape Grysbok (IS) are endemic to this habitat, and occur alongside the more widespread Common Duiker,

Steenbok, and Gray Rhebok. Cape Mountain Zebra and Leopard are much reduced in range but still occur locally. Smaller predators include Cape Fox, Cape Gray Mongoose, and Caracal.

Although it is bird-poor overall, the fynbos is of great interest to birders because of its endemic and near-endemic birds. These include some great-looking species like White-quilled Bustard and Black Harrier. Other typical birds include Gray-winged and Cape Francolins, Bokmakierie, Cape Grassbird, Victorin's Warbler, Karoo Prinia, Chestnut-vented Warbler, Cape White-eye, Karoo Scrub-Robin, Cape Robin-Chat, Southern Double-collared and Malachite Sunbirds, and Cape Siskin.

The fynbos has a rich set of reptiles comprising more than 100 species, of which more than a dozen are endemic. Endemics include Southern Adder, Southern Rock Lizard, Cape Mountain Lizard, Black and Oelofsen's Girdled Lizards, and False Girdled Lizard. This is also one of the most diverse habitats in the world for tortoises, home to such species as Parrot-beaked and Angulate Tortoises.

Endemism: The fynbos encompasses countless areas of local interest for botanists. For example, 30 separate and discrete botanical complexes have been identified within the mountain fynbos alone. But none of these nodes of plant endemism have produced a striking change in the mammal and bird assemblages. There is a slight difference between the wildlife of mountain and lowland fynbos, with lowland fynbos sharing more species with the succulent KAROO. Endemic birds include the Cape Rockjumper, Cape Sugarbird, and Orange-breasted Sunbird (IS). Half of the fynbos's 38 amphibian species are endemic.

DISTRIBUTION: Fynbos is found both along the coast and on the mountains of sw. South Africa. There are some isolated montane fynbos communities that are completely surrounded by KAROO habitat; although there are some similarities between fynbos and Karoo, especially the succulent Karoo, their plant and wildlife communities are very different.

WHERE TO SEE: Cape of Good Hope Nature Reserve, South Africa.

Left: **Cape Sugarbird is a common fynbos specialist preferring nectar of plants of the Protea family (photo: Cape Town, South Africa).**
© KEITH BARNES, TROPICAL BIRDING TOURS

Below: **Orange-breasted Sunbird, one of several birds endemic to the fynbos, is an erica specialist.**
© KEN BEHRENS, TROPICAL BIRDING TOURS

AFROTROPICAL MONTANE HEATH

IN A NUTSHELL: A montane heath-dominated shrub habitat that is fire-prone and grows on poor soils, mainly at high elevations but locally down to moderate elevations. **Habitat Affinities:** AUSTRALASIAN ALPINE HEATHLAND; EUROPEAN HEATHLAND AND MOORLAND. **Species Overlap:** KAROO; AFROTROPICAL MONTANE GRASSLAND; AFROTROPICAL MONTANE FOREST; AFROTROPICAL MOIST MIXED SAVANNA.

DESCRIPTION: Heath is a shrubland dominated by members of the heath family (Ericaceae), which includes heathers. These plants tend to have fine, leathery leaves and a brownish or off-green coloration, giving the habitat a distinctive character, even at a distance. Heath usually grows on poor, acidic, well-drained soil and is highly prone to fire. In the Afrotropics, the most diverse heath habitat is the FYNBOS of sw. South Africa. The continent's only other heath, the montane heath covered here, seems to act as a buffer between the frequently burned AFROTROPICAL MONTANE GRASSLAND on mountaintops and the lower-lying and generally unburned AFROTROPICAL MONTANE FOREST. At the

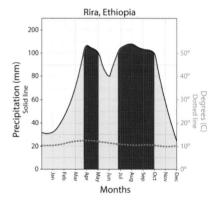

transition from montane forest to heath, there can be large ericas, up to 24 ft. (7m) tall, heavily swathed in mosses and other epiphytes. Higher up, the moorland (a wetter kind of heath) gradually becomes shorter, down to 3–7 ft. (1–2m), then gives way to grassland.

WILDLIFE: Montane heath is a poor habitat for wildlife. In general, it has a small subset of the species of adjacent AFROTROPICAL MONTANE FOREST and MONTANE GRASSLAND habitats. Black Rhinoceros is one mammal that readily uses heath, though it has been eliminated from most of its former range.

Typical birds include Abyssinian and Rwenzori Nightjars, Cape Eagle-Owl, African Stonechat, Moorland Chat, Cinnamon Bracken-Warbler, and a variety of sunbirds, including Tacazze and Malachite Sunbirds and several double-collared sunbirds. Ethiopia has perhaps the richest montane heath in the Afrotropics. The Indian Ocean island of Grande Comore has the distinction of supporting the only Afrotropical species that is entirely restricted to heath habitat: Comoro White-eye. This habitat is also used by the island endemic Grand Comoro Brush-Warbler. On Réunion, montane heath is inhabited by the endemic Réunion Stonechat, while in Madagascar, heath species include Madagascar Flufftail, Madagascar White-eye, and Madagascar Sunbird.

Above: **Though habitats like this, from the Bale Mountains of Ethiopia, are considered Montane Grassland in this book, they have wildlife similar to Montane Heathland, and include some scrubby ericas.** © NICK ATHANAS, TROPICAL BIRDING TOURS

Left: **Ethiopian Wolf crossing high-elevation habitat of the Sanetti Plateau, Ethiopia.** © KEN BEHRENS, TROPICAL BIRDING TOURS

Endemism: In montane heath habitats on the Indian Ocean islands, which have very different wildlife from those on mainland Africa, endemism ixs driven more by the high endemism of adjacent forest habitats rather than by the heath itself. On the continent, this habitat is readily used by several of the endemic mammals and birds of the Ethiopian Highlands: Gelada, Mountain Nyala, Ethiopian Wolf, and Moorland and Chestnut-naped Francolins. Elsewhere, it is an important habitat for the localized Albertine Rift endemic Rwenzori Red Duiker and for the Kenyan specialties Jackson's Francolin and Aberdare Cisticola.

DISTRIBUTION: Found exclusively in the higher mountains of the Afrotropics, this habitat typically occurs in a narrow elevational range, above AFROTROPICAL MONTANE FOREST but below AFROTROPICAL MONTANE GRASSLAND. Near the equator, this zone is between 10,000 ft–11,500 ft. (3,000–3,500m). Away from the equator, as in c. and s. Madagascar, heath can occur down to 6,500 ft. (2,000m). Some of the most extensive areas of this habitat are in the Bale Mountains of Ethiopia. This habitat can also be produced at lower elevations by frequent, human-caused fires, which transform montane forest into heath. Such artificial heaths are common in Madagascar.

WHERE TO SEE: Bale Mountains National Park, Ethiopia; Aberdare National Park, Kenya.

AFROTROPICAL FRESHWATER WETLAND

IN A NUTSHELL: Habitat that is permanently or seasonally flooded with fresh water, vegetated by a diverse set of uniquely adapted hydrophilic plants. **Habitat Affinities:** AUSTRALASIAN TROPICAL WETLAND; AUSTRALASIAN TEMPERATE WETLAND; NEOTROPICAL FLOODED GRASSLAND AND WETLAND; INDO-MALAYAN FRESHWATER WETLAND; NEARCTIC FRESHWATER WETLAND; PALEARCTIC TEMPERATE WETLAND. **Species Overlap:** AFROTROPICAL TROPICAL GRASSLAND; AFROTROPICAL SALT PAN; AFROTROPICAL MANGROVE; AFROTROPICAL TIDAL MUDFLAT AND SALT MARSH.

Shoebill is found mainly in Papyrus swamp. This oddity may be Africa's most sought-after species among birders. © IAIN CAMPBELL, TROPICAL BIRDING TOURS

DESCRIPTION: Freshwater wetlands are among the most productive and diverse habitats in Africa and are critical for a vast array of wildlife. They are also crucial to humans, watering most of the food they consume, such as rice, which is the staple food for half of earth's human population. This habitat encompasses a vast and complex array of microhabitats. These include rivers, streams, lakes (both dammed and natural), ponds, bogs, coastal and inland deltas, shrub swamps, freshwater marshes, and inland mudflats. Freshwater AFROTROPICAL SWAMP FOREST is covered separately, as its primary function is that of a thickly forested environment, though it also supports some wetland species. Many areas of TROPICAL GRASSLAND transform into wetlands when they are seasonally flooded. Saltwater habitats (AFROTROPICAL TIDAL MUDFLAT AND SALT MARSH, AFROTROPICAL MANGROVE, and AFROTROPICAL SALT PAN) are also covered in separate sections.

Freshwater wetlands support plants that can grow in the anaerobic conditions produced by temporary or perennial flooding. These plants have a diverse set of forms, making this an endlessly fascinating habitat to explore. Deep or fast-flowing water may largely lack emergent plants. The richest wetlands for wildlife are those in slow-flowing or shallow areas. Grasses are some of the most important and prominent wetland plants. Widespread and common grasses include the common reeds, cattails, wild rice, and Hippo Grass. Africa's most famous wetland grass is Papyrus, which can form vast stands, especially between Botswana and South Sudan. Floating plants include water lilies and Water Cabbage. Although Water Cabbage is native to the continent, it has spread far beyond its original range and is a serious invasive species.

Wetlands in well-watered areas such as the Congo Basin fluctuate little throughout the year, while wetlands in more seasonal environments, such as the savannas that cover much of the continent, change drastically. They can expand across vast floodplains in the rainy season, or when floods arrive from upstream, and then contract drastically or disappear completely by the end of the dry season.

Although Africa still contains far more pristine wetlands than most of the other continents, this habitat has nonetheless been heavily impacted by humans. Marshes are often destroyed to make way for agriculture or human settlement. The endemic-rich marshes of e. Madagascar have been so widely destroyed that they probably deserve to be considered the island's most-threatened habitat, despite the better-known plight of the island's forests. Throughout the Afrotropics, the water in rivers and lakes can be overutilized, causing them to slowly shrink. In some areas, such as e. South Africa, extensive planting of exotic tree plantations in the highlands has led to a drastic reduction of the water flowing into the rivers below. Wetlands easily collect pollutants, as contaminants naturally flow downhill. Invasive plants such as Common Water Hyacinth, an Amazonian species, now clog huge areas that were formerly far more diverse. Damming of rivers for irrigation or to generate hydroelectric power can drastically impact fish populations. One of the continent's greatest conservation tragedies occurred in the 1950s when the Nile Perch was introduced into Lake Victoria, resulting in the extinction or near-extinction of hundreds of endemic cichlids. Conserving wetlands has to be considered one of the continent's greatest conservation challenges, and this uniquely vulnerable habitat may serve as a sort of litmus test for the wider state of conservation.

WILDLIFE: This is one of the most exciting habitats for wildlife-watching, and any safari will spend much time in and around wetland habitats. The water draws wildlife like a magnet, and there is no surer way to liven up a dull midday in the savanna than to head for a river or marsh. While virtually all larger mammal species visit wetlands in order to drink, only a small set of species use this as their primary habitat. West African Manatees are found in the rivers of West Africa and in some cases have been isolated from the sea by dams or stranded in oxbow lakes. Wetlands are prime habitat for several species of otters. Hippopotamuses spend their days in wetlands, then range widely through surrounding habitats at night in search of fodder. The Sitatunga is a remarkable marsh-dwelling antelope that has long, narrow, widely splayed hooves that allow it to traverse boggy ground. Several other antelopes are found mainly at the edges of wetlands. These include reedbucks, Puku, lechwes, and Waterbuck. Other mammals often found in and around wetlands are Serval, Side-striped Jackal, Marsh Mongoose, and African Buffalo. In Madagascar, one remarkable species of lemur has adapted to life in reed beds: the critically endangered Lac Alaotra Bamboo Lemur.

While wetlands are relatively poor in reptiles, they do host the continent's biggest and most fearsome reptile, Nile Crocodile. They're also

Papyrus Gonolek is another Papyrus-restricted specialist of East Africa (photo: Lake Victoria, Uganda). © IAIN CAMPBELL, TROPICAL BIRDING TOURS

prime habitat for Nile Monitor and several water-adapted snakes. Wetlands are the primary habitat of many of continental Africa's 800-some species of amphibians and some of Madagascar's hundreds of endemics. Africa is home to around 20% of the world's approximately 15,000 freshwater fish species; the Congo River alone has more than 700 species.

Wetlands are paradise for birds and birders alike. Anyone with ambitions of checking off a long bird list must spend lots of time in this habitat. Many entire groups of birds are found almost exclusively in wetlands. These include herons, egrets, ibises, geese, jacanas, shorebirds, terns, and gulls. This is the habitat of some of the continent's most superb birds, such as Goliath Heron, Yellow-billed Stork (IS), Saddle-billed Stork, Hamerkop, Wattled Crane, African Finfoot, and Giant Kingfisher. Shoebill, which may be Africa's most sought-after bird, makes its home in Papyrus swamps, along with a range of other specialists, like Papyrus Gonolek, Papyrus Yellow-Warbler, and White-winged Swamp Warbler. Open water can hold grebes, cormorants, pelicans, and ducks. Certain species, such as African Skimmer, Egyptian Plover, pratincoles, and lapwings, are found mainly on riverine sandbars. Wetlands are hunted by African Fish-Eagle, harriers, and African Grass-Owl. Although many of this habitat's birds are big and flashy, it also supports smaller birds, including swallows, wagtails, stonechats, swamp warblers, reed warblers like the widespread African Reed Warbler (IS), cisticolas, prinias, weavers, bishops, widowbirds, and waxbills. Marsh skulkers include White-backed Night-Heron, flufftails, and a variety of rails.

Madagascar wetlands are the exclusive habitat of several endemic Malagasy birds: In the west are found Madagascar Fish-Eagle, Madagascar Jacana, and Sakalava Rail. The east holds Meller's Duck, Madagascar Rail, Slender-billed Flufftail, and the recently rediscovered Madagascar Pochard.

Most wetland species are widespread, probably due to the far-flung nature of this habitat, as only species that are able to find new habitat when local conditions change will survive. Indeed, many wetland birds are migratory or nomadic and move around to take advantage of changing water conditions across the continent. The richest wetlands are in the vast swath of Africa between Botswana and e. South Africa in the south to Ethiopia and South Sudan in the north.

Hippos remain common in protected areas of sub-Saharan Africa.
© KEITH BARNES, TROPICAL BIRDING TOURS

Endemism: Within the Afrotropics' wetland zone, there are minor centers of bird endemism in Botswana–Zambia, and around Lake Victoria. Many of the localized endemic birds are species virtually confined to pure stands of Papyrus, such as Papyrus Gonolek and Papyrus Canary.

DISTRIBUTION: Wetlands are found locally across the Afrotropics, usually nested within other habitats, ranging from AFROTROPICAL DESERT to AFROTROPICAL LOWLAND RAINFOREST. They often occur in very small patches and might easily be dismissed as simply another microhabitat if not for their immense importance to wildlife. Africa's four great rivers are the Niger, Nile, Congo, and Zambezi. The Nile provides water to the great Sudd swamps of South Sudan, and the Niger waters the Inner Niger Delta. A string of rich lakes and other wetlands runs down the two arms of the Great Rift Valley, from Ethiopia to Tanzania and from Uganda to Malawi. Western Tanzania and Zambia hold some important wetlands, including the Bangweulu Wetlands and the Kafue Flats. The continent's most famous wetland is the Okavango Delta, where a broad river flows out of the mountains of Angola and disappears into the sands of the dry Kalahari in n. Botswana. Madagascar's wetlands include many broad but shallow rivers in the west and vast marshes at Lac Kinkony and Lac Alaotra.

WHERE TO SEE: St. Lucia, South Africa; Okavango Delta, Botswana; Lake Awassa, Ethiopia.

AFROTROPICAL SALT PAN

IN A NUTSHELL: Sparsely vegetated or unvegetated habitat on highly alkaline soils, usually evaporation pans. **Habitat Affinities:** NEARCTIC SALT MARSH; AUSTRALASIAN TIDAL MUDFLAT AND SALT MARSH; INDO-MALAYAN TIDAL MUDFLAT AND SALT PAN. **Species Overlap:** AFROTROPICAL TIDAL MUDFLAT AND SALT MARSH; AFROTROPICAL ROCKY COASTLINE AND SANDY BEACH; AFROTROPICAL FRESHWATER WETLAND.

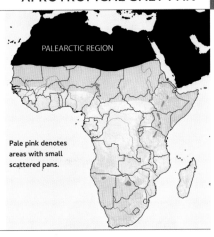

PALEARCTIC REGION

Pale pink denotes areas with small scattered pans.

DESCRIPTION: An extremely hostile environment in which very few plants will grow, this habitat is created and defined by alkaline soil that is incredibly rich in sodium carbonate. Pans form when large lakes evaporate away, when seawater is trapped in a coastal lagoon and then evaporates, or when salt-rich deposits of volcanic material accumulate in lake basins and along river courses. This might be considered little more than a microhabitat within arid environments if not for its huge extent in the Etosha Pan in Namibia, Makgadikgadi Pans in Botswana, and a scattering of large soda lakes in the Great Rift Valley, including Tanzania's Lakes Natron and Manyara, Kenya's Lake Nakuru, and Ethiopia's Lake Abiata. The life of this habitat is water, which can be direct rainfall or inflow from rivers and streams. Although dry pans are virtually barren, blue-green algae will grow and small shrimp quickly hatch after a pan is flooded, providing a source of food in the ecosystem. Despite their apparent inhospitable nature, soda lakes are highly productive, with high rates of photosynthesis enabled by the huge amount of dissolved carbon dioxide. Flooded salt pans become wetlands bustling with feeding and nesting birds, most famously flamingos. There is a small set of halophytic plants that will grow on salt pans, usually along the edges. These include Salt Grass, Smooth Flatsedge, *Odyssea paucinervis*, and seepweeds. Away from the pans, the typical habitat is AFROTROPICAL DRY THORN SAVANNA or MOPANE SAVANNA.

WILDLIFE: Most mammals rarely frequent this habitat, except when forced to cross it on the way to somewhere else. Scavengers like Black-backed Jackals occasionally raid waterbird colonies in the pans.

Birds are the most abundant and conspicuous form of wildlife by far. Most famous among these are flamingos, which breed exclusively in this habitat. While Greater Flamingo breeds more widely, the only historically consistent breeding site for Lesser Flamingo (IS) on the whole continent has been Tanzania's Lake Natron. When not breeding, flamingos move all around the continent. The creation and enhancement of Kamfers Dam in South Africa has boosted the declining Lesser Flamingo population, providing another dependable breeding site. When Namibia's Etosha Pan floods, every 7–10 years, Greater and Lesser Flamingos flock there to breed. But oftentimes, the pan dries too quickly to allow the young birds to be ready to migrate, stranding huge numbers of them, which either die or require a massive rescue operation. Another bird that nests colonially on flooded salt pans is Great White Pelican. The sight of many thousands of flamingos and pelicans standing virtually shoulder to shoulder is one of Africa's great wildlife spectacles.

Chestnut-banded Plover (IS) is virtually restricted to salt pans. The pans are frequented by a range of shorebirds, both residents such as Pied Avocet and Kittlitz's Plover, and migrant species like Ruff and Curlew Sandpiper. Other birds partial to this habitat include terns and gulls, Eared Grebe, and Cape Teal. One odd sight that may greet a visitor to a dry salt pan is Common Ostrich in the distance, distorted by the heat haze. Amazingly, these hardy birds sometimes choose to nest in this habitat due to the absence of predators.

There are a few reptiles that have adapted to this harsh habitat; Etosha and Makgadikgadi Pans each have an endemic agama.

DISTRIBUTION: Salt pans are found in arid environments where evaporation generally outpaces precipitation, so they are confined to sw., ne., and n. Africa. The two largest pans are in the Kalahari of the southwest, while the Great Rift Valley has a string of soda lakes, the largest of which is Lake Natron. A scattering of smaller salt pans may be found elsewhere, both inland and along the coast.

WHERE TO SEE: Etosha National Park, Namibia; Lake Manyara, Tanzania; Lake Abiata, Ethiopia.

Salt pans are the breeding habitat for both of Africa's flamingos: Lesser and Greater. © KEN BEHRENS, TROPICAL BIRDING TOURS

AFROTROPICAL MANGROVE

IN A NUTSHELL: Forest of salt-tolerant trees that grow coastally in the intertidal zone. **Habitat Affinities:** NEARCTIC MANGROVE; NEOTROPICAL MANGROVE; AUSTRALASIAN MANGROVE; INDO-MALAYAN MANGROVE FOREST. **Species Overlap:** AFROTROPICAL SWAMP FOREST; AFROTROPICAL LOWLAND RAINFOREST; AFROTROPICAL MONSOON FOREST; AFROTROPICAL TIDAL MUDFLAT AND SALT MARSH.

Mangroves line the banks of Mida Creek, Kenya.
© ORDERCRAZY, CC0, VIA WIKIMEDIA COMMONS CREATIVE COMMONS CC0 1.0

DESCRIPTION: Mangroves are trees that have a high tolerance for salt, allowing them to grow in the intertidal zone along the seashore and on brackish estuaries. Although the species composition of mangrove forest varies throughout the world, its structure and function remain much the same. The world's richest mangroves are found in Australasia, and the AUSTRALASIAN MANGROVE habitat account in that chapter includes the most thorough treatment. Among the regions covered in this book, the Afrotropics holds the third-most extensive mangrove habitat, after Australasia and Indo-Malaysia. Despite this, Afrotropical mangroves, unlike those of Indo-Malaysia and especially Australasia, do not support a very diverse or specialized set of terrestrial wildlife.

Trees that qualify in the broad sense as mangroves are found in many different families. The pressures of living in an exacting environment have forced these unrelated trees to evolve a set of similar adaptations, an example of converging evolution. One is the ability to survive in low-oxygen soil. Some species have stilt roots to lift them out of the water, and others push "breathing roots" (pneumatophores) out of the mud to give them access to oxygen. Another mangrove adaptation is thick and leathery leaves to limit water loss. Botanically, the East African, Arabian Peninsula, and Malagasy mangroves are similar to those elsewhere around the Indian Ocean and are completely different from the mangroves of West Africa, which are closely allied to those along the Atlantic coast of the Americas. The most common and widespread species in the east are Red Mangrove (*Rhizopora mucronata*), Black Mangrove (*Bruguiera gymnorrhiza*), Indian Mangrove, Cannonball Mangrove, Gray Mangrove, Mangrove Apple, and White-flowered Black Mangrove. In the poorer west, there are only six major species: three known as Red Mangrove (*Rhizophora racemosa, R. harrisonii*, and *R. mangle*), Black Mangrove (*Avicennia germinans*), White Mangrove, and the introduced but quickly spreading Nipa Palm. Typical mangrove forest is around 30 ft. (10m) tall, though well-developed stands can reach 100 ft. (30m). The trees are normally stunted along the shoreline but become taller inland, where they give way to AFROTROPICAL SWAMP FOREST, AFROTROPICAL LOWLAND RAINFOREST, or savanna habitat. This habitat is not directly dependent on rain for its sustenance, so mangroves grow in areas with a wide range of precipitation. However, areas without either sufficient rainfall or fresh groundwater, such as the coast of the Namib Desert, lack mangroves.

Human populations tend to be high along the coast, in the same areas that hold mangroves, so this habitat has been heavily impacted. Although it may seem inaccessible at first glance, mangroves are cut for firewood, building material, and charcoal. Areas of mangroves are often converted into shrimp farms, saltworks, or rice paddies. Their coastal location also makes them susceptible to oil spills and river pollution.

WILDLIFE: Considering how long mangroves must have been present in the Afrotropics, this is a remarkably poor habitat for terrestrial birds and mammals compared to the mangroves of the Indo-Malayan or Australasian regions. For the most part, mangroves host a much-reduced subset of the species of adjacent habitats. Of course, mangroves are extremely important to marine environments, harboring a huge and diverse array of species, and are especially crucial as a nursery for developing fish. This habitat is used by charismatic aquatic beasts like West African Manatee, Dugong, Nile Crocodile, and marine turtles.

Although virtually no species are restricted to this habitat, the assemblages of wildlife in West Africa, East Africa, the Malagasy region, and around the Red and Arabian Seas are all quite different. West African mangrove is botanically poorer but richer in wildlife, perhaps because it occurs alongside freshwater AFROTROPICAL SWAMP FOREST and AFROTROPICAL LOWLAND RAINFOREST. The muddy substrate covered in dense roots seems to preclude the presence of most ground-dwelling mammals, but the canopy holds primates including Western Red Colobus, Angolan and Gabon Talapoins, Sooty Mangabey, and Mona and Campbell's Monkeys.

A fair number of West African rainforest birds will use mangroves, and this is the primary habitat for Mouse-brown (IS) and Carmelite Sunbirds. The East African mangroves are much poorer in terms of terrestrial wildlife. The one bird strongly associated with this habitat is Mangrove Kingfisher. Around the Red and Arabian Seas, Collared Kingfisher and Clamorous Reed Warbler (IS) are found exclusively in mangroves.

In Madagascar, several endemic species such as White-throated Rail, Madagascar Swamp Warbler, and Malagasy Kingfisher use this habitat. It's also an important refuge for the critically endangered Madagascar Fish-Eagle and provides daytime refuge for Madagascar Flying Fox.

DISTRIBUTION: Mangrove forest is found along coasts in warm tropical and subtropical regions. In the Afrotropics, the largest tracts are at river mouths, including the Zambezi, Rufiji, and Niger Rivers, where the presence of brackish water allows mangroves to penetrate up to 30 mi. (50km) inland. In West Africa, mangroves are found from Senegal south to Angola. In East Africa, they occur locally from Somalia to South Africa. In the Indian Ocean, mangroves are common along the west coast of Madagascar and very locally on the east coast and on other islands. Mangroves can also be found locally around the Red and Arabian Seas.

WHERE TO SEE: Mida Creek, Kenya; between Tulear (Toliara) and Ifaty, Madagascar.

AFROTROPICAL TIDAL MUDFLAT AND SALT MARSH

IN A NUTSHELL: Coastal intertidal areas with rich soil that is sometimes unvegetated and sometimes supports low, salt-tolerant vegetation. **Habitat Affinities:** NEARCTIC SALT MARSH; NEARCTIC TIDAL MUDFLAT; AUSTRALASIAN TIDAL MUDFLAT AND SALT MARSH; INDO-MALAYAN TIDAL MUDFLAT AND SALT PAN. **Species Overlap:** AFROTROPICAL ROCKY COASTLINE AND SANDY BEACH; AFROTROPICAL SALT PAN; AFROTROPICAL FRESHWATER WETLAND; AFROTROPICAL OFFSHORE ISLANDS; AFROTROPICAL PELAGIC WATERS.

DESCRIPTION: Coastal mudflats mainly occur where rivers flow into the ocean, carrying sand, silt, and clay, which are deposited along the estuary. These soils are much richer than the nutrient-poor substrates of sandy and rocky coastlines found away from estuaries and support a bounty of invertebrates, which in turn support other wildlife, mainly migratory birds. Some mudflats are tidal, exposed at low tide. Others are exposed only when the winds correctly align. Although this is a harsh environment, due to wave action, constantly changing water levels, and high salt content of the ocean water, some specialized plants can grow in more sheltered areas, forming salt marshes. The halophytic plants of this environment are similar to those of AFROTROPICAL SALT PAN.

WILDLIFE: This habitat is very poor in mammals, although it is an important component of the habitat of Dugong and West African Manatee. Mudflats are a birder's paradise due to the range of shorebirds they attract. Many widespread species such as Gray Heron, Little Egret, Whimbrel, Bar-tailed Godwit (IS), and Black-winged Stilt are found along coasts throughout the Afrotropical region. Some shorebirds are more localized to the Indian or Atlantic Ocean side of the continent:

A typical coastal wetland, Winneba Lagoon, Ghana.
© KEN BEHRENS, TROPICAL BIRDING TOURS

Indian Ocean shorebirds include Crab-Plover, sand-plovers, and Terek Sandpiper, while Red Knot is largely restricted to Atlantic coasts. Afrotropical salt marshes are not particularly rich in larger wildlife but do support a few widespread species like African Pied Wagtail.

DISTRIBUTION: Found locally around all coasts of the Afrotropics, including on the Indian Ocean islands.

WHERE TO SEE: West Coast National Park, South Africa; Walvis Bay, Namibia.

AFROTROPICAL ROCKY COASTLINE AND SANDY BEACH

IN A NUTSHELL: Coastal areas with poor rocky or sandy soil, which support grass or low scrub or simply lack vegetation. **Habitat Affinities:** NEARCTIC ROCKY COASTLINE AND SANDY BEACH; AUSTRALASIAN ROCKY COASTLINE AND SANDY BEACH; INDO-MALAYAN ROCKY COASTLINE AND SANDY BEACH. **Species Overlap:** AFROTROPICAL TIDAL MUDFLAT AND SALT MARSH; AFROTROPICAL SALT PAN; AFROTROPICAL FRESHWATER WETLAND; AFROTROPICAL OFFSHORE ISLANDS; AFROTROPICAL PELAGIC WATERS.

DESCRIPTION: Most of Africa's coastline consists of sandy beaches and rocky areas that can range from flat rock to abrupt cliffs. Sand dunes accumulate in some areas, creating a localized desertlike environment. Rare natural saline lagoons, along with more common human-created ones, can also be found (see AFROTROPICAL SALT PAN).

WILDLIFE: Although beach is the favorite habitat of sun-seeking vacationers around the world, it is generally a poor habitat for wildlife. In the Afrotropics, the richest shoreline habitats are those of w. South Africa and Namibia, where the cold waters of the Benguela Current bring abundant nutrients close to the shore. This area has some massive Cape Fur Seal colonies, some on the mainland and some on offshore islands. These colonies attract scavengers like Brown Hyena and Black-backed Jackal.

Birds of the cold coastline of sw. Africa include African Oystercatcher and Bank, Cape, and Crowned Cormorants. More widespread birds found along much or all of the continent's shorelines are Western Reef-Heron, White-fronted Plover, Great Crested Tern, and Sandwich Tern. Some species, like Lesser Crested Tern, are restricted to the Indian Ocean coast, while others, like West African Crested and Damara Terns, are found mainly along the Atlantic. There are many generalist coastal birds that use both sandy and rocky shorelines and AFROTROPICAL TIDAL MUDFLATS. But even these species tend to prefer mudflats due to the greater abundance of food. Although sandy and rocky shoreline is far poorer than mudflats for shorebirds, it does attract some migrant species

Most people are surprised to see African Penguins on a South African beach. © KEN BEHRENS, TROPICAL BIRDING TOURS

like Sanderling (IS) and Ruddy Turnstone. Coastal cliffs provide breeding habitat for White-tailed Tropicbirds in the Gulf of Guinea and on the Indian Ocean islands.

Sandy beaches provide breeding habitat for Green, Loggerhead, Olive Ridley, Hawksbill, and Leatherback sea turtles, which are found throughout most of the region but are classified by the International Union for Conservation of Nature (IUCN) as vulnerable, endangered, or critically endangered. Their decline is a result of human activities including beach development, hunting, and dumping trash into the ocean.

DISTRIBUTION: Rocky coastline and sandy beach make up the most common coastal habitat, though it is periodically broken by AFROTROPICAL TIDAL MUDFLAT AND SALT MARSH in estuarine areas.

WHERE TO SEE: Pemba, Tanzania.

AFROTROPICAL PELAGIC WATERS

IN A NUTSHELL: The deep waters lying off the continental shelf surrounding Africa and the Indian Ocean islands. **Habitat Affinities:** NEARCTIC PELAGIC WATERS; INDO-MALAYAN PELAGIC WATERS; AUSTRALASIAN PELAGIC WATERS. **Species Overlap:** AFROTROPICAL OFFSHORE ISLANDS; AFROTROPICAL ROCKY COASTLINE AND SANDY BEACH; AFROTROPICAL TIDAL MUDFLAT AND SALT MARSH.

DESCRIPTION: The Atlantic Ocean lies west of Africa, and the Indian Ocean lies to its east. Cape Agulhas, South Africa, the continent's southernmost point, is somewhat arbitrarily considered the border between the two. Although the waters surrounding Africa are generally poorly known, both oceans are locally rich in marine mammals and seabirds. The edges of the continental shelf, where there is a drop-off between a relatively shallow seabed and much deeper waters, tend to have nutrient-rich upwellings and thus to be the best areas for seeing oceanic wildlife. In

The cold waters off the Cape of Good Hope, South Africa, host many bird species, including Atlantic Yellow-nosed Albatross, and marine mammals such as Short-beaked Common Dolphin.
© LISLE GWYNN, TROPICAL BIRDING TOURS

some places, such as off the island of Mauritius and off Dakar, Senegal, this drop-off is very close to shore. In general, the n. Indian Ocean is fairly poor in oceanic wildlife, and the Atlantic is a bit richer. The richest portion of the Atlantic is off sw. South Africa and Namibia, where the cold Benguela Current brings abundant nutrients from the Southern Ocean. Cape Town, South Africa, is the one place in the continent where boat trips regularly go offshore in search of pelagic birds.

Masses of seabirds congregate behind a fishing trawler in the productive waters of the Benguela Current off sw. South Africa. © KEITH BARNES, TROPICAL BIRDING TOURS

WILDLIFE: Some widespread marine mammals such as Humpback Whale, Blue Whale, Killer Whale, and Common Bottlenose Dolphin are found throughout the oceans of the region. Other species are divided between the warmer, more tropical waters (Short-finned Pilot Whale, Pygmy Killer Whale, Melon-headed Whale, Spinner Dolphin, and Pantropical Spotted Dolphin) and the cooler waters around the southwest side of Africa (Southern Right Whale and Heaviside's Dolphin). The austral winter is the best time for whale-watching off s. Africa, as many species of the Southern Ocean move northward to escape the frigid Antarctic winter.

For birds, the major divide is between the Atlantic and Indian Oceans, though a few common birds like Wilson's Storm-Petrel are found in both. Widespread Atlantic species include Cory's and Sooty Shearwaters, and European and Leach's Storm-Petrels. Typical Indian Ocean species are Wedge-tailed, Tropical, and Flesh-footed Shearwaters; and Great and Lesser Frigatebirds. The richest and best-known African waters for seabirds are those off of Cape Town, South Africa. Here the rich, cold waters of the Benguela Current attract a great diversity of pelagic birds. Different seasons bring strikingly different birds. Many species of the far Southern Ocean move north during the austral winter, roughly between May and October; these include Northern Royal and Wandering Albatrosses, giant-petrels, and Cape Petrel. During the austral summer, different birds visit from the north during their nonbreeding season; these include European Storm-Petrel, Sabine's Gull, and Parasitic, Pomarine, and Long-tailed Jaegers.

WHERE TO SEE: Pelagic birding boat trip out of Cape Town, South Africa.

AFROTROPICAL OFFSHORE ISLANDS

IN A NUTSHELL: Small offshore islands, usually covered in grass or low scrub, that provide refuge for oceanic wildlife and a small selection of other species. **Habitat Affinities:** NEARCTIC OFFSHORE ISLANDS; INDO-MALAYAN OFFSHORE ISLANDS; AUSTRALASIAN SANDY CAYS. **Species Overlap:** AFROTROPICAL PELAGIC WATERS; AFROTROPICAL ROCKY COASTLINE AND SANDY BEACH; AFROTROPICAL TIDAL MUDFLAT AND SALT MARSH.

DESCRIPTION: While small offshore islands look insignificant on a map, they provide very important habitat for breeding seabirds, sea turtles, and seals, because they are generally free of the predators of the mainland and are also less likely to have been pillaged by humans. They also give refuge to migrating birds such as shorebirds, gulls, and terns, which also use beaches and mudflats along the mainland. Most small offshore islands are sandy, formed by the breakdown of coral reefs, though some are sedimentary, produced by the outflow of sediment from large rivers; some are volcanic; and others, such as the main islands of the Seychelles, are granitic fragments of Gondwana. Despite their remoteness, most offshore islands have still been heavily impacted by hunting and egg collection, but the most drastic impact has been the introduction of mammals such as rats, goats, and rabbits. The rats prey on seabird eggs, especially those of the burrow-nesting shearwaters and petrels. Goats and rabbits overgraze the vegetation and drastically alter the natural landscape. In some places, such as Round Island, off Mauritius, there has been a concerted effort to rid islands of exotic animals and to restore them to their natural state.

Cape Gannets breed on a handful of offshore islands in South Africa and Namibia (photo: Lambert's Bay, South Africa). © KEN BEHRENS, TROPICAL BIRDING TOURS

WILDLIFE: Although there are only a few islands off the coast of w. South Africa and Namibia, they are heavily used by wildlife, as they provide secure breeding sites in the nutrient-rich, cold waters of the Benguela Current. Sites such as South Africa's Dassen and Dyer Islands and Namibia's Penguin Islands provide breeding habitat for most of the world's Cape Fur Seals, African Penguins, Cape Gannets, Kelp and Hartlaub's Gulls, and Cape, Crowned, and Bank Cormorants.

Islands in warm tropical waters of both the Atlantic and Indian Oceans support breeding colonies of tropicbirds, frigatebirds, Brown Booby, Brown Noddy, and several terns, including Sooty Tern (IS). Aldabra, in the Seychelles, has the world's second-largest frigatebird colony, hosting both Great and Lesser Frigatebirds. Bird Island, also in the Seychelles, hosts more than a million breeding Sooty Terns. Black Noddy breeds on islands only in the tropical Atlantic portion of the Afrotropical region. Red-tailed Tropicbirds breed on Nosy Be off Madagascar. Breeding colonies of Wedge-tailed and Tropical Shearwaters, Red-tailed Tropicbird, Masked and Red-footed Boobies, Lesser Noddy, and Roseate Tern are found only on islands on the Indian Ocean side of the continent. Islands of the n. Indian Ocean and Red Sea are home to White-eyed Gull, and Lesser Crested, Sandwich, and White-cheeked Terns. Round Island is one of the few relatively high-elevation (rather than low and sandy) tropical islands that lack rats, and as such is a critical breeding refuge for several seabirds such as Trindade, Kermadec, and Herald Petrels, and also holds Bulwer's and Black-winged Petrels and Wedge-tailed Shearwater.

Sandy offshore islands provide crucial breeding habitat for Green, Loggerhead, Olive Ridley, Hawksbill, and Leatherback sea turtles, all of which are threatened or endangered.

Endemism: Although this habitat doesn't harbor many endemic birds, there are major differences between the species breeding in the tropical Atlantic, those off the cold southwest side of Africa, and those in the tropical Indian Ocean. A few species and subspecies of lizards are endemic to small offshore islands. Round Island has an exceptional assemblage of endemic reptiles: Round Island Keel-scaled Boa, Round Island Day Gecko, and Round Island Ground Skink.

DISTRIBUTION: The largest concentrations of small islands are found in the Red Sea, in the Seychelles, and off Madagascar, Mauritius, and Rodrigues. There are a few offshore islands in the cold water of the Benguela Current off sw. Africa. The coast of West Africa has relatively few islands, though there are some in the Gulf of Guinea.

WHERE TO SEE: Nosy Ve, Madagascar.

AFROTROPICAL CROPLAND

IN A NUTSHELL: Areas dominated by human-planted crops. **Habitat Affinities:** Tropical croplands are pretty generic worldwide. **Species Overlap:** AFROTROPICAL TROPICAL GRASSLAND; AFROTROPICAL MONTANE GRASSLAND; Afrotropical savannas (six different habitats).

DESCRIPTION: Crop cultivation takes many different forms in the Afrotropics, running the gamut from small-scale farming that might even enhance a habitat for some species of wildlife to vast industrial monocultures that completely replace the indigenous habitat and most of its wildlife. Examples of the former are the small Teff Grass farms in the Ethiopian Highlands and the polycultures of much of Uganda and w. Tanzania and Kenya, which are rich in birds and smaller mammals. Examples of the latter are the wheat fields of sw. South Africa and the soybean farms along the Kavango River in n. Namibia, which attract some wildlife but are completely different from the habitats they replaced. Grainfields are a sort of grassland and attract a subset of the species of natural AFROTROPICAL TROPICAL GRASSLAND and AFROTROPICAL MONTANE GRASSLAND. Most of the cultivation in Africa is still small scale; vast monocultures are rare. But the continent is rich in arable land, and with the economic rise of Africa and the global demand for food, industrial farms are certain to become increasingly common. The impact of this remains to be seen but is likely to be negative for wildlife. Tree farms growing exotic species, primarily eucalyptus and pine, are common throughout Africa, especially in the mountains. In most cases, these plantations directly replace natural forest, resulting in a massive loss of biodiversity.

Cultivated areas on the Indian Ocean islands tend to be much poorer in wildlife than those on the continent, as these originally heavily forested islands lack a reservoir of indigenous species that are adapted to open habitats. The starkest example of this is the island of Mauritius, the refuge of some fabulous forest birds, the most famous of which was the Dodo. Today Mauritius is nearly deforested, the Dodo and many other forest species are extinct, and about one third of the island is covered in vast sugarcane monocultures.

The wheat fields and other agricultural lands of South Africa's Overberg provide habitat for some open-country birds like **Blue Crane.** © KEN BEHRENS, TROPICAL BIRDING TOURS

WILDLIFE: In general, cultivated areas are used only fleetingly by larger mammals that venture out of adjacent natural habitats. Human conflict with wildlife is becoming increasingly common in places where cleared and cultivated land lies adjacent to natural habitats. Examples are c. Kenya, where African Bush Elephants venture out of Mt. Kenya and Aberdare National Parks; and the mountains of Rwanda and Uganda, where Mountain Gorillas sometimes raid farms.

Cultivated areas can be good for birds and are actually the best places to look for certain species. Grainfields are used by Common Quail, cranes, some bustards, and many larks and pipits. Various sorts of cultivation are favored habitats for Cape Crow, African Stonechat, and bishops. African Oil Palm plantations attract frugivores like hornbills and Gray Parrot. Fallow fields are suitable for a wide variety of open-country birds such as lapwings, migrant wheatears, and Quailfinch. Exotic tree plantations are fairly sterile overall but do function as habitat for a subset of forest species. Raptors such as Black Goshawk, Rufous-breasted Sparrowhawk, and Forest Buzzard readily nest in the tall trees provided by tree plantations. Beehives are often placed in eucalyptus plantations, making them attractive to Greater and Lesser Honeyguides. Other birds with an affinity for tree plantations include several species of cuckoos, African Wood-Owl, Fiery-necked Nightjar, and Rufous-necked Wryneck.

AFROTROPICAL GRAZING LAND

IN A NUTSHELL: Areas grazed by domestic animals. **Habitat Affinities:** Intense grazing is generic worldwide. Nomadic grazing lands are similar to PALEARCTIC MAQUIS; NEOTROPICAL MONTE; and AUSTRALASIAN OPEN EUCALYPT SAVANNA. **Species Overlap:** Afrotropical savannas (six different habitats); AFROTROPICAL TROPICAL GRASSLAND; AFROTROPICAL MONTANE GRASSLAND.

DESCRIPTION: Grazing lands with minimal impact on the original habitats include those of the AFROTROPICAL DRY THORN SAVANNA of the Horn of Africa, virtually all of which is grazed, and the KAROO shrublands and the AFROTROPICAL MONTANE GRASSLAND of the Highveld of s. Africa. The story has been quite different on the Indian Ocean islands, especially Madagascar. There, vast areas of forest were burned to create cattle pasture, which is shockingly devoid of indigenous plants or wildlife. The major reason for the poverty of these habitats seems to be the lack of a reservoir of Malagasy savanna- or grassland-adapted species that can move into them. On the African continent, forested areas that are destroyed to create pasture or for cultivated crops are quickly colonized by the species of adjacent grassland and savanna habitats.

WILDLIFE: Grazing lands generally have far fewer big wild mammals than natural habitats, due to the competition for food resources with domestic animals. When you see a herd of 100 goats somewhere in the DRY THORN SAVANNA, you can probably assume that they're directly replacing the natural herbivores of that habitat. While herbivorous big mammals usually remain in reduced numbers, except in areas with lots of hunting, larger predators are rarely tolerated. Lion, Cheetah, Leopard, and hyenas have been wiped out across much of Africa in order to protect domestic animals. Even the smaller predators like Caracal, foxes, and jackals are often killed.

The situation is much better with birds. Grazing lands generally support a nearly complete set of the birds of the habitat in which they lie. A few species even closely associate with domestic animals, often in lieu of the presence of wild mammals. Cattle Egret, Fork-tailed and Glossy-backed Drongos, Piapiac, and Wattled Starling all follow domestic animals, sometimes even perching on them and catching the insects that they disturb. Hornbills eat insects such as dung

Oxpeckers don't seem to discriminate between wild and domestic big mammals. © KEN BEHRENS, TROPICAL BIRDING TOURS

beetles that they extract from the dung of both wild and domestic animals. Oxpeckers eat ticks and other parasites off their host animals, drink their blood, and even use their hair to line their tree cavity nests. Certain groups of birds seem to thrive in grazed areas, occurring perhaps at even higher densities than in natural areas. These include some species of starlings, waxbills, and weavers. Two unusual birds, Stresemann's Bush-Crow and White-tailed Swallow, are restricted to a tiny and heavily grazed area of s. Ethiopia, where they remain in good numbers.

DISTRIBUTION: Vast areas of the Afrotropics are grazed at least occasionally by domestic animals: goats, sheep, cattle, donkeys, and camels. Thankfully, most of these grazing areas retain much of their natural character and wildlife, and still function as savanna or grassland habitats.

AFROTROPICAL CITIES AND VILLAGES

IN A NUTSHELL: Areas of human habitation. **Habitat Affinities:** INDO-MALAYAN CITIES AND VILLAGES. **Species Overlap:** Smaller villages retain some similarity to surrounding natural habitats, but cities have very little in common with any natural areas.

DESCRIPTION: Humans have heavily altered the natural landscape of Africa in countless ways but nowhere more profoundly than in and immediately surrounding their settlements, from sprawling cities to tiny and remote villages. These areas contain a bounty of different microhabitats that provide different niches for wildlife, including gardens of exotic flowers, stands of trees planted for shade and shelter, granaries, trash dumps, ornamental ponds, urban parks, manicured lawns, and concrete canyons between skyscrapers. Indeed, humans have created such a diverse and complex set of wildlife habitats that a whole book could easily be devoted just to them. Grazing lands and cultivated areas are covered in the preceding sections (AFROTROPICAL CROPLAND; AFROTROPICAL GRAZING LAND); this section is devoted to the areas immediately associated with human habitation.

In forests, human settlements tend to be cleared to some extent and thus more open than the surrounding natural habitat. The opposite is true in more open environments, which characterize the majority of the Afrotropics; there areas of human habitation tend to be lusher than their environs, as people access water in order to grow ornamental plants and soften their surroundings, creating islands of lusher habitat. In areas of AFROTROPICAL DRY THORN SAVANNA, towns tend to take on the character of AFROTROPICAL MOIST MIXED SAVANNA, while in an area of moist mixed savanna, a well-established town can have a character similar to AFROTROPICAL MONSOON FOREST. Habitat islands like this can be very attractive to birds migrating through dry country. In some cases, human-modified habitats have allowed large range extensions of certain wildlife species. The most striking example of this is in the Cape area of South Africa, where many bird species of moist mixed savanna have extended their ranges hundreds of miles to the west by using the new habitats created by humans.

WILDLIFE: An amazing variety of larger mammals can survive alongside humans in areas where their presence is tolerated, as occurs in most of Ethiopia. Lush gardens can hold Guereza Colobus and other monkeys, squirrels, and Bushbuck. Treed urban areas frequently provide daytime refuge for Straw-colored Fruit Bat and other large bats. Predators like jackals, African Wildcat, Common Genet, and even Leopard can survive alongside large human populations. Spotted Hyenas are even known to live in the sewers of Addis Ababa, Ethiopia. In parts of South Africa, Hippopotamuses live in ponds and wetlands adjacent to towns and emerge to "mow" the lawns at night. In parts of s. and e. Africa, herbivores like Impala, Thomson's Gazelle, Blesbok, and Common Zebra have been introduced to the grounds of gated estates, where they munch the lush lawns. In contrast, the towns and cities of heavily hunted West Africa are virtually devoid of bigger mammals.

Urban areas can provide excellent habitat for birds and are the main habitat of several introduced species, such as feral Rock Pigeon, Rose-ringed Parakeet, House Crow, European Starling, Common Myna, and House Sparrow (IS). They are favored in Ethiopia by White-collared Pigeon and throughout the continent by several species of indigenous sparrows. Some of these species are now so adapted to human-created habitats that it's difficult to imagine where they lived before the ascendancy of human beings. In many parts of Africa, scavenging Black Kites and Marabou Storks are trash collectors and can be found in large numbers around dumps or perched ghoulishly on streetlights. Lawns and artificial wetlands are favored by Hadada Ibis and Egyptian Goose. Mowed grass is attractive to Helmeted Guineafowl and several species of thrush. Tall buildings seem to function much as cliffs for some birds and even provide nesting habitat for Eurasian and Rock Kestrels and Red-winged Starling. Highway underpasses and culverts are colonized by Little and White-rumped Swifts and many species of swallows. As suggested by their names, Village Indigobirds and Village Weavers are often seen in towns. Some species that favor the AFROTROPICAL MOIST MIXED SAVANNA–analogous habitat of lush gardens include Red-eyed and Laughing Doves, Dideric and Klaas's Cuckoos, bulbuls, and robin-chats. These habitats also attract mousebirds, white-eyes, and many species of sunbirds, which are especially partial to the flowers of introduced Australian bottlebrushes.

Urban habitats support many reptiles, including smaller snakes, skinks, and geckos. The most common gecko is the Tropical House Gecko (IS), which is easily found at night on walls and around lights across most of the Afrotropics. In reptile-rich Madagascar, human structures are often inhabited by Common House Gecko, several beautiful day geckos (*Phelsuma*), fish-scale geckos (*Geckolepis*), and sometimes even huge Madagascar velvet geckos (*Blaesodactylus*).

Human settlements in Madagascar can be great for gecko-watching. Shown here *(clockwise from left)* are **Northern Madagascar Velvet Gecko, Madagascar Giant Day Gecko eating a Common House Gecko, Peacock Day Gecko, and fish-scale gecko.** © KEN BEHRENS, TROPICAL BIRDING TOURS

HABITATS OF THE PALEARCTIC

(EUROPE, NORTHERN ASIA, AND NORTH AFRICA)

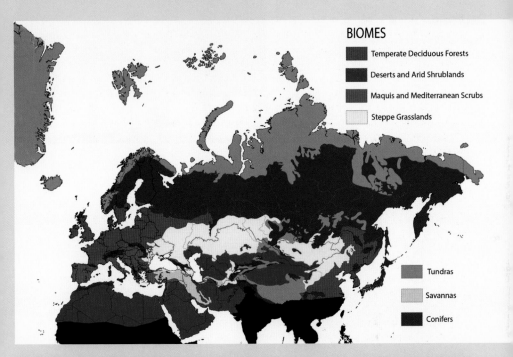

BIOMES

- Temperate Deciduous Forests
- Deserts and Arid Shrublands
- Maquis and Mediterranean Scrubs
- Steppe Grasslands
- Tundras
- Savannas
- Conifers

EURASIAN SPRUCE-FIR TAIGA

IN A NUTSHELL: These are the endless forests of conifers that people imagine when they think of Norway or Russia's Kamchatka Peninsula. **Habitat Affinities:** NEARCTIC SPRUCE-FIR TAIGA. **Species Overlap:** SIBERIAN LARCH FOREST; EURASIAN MONTANE CONIFER FOREST.

DESCRIPTION: From Norway and most of Sweden to the coast of Kamchatka and n. Japan, there is a massive belt of taiga (often called boreal forest) dominated by conifers—spruces, firs, and pines. There are numerous subzones within this huge region, but with the exception of the eastern SIBERIAN LARCH FOREST, which is treated separately, most of this forest is rather uniform, with few conifer species, and very few broadleaf tree species. Because of their great shade tolerance, the dominant trees of this habitat are Scots Pine and Ayan Spruce in the east and Norway Spruce in the west; they are replaced in the boggiest areas by willows and birches; in the rockiest zones with lithosols by pines and cedars; and close to the tree line by alders. The eastern taiga

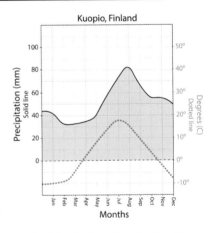

Kuopio, Finland

forests to the east and south of the Siberian larch forest are far richer in plant diversity than the western taiga forests found in Norway. One explanation is that during the last glacial maximum (25,000–16,000 YBP) and the Younger-Dryas mini-glaciation (12,900–11,700 YBP), the European boreal forests were essentially eliminated, and e. Asia acted as the refugium from which boreal forests recolonized the n. Palearctic.

When you visit this kind of forest, it feels very monotonous, certainly much more so than grassland or desert, and seems endless. It has a heavy canopy cover between 65 ft. and 165 ft. (20–50m) tall, though usually under 100 ft. (30m), and little undergrowth. The forest floor is covered in a litter of conifer needles, decaying trunks, and lichens. You can easily walk through the drier parts of this forest without a trail, as there are few shrubs to impede you. In boggy areas, many sphagnum and feather mosses, shrubs like Bilberry and Cranberry, and grasses grow. Even though precipitation is not high, the low temperatures in this region mean that precipitation

Spruce-fir taiga with deciduous birch, south of Kirkenes, Norway.
© IAIN CAMPBELL, TROPICAL BIRDING TOURS

exceeds evaporation, producing an abundance of water. Glacial deposits block drainage, generating many potholes and depressions, and in combination with the positive water balance, this means that the whole region is dotted by lakes, bogs, and seasonally flooded areas. There is variation due to slope, drainage, and bedrock type, and on a broader scale the forests change from west to east, with the east being far richer. From Fennoscandia to the Ural Mountains, the region is dominated by just two associations: stands of Scots Pine that dominate in well-drained and sandy soils, along with areas with more frequent fires; and forests of spruce and fir that cover the wetter and flatter areas, which have fewer fires. Pine forests have mature trees of varying ages (up to 700 years) as a result of their diversified life cycles, while spruce forests tend to have a canopy consisting of similarly aged trees (up to 400 years) and copious younger trees because of their harmonized circuition, where trees in close proximity have a synchronized life cycle. At their elevational limits, these forests can have patches dominated by birch. Plant diversity gradually increases as you move east, and Siberian Fir, Dwarf Siberian Pine, and Siberian Spruce become common.

WILDLIFE: As well as resident passerines such as Siberian Tit (IS), the bird assemblage contains many woodpeckers such as the Black Woodpecker, corvids like the Siberian Jay (IS), and grouse such as Western Capercaillie. Most resident species found in the Eurasian spruce-fir taiga such as Eurasian Three-toed Woodpecker, Pine Grosbeak, Bohemian Waxwing, and Red Crossbill are found in both blocks of this habitat in Europe and e. Asia. Most of the breeding migrant raptor species such as Eurasian Kestrel, Eurasian Hobby, Hen Harrier, and Northern Goshawk occur throughout the taiga. Passerine migrant breeders of many species are found across the whole region, but the western parts of this habitat have birds that winter in Africa, and the eastern block is dominated by species that winter in s. and se. Asia. African passerine migrants to Europe include Common

Eurasian Brown Bears are confiding where not persecuted.
© VIKRAM PODKAR

Cuckoo, Spotted Flycatcher, Siberian Blue Robin, European Pied Flycatcher, and Willow Warbler. Breeding migrants to e. Asia include Japanese Waxwing, Asian House-Martin, Gray's Grasshopper-Warbler, Dusky Warbler, and Dusky Thrush.

Herbivorous mammals of this habitat include Moose (known as Elk in Eurasia) and Mountain Hare. Carnivores include Wolverine, Red Fox, and Brown Bear.

DISTRIBUTION: This is the predominant habitat from the coast of Norway, Sweden north of Stockholm, and n. Estonia, east across the Ural Mountains to the Yenisei River in c. Siberia, where it is replaced by SIBERIAN LARCH FOREST. It replaces larch forest again in se. Russia, n. Japan, and n. China. It is bordered to the south by EUROPEAN TEMPERATE DECIDUOUS FOREST in Europe and by PALEARCTIC FOREST-STEPPE in c. Russia. In the north it blends into BERINGIAN TAIGA SAVANNA or directly into EURASIAN ROCKY TUNDRA and BOGGY TUNDRA.

WHERE TO SEE: Oulu and Kuusamo regions of Finland, and Sakhalin in Russia.

Siberian Jay can be found in spruce-fir taiga as well as Siberian larch forest. © KEITH BARNES TROPICAL BIRDING TOURS

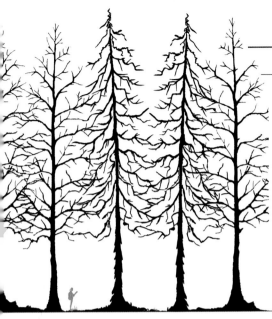

SIBERIAN LARCH FOREST

IN A NUTSHELL: A very uniform, deciduous conifer forest. **Habitat Affinities:** There are no extensive larch forests anywhere else in the world. **Species Overlap:** EURASIAN SPRUCE-FIR TAIGA; EURASIAN MONTANE CONIFER FOREST.

DESCRIPTION: The first impression you might have in a larch forest in summer is the uniformity of the coniferous vegetation; this is because it usually grows in uniform stands of the same age and height. Much of c. Siberia is covered with this deciduous forest, which grows where conditions are too extreme for the more widespread EURASIAN SPRUCE-FIR TAIGA. Although the same species, Dahurian Larch, occurs here in a wide variety of situations, its structure varies depending on where it's growing. In the north, in the coldest parts of the Northern Hemisphere, there is permafrost and even tundra does not grow. In this area, larch trees are stunted, reaching only 4 ft. (1.2m) tall. Farther south, larches become true trees and can grow to 150 ft. (45m) tall. In winter, the difference between larch forests and spruce-fir-pine forests is even starker. Larch is deciduous, and a winter larch grove gives the impression of a conifer forest after a fire, with bare-branched trees and the ground strewn with shed needles. The deciduous nature of the canopy, and the thin understory of broadleaf deciduous dwarf birches and willows, allows for a layer of shrubs such as Lingonberry and many mosses and lichens on the frequently boggy ground. Interestingly, this forest shares much of its ground-cover plant assemblage with the EURASIAN ROCKY TUNDRA and BOGGY TUNDRA to the north, a connection that is not as strong between the spruce-fir forests and the tundra.

Groves of larch cover a wide variety of environments, from uplands and wide valleys to the Arctic and alpine tree lines. Within cold areas, larch grows in both humid and dry microenvironments and in a wide variety of soils. Within these forests, larch forms nearly uniform, similarly aged, and same-height canopies without many other trees. Although some more favorable microenvironments do allow for various birches, willows, and Mongolian Poplar to grow, larch rarely forms long-term large mixed stands with them, due to the fact that it is intolerant of shade.

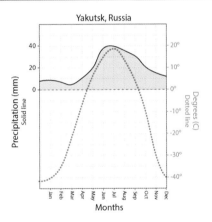

Yakutsk, Russia

Larch forests are easily identified in winter because their trees are uniform in size and deciduous.
ANNELI SALO, CC BY-SA 3.0, HTTPS://CREATIVECOMMONS.ORG/LICENSES/BY-SA/3.0, VIA WIKIMEDIA COMMONS

When mixed canopies begin to form, new larch cannot regenerate (even under its own canopy) and is outcompeted by the other conifers and birches. It seems that larch stands tend to colonize, grow, and die together. There are two things that make Dahurian Larch so successful in this region. The first is its ability to colonize and hold onto new areas, which is due to its very rapid growth at young ages and its longevity—some specimens having been dated as older than 540 years. The second is its ability to take up water even from very cold soil and to transpire and photosynthesize faster than other conifers, combined with its tolerance of exceedingly frigid winter temperatures. These adaptations are due to larch being both very shallow-rooted, with no taproots, and a deciduous conifer, which means it does not need to absorb water from the frozen ground during the winter when it has no leaves.

WILDLIFE: Birdlife in larch forests differs from that of the EURASIAN SPRUCE-FIR TAIGA farther west and east, in that it is even more depauperate in the colder months. The deciduous nature of the forest in winter does not provide as much shelter as other conifers for common groups of birds such as woodpeckers and grouse. However, larch forests are richer in birds than other conifer forests during summer, because the breeding birds have no competition from resident species. Larch forests are important breeding grounds for many of e. Asia's migrant passerines, such as Siberian Thrush, Pallas's Grasshopper-Warbler, and Red-flanked Bluetail.

There is a lack of resident mammals for the same reason there aren't winter birds. Because the larches cannot hold snow, like spruce, fir, or hemlock, deer cannot find shelter from the wind, and therefore reside in much lower numbers than farther west or south.

DISTRIBUTION: The larch forest of e. Siberia is one of the largest homogeneous forest blocks in the world, covering 950,000 sq. mi. (2.5 million km^2). It stretches from the Yenisei River in the west all the way to the east coast of Siberia, and from EURASIAN ROCKY TUNDRA and BOGGY TUNDRA or BERINGIAN TAIGA SAVANNA in the north to PALEARCTIC FOREST-STEPPE of northernmost Mongolia in the south.

WHERE TO SEE: Yakutsk, Sakha, Russia.

Hazel Grouse is another species that uses larch forest sparingly in winter, but it will move in to breed in summer. © PETE MORRIS, BIRDQUEST

Wolverine is the largest of the terrestrial weasels. © KEITH BARNES, TROPICAL BIRDING TOURS

EURASIAN MONTANE CONIFER FOREST

IN A NUTSHELL: These are the moist, open, coniferous forests in the mountains of most of Europe. **Habitat Affinities:** NEARCTIC MONTANE SPRUCE-FIR FOREST. **Species Overlap:** EURASIAN SPRUCE-FIR TAIGA.

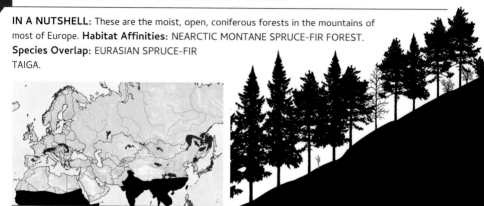

DESCRIPTION: The montane conditions at high elevation in the temperate zone are similar to the cold conditions found farther north in the taiga (or boreal) zone. The conifer forests of these areas are related to (and can superficially resemble) EURASIAN SPRUCE-FIR TAIGA forests to their north, and in areas of c. and e. Asia, the forests blend together. Relative to their boreal cousins, the mid-latitude conifer forests have longer growing seasons and greater daily temperature fluctuations.

Pearl Shoal Waterfall in Jiuzhaigou, China, is surrounded by montane conifer forests.
© KEITH BARNES, TROPICAL BIRDING TOURS

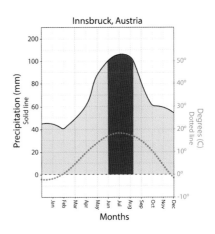

Innsbruck, Austria

Black Woodpecker prefers conifer forests in Japan and e. Asia but is more catholic in habitat choice in Europe, where it is more common in mixed conifer-broadleaf forest.
© PETE MORRIS, BIRDQUEST

Temperatures can be balmy through the day only to plunge well below freezing after nightfall. This regular freeze-thaw activity during many months of the year results in large amounts of mechanical rock weathering and unstable ground. As with the taiga, this forest tends to be locally dominated by one or two tree species. However, the rugged terrain and large variations in sunlight, rainfall, drainage, and soil types allow for many microclimates in which several species find their own niche. The continual weathering and erosion of the mountain ranges, paired with the continual opportunities for new colonizing forest communities, create a mosaic of forest variations that are far more interesting than taiga.

The European Alps are dominated by spruce forests, the majority of which have been replanted. Pine forests are also found, at higher elevations. Although there is much variation, the general pattern is that the lower conifer forests are dominated by tall spruce trees, above which are similarly sized pines and some larches. Above these larger conifers, on the scree slopes, is a band of small bushy pines that form dense groves that are very difficult to walk through. The Pyrenees have a pine and fir mix.

Eastward, the mountain ranges generally become drier, and the Carpathian and Caucasus Mountains have spruce forests on the lower slopes and pines at higher elevations. The species change, but the underlying theme is fairly constant. Pine is found on sandier and poorer soils or in areas with more influence from fire, while spruce and fir are found on wetter, more nutrient-rich soils. The Altai Mountains in e. Asia differ in that they have a closer relationship with the taiga, since they merge directly into that habitat. The Tian Shan range of China is dominated by spruce, while the Altai range is dominated by larch and fir lower down and pine at higher elevations.

Chamois is a European mountain goat that spends summers in alpine meadows and winters in coniferous forests. © DALE FORBES, SWAROVSKI OPTIK

WILDLIFE: As in many n. Palearctic habitats, the birdlife in montane conifer forest is dominated by migrant species that spend the winter in Africa or s. Asia, and migrate north to breed, such as Willow Warbler, Common Chiffchaff, Goldcrest, and Mistle Thrush. These forests do retain some resident species such as Boreal and Pygmy Owls, Eurasian Nutcracker, Red Crossbill, and Black Woodpecker, found through the European Alps.

Larger herbivores such as Chamois are still found in the higher levels of these mountain forests. Other than Red Fox, carnivores are now rare, but Pine Martens occur in most of these forests.

DISTRIBUTION: These montane conifer forests occur in all the humid and semi-humid mountain ranges from Scotland south to the Pyrenees and European Alps, east to e. Europe, across to the Altai Mountains of Mongolia, the Dzhagdy Mountains in ne. Russia, and the peaks of Hokkaido, Japan.

WHERE TO SEE: Tyrol, Austria; Pyrenees, Spain; Tian Shan, China; Jiuzhaigou, China.

MEDITERRANEAN AND DRY CONIFER FOREST

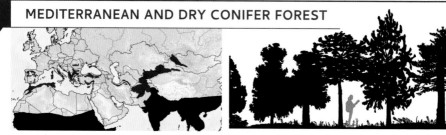

IN A NUTSHELL: The cypress, pine, and juniper forests of the drier parts of Eurasia. **Habitat Affinities:** NEARCTIC MONTANE MIXED CONIFER FOREST; NEARCTIC PINYON-JUNIPER WOODLAND. **Species Overlap:** EURASIAN MONTANE CONIFER FOREST.

DESCRIPTION: The conifer forests of the Mediterranean, Middle East, and east to Pakistan are an extension of the mid-latitude EURASIAN MONTANE CONIFER FORESTS, and therefore, where they overlap (in places such as Spain), they share some of the same tree species. However, the growing conditions change, and these forests experience a harsher environment, since in addition to cold winters, they also have to deal with very dry conditions for much of the year. In the north, farther away from the Mediterranean Sea, firs and Scots Pine dominate. Umbrella Pine becomes the dominant tree of the w. Mediterranean, where it grows alongside Maritime Pine, Wild Olive, and oaks. The eastern shore of the Mediterranean and the mountains of Lebanon used to be covered in vast forests of Lebanon Cedar and Aleppo Pine, though these were cut during biblical times. In the Maghreb region of nw. Africa and along the e. Mediterranean shores, oaks and other broadleaf trees are present but a minor component of any forest, while Aleppo Pine and fir trees form extensive forests. The characteristic all these forests have in common, that sets them apart from the forests of higher latitudes and elevations, is that they are much drier, and although there

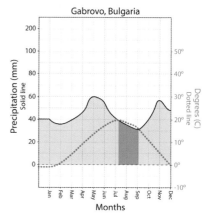

Gabrovo, Bulgaria

Right: **Common Firecrest is primarily a conifer species in the Mediterranean but is also common in the region's oak forests.**
© SIMON BUCKELL

Below: **Pine forest, Andalusia, Spain.**
© IAIN CAMPBELL, TROPICAL BIRDING TOURS

is leaf litter, the soil has much less humus than in wetter environments. It is easy to walk through dry conifer forest, as shrub and ground cover are limited. Shrubs that do grow here are the same species found in the MAQUIS and GARRIGUE habitats, sclerophyllous species with waxy leaves, and aromatic herbs occur as well.

WILDLIFE: The eastern range of this forest has more bird species that are closely tied to this habitat. Examples from North Africa include Algerian Nuthatch and Atlas Flycatcher, while to the east are Kashmir Nutcracker, Krüper's Nuthatch (in the Caucasus), and the rare Syrian Serin. Overall, this wide-ranging habitat is used by many bird species, including Hawfinch (in eastern parts of the bird's range); Black-billed Capercaillie (IS); Northern Goshawk (IS); Golden Eagle; Eurasian Pygmy-Owl; Long-eared and Boreal Owls; Levaillant's Woodpecker; Sombre, Coal, and Crested (IS) Tits; Eurasian Jay; Willow Warbler; Masked Shrike; Common Chaffinch; European Robin; White-winged Redstart; and Red Crossbill (IS).

Among the mammals that occur are Pine Marten and Eurasian Red Squirrel, which are both strongly tied to this habitat, and Barbary Macaque in North Africa. More widespread species include Red Deer, Chamois, Wild Boar, Brown Bear, Gray Wolf, and both European species of lynx, Eurasian and Iberian.

DISTRIBUTION: These conifer forests range from the Iberian Peninsula to the northern shores of the Mediterranean Sea to the highlands of Lebanon, Jordan, and Israel. On the African side, they occur in the Atlas Mountains from sc. Morocco to the very north of Algeria to northernmost Tunisia. In the Caucasus Mountains, this forest type is found at lower elevations than spruce forests. It continues through Syria to n. Iran and s. Turkmenistan. It forms another solid expanse from n. Afghanistan through Tajikistan and most of Kyrgyzstan, where it blends into the EURASIAN MONTANE CONIFER FOREST in the Tian Shan of c. Asia. An outlier block exists in sw. Pakistan in the mountains of Balochistan. This forest also occurs high in the Al Hajar Mountains of e. Oman.

WHERE TO SEE: Ifrane, Morocco; Tannourine Cedars Reserve, Lebanon; Dibbeen Forest Reserve, Jordan.

BERINGIAN TAIGA SAVANNA

IN A NUTSHELL: This habitat is a hybrid of tundra and taiga, with low spongy ground and a few small conifers. **Habitat Affinities:** None. **Species Overlap:** EURASIAN SPRUCE-FIR TAIGA; EURASIAN ROCKY TUNDRA.

DESCRIPTION: A common misconception is that conifers are the northernmost trees. But on the southern edge of the tundra, an Arctic scrub of birch and aspen groves, not evergreen conifers, grows in the most protected areas. These groves expand in favorable years and contract in more severe years. In the toughest conditions, conifers such as spruce and fir are unable to survive through the winter, due to the fact that they are evergreen. Any tree that carries leaves at all, and that needs to eke out water from the permafrost during the winter months, finds it difficult to survive.

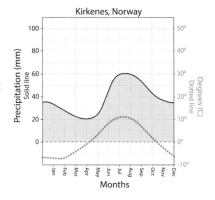

Kirkenes, Norway

South of the tundra, average midsummer temperatures rise to 54°F (12°C), warmer than the 50°F (10°C) limit for tundra. Birch and aspen groves become widespread, and the tundra starts to become dotted with stunted spruces, pines, and larches. This is the Beringian taiga savanna ecotone, a zone between the more typical tundra and the forest proper. In some areas the transition from EURASIAN ROCKY TUNDRA to taiga is rapid, but in other areas the transition is gradual. Beringian taiga savanna has a mosaic of individual trees, evergreen conifer groves, patches of larches (deciduous conifers), birches, and aspens, and mires of rocky or BOGGY TUNDRA. In these bogs, large hummocks of sphagnum mosses grow, with marsh grasses in the channels between them.

WILDLIFE: Beringian taiga savanna is critical habitat for nesting shorebirds. Many of the Old World shorebirds that winter in areas as remote as Africa, India, se. Asia, and Australia, come to this habitat

Beringian taiga savanna (photo: northernmost Finland). © IAIN CAMPBELL, TROPICAL BIRDING TOURS

to nest, arriving in late May as the snow first begins to melt, uncovering the marshy areas beneath it. In Fennoscandia, Spotted and Common Redshanks, Purple Sandpiper, and Black-tailed Godwit nest here. In e. Asia this habitat is used by Curlew Sandpiper, Spoon-billed Sandpiper, and Red-necked Stint. Species shared with the EURASIAN SPRUCE-FIR TAIGA to the south include ground birds such as Willow Ptarmigan and passerines such as Bluethroat.

The few mammals include some very intriguing ones, like Wolverine, Moose (known as Elk in Eurasia), and Reindeer (aka Caribou).

DISTRIBUTION: From Norway to the eastern tip of Siberia, there is always a zone of Beringian taiga savanna between true tundra and true taiga, though the width varies greatly. In the very north of Finland, just south of the Norway border, the band is only around 3 mi. (5km) wide, while it can be over 30 mi. (50km) wide in c. Siberia.

WHERE TO SEE: Kirkenes, Norway.

Willow Ptarmigan turns all white in winter. In the breeding season, the female is mottled brown all over, but the male (pictured) retains its white belly.
© IAIN CAMPBELL, TROPICAL BIRDING TOURS

PALEARCTIC HOT DESERT

IN A NUTSHELL: Desolate deserts with almost no vegetation that are hot in summer, warm in winter. **Habitat Affinities:** NEOTROPICAL DESOLATE DESERT. **Species Overlap:** PALEARCTIC HOT SHRUB DESERT; PALEARCTIC SEMIDESERT THORNSCRUB.

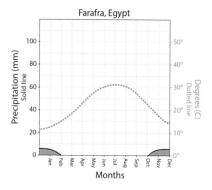

Farafra, Egypt

DESCRIPTION: From the nw. African coast of nw. Mauritania through the Middle East and all the way to far w. India, stretches an absolutely massive area of very hot deserts: the Sahara, Arabian, Iranian, and Thar. These deserts string together to form a continuous band of sandy and rocky habitats broken only by rivers such as the Nile, bodies of water like the Red Sea and Persian Gulf, and small mountain ranges like the Zagros Mountains of Iran and the Kirthar Range of Pakistan. Each of these deserts can be further broken down into smaller deserts, but they all hold a similar set of microhabitats and are therefore treated as one habitat.

Desert landscape with massif and sandy plain, se. Jordan. © IAIN CAMPBELL, TROPICAL BIRDING TOURS

Within the hot desert, there are vegetated mountains, such as the Tibesti Mountains of the Sahara, and isolated wadis (gullies), as well as bare rock mountains and plateaus, called *hammadas*. Surrounding these uplands are areas of colluvial outwash with stony desert and desert pavement, called reg, very similar to the gibber plains of c. Australia (see CHENOPOD AND SAMPHIRE SHRUBLAND in the Australasia chapter). But when most people imagine the Sahara, they think of seas of moving sand, known as ergs. Actually the least common of the desert landscapes, ergs make up only 20% of the Sahara, but that is still a massive area. These dunes can reach spectacular heights, up to 820 ft. (250m). Ergs are most common in the Arabian Desert, where the Rub' al Khali dune field covers 250,000 sq. mi. (650,000km²). When these dunes are moving, they are unable to sustain any real vegetation, but when they stabilize, water does accumulate beneath them over hundreds of years, and this water can be enough to sustain trees such as Gum Acacia, Babul, Desert Date, and Ghaf. In some areas Ghaf can concentrate in groves where the trees are as close as 12 ft. (4m) apart and form a closed canopy. These Ghaf forests are superficially similar to the casuarina forests of c. Australia or the Tamarugo groves in the Atacama Desert of Chile. Plants that grow in swales between dunes include Christ's Thorn Jujube, Hackenkopf, and grasses including panic grasses.

Reg (see sidebar 1.1 for a full description) is the most widespread landform, making up about half of the Sahara and much of the Iranian and Thar Deserts. Here the landscape is barren for most of the time, the vegetation limited to artemisia shrubs and grasses that sprout whenever the extremely erratic rains come. Hammadas have plants only in cracks in the rock, resembling the colonization of basalt flows in the Galápagos. The wadis support more vegetation and may resemble the desert shrub that surrounds these deserts, holding tamarisks, small acacias, Wild Jujube, and Joint Pine. Oases, supporting palm groves, made up a minuscule percentage of original desert, and almost all of these have been altered beyond recognition by humans. These were the only locations for permanent settlements and are almost all now towns and cities. In some areas, oases have not been covered in concrete and are still farmed, and although a human-altered habitat, remain vitally important stopover locations for migrating birds. The expanse of Palearctic hot desert encompasses some areas with widespread shrubland, called PALEARCTIC HOT SHRUB DESERT, which is treated as a separate habitat.

These places are some of the hottest in the world. Areas away from the moderating effect of the coast have average summer temperatures up to 113°F (45°C) and maximum temperatures topping 130°F (50°C). In winter, temperatures average 50–55°F (10–13°C), though they go well below freezing in the Iranian Desert. The daily temperature range can be extreme, and daily fluctuations of 70°F (21°C) have been recorded. The moistest area that holds this habitat is the Thar Desert, which gets up to 12 in. (300mm) of rain per year. But many areas receive less than 1 in. (25mm) per year and even go years without rain. The average annual rainfall is 4 in. (100mm), which, when combined with evapotranspiration rates as high as 15 ft. (4.5m) annually, makes for hyperarid conditions.

WILDLIFE: Typical hot desert birds include ground-dwelling species like Cream-colored Courser; Spotted, Lichtenstein's, and Pin-tailed Sandgrouse; Arabian and Houbara Bustards; Pharaoh Eagle-Owl; Desert Owl; Desert and Temminck's Larks; Greater Hoopoe-Lark; Iranian Ground-Jay; and Desert, White-crowned, Hooded, and Mourning Wheatears.

The Sahara and the Arabian Desert used to be laden with savanna animals (see sidebar 5.1), but today most large herbivores such as Arabian Sand, Mountain, and Goitered Gazelles; Nubian Ibex; Arabian Tahr; and Arabian Oryx are rare outside reserves. In the Thar Desert, Blackbuck is still common in reserves. Smaller animals have fared much better in the deserts, and hamsters, gerbils, mice, and shrews are common. Smaller carnivores include the Rüppell's Fox in the Arabian Desert, and Fennec Fox in the Sahara. Larger carnivores include Striped Hyena and Leopard. Asiatic Lions

The very distinctive Greater Hoopoe-Lark. © LISLE GWYNN, TROPICAL BIRDING TOURS

used to occur from nw. Africa to India, but today their distribution in Asia is limited to small reserves in Gujarat.

DISTRIBUTION: The Sahara alone stretches across the African continent from the Atlantic coast of Western Sahara and Mauritania to the Red Sea, 3,300 mi. (5,300km) in total. North to south, it runs from the southern edge of the Atlas Mountains in Morocco and Algeria and the Mediterranean coast of Libya, 1,200 miles (1,900km) south to the Sahel (AFROTROPICAL DRY THORN SAVANNA). The Arabian Desert stretches from se. Jordan, Syria, and Iraq south, over almost all of the Arabian Peninsula to the mountains of Yemen. The Iranian Desert covers most of s. Iran and parts of Afghanistan. The Thar Desert covers most of s. Pakistan and parts of w. India, where it merges into the PALEARCTIC SEMIDESERT THORNSCRUB of Asia.

WHERE TO SEE: Douz-Redjim Maaloug, Tunisia; Erg Chebbi (dunes), Morocco; Negev Desert, north of Eilat, Israel; Azraq, Jordan.

SIDEBAR 5.1 ▶ EXTINCTION OF PEOPLE, PLANTS, AND ANIMALS OF THE SAHARA

In the driest and coldest recent period of the late Pleistocene glaciation (around 20,000 –15,000 YBP), the Sahara was much more extensive than it is today. Moving seas of sand extended around 300 mi. (450km) south of where they are now, and the Palearctic biogeographic region extended far into Ghana and Nigeria. The rewarming began at the end the Younger-Dryas mini-glaciation (around 11,700 YBP) and by 9,000 YBP, almost all of the Sahara west of the Nile became lush with savanna; wet GUINEA SAVANNA, which is actually an evergreen woodland, occurred from Ghana right up through Senegal, and moist savanna in Mauritania merged with drier Mediterranean-type shrubland in what is now hyperarid reg. Drainage of most of n. Africa was internal, and where vast lake systems existed, people rapidly colonized this new lush land, adapting to an aquatic diet of fish and snails but also hunting the Giraffes, hippos, and antelopes of the wet savanna. The desert plants, and animals such as Dorcas Gazelle, now found refuge in the Egyptian desert and the Arabian Peninsula. From 7,000 to 5,700 YBP, much of the Sahara changed to AFROTROPICAL DRY THORN SAVANNA AND THORNSCRUB, and the lakes became drastically reduced. By 5,000 YBP, the game was over, and most of the region was the very desolate place it is now, and all previous settlements were abandoned. Whether the people of the aquatic/hunting culture died out, moved out, or became cattle herders remains unknown.

EAST ASIAN COLD DESERT

IN A NUTSHELL: Bleak deserts of w. China. **Habitat Affinities:** NEOTROPICAL DESOLATE DESERT. **Species Overlap:** CENTRAL ASIAN COLD DESERT.

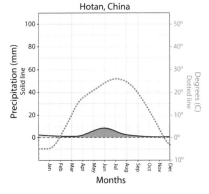

DESCRIPTION: The geomorphological terrain in the southern part of c. Asia is some of the most complex on earth, with east–west and north–south internally drained basins surrounded by massive mountains creating rain shadows. In these zones the Dzungarian and Taklamakan Deserts have formed. By comparison, the Gobi Desert is in a much simpler zone and is an arid extension of the TEMPERATE DESERT STEPPE to the north. In all three of these deserts, a series of high mountains and valleys, an array of internally drained basins, and drastically contrasting weather patterns, including extreme rain shadows, result

The Taklamakan Desert, China.

in desolate habitats, such as loess badlands (unstratified wind-derived sand deposit), hammadas (rock), reg (gibber plains; see sidebar 1.1), salt pans, and seas of sand and dunes (otherwise known as erg). The climatic conditions in this habitat are some of the harshest, and the scenery the starkest, on the planet. Cold desert diverges from both the PALEARCTIC HOT DESERT, like the Sahara, and the ARCTIC POLAR DESERT in its extremes, in that it seems to have the harshest of both worlds. In winter, temperatures can drop to -40°F (-40°C), while in summer it can be scorching hot, reaching 100°F (38°C). Annual precipitation is around 5 in. (125mm) over most of the region, though decades of drought sometime occur in the Taklamakan Desert. The bitterly cold winds are also dry, unlike those of the Kyzylkum and Karakum Deserts farther west (see EAST ASIAN COLD DESERT), and most of the rain falls in the summer. The freezing, dry winters inhibit any growth, even in spring, so plants can grow only during a short period after the summer rains and before the cold returns.

The northern part of the cold desert blends into temperate desert steppe and EASTERN GRASS STEPPE. For a naturalist, the general distinction between steppe and cold desert is in the ground cover, with the steppes always having over 50% ground cover and cold deserts always having less than 50% ground cover and usually less than 20%. The southern portions of the cold desert are higher and moister, and blend into the EURASIAN ALPINE TUNDRA of the Tibetan Plateau.

Much of this habitat consists of erg (sandy desert), though here it is not as mobile as the erg of the Sahara, as it is sometimes stabilized and covered in saksaul trees. Because much of this environment has internal drainage and can be hypersaline, almost all plants are halophytic. In most of the region, short White Saksaul trees are found in stabilized sandy areas between sand dunes. Other common plants include artemisias, salvias, saltbushes, Russian Thistle, and tussock grasses. The vegetation is rarely taller than 12 in. (30cm) and is very widely spaced, rarely reaching more than 20% ground coverage. In more saline areas, Black Saksaul and tamarisk grow in groves that are up to 24 in. (60cm) high, alongside chenopods such as saltbushes in saline areas.

Above: **Mongolian Ground-Jay is a target species for birders in c. Asia.** © KEITH BARNES, TROPICAL BIRDING TOURS

Below: **Wild Bactrian Camels are being reintroduced to protected areas in Mongolia and China as well as steppes out of their traditional range in Russia.** © ANDREW SPENCER

WILDLIFE: Because this is such an inhospitable environment, very few birds live here year-round, most just passing through on migration. There are some endemics though, such as the Xinjiang Ground-Jay (IS) of the Taklamakan Desert, which specializes on tamarisk and poplar bushes. Other birds include Henderson's Ground-Jay, Pallas's Sandgrouse, Desert Finch, Tarim Babbler, Mongolian Ground-Jay, Isabelline Wheatear, and Blanford's Snowfinch.

Mammals in these deserts include the wild Bactrian Camel, Asiatic Wild Ass, and Goitered Gazelle. However, small burrowing species make up most of the mammalian fauna, such as gerbils, jirds, hamsters, jerboas, pikas, and hares.

DISTRIBUTION: The East Asian cold deserts occur from the China–Mongolia border on the edge of the Gobi south to the Tibetan Plateau in Qinghai, China. In the east they are bounded by TEMPERATE DESERT STEPPE, and in the west by the CENTRAL ASIAN COLD DESERT in e. Kazakhstan. The Taklamakan Desert extends from the north of the Tibetan Plateau all the way west to China's border with Tajikistan, where it is wedged between the Tian Shan mountains and the Tibetan Plateau. The Dzungarian Desert is wedged between the Tian Shan to the south and the Altai Mountains to the north.

WHERE TO SEE: Hotan, Xinjiang, China.

CENTRAL ASIAN COLD DESERT

IN A NUTSHELL: Desolate deserts with freezing winters and hot summers. **Habitat Affinities:** NEOTROPICAL DESOLATE DESERT. **Species Overlap:** PALEARCTIC SUBTROPICAL SAVANNA; EAST ASIAN COLD DESERT.

DESCRIPTION: The Kyzylkum and Karakum Deserts are areas of bare plateau, stabilized dune systems, erg (moving sand dunes that support little vegetation), reg (desert pavement, similar to the gibber plains of Australia; see sidebar 1.1), and clay or salt pans that have almost no vegetation. Generally, the Kyzylkum is dominated by red, less fertile sand, and the Karakum has black, more fertile sand. Despite its more fertile soil, however, the Karakum is extremely dry over 90% of its area and has very little vegetation. These deserts have an intense version of the Mediterranean climate that prevails farther west, with winter wetter than summer, which is a period of intense drought. Unlike the Mediterranean areas though, here the soils are sometimes frozen, and the plants are covered in a thin blanket of snow. Together, these conditions mean that the majority of growing must occur in the short period after the thawing of the ground and melting of the snow but before the blistering heat of summer.

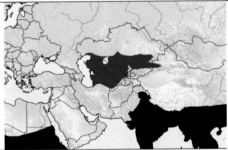

Bukhara, Uzbekistan

The Kyzylkum and Karakum seem devoid of plant life for most of the year because of drought and temperatures as low as -15°F (-26°C), but they burst alive for a short period in March, when there may still be frost, but daily high temperatures reach 68°F (20°C), and monthly rainfall is 1 in. (25mm). The good times are very short-lived, and by May, temperatures will average 82°F (28°C), and monthly rainfall drops to 0.5 in. (12mm). By June, the month's rainfall will be less than 2mm, average daytime

Right: **The Charyn Canyon in Kazakhstan is now part of the Central Asian Cold Desert, but has remnant links to vegetation of the nearby Tien Shan mountains.** © KEITH BARNES, TROPICAL BIRDING TOURS

Below: **Zarudny's Sparrow is a specialist of saksaul trees.** © JAMES EATON, BIRDTOUR ASIA

temperatures are 95°F (35°C) and on many days exceed 100°F (40°C). These conditions desiccate and kill most of the vegetation that sprouted in late March.

Plants have different growth strategies to eke out an existence. Some plants are evergreen but microphyllous (having tiny leaves), such as the trees and shrubs in the *Calligonum* genus and saksaul trees. Some plants have no leaves whatsoever and photosynthesize through their stems. Other plants, such as tulips and irises, have bulbs to store nutrients through the nine harsh, nongrowing months of the year. There are grasses with thick rhizomes (roots that store food), all of which grow in spring and then die during summer. White Saksaul is the most widespread tree of these deserts, especially in the south, where it can exceed 20 ft. (6m) tall. There are also acacias, which usually grow in widely dispersed small groves, with a canopy lower than 12 ft. (4m), in a growth form similar to the trees of AUSTRALASIAN MALLEE WOODLAND AND SCRUBLAND but with branches growing from a short trunk just above the ground rather than from a central node at ground level. Acacias can rarely grow as tall as 30 ft. (9m). Between these scattered trees, there are low shrubs like artemisias, plus halophytic shrubs and herbs in the saline plains.

WILDLIFE: Birds are limited in these extremely harsh environments, but there are some specialties, such as Zarudny's Sparrow, and Turkestan Ground-Jay. More widespread species include Pallas's Sandgrouse, Scrub Warbler (also found in PALEARCTIC HOT DESERT and HOT SHRUB DESERT), Brown-necked Raven, Rufous-tailed Scrub-Robin, and Great (Turkestan) Tit.

To survive in these extremely harsh conditions, most small mammals need to take cover underground, so it is no surprise there are a number of burrowing species. The most common of the larger burrowers are Long-eared Hedgehog and Tolai Hare. There are many smaller rodents such as gerbils, jirds, and jerboas, which resemble the Bilby of Australian deserts. Larger mammals such as Saiga Antelope, Asiatic Wild Ass, and various species of wild sheep are native to these deserts, though are now only in protected areas. Larger carnivores include Sand Cat and Corsac Fox.

DISTRIBUTION: The deserts of Karakum and Kyzylkum are bounded by the Tian Shan mountains to the east, the Caspian Sea to the west, the TEMPERATE DESERT STEPPE and WESTERN FLOWER STEPPE of Kazakhstan to the north, and by the forest-steppe savannas to the south.

WHERE TO SEE: Kyzylkum Nature Reserve, Uzbekistan.

TEMPERATE DESERT STEPPE

IN A NUTSHELL: The areas on the edges of Asian cold deserts that have extensive grass blooms in good years but spend most of the year barren. **Habitat Affinities:** NEOTROPICAL PATAGONIAN STEPPE. **Species Overlap:** EASTERN GRASS STEPPE; WESTERN FLOWER STEPPE.

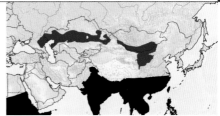

DESCRIPTION: Around the fringes of the EAST ASIAN COLD DESERT and the CENTRAL ASIAN COLD DESERT, and separating them from the EASTERN GRASS STEPPE and WESTERN FLOWER STEPPE of Eurasia, arid grasslands blend with scrublands in the temperate desert steppe. These steppes have many of the same species as the deserts in the southwest, such as Mongolian Wormwood. However, after the summer rains, these desert plants are joined by many of the drought-tolerant tuft grasses and perennial forbs that also grow in the steppes to the north. So in winter

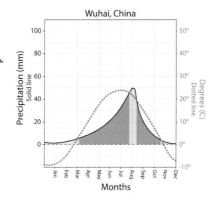

Desert steppe, c. Kazakhstan.
© KEITH BARNES, TROPICAL BIRDING TOURS

and early spring, these expanses appear like those deserts, while in summer and fall they look like the steppes to the north. Temperate desert steppe has very few trees, and around 50% shrub ground cover that grows to only about 1 ft. (0.3m) high, the exception being the occasional saksaul bush that gets to 10 ft. (3m) tall. These shrubs grow in widely spaced clumps, so that for the vast majority of the year there is extensive bare ground. After rains, grasses grow to around 10 in. (25cm) tall.

Midday Jirds live in large colonies and are hunted by various species of polecats and foxes.
© KEITH BARNES, TROPICAL BIRDING TOURS

WILDLIFE: Birdlife in this zone is a mixture of typical steppe species and those that prefer more arid environments, and the mix of species present at any moment is largely governed by the recent season's rainfall and summer grass growth. In summer, typical species include Lesser Kestrel, Oriental Plover, Upland Buzzard, Desert Wheatear, Egyptian Nightjar, Sociable Lapwing, Macqueen's Bustard, Pallas's and Pin-tailed Sandgrouse, Great Gray (Steppe) Shrike, Henderson's Ground-Jay, Bimaculated Lark, and Pere David's Snowfinch.

Larger mammals of the desert steppe include the incredibly rare wild Bactrian Camel, now mainly restricted to the Gobi Strictly Protected Area of Mongolia, and a subspecies of Brown Bear called the Gobi Bear, which is on the brink of extinction. It also holds the Asiatic Wild Ass and Goitered Gazelle. Smaller burrowing mammals are widespread, even in unprotected lands, including Midday Jird, gerbils, and jerboas.

DISTRIBUTION: Much of the n. and e. Gobi Desert, while being a true desert because of limited precipitation, has substantial plant growth (at least compared to the barren EAST ASIAN COLD DESERT and CENTRAL ASIAN COLD DESERT farther south and west) and is classified here as desert steppe. Temperate desert steppe covers the southern half of Mongolia, where it merges with the true desert of the Gobi around the Mongolia-China border. This habitat can also be found in the Mu Us Desert, east of the Gobi, and as far east as Yulin, Inner Mongolia. It also extends around the north of the Central Asian cold desert, west to north of the Caspian Sea and into e. Ukraine.

WHERE TO SEE: Khongoryn Els, Mongolia; Gobi Gurvan Saikhan National Park, Mongolia.

PALEARCTIC HOT SHRUB DESERT

IN A NUTSHELL: Low shrublands that surround the Sahara and other hot deserts. **Habitat Affinities:** NEOTROPICAL CHIHUAHUAN DESERT SHRUBLAND. **Species Overlap:** PALEARCTIC HOT DESERT; PALEARCTIC SEMIDESERT THORNSCRUB.

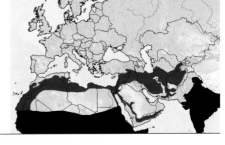

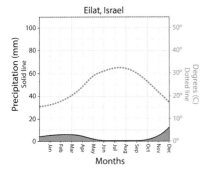

Eilat, Israel

DESCRIPTION: Hot shrub deserts are similar to some other arid lands, such as the Simpson Desert of Australia, in that they do not seem desolate enough to be regarded as desert, having seemingly too many trees or shrubs. But despite having plant cover at over 50%, and being much more lush than the barren ergs, regs, and hammadas of the Sahara and the Arabian Peninsula, they still have an evaporation level of over 10 ft. (3m) annually. This vastly exceeds the level of precipitation, at around 16 in. (400mm), so this shrub habitat is regarded as true desert rather than semidesert.

In se. Morocco, the hot shrub desert merges into barren desert.
© LISLE GWYNN, TROPICAL BIRDING TOURS

This habitat grows mainly on nutrient-deficient bedrocks like granites, rhyolites, and quartzites. The vegetation is a shorter, more widely spaced, scrubbier version of PALEARCTIC SEMIDESERT THORNSCRUB habitat, and includes trees such as Gum Acacia and Mimosa, spaced around 165 ft. (50m) apart and growing to 8 ft. (2.5m) high. Bushes include tamarisk, but also Ragwort and halophytic saltbushes. Between these ground shrubs are perennial grasses that remain low, generally around 15 in. (38cm) high. After the seasonal rains, these grasses are augmented by a burst of ephemeral grasses and shrubs such as Prickly Clover and Had. Even on relatively more nutrient-rich soils, vegetation is still sparse, with scattered small trees and shrubs. In n. Iraq and n. Iran, Wild Jujube and sumac shrubs dominate, giving the appearance of a shrub savanna rather than a desert. The more arid areas have artemisia with Bulbous Bluegrass. In Afghanistan, the semidesert communities are dominated by Hackenkopf and saksaul shrubs, other chenopods, and a few grasses. Southern areas of Iraq, Iran, and Afghanistan have Hackenkopf, saksaul, and heliotrope, and through the coastal plains from Qatar to Pakistan there are very low shrubs of Hackenkopf and Bean-caper.

Egyptian Nightjar is a migratory species that winters in similar habitat within the Sahel zone, on the southern edge of the Sahara. © JAMES EATON, BIRDTOUR ASIA

WILDLIFE: Because this habitat is contained within the broader PALEARCTIC HOT DESERT zone of the Sahara, Arabian, Iranian, and Thar Deserts, there is overlap in many of the arboreal bird species, such as vultures, eagles, and falcons. However, many ground birds prefer this shrubbier habitat to more desolate desert. Examples include Houbara Bustard, Egyptian Nightjar, Chestnut-bellied and Black-bellied Sandgrouse, Sykes's Nightjar, Scrub Warbler, Graceful Prinia, Fulvous Chatterer, Arabian Babbler, Palestine Sunbird, Masked Shrike, Brown-necked Raven, Lesser Short-toed Lark, Red-tailed Shrike, Blackstart, and Isabelline, Finsch's, and Hooded Wheatears.

Mammals are very similar to those of the surrounding

Once extinct in the wild, the Arabian Oryx is being widely reintroduced in the Middle East. © LISLE GWYNN, TROPICAL BIRDING TOURS

desert, though protected areas usually contain much greater numbers of mammals than those in more barren areas. Mammals of hot shrub desert include Dromedary (aka One-humped Camel), Blackbuck, Goitered Gazelle, Nubian Ibex, Arabian Tahr, and Arabian Oryx. Smaller animals are mainly hamsters, gerbils, mice, and shrews.

DISTRIBUTION: Hot shrub desert occurs in coastal areas of North Africa, separates the Atlas Mountains from the Sahara in Morocco and Algeria, blends with the Sahel of the Afrotropics (AFROTROPICAL DRY THORN SAVANNA), and dominates the landscape of Syria, Iraq, and coastal Iran. It is widespread on the west side of the Arabian Peninsula and extends into Yemen and Oman. This habitat develops a distinct Indian feel when it blends into INDO-MALAYAN THORNSCRUB in Pakistan; in n. Iran and Afghanistan, this scrubland merges with the Karakum Desert (see CENTRAL ASIAN COLD DESERT) of Turkmenistan.

WHERE TO SEE: Negev Desert, north of Eilat, Israel.

SIDEBAR 5.2 ▸ HABITATS IN FLUX

All habitats are in a constant state of flux due to an ever-changing climate and the effects of humans. The transition from humid, more complex habitats to drier, more depauperate habitats is usually rapid, but the change back takes a very long time, usually much longer than a human lifetime. Savanna can go back to rainforest, and desert can go back to savanna, but not without transitioning through intermediary habitats with plants that are then replaced. These are trends not absolute rules, and many habitats such as FYNBOS are in no way a depauperate version of humid forests.

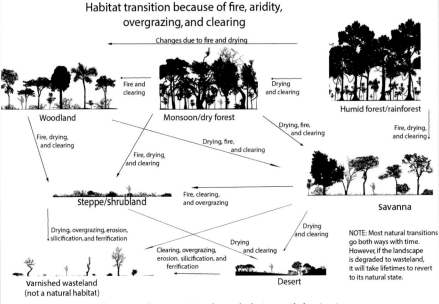

Habitat transition because of fire, aridity, overgrazing, and clearing

The general progression through drying and clearing is
humid forest→ monsoon forest→ woodland→ savanna→ steppe→ desert→ varnished wasteland

PALEARCTIC SEMIDESERT THORNSCRUB

Transition to Afrotropical thornscrub

IN A NUTSHELL: Arid scrub of bushes with small leaves and spiny branches. **Habitat Affinities:** AFROTROPICAL DRY THORN SAVANNA AND THORNSCRUB. **Species Overlap:** PALEARCTIC HOT SHRUB DESERT.

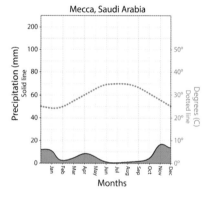

DESCRIPTION: This habitat connects the INDO-MALAYAN THORNSCRUB of India and the AFROTROPICAL DRY THORN SAVANNA of the Sahel, and blends into the PALEARCTIC HOT DESERT or HOT SHRUB DESERT. Semidesert thornscrub is produced not only by low rainfall but also by the number of months of intense drought. In drier areas with rainfall as low as 4 in. (100mm) per year, its flora can grow as long as the rain is spread throughout the year, whereas in areas with prolonged drought of six months or more, the thornscrub needs up to 2 in. (50mm) of precipitation in the wet months to survive. Thornscrub is most prevalent in areas where the annual rainfall is around 12 in. (300mm) and the drought lasts less than four months. The underlying geomorphology may be as remarkable as that of the deserts, but here the rocky terrains, regs, and stabilized dunes are obscured by vegetation. Except in the most barren of rocky outcrops, the trees and shrubs remain rather constant over most of the underlying soil materials on all terrains, including stabilized ergs, regs, and alluvial and colluvial washes, giving the habitat a uniform appearance from a distance.

This habitat consists of individual acacia groves joined with other trees, succulents, and euphorbias, in stands that are usually 6–10 ft. (2–3m) high and closely spaced to form a canopy of around 70% cover. The ground cover is shrubs including Hackenkopf, saksauls, and wormwoods. In drier regions where this habitat merges into hot shrub desert, the tree cover declines, and the low shrubs become more widely spaced; rather than a mosaic of the two habitats, there is a very gradual ecotone, which makes delineating the habitat boundary very difficult.

Because rocky outcrops of igneous rhyolites and metamorphic quartzites are very resistant to chemical or mechanical weathering in this climate, they do not support the development of even lithosols, so large areas in this habitat remain barren rock with little more than the occasional shrub. However, around these outcrops, where a break in slope allows soil development and water retention, thorny bushes grow well and can be thicker than the shrubland found just a little farther away from the outcrops.

Above: **Thornscrub in Oman.** © KEN BEHRENS, TROPICAL BIRDING TOURS

Inset: **Arabian Warbler is typical of this habitat in the Middle East.** © KEN BEHRENS, TROPICAL BIRDING TOURS

Left: **Hypocolius represents a monotypic family found in the thornscrub of the Middle East.** © JAMES EATON, BIRDTOUR ASIA

WILDLIFE: The bird life of the semidesert thornscrub is a combination of resident species, desert species that use this habitat in dry periods, and species from the more humid MAQUIS and GARRIGUE shrublands that use it during wetter times. Typical resident birds include Gray Francolin, Peregrine (Barbary) Falcon, Lichtenstein's Sandgrouse, Sykes's Nightjar, Namaqua Dove, Sind Woodpecker, Arabian Woodpecker, Hypocolius, Arabian Babbler, Common Babbler, Arabian Warbler, Nile Valley Sunbird, Arabian Golden Sparrow, and Pale Rockfinch.

There are many more mammals in the thornscrub than in barren desert, including such larger species as Striped Hyena, Dromedary (aka One-humped Camel), Blackbuck, Nilgai, and Goitered Gazelle. Smaller species that live here include jirds and gerbils.

DISTRIBUTION: This habitat occurs along both sides of the Red Sea, on the s. Sinai Peninsula, in s. Israel, s. Jordan, n. Syria, and s. Iraq, around the Persian Gulf, in much of Iran, Afghanistan, and most of Pakistan, to westernmost Rajasthan and Gujarat in India.

WHERE TO SEE: Al Saleel National Park, Oman.

SIDEBAR 5.3   DESERT REFUGIA

Birdlife has varied markedly in the Palearctic deserts over the Holocene, as the deserts have sometimes been much larger and at other times much smaller, shrinking to about 20% of their current size. During these times of contraction, desert birds retreated to desert refugia such as the Horn of Africa, the Arabian Peninsula, and the w. Sahara, and from these population centers they recolonized the deserts when they again expanded. These desert refugia remain the centers of highest endemism for hot desert species even today.

EAST ASIAN MOIST MIXED FOREST

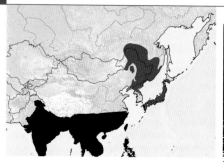

IN A NUTSHELL: These are the highly diverse closed-canopy broadleaf and conifer forests of Japan and that used to occur throughout n. China. **Habitat Affinities:** NEARCTIC TEMPERATE MIXED FOREST. **Species Overlap:** EAST ASIAN TEMPERATE DECIDUOUS FOREST.

DESCRIPTION: These forests of se. Russia, ne. China, Korea, and Japan are a fascinating blend of EURASIAN SPRUCE-FIR TAIGA and broadleaf deciduous forest. This habitat is much richer than its European equivalent (EUROPEAN MOIST MIXED FOREST); and instead of that forest's mosaic of stands of conifers and stands of broadleaf trees, the forest types merge here on a finer scale. At the upper elevations on hills, this forest merges with both taiga and EURASIAN MONTANE CONIFER FOREST, while at lower elevations and latitudes it merges with EAST ASIAN TEMPERATE DECIDUOUS FOREST. In the north of its range, moist mixed forest covers areas from sea level to the tree line, and in the south exists up to 4,000 ft. (1,200m) elevation. This forest stands in stark contrast to the SIBERIAN

Moist mixed forest in winter in Japan.
© KEITH BARNES, TROPICAL BIRDING TOURS

LARCH FOREST, just slightly to its north, in how varied the canopy is, with a mix of trees of varying sizes and ages growing together. Emergent Korean Pine and Korean Fir trees can reach 150 ft. (45m), but generally the closed canopy is around 70–115 ft. (22–35m) high and consists of oaks, beeches, ashes, birches, lindens, and Kalopanax. In well-watered areas, Korean Pine, elms, and ashes dominate. In Japan, the conifers are represented by Japanese Cypress, firs, and Japanese Cedar, which blend with maples, oaks, beeches, birches, and poplars. The understory can be fairly open, with a lower layer of deciduous shrubs to around 17 ft. (5m) tall and good visibility at head height. However, in some areas it can have dense stands of bamboo, which rise to 20 ft. (6m), impede visibility, and make walking difficult.

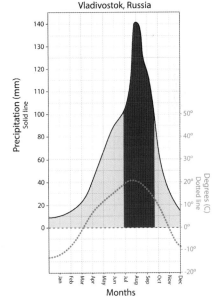

WILDLIFE: In its remaining protected areas, this habitat is surprisingly rich for such high latitudes. It holds Golden Snub-nosed Monkey, Siberian Tiger, Raccoon Dog, Sika Deer, and Giant Panda.

White-backed Woodpecker and Koklass Pheasant are resident. Chinese Thrush and Japanese Robin migrate to breed here. Smaller passerines include Brown-eared Bulbul; Mugimaki, Yellow-rumped,

Above: **Sika Deer populations are rising in e. Asia where not persecuted.** © KEITH BARNES, TROPICAL BIRDING TOURS

Left: **Siberian Thrush winters in se. Asia but breeds in the moist mixed forests of ne. Asia.** © PETE MORRIS, BIRDQUEST

Below: **Varied Tit is a species of Japan and Korea.** © SAM WOODS, TROPICAL BIRDING TOURS

and Narcissus Flycatchers; Eastern Crowned Warbler; Ashy Minivet; Japanese Paradise-Flycatcher; and the resident Varied Tit.

DISTRIBUTION: Moist mixed forests occur from about 300 mi. (480km) north of Vladivostok in the Primorsky province of Russia through the lowlands of n. Manchuria in far ne. China, and south to Jinzhou and the n. Korean peninsula. It occurs extensively in the lowlands of the Japanese archipelago south of Hokkaido.

WHERE TO SEE: Sapporo, Hokkaido, Japan; Karuizawa, Nagano, Japan; Gapcheon Stream and Forest, South Korea.

Until recently, Europe was covered in fairly uniform deciduous forests that resembled the EAST ASIAN TEMPERATE DECIDUOUS FOREST and the NEARCTIC TEMPERATE DECIDUOUS FOREST of e. North America. These forests were mostly eradicated during the Pleistocene ice ages, then partially recovered in the early Holocene, only to be severely restricted during the Younger-Dryas climate reversal (12,900–11,700 YBP). The forests in nw. Europe had no time to adapt as the climate dropped back to glacial temperatures, causing the extinction of many tree species. As a result, the modern EUROPEAN TEMPERATE DECIDUOUS FOREST of these regions lacks diversity. The later warming that occurred 11,700 YBP likely happened within a few human lifetimes, so the rapid demise of the tundra and then the arboreal recolonization of the n. European plain began from the limited refugia, primarily along the se. Black Sea coast (what is now the COLCHIC DECIDUOUS RAINFOREST), along the western edge of the Caucasus, where forest remained right through the Pleistocene, and the large remaining tracts of deciduous forest farther east in Asia.

The northern and western recolonization was fast for some species, which established some form of deciduous forest between 11,700 and 9,000 YBP. This recolonization happened patchily, rather than in a solid block; in some places, like e. Hungary, the environment rapidly turned over from a coniferous forest steppe to deciduous forest around 9,500 YBP.

The composition of the forest changed over time. Birch, hazel, and elm arrived within 500 years of the warming, forming open forest. They were followed much more slowly by alder and ash, which formed closed-canopy forests. It was the arrival and dominance of Pedunculate Oak that produced most of the lowland forests of w. Europe by Mesolithic times. Beech was one of the last of the trees to arrive in w. Europe and occurred mainly along with ash on calcareous soils or fertile, well-drained soils. Where the Pedunculate Oak was once dominant, beech became the most successful tree, through human thinning of oak, hazel, elm, and ash.

EUROPEAN MOIST MIXED FOREST

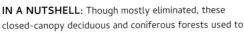

IN A NUTSHELL: Though mostly eliminated, these closed-canopy deciduous and coniferous forests used to cover most of e. Europe. **Habitat Affinities:** NEARCTIC TEMPERATE DECIDUOUS FOREST. **Species Overlap:** EUROPEAN TEMPERATE DECIDUOUS FOREST.

DESCRIPTION: The moist mixed forests of Europe are one of the most heavily impacted forest types of the continent and have been almost completely replaced by agriculture. As opposed to EAST ASIAN MOIST MIXED FOREST, the mixed forests of Europe are a broad mosaic of pine stands alongside stands

Moist mixed forest, Belarus.
© LISLE GWYNN, TROPICAL BIRDING TOURS

of broadleaf beech and oak. The basic difference from EURASIAN SPRUCE-FIR TAIGA or MONTANE CONIFER FOREST is that spruce and fir are replaced here by oak on the heavier clay-rich soils, while pines remain the dominant tree of sandier soils in areas with more fires. The main plant assemblages are pine, spruce, fir, oak, birch, and beech. All are usually found in monotypic clusters. There is a very weak understory of mature trees, with a canopy cover of less than 20%. A lot of light makes it to the forest floor, but the shrub layer is surprisingly thin. The understory is dominated by forbs rather than grasses, though there are extensive patches where the accumulation of pine needles prohibits the growth of any plants at all. The climate of these forests tends to be much milder than that of the conifer forests to the east and north, with temperature ranges of 23–65°F (-5–18°C) and fairly uniform precipitation throughout the year.

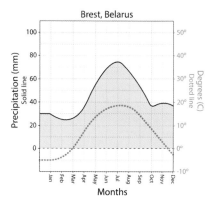

Farther east, away from the Atlantic coast, in c. Europe, the climate becomes much more severe, with hotter summers and colder winters. Here, beech becomes far less common and has its limits in e. Poland and Moldova. From Poland and Transylvania east to the Urals, lacking beech, the forest becomes much more diverse and is referred to as Sarmatic mixed forest; Pedunculate Oak and Common Hornbeam dominate but are joined by many other canopy trees, such as Common Maple, Norway Spruce, Scots Pine, ashes, aspens, alders, and birches.

WILDLIFE: These forests are much richer than the EUROPEAN TEMPERATE DECIDUOUS FORESTS. Many bird species that breed in the EURASIAN SPRUCE-FIR TAIGA to the north use these forests as wintering grounds or as stopover points during spring and fall migration; examples include Hen Harrier, Rough-legged Hawk, Hoary Redpoll, Brambling, and Arctic Warbler. Breeding migrants include Red Kite, Eurasian Sparrowhawk, Lesser Spotted

Rough-legged Hawk winters in moist mixed forests but breeds much farther north in the tundra. © IAIN CAMPBELL, TROPICAL BIRDING TOURS

Fieldfare winters and sometimes breeds in Eurasian moist mixed forest. © PETE MORRIS, BIRDQUEST

Eagle, Mistle Thrush, Eurasian Blackcap, Greenish Warbler, and European Serin. As most bird species are summer-breeding migrants, these forests have few residents, and the resident bird assemblages are dominated by woodpecker species, like Gray-headed, Black, and Lesser Spotted Woodpeckers.

Most of the charismatic European mammals such as Wild Boar, European Bison, European Wildcat, Gray Wolf, and Brown Bear occur in this habitat, though they are limited to dedicated forest reserves, as the remaining forest corridors are not extensive enough to support larger mammals. More common and widespread mammals include various deer, rabbits, and rodents, as well as Pine Marten and European Badger.

DISTRIBUTION: European moist mixed forest habitat is found from around 60°N, in far s. Finland, s. Sweden, and n. Estonia, south into n. Germany, across to the Ural Mountains. There are smaller areas of this forest (not mapped) in the Scottish Highlands, n. Spain, and other mountainous areas of Europe.

WHERE TO SEE: Bialowieza National Park, Poland.

European Bison can still be found in good numbers in the Belovezhskaya forests, Belarus.
© NICK ATHANAS, TROPICAL BIRDING TOURS

EAST ASIAN TEMPERATE BAMBOO FOREST

IN A NUTSHELL: Forest in which bamboo stands dominate the understory and can form parts of the canopy. A common setting for movies set in China. **Habitat Affinities:** This type of forest does not occur anywhere else in the world. **Species Overlap:** EAST ASIAN MOIST MIXED FOREST; EAST ASIAN TEMPERATE DECIDUOUS FOREST.

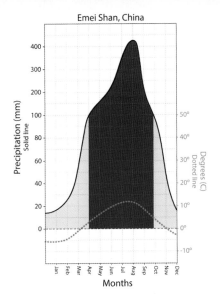

Emei Shan, China

DESCRIPTION: Most forest habitats of the world are defined by their canopy trees, which determine the kinds of birds and other animals that are most common in them. This is not the case with e. Asian bamboo forests, which can grow under a variety of conditions, including in monotypic stands with no canopy cover, in deciduous forests of beech and oak, in forests with spruce and fir, and in the mixed forests of n. China. In these forests, bamboo dominates all and gives the habitat a distinctive feel, regardless of the canopy cover. This habitat supports a whole suite of specialist species, including an entire group of birds, the parrotbills, that evolved to use this niche. The canopy birds reflect the type of canopy, such as conifer, broadleaf, or mixed and may have no relation to the underlying bamboo.

WILDLIFE: The bamboo stands within coniferous forests are the core habitat for the critically endangered Giant Panda, Red Panda, and Golden Snub-nosed Monkey, as well as birds such as Golden-breasted Fulvetta, Barred Laughingthrush, and Blackthroat. In the forests of Sichuan, China, Golden, Brown, and Rusty-throated Parrotbills all feed on bamboo. Temminck's Tragopan, Golden Pheasant, and Lady Amherst's Pheasant are three pheasant species typical of these forests.

DISTRIBUTION: East Asian temperate bamboo forests are found as isolated patches around China, although the most extensive stretches are north and west of Chengdu, in the area that separates the

Above: **Bamboo forest mixed with deciduous broadleaf trees and conifers, e. China.** © ALEX ANSLEY, CC BY-SA 2.0, HTTPS://CREATIVECOMMONS.ORG/LICENSES/BY-SA/2.0, VIA WIKIMEDIA COMMONS

Right: **Giant Panda is so rare in the wild that most naturalists see it only at captive breeding centers, like this one in Chengdu, Sichuan, China.** © SAM WOODS, TROPICAL BIRDING TOURS

Tibetan Plateau from the densely populated Sichuan basin. They also occur on the Korean peninsula and in s. Japan.

WHERE TO SEE: Emei Shan, Wawu Shan, and Longcanggou, all in Sichuan, China.

When food supply is high, Golden Snub-nosed Monkeys can form very large groups. When supplies dwindle, the groups break up into smaller, competing groups. © KEITH BARNES, TROPICAL BIRDING TOURS

COLCHIC DECIDUOUS RAINFOREST

IN A NUTSHELL: Europe's only rainforest; a bizarre mixture of deciduous canopy and evergreen undergrowth. **Habitat Affinities:** NEOTROPICAL MAGELLANIC RAINFOREST. **Species Overlap:** EUROPEAN MOIST MIXED FOREST.

DESCRIPTION: This habitat includes the Euxine forests, stretching from southeasternmost Bulgaria east to the c. Black Sea coast of Turkey, and also encompasses one of the most interesting forest blocks in the world, comprising a myriad of micro-forest types, tucked away on the e. Black Sea coast of ne. Turkey and sw. Georgia. As well as being fascinating now, this habitat was the refugium (see sidebar 5.4) for most of the trees in the EUROPEAN TEMPERATE DECIDUOUS FOREST, EUROPEAN MOIST MIXED FOREST, and EURASIAN MONTANE CONIFER FOREST during the Pleistocene ice ages. When the rest of Europe was converted to tundra or steppe, this area stayed forested, and it was from here that the rest of the deciduous forest recolonized.

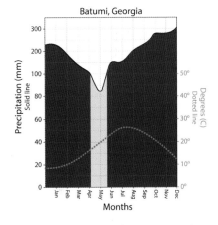

Batumi, Georgia

Some people regard this as a temperate rainforest, but most of its broadleaf (non-coniferous) canopy trees are deciduous. Temperate rainforests in the Nearctic are coniferous, and those in the Southern Hemisphere are dominated by broadleaf evergreen *Nothofagus* and *Eucalyptus* canopy trees. In southernmost South America, the MAGELLANIC RAINFORESTS have a few deciduous trees, but deciduous trees do not dominate the canopy as they do here, where the canopy comprises deciduous oaks, hornbeams, and beeches, with some conifers.

The makeup of this forest changes from sea level to the tree line, but the extreme humidity remains throughout. At the lower parts of the range, up to 1,640 ft. (500m) elevation, the rainforest is dominated by European Alder, Caucasian Walnut, Strandzha Oak, Common Hornbeam, Sweet Chestnut, and Oriental Beech. The understory is dominated by many rhododendrons and *Ficus* trees that do not extend to elsewhere in Europe. The ground is laden with a thick fern layer. Above this zone, to 3,300 ft. (1,000m), the canopy loses biodiversity and becomes dominated by Oriental Beech and Sweet Chestnut, while other canopy species from below become minor components. In contrast, the 9–12 ft. (3–4m) tall understory becomes very diverse, with many species of evergreen bushes that are more typical of a Himalayan moist forest than a European one, such as many rhododendron species, hollies, and Cherry Laurel. Even higher up, the forests become straight beech forests with rhododendron understories, and then transition into wet conifer forests.

This region has a wet, warm-temperate climate unlike any other in Europe. Summers are warm for Europe, averaging 70°F (21°C), and winters cool, averaging 45°F (7°C), much like the climate of Barcelona, yet with much more rain here, 95–177 in. (2,400mm–4,500mm) annually, compared to just 24 in. (600mm) for Barcelona. Precipitation is not only much higher compared to other locations in the region but also distributed relatively evenly throughout the year. The winds in this region are predominantly westerlies. As the warm winds cross the Black Sea, they pick up a lot of evaporation, and these moisture-laden air masses are funneled between mountain ranges on the southern and northeastern shores of the Black Sea. When they hit the steep mountains of the southeastern shore, they drop their rain as orographic rainfall (see sidebar 3.3).

WILDLIFE: It seems incongruous that a forest that is botanically so distinctive from any other in the region does not have an equally distinctive animal assemblage. The forest is stark in its lack of resident bird species. The majority of European birds are migratory, and there are few resident bird species in the rainforest. The endemic Caucasian Chiffchaff (a subspecies of Mountain Chiffchaff) occurs here along with Green Warbler. The Colchic forest zone is an important funnel for raptor migration, and more than 1 million birds of 35 species use it, including over half the world's populations of European Honey-buzzard, Levant Sparrowhawk, and Booted Eagle.

Amphibians seem much more indicative of the special nature of the Colchic rainforests: Caucasian Toad, Caucasian Parsley Frog, Caucasian Salamander, and Caucasian Northern Banded Newt are all restricted to this habitat.

At middle elevations, the understory of Colchic deciduous rainforest becomes dense with rhododendrons and other evergreens and feels more like a forest of the Himalayas than Europe.

DISTRIBUTION: True Colchic deciduous rainforest, found around the southeastern corner of the Black Sea, is surrounded by dry habitats and water. It ranges from sea level to 6,500 ft. (2,000m) and mainly occurs within Georgia but also in far ne. Turkey. A drier form of Colchic deciduous rainforest, called Euxine forest, extends from the Turkish-Bulgarian border to easternmost Turkey; it resembles EUROPEAN MOIST MIXED FOREST.

WHERE TO SEE: Mtirala National Park, Georgia.

EUROPEAN TEMPERATE DECIDUOUS FOREST

IN A NUTSHELL: The rather uniform deciduous forests of Europe, dominated mostly by beech trees and a few other species. **Habitat Affinities:** NEARCTIC TEMPERATE DECIDUOUS FOREST. **Species Overlap:** EUROPEAN MOIST MIXED FOREST.

DESCRIPTION: European temperate deciduous forest (also known as summer deciduous forest because it has foliage during the summer months) has closely spaced trees that create a canopy at 55–80 ft. (17–25m), with some emergent trees reaching up to 130 ft. (40m). The average age of the canopy trees is between 40 and 60 years old, making these forests very young by global standards. In winter, the majority of the canopy trees lose their leaves, though some of the underlying shrub species keep their leaves throughout the cold months. Snow usually does not stay on the ground as often as it does in EURASIAN SPRUCE-FIR TAIGA or EUROPEAN MOIST MIXED FOREST. In many areas and for most winters, there is no snow accumulation at all.

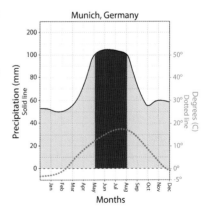

Beech was one of the last of the trees to recolonize the landscape of w. Europe after the last glaciation event (sidebar 5.4) and mainly occurred along with ash on calcareous soils or fertile, well-drained soils. Where the Pedunculate Oak was once dominant, beech became the most successful tree, through human thinning of oak, hazel, elm, and ash. Now beech is the pervasive canopy tree

Young European temperate deciduous forest. © SIMON BUCKELL

European Robin *(foreground)* and Eurasian Blue Tit are common birds of this habitat.
© LISLE GWYNN, TROPICAL BIRDING TOURS

European Badger remains common in some parts of the United Kingdom.
© KEITH BARNES, TROPICAL BIRDING TOURS

of w. Europe, and most forest plant assemblages are defined as "beech plus *x*," the other trees being mainly different oaks (sidebar 5.5). Recently, cultivation has made Scots Pine much more common in the landscape, and it is becoming the dominant forest tree in many areas of w. Europe.

WILDLIFE: Because these forests have been logged extensively, the tree diversity is limited, structure is simple, and most forests do not have highly varied tree ages. There are far fewer niche habitats than in the EUROPEAN MOIST MIXED FOREST. These factors result in a less diverse set of birds than in the other European forests or in EAST ASIAN TEMPERATE DECIDUOUS FOREST. There are a few resident species here, such as Great Spotted, Lesser Spotted, and White-backed Woodpeckers; Eurasian Blue and Willow Tits; Wood Warbler; Eurasian Bullfinch; Common Chiffchaff; and Goldcrest. No breeding species are confined to this habitat; most of its breeding species are found widely in other habitats as well. A few examples include Eurasian Sparrowhawk, Tree Pipit, European Robin, Song Thrush, and Garden and Wood Warblers.

Mammals that now live in this forest include Wild Boar, European Badger, Red Deer, and Fallow Deer, as well as various rabbits and rodents.

DISTRIBUTION: The range of European temperate deciduous forest begins south of the EUROPEAN MOIST MIXED FOREST and EURASIAN SPRUCE-FIR TAIGA of Scandinavia in the north and extends south to the MEDITERRANEAN OAK FOREST in the n. Iberian Peninsula. It runs from the British Isles and Atlantic coast in the west to w. Poland where it merges into European moist mixed forest. European temperate deciduous forest has been reduced and depleted by widespread agriculture, and few large stands remain.

WHERE TO SEE: White Wood, Dartmoor National Park, Devon, UK; Bavarian Forest National Park, Germany.

SIDEBAR 5.5 **ALL THOSE BEECHES**

Common Beech (*Fagus sylvatica*) is well adapted and highly successful in most European climates. This compatibility, in conjunction with the elimination of potential Asian competitors during the last ice age, has produced the uniform forests we see today. Regardless of the local climate, most forests in Europe vary only in the types of secondary trees that grow beside beech.

Dominance of beech in deciduous and mixed forests of Europe

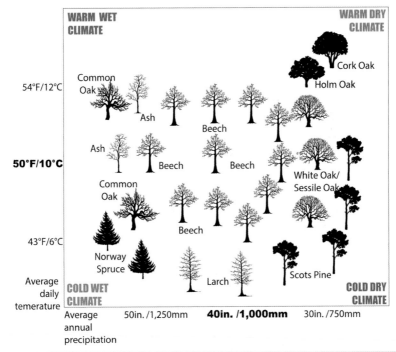

EAST ASIAN TEMPERATE DECIDUOUS FOREST

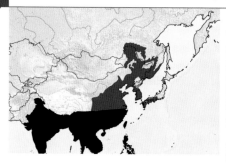

IN A NUTSHELL: A very biodiverse moist temperate deciduous forest of e. Asia. **Habitat Affinities:** NEARCTIC TEMPERATE DECIDUOUS FOREST. **Species Overlap:** EAST ASIAN MOIST MIXED FOREST; INDO-MALAYAN SUBTROPICAL BROADLEAF FOREST.

DESCRIPTION: The East Asian temperate deciduous forests once formed a solid block that spread across the central latitudes of e. Asia. They generally resembled the EUROPEAN TEMPERATE DECIDUOUS FORESTS but were far richer in plant, mammal, and especially bird diversity. Here, rather than a monotonous canopy of beeches with a few oaks, the forest is a medley of maples, hornbeams, alders, oaks, walnuts, Hackberry, ashes, and lindens, with poplars forming stands

Temperate deciduous forest around a lake edge, with montane conifer forests on the hills, in Jiuzhaigou, Sichuan, China. © KEITH BARNES, TROPICAL BIRDING TOURS

along rivers. In common with the NEARCTIC TEMPERATE DECIDUOUS FORESTS of North America, there is usually a secondary layer of younger canopy trees, a shrub layer of plants like dogwoods, and a ground cover of herbs. Many of these trees are the same species that co-dominate with conifers in the EAST ASIAN MOIST MIXED FORESTS to the north, but here they dominate without the conifers, mixing instead with more deciduous trees from farther south.

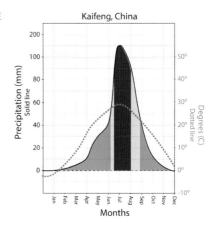

There is a north–south gradation of these forests corresponding to precipitation and temperature: annual precipitation in the forests in the north and northwest is around 12 in. (300mm) and in southern areas is around 55 in. (1,400mm). Surprisingly, the highest tree diversity is at higher latitudes, which is not the case in most of the forests of the world, including the deciduous forests of North America and Europe. Here, in the e. and ne. Chinese provinces and extreme se. Russia, the forest is very diverse, with over 20 genera of deciduous trees. To the south (at about 43°N), oaks begin to dominate the canopy. In the e. Yangtze Valley, the environment becomes milder and more mesophytic (not too hot, too cold, too dry, or too wet); here the canopy is dominated by oaks, Sweet Gum, hornbeams, Sassafras, and Chinese Chestnut. The canopy remains predominantly deciduous, and the sub-canopy and shrub layers have many more broadleaf evergreen plants, mostly species shared with the INDO-MALAYAN SUBTROPICAL BROADLEAF FOREST to the south. In Japan, this forest is dominated by a very closed canopy of beeches and large oaks, and there is a less defined sub-canopy.

With the exception of Japan, this part of Eurasia has been tectonically inactive for many millions of years, but it retains many isolated mountain ranges, as well as floodplains and coastal plains, creating different environments and promoting the development of high plant and animal diversity. This region avoided the scrubbing of the landscape and resulting extinctions that Europe and North America faced during the Pleistocene glaciations and therefore re-creates the high diversity that must have existed across Europe prior to the ice ages.

WILDLIFE: Pheasants have their highest diversity in this part of Asia; among the species that are associated with, but not limited to, these forests are Reeves's, Copper, and Ring-necked Pheasants,

Central China is the world's hotspot for native pheasants such as the Reeves's Pheasant. © JAMES EATON, BIRDTOUR ASIA

Japanese Macaques are nicknamed Snow Monkeys where they take to thermal springs in winter.
© KEITH BARNES, TROPICAL BIRDING TOURS

the last of which has been introduced worldwide. Other resident bird species are Pygmy and Japanese Woodpeckers, Lesser Cuckoo, and Varied Tit. Notable migrant species include Fairy Pitta, Black-naped Oriole, Ashy Minivet, Yellow-rumped Flycatcher, and Japanese Paradise-Flycatcher.

Mammals of this habitat include Japanese Badger, Raccoon Dog, Leopard Cat, Asiatic Black Bear, and Japanese Macaque (Snow Monkey), one population of which may be the most photographed monkeys in the world.

DISTRIBUTION: This forest type originally existed as a continuous block at lower elevations from far n. China and extreme se. Russia to Shanghai, China. In the northeast, it bordered the grasslands of Inner Mongolia, which begin 100 mi. (160km) west of Beijing, and to the southwest it ranged to the coniferous mountain forests of w. Sichuan, where it still occurs on the lower elevations of the mountain ranges around Chengdu. The vast majority of these forests have been converted to cultivation, with few significant tracts remaining, mainly in Manchuria, in ne. China.

WHERE TO SEE: Beidaihe, China; lower parts of Jiuzhaigou, Sichuan, China.

PALEARCTIC FOREST-STEPPE

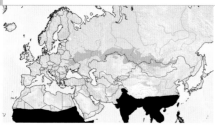

IN A NUTSHELL: This is where the temperate coniferous and broadleaf forests thin out to steppe, creating a mélange of forest groves and meadows. **Habitat Affinities:** NEARCTIC MADREAN PINE-OAK WOODLAND. **Species Overlap:** EURASIAN SPRUCE-FIR TAIGA; WESTERN FLOWER STEPPE; EASTERN GRASS STEPPE.

DESCRIPTION: The forest-steppe is the mixed habitat between forests and steppe grasslands. Within this transitional zone, the vegetation changes as precipitation decreases, summer temperatures rise, and fires become a dominant force in the landscape. On the northern edge of this habitat, adjacent to the forest, the landscape is a mosaic of complex forest patches, meadows, and marshes. On the southern edge, closer to true steppe, the patches change to isolated individual trees within a much

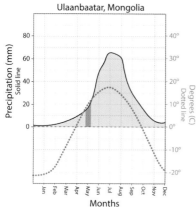

Ulaanbaatar, Mongolia

Forest-steppe, Kazakhstan.
© KEITH BARNES, TROPICAL BIRDING TOURS

broader expanse of grasslands and shrublands. This inner band superficially resembles the savanna of more tropical environments and has been described as a boreal savanna. This savanna-like habitat has been partly created by humans through fires, clearing, and grazing, and most of the meadows have been turned over to agriculture. But although these habitats must have been much reduced in early Holocene times, they did occur naturally.

The trees in this habitat tend to be more drought- and fire-tolerant than the dominant trees in the nearby forest habitats. The prominent exception to this rule is Eurasian Aspen, which isn't fire tolerant. Aspen groves are the first colonizers to invade the steppe when fire and grazing are limited. The boundaries of the forest-steppe with broadleaf or mixed forest zones are dominated by species like Mongolian Oak and birches, though in drier areas, such as around the Tian Shan mountains in Kazakhstan, conifers like cypress (*Cupressus*) and junipers dominate. The meadows within the forest-steppe habitat are much lusher than those in the steppes proper, with grasses and herbs here growing to 4 ft. (1.2m) tall in summer. In the drier versions of this habitat, moisture is concentrated on the northern faces of hillsides and there promotes forest growth. The southern hillsides are prone to desiccating winds and tend to have far fewer trees and shorter pastures. In this habitat the vegetation type often does not seem to be in sync with the underlying soil, with meadow soils often under forest patches, and forest soils under meadows, suggesting that this system is always in flux (see sidebar 5.2), with constant boundary movements between grasslands and forests.

Demoiselle Cranes winter mainly in the n. Indian plains but nest in the forest-steppe of Kazakhstan and Mongolia. © IAIN CAMPBELL, TROPICAL BIRDING TOURS

WILDLIFE: Birdlife in this habitat includes both forest and steppe birds. The forest birds tend to be species that use the forest edge in this habitat's more closed environments, such as Northern Hawk Owl, Hazel Grouse, Booted Eagle, Pallas's Rosefinch, Daurian and White-cheeked Starlings, and Long-tailed Tit. Examples of steppe birds are Demoiselle Crane, Siberian Stonechat, and Richard's Pipit. As in steppe, raptors are prominent, especially harriers, including Montagu's, Pallid, and Hen Harriers. In slightly wetter areas, migratory shorebirds such as Asian Dowitcher, Swinhoe's Snipe, and Green Sandpiper are among a host of species nesting here in summer.

Roe Deer occurs in forest-steppe from Bulgaria to w. Ukraine.
© KEITH BARNES, TROPICAL BIRDING TOURS

Most of the mammals of this habitat have long since been extirpated from most of the region. In well protected areas you can still find Altai Wapiti, Roe Deer, Wild Boar, Gray Wolf, Eurasian Lynx, Red Fox, and European Badger.

DISTRIBUTION: Forest-steppe occurs as a patchy ribbon from Bulgaria through Ukraine to s. Russia, far n. Kazakhstan, n. Mongolia, and n. China.

WHERE TO SEE: Gorkhi-Terelj National Park, Mongolia.

PALEARCTIC SUBTROPICAL SAVANNA

IN A NUTSHELL: This is a temperate and subtropical savanna with a mosaic of tree stands and grasslands. **Habitat Affinities:** AFROTROPICAL GUSU WOODLAND; NEARCTIC OAK SAVANNA. **Species Overlap:** CENTRAL ASIAN COLD DESERT; MEDITERRANEAN AND DRY CONIFER FOREST.

DESCRIPTION: The Palearctic subtropical savanna is the moderate-climate equivalent of the PALEARCTIC FOREST-STEPPE. It occurs where the DRY CONIFER FOREST, MEDITERRANEAN OAK FOREST, and MAQUIS all merge with the CENTRAL ASIAN COLD DESERT or with the PALEARCTIC HOT DESERT of the Middle East. Over much of the western part of this habitat, juniper and oak assemblages form a very open woodland/tree savanna. In the east of the range, the juniper trees blend with different species of pistachio trees to form a savanna that feels like NEARCTIC OAK SAVANNA of California and GUSU WOODLAND of Africa, with trees reaching 30 ft. (10m) and spaced between 25 ft. (8m)

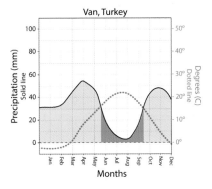

and 65 ft. (20m) apart. In the drier and/or well-drained parts of these pistachio savannas, the pistachio trees are mixed with the smaller White Saksaul trees and desert sedges that extend into the Karakum Desert (see central Asian cold desert). The grasses and forbs tend to be more annual in eastern and southern parts of this habitat and perennial in the western regions. In the slightly moister and/or cooler areas of the pistachio savanna, the pistachio trees mix with the juniper trees that extend into the dry conifer forests of the Hindu Kush mountain range.

This habitat grows in a climate with a wet winter and/or early spring and a hot summer. Annual precipitation is in the range of 9–16 in. (240–400mm), though the majority of this falls between December and April, and a prolonged drought occurs from May through October. Temperature extremes are the norm here, with temperatures hitting 122°F (50°C) in summer and -26°F (-32°C) in winter. These dry savannas can be regarded as one of the hardiest of all the world, able to tolerate drought, extreme temperature ranges, and fire.

WILDLIFE: This savanna environment has a carnivore assemblage similar to that of n. Africa, with Honey Badger, Corsac Fox, Jackal, Striped Hyena, Leopard, and a subspecies of Caracal known as Sand Lynx, as well as more temperate species like Gray Wolf, Red Fox, and Pallas's Cat. These carnivores prey on large herbivores such as Goitered Gazelle, Urial, and a subspecies of Asiatic

Wild Ass called Onager, as well as smaller mammals like gerbils, ground squirrels, hamsters, voles, and desert hedgehogs.

Birdlife resembles a depauperate version of warmer savanna habitats. The avian assemblage is dominated by aerial birds like Barbary Falcon, Chukar (a partridge), Saker Falcon, Cinereous Vulture, and European Bee-eater. Ground birds

Right: **The Corsac Fox lives in the subtropical savanna and semidesert shrubs of Turkmenistan but does not cross the mountains into c. Iran.** © KEITH BARNES, TROPICAL BIRDING TOURS

Below: **In subtropical savanna, Pin-tailed Sandgrouse lives in the grassier areas between tree groves.** © MIKE WATSON, BIRDQUEST

Bottom: **Subtropical savanna in Vashlovani National Park, s. Georgia.** © PAATA VARDANASHVILI FROM TBILISI, GEORGIA, CC BY 2.0, HTTPS://COMMONS.WIKIMEDIA.ORG/W/INDEX. PHP?CURID=2809070

include Eurasian Thick-knee, Collared Pratincole, Pin-tailed Sandgrouse, Calandra and Crested Larks, Rufous-tailed Scrub-Robin, Black Redstart, and Finsch's Wheatear. The few trees in this terrain host Long-tailed and Lesser Gray Shrikes and Upcher's Warbler.

DISTRIBUTION: This habitat has two main blocks. The western block extends from w. Turkey through c. Turkey into Georgia, Armenia, and Azerbaijan, and south through w. Iran. The eastern block has small stands in northernmost Iran and Afghanistan and becomes extensive in s. Turkmenistan. Much of this habitat has been relinquished to agriculture throughout the region.

WHERE TO SEE: Badhyz State Nature Reserve, Turkmenistan; Vashlovani National Park, Georgia.

MEDITERRANEAN OAK FOREST

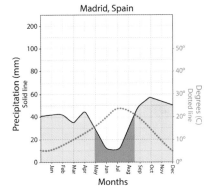

IN A NUTSHELL: The deciduous and evergreen broadleaf forest of the warmer, drier parts of Eurasia.
Habitat Affinities: NEARCTIC OAK SAVANNA.
Species Overlap: MAQUIS.

DESCRIPTION: In most of s. Europe, and the lowlands of c. Europe, beech-oak assemblages are replaced by these much more interesting oak-dominated forests. Both deciduous and evergreen oaks occur, with evergreen species becoming more dominant farther south. The oaks share the forest canopy with hornbeams, Sycamore Maple, chestnuts, ashes, boxwoods, and elms. Oak forests with species like Sessile, White, and Cork Oaks generally occur in drier areas of Europe where beeches cannot dominate, though some oak forests form where it is too moist for beech to dominate, and this moister forest often includes Common Oak. This habitat blends in with beech forest on the mountain slopes of Europe, whereas in the Atlas Mountains, beech is not significant, so the oak forests transition to MONTANE CONIFER FOREST dominated by Atlas Cedar.

In the early Holocene, oak and conifer forests would have extended down to the Mediterranean coast in areas that are now MAQUIS. Where large oak forests still exist with a closed canopy, the sub-canopy is very open, and there is a significant low shrub layer. Walking through these forests is not difficult. The soils remain moist and do not have the intense red staining from iron oxyhydroxides that occurs in the soils of the surrounding maquis. This suggests that the original forests maintained a more humid microclimate in the sub-canopy and undergrowth than surrounding habitats, despite the intense summer droughts common in this region.

WILDLIFE: Oak forests are much richer than the beech-dominated EUROPEAN TEMPERATE DECIDUOUS FOREST found to the north and upslope on mountain ranges. Many species that breed in the EURASIAN SPRUCE-FIR TAIGA forests to the north use this habitat in the winter; among them are Hen Harrier and Long-eared Owl. There are a few resident birds, such as Spanish Eagle, Middle Spotted and White-backed Woodpeckers, Short-toed Treecreeper, Eurasian Blue and Great Tits, and Eurasian Nuthatch. Most species of this habitat are summer-breeding migrants, such as Booted Eagle, Common Nightingale, Eastern and Western Orphean Warblers, Olive-tree Warbler, Spotted Flycatcher, and Ortolan Bunting. There are some endemic species that primarily use this habitat along with surrounding MAQUIS, such as Cyprus Warbler and Cyprus Wheatear.

Most of the charismatic European mammals such as Iberian Lynx, European Wildcat, Wild Boar, European Badger, and Brown Bear occur in oak forest.

Above: **Cork Oak forests are well protected in Spain.** © IAIN CAMPBELL, TROPICAL BIRDING TOURS

Left: **The critically endangered Iberian Lynx prefers Cork and Holm Oaks in Andalusia with significant grassy patches and brambles.** © SAM WOODS, TROPICAL BIRDING TOURS

DISTRIBUTION: There are some oak-hornbeam forests in se. England and nw. France, but the true Mediterranean oak forest starts in Portugal, with Cork Oak and Holm (aka Holly) Oak forests, and extends eastward. The forest grows south of the European Alps but north through the Balkans to extreme sw. Ukraine. It occurs in w. Turkey and along the s. Black Sea coast, as well as along the southern edge of the Caspian Sea, then stretches northeast again to w. Ukraine and s. Belarus. Oak forests also occur in limited patches in the Atlas Mountains of Morocco.

WHERE TO SEE: Douro International Natural Park, Portugal; Monfragüe National Park, Spain; Sierra de Andújar Natural Park, Spain.

EUROPEAN HEATHLAND AND MOORLAND

IN A NUTSHELL: An open, low shrubby habitat of exposed areas of the cooler parts of the region. **Habitat Affinities:** NEARCTIC ALPINE TUNDRA. **Species Overlap:** EURASIAN ALPINE TUNDRA; GARRIGUE.

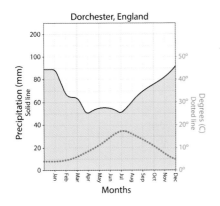

Dorchester, England

DESCRIPTION: The chalk heathlands common in s. England were long thought to be natural grasslands but are more likely to have been human-derived and can be viewed as a wetter version of the human-derived MAQUIS found around the Mediterranean and Middle East. Climax heathlands are low fields of herbs and grasses with isolated Common Yew trees and thickets of hawthorn and juniper. Ground cover is near universal; the dominant plant over many drier heathlands is Common Heather, which occurs with gorse, milkwort, and bracken. In Europe, heathlands and moorlands (boggy, wetter heaths) exist primarily as a colonizing subclimax community—that is, a plant community that would change into something else over time; however, they can be climax vegetation in well-watered areas where the strong, desiccating winds or extremely nutrient-deficient soils make the development of trees impossible. Heaths predominately form on extremely alkaline soils derived from limestone or on exceedingly acidic soils formed where weathering of silicates results not in clays but rather in inert sandy soils; this seems counterintuitive, but they are habitats of soil conditions that are less than favorable for other plant assemblages. In moorlands, a type of wet heath more common in higher areas, the plants form thick peat layers. Heath plants that grow in well-drained lowland heathlands receive very little from the soil and return very little, and thus there is rarely a significant humus layer.

WILDLIFE: Dartford Warbler is the bird most associated with heathland habitat, though many other birds use the habitat including Eurasian Hobby, Eurasian Nightjar, European Stonechat, and Wood Lark.

Many larger grazing animals use heathland habitat along with surrounding forest, including ungulates like Fallow, Red, Roe, and Sika Deer, though the Sika is an introduced species from Asia.

DISTRIBUTION: Heathland is found around coastlines, on chalk plains, and in highlands, where it merges into EURASIAN ALPINE TUNDRA above about 3,000 ft. (900m). This habitat is more common in the British Isles than anywhere else in the Palearctic. About 70% of the heathland that occurred in preindustrial times has been lost to intense agriculture, forest growth through planting and succession, or urban development, and the remaining heathland is maintained by grazing to stop colonization by trees.

WHERE TO SEE: Thursley Common, Surrey, UK.

Left: **Dartford Warbler is expanding its range north into the heathlands of the British Isles.**
© LISLE GWYNN, TROPICAL BIRDING TOURS

Below: **Heathland in s. England.**
© SIMON BUCKELL

MAQUIS

IN A NUTSHELL: Short, dry thickets around the Mediterranean with trees that are probably not in their climax state. **Habitat Affinities:** NEARCTIC PACIFIC CHAPARRAL. **Species Overlap:** MEDITERRANEAN OAK FOREST; GARRIGUE.

DESCRIPTION: The original early Holocene to Mesolithic habitat of s. Europe, n. Africa, and the Middle East is still open to dispute. Some scientists suggest that it was predominantly an evergreen mixed woodland of cypresses, junipers, and oaks. Other scientists suggest that it was a fire-adapted wooded savanna, while others believe that the maquis that exists now is a natural response to the Mediterranean climate. Although maquis is often regarded as a human-induced, disturbed (plagioclimax) habitat, it is likely that in the early Holocene, a habitat similar to maquis existed in gaps of mixed woodland, oak woodland, conifer woodland, or woodland savanna. The increased fire regime, overgrazing by domesticated sheep and goats, clearing for agriculture, and the development of olive

Maquis is one of the main habitats in the area of Tarifa, Spain.
© IAIN CAMPBELL, TROPICAL BIRDING TOURS

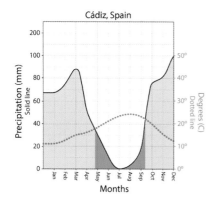

Cádiz, Spain

Eurasian Hoopoe migrates from Spain to Africa in winter. © IAIN CAMPBELL, TROPICAL BIRDING TOURS

groves has all but eliminated the oaks and conifers. The plants native to the original, completely natural maquis have a massive advantage in this human-altered environment, and in modern times have become the dominant vegetation around the Mediterranean basin and through the western parts of the Middle East.

Maquis is a dense shrub or small tree community with a canopy between 3 ft. and 20 ft. (1–6m) high, most often around 10 ft. (3m). It consists mostly of evergreen canopy plants like junipers, Aleppo Pine, Carob, Strawberry Tree, and Wild Olive. Oleander and Rock Rose form a shrub layer, while in the open ground cover, many annuals grow until they are shaded out by sclerophyllous plants. Walking through maquis is not an easy task, as canopy cover at eye level is nearly solid. The canopy is usually not high enough to be able to walk under comfortably, and almost all canopy plants have shrubby growth forms, with small branches growing up the trunk and multiple trunks growing from one base. Because this environment gets most of its rain during the winter months, and the summer drought months are brutal on most plant life, many plants are sclerophyllous and have developed leathery leaves with a waxy coating that helps limit transpiration. Plants also tend to have narrower leaves than those in broadleaf deciduous forest. Some non-sclerophyllous shrubs are dry-deciduous species, losing their leaves in summer to avoid desiccation.

WILDLIFE: Animal assemblages in the maquis have been drastically affected by 4,000 years of intense human pressure, resulting in the extinction of many mammals. The Iberian Lynx still holds on in Andalusia, the Barbary Macaque is found in n. Africa and Gibraltar, and the Nubian Ibex can still be found in Israel.

Maquis has many more reptiles than other habitats of Europe, with chameleons as well as many other lizards and snakes.

This area holds many resident birds, like Red-legged Partridge and Sardinian Warbler, and some restricted-range ones such as the Cyprus Warbler and Cyprus Wheatear, in their namesake island. It is most important as wintering grounds for species that breed farther north, such as Little Owl and Song Thrush. In spring and summer, maquis hosts such breeding migrants as Red-necked Nightjar, Rufous-tailed Scrub-Robin, Woodchat Shrike, Western Orphean Warbler, Western Olivaceous Warbler, and Marmora's Warbler. The habitat also becomes vitally important as the first feeding place for the many spring migrant passerines that have crossed the Sahara on their way to the mixed forests farther north. Following these birds are numerous accipiters and falcons, along with other raptors. This is the only habitat in Europe where vultures are common. Birding in the spring can be amazing, and April is a great time to be in the Iberian Peninsula.

DISTRIBUTION: This habitat is very widespread and common below 3,500 ft. (1,100m) on the Mediterranean islands, around the s. Mediterranean coast from Morocco to Tunisia, from Portugal and around the n. Mediterranean coast to Greece, and from Turkey around to Israel. It extends west through much of s. Turkey into northernmost Syria. Some of the best areas of maquis are along the Andalusian coast in s. Spain.

WHERE TO SEE: Doñana National Park, Andalusia, Spain; Tarifa, Andalusia, Spain; Arrábida Natural Park, Portugal.

GARRIGUE

IN A NUTSHELL: The open scrub around much of the Mediterranean Sea; akin to a depauperate MAQUIS. **Habitat Affinities:** None. **Species Overlap:** MAQUIS.

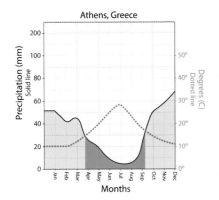

Athens, Greece

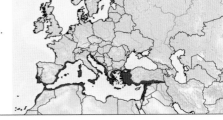

DESCRIPTION: Garrigue is a very depauperate, short counterpart of MAQUIS, and forms of it are usually found in parts of this region that are hotter and drier than maquis or on soils developed over limestone. It also exists as the end result of overburning and overgrazing of maquis and can be regarded as a degraded version of that habitat. This shrubland is less than 3 ft. (1m) high, with very scattered trees (if any), such as Wild Olive or Strawberry Tree. Many of the common household herbs originated in this environment, such as sage, thyme, lavender, and rosemary. Given that this habitat is the end result of human overuse, it is no wonder that it is found in all disturbed areas around the Mediterranean.

WILDLIFE: Birdlife in this man-made, depauperate habitat is similar to that of the MAQUIS from which it derives but is much reduced, both in species and numbers. The more open nature of garrigue has allowed species from drier and sparser habitats to become more widespread in this region. Birds that take advantage of garrigue include Western Rock Nuthatch, Dupont's Lark and a variety of other larks and pipits, Black-eared and Black Wheatears, Spectacled Warbler, and Masked Shrike. Northern Bald Ibis has been extirpated but has been recently reintroduced in s. Spain in garrigue and surrounding farmland.

DISTRIBUTION: This habitat is very widespread on Mediterranean islands and all around the Mediterranean coast. It extends east through much of Turkey into Syria, where it merges into PALEARCTIC SEMIDESERT THORNSCRUB.

WHERE TO SEE: Tarifa, Andalusia, Spain.

Top: **Western Rock Nuthatch lives around rocks within garrigue.**
© JAMES EATON, BIRDTOUR ASIA

Above: **Garrigue near Cádiz, Spain.**
© IAIN CAMPBELL, TROPICAL BIRDING TOURS

Northern Bald Ibis was close to extinction but has been reintroduced in Spain. © IAIN CAMPBELL, TROPICAL BIRDING TOURS

WESTERN FLOWER STEPPE

IN A NUTSHELL: The large area of dry grasslands through most of c. Asia in which flowers are very common in spring. **Habitat Affinities:** NEARCTIC TALLGRASS PRAIRIE. **Species Overlap:** EASTERN GRASS STEPPE.

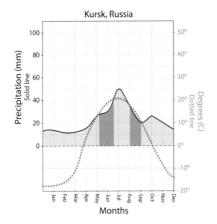

Kursk, Russia

Western flower steppe, Kazakhstan.
© KEITH BARNES, TROPICAL BIRDING TOURS

DESCRIPTION: This habitat is the Palearctic equivalent of the NEARCTIC TALLGRASS PRAIRIE of North America and has much in common with it. The main factor in the creation of this habitat is a lack of precipitation, rather than temperature, as it can be found in hot parts of the Iberian Peninsula as well as frigid zones of Kazakhstan. But throughout its range, annual precipitation remains below 30 in. (750mm) per year, too dry for temperate trees or large bushes to grow, when combined with fire. Many of the grasslands classified as steppe (such as the Puszta in Hungary) are actually not a climax community (one that is in complete harmony with precipitation and temperature controls) but have been kept in a plagioclimax state by fire and more recently by human-caused fire and grazing by domestic animals. If fire or grazing were to cease, these grasslands would revert to forest-steppe at the least and most likely to full deciduous and mixed forest.

Western flower steppe has many species of grasses and flowering plants, and it is the abundance of flowering plants that distinguishes the European steppes from those east of the Ural Mountains (EASTERN GRASS STEPPE). Many of the flowers now seen in anthropogenic steppes of Europe are originally from alpine meadows and have colonized the (geologically) recently cleared areas. In spring, the grasses start to grow, reaching around 20 in. (50cm) tall, and flowers can grow slightly taller. The growth cycles vary among species, with some flowering early in the spring, and smaller numbers through summer and into the fall. The grasses are dominated by feathergrasses, mostly *Stipa*, and flower mainly in midsummer, giving the whole landscape a much browner appearance than in spring. The steppes are underlain by extremely fertile chernozems, which can get to 5 ft. (1.5m) deep; these rich soils have meant heavy agricultural use, resulting in most of the natural prairie being converted to monoculture.

Great Bustard is a major target for naturalists in Extremadura, Spain.
© KEN BEHRENS, TROPICAL BIRDING TOURS

WILDLIFE: Smaller mammals, such as Bobak Marmot, Thick-tailed Three-toed Jerboa, Steppe Pika, and Tolai Hare, are common in western flower steppe. The bizarre-looking Saiga Antelope is now severely restricted to well-protected areas. These grasslands originally supported Wild Horse and Asiatic Wild Ass, but both have long been eliminated. Carnivores of this habitat include Asiatic Wildcat, Corsac Fox, and Steppe Polecat.

The steppe has a very different set of birdlife from forested Eurasian habitats. Rosy Starling

Black Lark is a resident and short-distance migrant in c. Asia. © JAMES EATON, BIRDTOUR ASIA

(IS) is a marquee bird of this habitat. You can also expect to see open-country birds like Montagu's Harrier, Eurasian Thick-knee, Corn Bunting, and Spanish Sparrow; larks such as Calandra, Black and Thekla's Larks; shrikes like Iberian Gray Shrike; and groups of pipits and wheatears. Along with these more widespread groups there are such sought-after birds as Great and Little Bustards, Sociable Lapwing, Black-winged Pratincole, and Black-bellied Sandgrouse.

DISTRIBUTION: Small outliers of these grasslands exist on the Iberian Peninsula in Spain and Portugal, and in easternmost Austria, Hungary, and Czech Republic. The southern boundary of the main tract of these grasslands begins in far n. Bulgaria and extends through s. Ukraine to the northern half of Kazakhstan and s. Russia. This steppe is separated from the EASTERN GRASS STEPPE by the Altai Mountains. In the north it blends into the PALEARCTIC FOREST-STEPPE of Russia, n. Ukraine, and Kazakhstan.

WHERE TO SEE: Cáceres, Spain; Kiskunság National Park, Hungary.

EASTERN GRASS STEPPE

IN A NUTSHELL: The expansive grasslands of Mongolia, which flower in summer (not spring) and have fewer flowers than the steppes farther west. **Habitat Affinities:** NEOTROPICAL PAMPAS AND CAMPO. **Species Overlap:** TEMPERATE DESERT STEPPE.

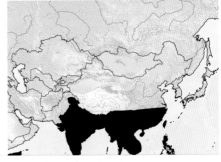

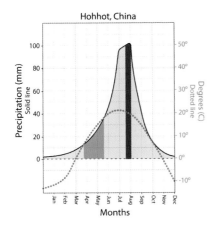

Hohhot, China

DESCRIPTION: The eastern grass steppe of Eurasia is an extension of the WESTERN FLOWER STEPPE but grows in conditions that are harsher, with much greater temperature fluctuations during the year and even during the day. There are long, bitterly cold (minimum temperature, –45°F/–43°C) and more importantly very dry winters and incredibly short summers (maximum temperature, 78°F/26°C), when

In summer, the eastern grass steppes of c. Mongolia look similar to human-made grasslands of Europe or e. **North America** © IAIN CAMPBELL, TROPICAL BIRDING TOURS

most of the rain occurs, with an annual precipitation of 10 in. (250mm). These extreme conditions prohibit the growth of herbs and shrubs and favor grass types that have dense, shallow roots. Most grasses and the very few shrubs bloom in summer, rather than in spring, giving the summer grass steppe a very different feel from the western flower steppe. The eastern regions of these steppes receive more rain, and there the grasses grow to 30 in. (80cm) tall. But over most of the habitat, perennial bunchgrasses and feathergrasses grow only 10–20 in. (25–50cm) high, and in the driest and coldest areas, at the transition into desert habitats, they reach only 8 in. (20cm).

Most of these grasslands completely lack trees, and the trees that do grow usually occur around an existing settlement. The very few bushes are usually less than 2 ft. (0.6m) tall and are widely spaced. Eastern grass steppes have had human intervention for an incredibly long time, with extensive grazing. More recently the moister areas, especially in China, have been mostly converted to monoculture agriculture, though the nature of the farming leaves large corridors of semi-natural habitat between fields. In Mongolia, there does not appear to be an obvious difference between protected and unprotected habitat, suggesting that the steppes are much more natural. Presumably this is because of the lower population density and more of a grazing rather than cultivating economy.

WILDLIFE: Birds have fared very well in the eastern grass steppe, and most species are still found in good numbers. As in the WESTERN FLOWER STEPPE, bird groups like larks, pipits, shrikes, and wheatears dominate the bird assemblage. Noteworthy birds include Oriental Plover, Isabelline Wheatear, Meadow Bunting, Mongolian Lark, Daurian Partridge, and Pallas's Sandgrouse.

In well-protected areas, such as in Mongolia, you can still find Przewalski's Horse, Mongolian Gazelle, Wild Boar, Asiatic Wild Ass, Pallas's Cat, and Steppe Polecat.

DISTRIBUTION: The boundary between the WESTERN FLOWER STEPPE and the eastern grass steppe is a diffuse one, though these eastern steppes become more prominent east of the Altai Mountains where Mongolia and Kazakhstan almost meet. They extend through the northern half of Mongolia into n. China, coming to within 100 mi. (160km) of Beijing.

WHERE TO SEE: Khustain Nuruu (aka Hustai) National Park, Mongolia.

Pallas's Cats have short legs and are not fast runners. They hunt their prey of pikas and gerbils through stalking and ambush. © KEITH BARNES, TROPICAL BIRDING TOURS

ARCTIC POLAR DESERT

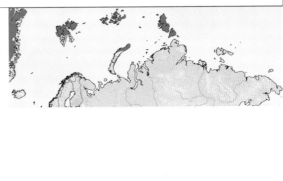

IN A NUTSHELL: This is the habitat at the very edge of where life is possible on earth. There is almost no vegetation of any kind. **Habitat Affinities:** None. **Species Overlap:** EURASIAN ROCKY TUNDRA; EURASIAN BOGGY TUNDRA.

DESCRIPTION: The northernmost vegetated area is one of both bitter cold and aridity. The polar desert receives less than 10 in. (250mm) of precipitation per year and never has a month when temperatures average over 50°F (10°C). What truly separates this climate regime from others is not the bitter winters, because the tundra to the south also has those, but rather the cold summers.

Crustose and fruticose lichens grow in protected areas of rock, n. Svalbard, Norway.
© IAIN CAMPBELL, TROPICAL BIRDING TOURS

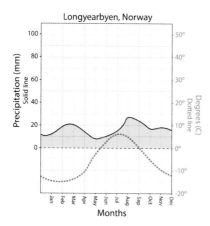

Longyearbyen, Norway

The ice may melt for a very short period of summer, allowing photosynthesis to start at about 36°F (2°C), but there is no prolonged period of growth before temperatures drop to 23°F (-5°C) and photosynthesis halts, so plant life is severely limited to the hardiest of species. The ground is sparsely covered, with no closed vegetation cover, and up to 80% of the surface is bare rock. Lichens exist in protected areas in cracks and hollows and consist mainly of crustose (flat-laying) species, though some freestanding (fruticose) lichens survive in the most protected areas. Mosses, fruticose lichens, and rare grasses occur only where fine soils have accumulated in the best-protected areas.

WILDLIFE: There are no endemic animals in this region. The few birds and other animals that do occasionally arrive here are all present in the tundras

to the south. Birds use cliffs for nesting but do not rely on the limited vegetation for food, and no resident bird species can survive. No land mammals reside here, though Walrus, seals, and Polar Bear use the fringes of the land as haul-outs. In the true polar desert environment, no insects or reptiles can exist; however, a few insects live in transitional areas between polar desert and EURASIAN ROCKY TUNDRA in s. Svalbard, Norway.

DISTRIBUTION: This habitat is limited to n. Greenland, Wrangel Island, and the islands of n. Russia. Most people encounter this habitat on the northernmost parts of a Svalbard cruise.

WHERE TO SEE: Cruises from Svalbard, Norway that go to the north of the Svalbard archipelago pass by this habitat.

EURASIAN ROCKY TUNDRA

IN A NUTSHELL: This low, mossy habitat spends most of the year under snow but has a spectacular burst of insects and birdlife in summer. **Habitat Affinities:** NEARCTIC ROCKY TUNDRA. **Species Overlap:** EURASIAN BOGGY TUNDRA; BERINGIAN TAIGA SAVANNA.

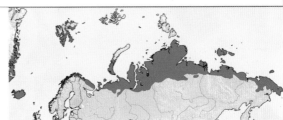

DESCRIPTION: There are multiple microhabitats within this broad classification, but generally rocky tundra vegetation is low-lying, one-layered ground cover with few bare patches over rock. The vegetation comprises mainly perennial plants no higher than 12 in. (30cm) aboveground, especially dwarf shrubs and mosses, and there are no larger shrubs or trees. There are bare patches where only lichens and mosses grow, but over most of the area, there is some soil, allowing the growth of low heath tundra. This very low heath is what most people think of when they imagine tundra and will primarily be described here. In well-protected areas and those with thicker soil, Arctic scrub can grow, reaching several feet high, but it is generally made up of taller versions of the same plant species that grow in more exposed areas.

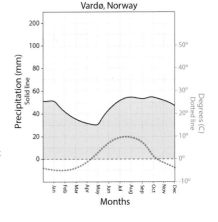

Vardø, Norway

One's first impression of tundra in late spring to midsummer is a blend of earthy oranges, olive greens, and browns of the mosses and lichens, along with the previous year's dead growth. Perennials and shrubs tend to flower early in the season, while annuals or plants without substantial biomass in their root systems flower later in the season and either wither away or spend the winter as seed.

Soils of rocky tundra are usually very shallow, with little weathered rock below the vegetation (photo: n. Norway). © IAIN CAMPBELL, TROPICAL BIRDING TOURS

Temperature alone would intuitively seem to be the main factor determining vegetation limitations in the tundra, but temperature varies little from one area to another in these regions, and the SIBERIAN LARCH FOREST is actually colder than tundra areas. However, these northern tundra areas can be very windy, and at extreme temperatures in areas of permafrost, these winds cause rapid transpiration and desiccation with no chance of replenishment, as all water is frozen. So it is the dry wind rather than temperature that causes most of the tundra to be treeless, except for tiny groves of birches that eke out an existence in nooks that are protected from the wind.

The growing season for most of the tundra is only four months a year, when there are up to 24 hours of sunlight, and the rest of the year the habitat remains dormant; for 200 days, it is buried under snow cover (which actually protects the plants from desiccating winds). Therefore, the dominant features of plants of this habitat are that, rather than simply being resistant to extremely cold conditions, they are adapted to a very short growing season with long periods of daylight, and they can withstand the intense drought (through desiccation) of the winter months as well as the exposure to winds when there is no snow cover.

WILDLIFE: Most of the mammals found on rocky tundra are not able to migrate and have had to find ways of dealing with this environment for the eight months of the year when it is inhospitable. The larger grazing mammals such as Reindeer and Muskox (and until recently Woolly Mammoth and Woolly Rhinoceros) do not hibernate, so they need a large fat layer beneath the skin, as well as two fur layers. Beneath the coarser outer layer is a much finer, denser undercoat that traps body heat. In the summer months, the larger mammals will molt their coat for a much cooler coat

Snowy Owl nests in rocky tundra throughout the Holarctic.
© KEITH BARNES, TROPICAL BIRDING TOURS

Many Polar Bears spend their summers in rocky tundra. © IAIN CAMPBELL, TROPICAL BIRDING TOURS

to prevent overheating in the mild temperatures of summer.

Rocky tundra is home to plenty of smaller mammals, such as lemmings and voles, which are widely distributed in the southern tundra. They spend the winter in snowbanks, where they can seek protection from the bitter winds, but do not hibernate. Rather, they continue to search for food throughout the winter, and some species store food in preparation for the harder times. Both voles and lemmings have fluctuating populations; vole numbers are controlled by predators whereas lemming numbers are controlled by food supply. Voles eat mainly fast-growing grasses, which continue to be replenished throughout the summer, and their populations rise and fall with predation. Lemmings feed mainly on mosses, and there is evidence to suggest that lemming numbers rise a little faster than vole numbers in areas where moss is abundant; then the lemmings overgraze, exhausting the moss supply, and their population drops suddenly. Population explosions of voles and lemmings trickle through the food chain and affect predators like Arctic Fox, Long-tailed and Pomarine Jaegers, and Snowy Owl.

The biggest change in animal numbers comes with the invertebrate ephemeral breeders such as mosquitoes and midges that burst into life in early summer, have phenomenal growth over a very short time, and provide the food source for the bulk of migratory birds. Many shorebirds migrate north to exploit this food source with its 24-hour feeding opportunities. The rocky tundra will come alive with species like Ruff, Purple Sandpiper, Black-bellied Plover, European Golden-Plover, and Whimbrel, all nesting on the higher rocky tundra. There are few passerines this far north, but a few species like Snow Bunting and Lapland Longspur can be very common.

DISTRIBUTION: If defined by an annual temperature below freezing and summer months that are below 50°F (10°C), then the tundra environment extends from Iceland, Norway, Sweden, and Finland (Fennoscandia) right across the top of Eurasia to e. Siberia, south to about 65°N. Many people also add the existence of permafrost to the definition, thereby discounting Iceland or Fennoscandia; however, permafrost also prevails through much of the SIBERIAN LARCH FOREST, so we do not see it as a defining feature.

WHERE TO SEE: Varanger area (around Vadsø and Vardø), n. Norway.

EURASIAN BOGGY TUNDRA

IN A NUTSHELL: The wetland areas of the northern tundra, where most of the waterbirds nest. **Habitat Affinities:** NEARCTIC BOGGY TUNDRA. **Species Overlap:** EURASIAN ROCKY TUNDRA.

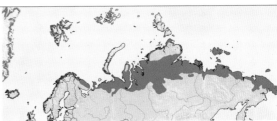

DESCRIPTION: This habitat is closely associated with EURASIAN ROCKY TUNDRA, occurring in the same zones with the same climate. The only significant difference between the two is drainage. Wet tundra, with its bogs, marshes, and lagoons, resembles the bogs that exist in the EURASIAN SPRUCE-FIR TAIGA to the south, but here this wet habitat covers vast areas of the tundra belt. Much of this landscape is very flat, therefore promoting the concentration of water. The bogs can have a grassy cover of sedges and cottongrasses, along with water-loving mosses. Raised areas of the bog,

called "pingos," are formed by the swelling of underlying permafrost, causing an elevated dome. These very small, isolated, water-surrounded areas tend to be drier than the surrounding boggy tundra, have a similar vegetation to the rocky tundra, and are where the boggy tundra shorebirds breed. The pingos support mosses, fruticose lichens, and dwarf shrubs, especially Dwarf Willow.

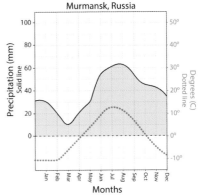

King Eider nests in boggy tundra (photo: n. Norway).
© IAIN CAMPBELL, TROPICAL BIRDING TOURS

WILDLIFE: These boggy and marshy areas are rich in nesting shorebirds and other waterbirds. Species include King and Steller's Eiders, Arctic and Yellow-billed Loons, and Sharp-tailed, Solitary and Green Sandpipers. The density of birdlife and the numbers of insects can be staggering.

Very few mammals live in this cold wetland environment because it is productive for only a short time of the year, and aquatic mammals cannot survive the winter here. Arctic Fox wanders into it from ROCKY TUNDRA to predate nesting birds, and Polar Bears roam here in summer but do not den in this marshy environment.

DISTRIBUTION: The broad distribution of this habitat follows that of the EURASIAN ROCKY TUNDRA, but it is more common closer to coastlines.

WHERE TO SEE: Varanger area, n. Norway.

The male Ruff has magnificent breeding plumage.
© IAIN CAMPBELL, TROPICAL BIRDING TOURS

Arctic Fox lives in both boggy and rocky tundra (photo: n. Norway).
© IAIN CAMPBELL, TROPICAL BIRDING TOURS

EURASIAN ALPINE TUNDRA AND HIMALAYAN MONTANE DESERT

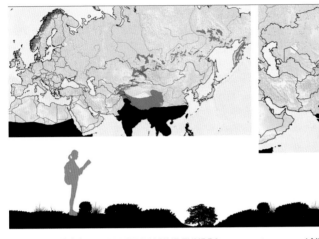

IN A NUTSHELL: The open spongy and mossy ground above the tree line in temperate parts of the region, and the desert above the tree line in the Himalayas.

Habitat Affinities: NEARCTIC ALPINE TUNDRA at one extreme, and NEOTROPICAL PUNA at the other.
Species Overlap: EURASIAN ROCKY TUNDRA; EASTERN GRASS STEPPE.

DESCRIPTION: Alpine tundra occurs above a variable tree line through most of Eurasia. The tree line can be low in the northern latitudes, such as s. Norway (almost at sea level), and it continues to rise southward, reaching as high as 12,000 ft. (3,650m) in the Himalayas. Above this tree line, the plants resemble those of the Arctic tundra (EURASIAN ROCKY TUNDRA), and indeed many of the Arctic tundra plants also occur in the alpine tundra. However, the alpine tundra is far richer vegetatively and has acted as a refuge for tundra plants during times of rapid climate change.

With the exception of the Tibetan Plateau and a few other locations, where extensive flat areas cause drainage impedance and large internal lakes, the topography of the alpine areas is steep and undergoes constant disturbances like rock slides and avalanches. With very little chemical weathering possible, there is little clay and iron-oxide development, so soils are mainly lithosols (from bare rock) and regosols (from scree), which are shallow and have little relationship to the underlying parent material. In more stable or protected areas, such as "snow crannies," chemical weathering of the parent material can occur. In some areas, deeper meadow soils develop, which are more conducive to shrub growth.

Alpine tundra has an even harsher growing environment than Arctic tundra, with a low median annual temperature, brutal winters (lows to -40°F/-40°C), and cool summers (up to 70°F/21°C). The growing season is very short, just 4–10 weeks. This is similar to the Arctic tundra, but alpine regions do not experience the months of continuous light during the growing season. The alpine tundra also has to deal with large daily temperature ranges, and temperatures that may dip below freezing at night, even during the growing season. The growing season is limited even further by the snow that buries most plants for much of the year. This protects them from the strong, desiccating winds of winter, yet the plants are able to photosynthesize and grow only once the snow has melted and they can receive sunlight. This really does put these plants in the very harshest of environments.

The Cairngorms of Scotland contain the most familiar area of alpine tundra in the British Isles.
© KEITH BARNES, TROPICAL BIRDING TOURS

With varied microenvironments within a small area, plant associations form a complex mosaic in these habitats. In the highest areas of the Tibetan Plateau and in the Himalayan montane desert above 15,000 ft. (4,500m), at first glance the area looks like ARCTIC POLAR DESERT, its limited ground cover consisting mainly of crustose and fruticose lichens. At lower elevations, plant communities on the poorly developed soils are specialized and are dominated by deep-rooted cushion plants and mosses, along with crustose and fruticose lichens. In more stable, moister areas, the cushion plants still dominate, but other plants include members of the rose, buckwheat, and saxifrage families that are more widespread in temperate areas. Perennials, which grow up to 2 ft. (60cm) in height, keep more biomass in their roots, ready for rapid growth at the onset of spring. While annuals may look prevalent when they are flowering in summer, they actually make up less than 5% of the total alpine flora in most areas.

In the snow crannies the shrubs are low but generally of the same types as those at the very edge of the tree line, where this habitat merges into conifer forest. In the moister snow crannies and protected valleys around the eastern edge of the Tibetan Plateau are extensive areas of deciduous shrubs with isolated conifers. Protected areas in the much drier western and southern (Himalayan montane desert) parts of this region, such as Ladakh in India, juniper shrubs form thickets, and some poplars and birches grow.

WILDLIFE: A fascinating difference between Arctic tundra (EURASIAN ROCKY TUNDRA and BOGGY TUNDRA) and alpine tundra in Eurasia is the comparative lack of shorebird and wildfowl migrants in the latter. Whereas the Arctic tundra bursts into life with insects and a mass of shorebirds that migrate north to make good use of the summer food supply, the alpine tundra does not seem to have the same attraction. While most of Asia's shorebirds pass overhead on their way to the Arctic, the Tibetan Plateau gets migrant waterfowl like Bar-headed Goose (which flies directly over the Himalayas rather than bother using valleys and passes), and cranes like the Black-necked Crane. In the low shrubs of the snow crannies, a whole different suite of resident and migratory birds exists, such as Himalayan Rubythroat, White-browed Tit-Warbler, Blue Eared-Pheasant, and Chinese Monal. Alpine tundra in Europe has nesting breeders like Eurasian Dotterel, as well as resident species like Rock Ptarmigan and Red-billed Chough.

The Himalayan montane desert has Ibisbill along river edges and White-winged Redstart in more protected areas. The higher, more exposed areas have Brown Dipper in rivers, Solitary Snipe in boggy areas, Robin and Brown Accentors in riparian vegetation, Cinereous Tit in the very small shrubs, Bearded Vulture (Lammergeier) soaring overhead, and the enigmatic Wallcreeper on the cliff faces.

Right: **White-browed Tit-Warbler is a resident of the shrubbery in the snow crannies of alpine tundra in China.**
© SAM WOODS, TROPICAL BIRDING TOURS

Below: **Ibisbill winters in the Indo-Malayan region but nests on river edges of alpine tundra in c. Asia.** © SAM WOODS, TROPICAL BIRDING TOURS

Left: **Many people visit the Himalayan montane desert to search for Snow Leopard.** © SAM WOODS, TROPICAL BIRDING TOURS

Below: **Snow Leopard is the apex predator in these habitats—and highly sought after by naturalists.** © MIKE WATSON, BIRDQUEST

Mammals are more prevalent in these montane habitats than in the High Arctic tundra; predators like Tibetan Wolf (a subspecies of Gray Wolf), Tibetan Sand Fox, Snow Leopard, and Pallas's Cat hunt sheep like Bharal and Argali, goats like Chamois, Siberian and Alpine Ibex, and small mammals like rabbits, hares, and pikas.

DISTRIBUTION: Alpine tundra is found on most of the mountain ranges in Europe, including the Highlands of Scotland, the Pyrenees, the Alps, the spine of Fennoscandia, and the Urals. It is extensive in the Tian Shan and Altai Mountains of c. Asia. In e. Asia it occurs in Kamchatka down to Japan. The largest block of alpine tundra is on the Tibetan Plateau and the high Himalayas between 12,000 ft. and 19,000 ft. (3,650–5,800m). The Himalayan montane desert is concentrated in the dry areas in Tibet, northernmost Nepal, Ladakh (India), and north though Tajikistan to Kyrgyzstan.

WHERE TO SEE: Alpine tundra: Qomolangma National Park, Tibet, China; the Cairngorms, Scotland. Himalayan montane desert: Ulley Valley, Ladakh, India; Jomsom, Kali Gandaki Valley, Nepal.

PALEARCTIC TEMPERATE WETLAND

IN A NUTSHELL: Permanent and ephemeral wetlands similar to those found throughout the world. **Habitat Affinities:** NEARCTIC FRESHWATER WETLAND. **Species Overlap:** There is almost no overlap with surrounding habitats.

DESCRIPTION: The broad term *wetland* includes a vast and complicated set of microhabitats in which water is the key determinant of the environment. Some of the habitats considered under wetlands include (but are not restricted to) lakes, ponds, bogs, reed islands and reed beds, mires, fens, rivers, marshes, and floodplains, and also includes artificial ones such as reservoirs, gravel pits, and dams. The extent of wetlands in this region can be seen, for one example, in Russia, where more than 120,000 rivers and 2 million lakes exist, including Lake Baikal, an immense water body holding 20% of the world's fresh water, with a surface area of 12,248 sq. mi. (31,722km²). In this book we treat mangroves and tidal flats separately, though almost all other wetland microhabitats are included here, from small farm dams and ponds to vast lakes and deltas, such as those of the Ebro (Spain) and Volga (Russia) Rivers. Wetlands have been under intense pressure in this region for centuries, from activities such as overexploitation of the water through large-scale irrigation projects, which affects water tables; drainage for other land use; abandonment of grazing, which leads wetlands to become overgrown and choked with other vegetation; and pollution from industrial and agricultural practices. Europe has seen some of the most dramatic declines in wetlands over the latter half of the 20th century, with France, Spain, Italy, and Greece in particular all suffering a 50% loss of wetland habitats during that time. However, plentiful wetland areas remain in the reserve systems throughout the region.

Wetlands are often highly seasonal, and the timing of congregations of birds is often determined by this. In this temperate region, there are both wetland areas that attract large numbers of wintering populations, such as ducks, shorebirds, and geese, which may disperse to breed over a wider area for the summer, and also areas that hold large colonies of breeding waterbirds too. This means the composition of species and the numbers of individuals are often in flux within these dynamic microhabitats. In general, the wetland birds of any area do not really relate to the surrounding vegetation types, so the same overall assemblage of bird species can be found in different temperate wetlands, regardless of whether they are surrounded by desert, steppe, or deciduous forest.

Typical wetland, Doñana National Park, Andalusia, Spain.
© IAIN CAMPBELL, TROPICAL BIRDING TOURS

WILDLIFE: Some of the largest concentrations of birdlife can be seen in wetland areas, which hold a wonderful variety of waterbirds, providing breeding sites for some, wintering sites for others, or critical stopover sites for migrating birds. In the deeper open waters, diving ducks, pelicans, loons, and grebes can be found; while shallow waters at the edges can be home to dabbling ducks, swans, and geese; and muddy edges may be inhabited by spoonbills, shorebirds, bitterns, crakes, and rails. Among reed beds are Bearded Reedling and a series of reed-bed specialist warblers, like Aquatic and Paddyfield Warblers. Rivers and channels

The natural breeding population of Ruddy Shelduck is decreasing in most of its range, but introduced populations are growing in w. Europe.
© IAIN CAMPBELL, TROPICAL BIRDING TOURS

Bearded Reedling represents a monotypic family from the Palearctic. © SIMON BUCKELL

can provide habitat for mergansers, dippers, herons, and egrets.

Many of the wetlands in the region are on important flyways, such as the East Atlantic Flyway, which links migratory birds from Siberia with their wintering grounds in nw. Europe and the west coast of Africa, in countries like Mauritania. The largest spectacles of birds at these sites often occur when birds gather to prepare to migrate or at large-scale winter feeding areas. The birds are typically from the following groups: shorebirds (stilts, snipes, sandpipers, avocets, plovers), pelicans, ducks, geese, cranes, rails, gulls, terns, swans, grebes, storks, ibises, herons, egrets, and cormorants. These include a long list of threatened species, like Lesser White-fronted and Red-breasted Geese, White-headed Duck, and Baikal and Marbled Teal in Europe, and Black-faced Spoonbill, Oriental Stork, Chinese Egret, Swan Goose, Scaly-sided Merganser, Baer's Pochard, and Red-crowned, Siberian, Hooded, and White-naped Cranes in e. Asia.

Many mammals are strongly associated with wetlands, including Eurasian Otter, voles and water voles, European Mink, Eurasian Beaver, Common Muskrat, and Moose (known as Elk in Eurasia), and a huge variety of mammals use them to some degree during their lives, including European Polecat, Arctic Fox, and Brown Bear.

DISTRIBUTION: Wetlands are found throughout this temperate region, as this category encompasses a wide range of natural and artificial habitats in a region with abundant rainfall and natural sites. An indication of the large number of globally important wetland sites in Europe can be gleaned from a list of Ramsar sites (wetlands designated to be of international importance); over 50% of the world's such sites are in Europe—more than 1,000 different sites. Taken together, these wetlands cover a land area equivalent to that of the United Kingdom and involve over 40 different countries. The United Kingdom has the world's highest number (175) of Ramsar sites, while six other European countries have more than 50 (Spain, France, Sweden, Norway, Italy, and the Netherlands), illustrating Europe's importance in terms of wetlands. In North Africa, Algeria, Tunisia, and Morocco hold the largest numbers of internationally important wetlands, while in the e. Asian part of the region, China, Japan, and South Korea hold the most.

WHERE TO SEE: Lake Baikal, Russia; Danube River delta, Romania and Ukraine; Ebro River delta, Catalonia, Spain; Doñana National Park, Andalusia, Spain; Biebrza Marshes, Poland; the Camargue, France; London Wetland Centre (artificial), England, UK; Minsmere-Walberswick Heaths and Marshes, England, UK; Souss-Massa National Park, Morocco; Nile Delta (e.g., Lake Burullus), Egypt; Kushiro Wetlands, Hokkaido, Japan.

PALEARCTIC TIDAL FLAT

IN A NUTSHELL: The muddy intertidal areas around the coastline. **Habitat Affinities:** NEARCTIC TIDAL MUDFLAT. **Species Overlap:** PALEARCTIC ROCKY COASTLINE AND SANDY BEACH.

DESCRIPTION: Tidal flats are nonvegetated, gently sloping sections of coastline that experience regular flooding and exposure by the tides. They are composed of fine sediments, such as mud, sand, gravel, and silt, which are deposited there by tides or rivers. Sandflats and mudflats come under this category; mudflats are usually organically richer, with finer sediment composed mostly of silt and clay, while sandflats are composed of larger-grained sediment, mostly quartz. Flats form in intertidal areas that are calm, low-energy systems, which therefore allow deposition and buildup of the sediment. While on the surface these flats can seem devoid of life, beneath the sediment, a high diversity of benthic microscopic organisms provides food for invertebrates, fish, and birds. Flats are situated between seas and landward habitats, making them a significant barrier against erosion.

One of the major conservation concerns with mudflats is they are frequently modified though land-reclamation projects, causing catastrophic loss of habitat and birds. The Saemangeum estuary system in South Korea was cut off from the Yellow Sea when a large seawall was built in 2006, which resulted in a more than 90% decline in bird numbers. Over a third of all Yellow Sea tidal flats have undergone large-scale land-reclamation projects. However, other areas of flats still exist on the Korean Peninsula and neighboring China. For example, the Yalu River estuary (within China, bordering North Korea) still hosts an estimated 70,000 birds a day during the peak of migration.

Tidal flats are best viewed as interconnected systems that in concert provide feeding grounds for massive migratory bird populations, which need multiple sites due to the great distances they cover. This habitat and its numerous sites perhaps offer one of the greatest conservation challenges of our times, requiring cross-nation support and coordination.

WILDLIFE: Tidal flats are critical sites for many migrant species of shorebirds, either as a wintering site or stopover site on migration. Some of the typical groups found on flats include sandpipers, plovers, and avocets among the shorebirds, as well as flamingos, terns, gulls, herons, pelicans, and other waterbirds. Among the species of interest are Great and Red Knots, Common and Spotted Redshanks, Common Greenshank, Broad-billed and Curlew Sandpipers, Little Stint, Bar-tailed and Black-tailed Godwits, Kentish, Common Ringed, and Black-bellied Plovers, Lesser and Greater Sand-Plovers, Pied Avocet, and Eurasian and Far Eastern Curlews. In the Yellow Sea area, flats are important feeding areas for two species of conservation concern, Nordmann's Greenshank and Spoon-billed Sandpiper.

Aside from birds, and the abundant aquatic life within the sediment, such as microscopic organisms, crabs, mollusks and fish, there are few other conspicuous animals in this habitat.

Sanderling are found throughout tidal flats of the Palearctic. © IAIN CAMPBELL, TROPICAL BIRDING TOURS

Crab-Plover, representing a monotypic family, winters on the East African coast and breeds on the shores of the Red Sea, Persian Gulf, and nw. Indian Ocean. © KEN BEHRENS, TROPICAL BIRDING TOURS

DISTRIBUTION: While widespread through the region, not all flats are the same. There are a number of key sites in the region that are particularly nutrient-rich and attract huge congregations of birds. These are typically low in diversity of breeding birds but are most noteworthy as wintering sites or stopover sites for migrant birds, particularly shorebirds.

The Wadden Sea UNESCO World Heritage Site, shared by the Netherlands, Germany, and Denmark, represents the largest system of connected tidal flats in the world. They span 300 mi. (500km) and cover an area of 3,900 sq. mi. (10,000km^2). This area is important for shorebirds that use two principal flyways, the East Atlantic Flyway and the Eurasian-African Flyway. These flyways cover a number of migration routes of bird species that breed in the Arctic, from North America to Siberia, and fly south for the winter to nw. Europe and or farther south to w. and s. Africa. For example, over 40 migratory bird species use the East Atlantic Flyway (mostly waterbirds and shorebirds), and the Wadden Sea itself attracts an estimated 12 million birds each year. Many of these are migrant species, some of which use the area for only a short period, while others winter there for months at a time.

Another critical area of tidal flats in Eurasia is in the Yellow Sea region of China, including the entire west coast of the Korean Peninsula. These flats are important for migratory shorebirds that use the East Asian–Australasian Flyway, which covers routes from the Arctic of North America and Siberian Russia south to Australia and New Zealand. For many species, during their spring, northward migration, the Yellow Sea area is the last stopover site before they reach their northern breeding grounds. About 90% of all Lesser Sand-Plovers are believed to pass through the Yellow Sea area, illustrating its global significance for that species. The importance and complexity of tidal flats as a conservation issue can be gleaned from the numbers concerning the East Asian–Australasian Flyway: it hosts over 400 species of migratory birds, spans across 37 nations, and covers an area of 32,728,000 sq. mi. (84,765,020km^2).

In the Middle East, the peninsula of Barr al Hikman in Oman is considered one of the most important areas for wintering waterbirds in w. Asia. The area holds more than 30,000 acres (12,000ha) of tidal flats, which are up to 4.5 mi. (7km) wide, and is recognized as an important bird area, mainly for wading birds, such as Crab-Plover.

In North Africa, Banc d'Arguin National Park in Mauritania protects 40% of the country's coastline. Within the park are extensive tidal flats, covering 135,000 acres (55,000ha). These are home to the largest number of nonbreeding waders in the world, estimated at more than 2 million individuals. This represents 30% of the birds using the East Atlantic Flyway, which migrate to the park for the winter from their breeding grounds in n. Europe, Greenland, and Siberia. While there are many tidal flats in parts of Europe that support these species too, what sets Banc d'Arguin apart is that it has relatively low disturbance from humans, with controls on local visitors and their activities within the park.

These are just some of the most significant tidal flat sites in the region, but there are many others (e.g., the Thames estuary and the Wash bay and estuary, both in England).

WHERE TO SEE: Morecambe Bay, England, UK; the Wash, England, UK; Wadden Sea, Netherlands, Germany, and Denmark; Yalu River estuary, China; Banc d'Arguin National Park, Mauritania; Barr al Hikman, Oman.

PALEARCTIC ROCKY COASTLINE AND SANDY BEACH

IN A NUTSHELL: Coastal headlands and beaches. **Habitat Affinities:** NEARCTIC ROCKY COASTLINE AND SANDY BEACH. **Species Overlap:** PALEARCTIC TIDAL FLAT.

DESCRIPTION, WILDLIFE, DISTRIBUTION: On shorelines surrounding the Palearctic from Japan to Morocco and including the Mediterranean coastline, there are many headlands and sandy beaches, which are important breeding areas for many seabirds and shorebirds. In the east, the headlands are particularly important for breeding alcids such as Horned and Tufted Puffins and Least, Parakeet, and Whiskered Auklets. Along the Atlantic coastline the colonies of Atlantic Puffin, Razorbill, and Thick-billed Murre are joined by gull colonies of Black-legged Kittiwake and Glaucous and Herring Gulls. While tubenose seabirds are not as varied as in the Southern Hemisphere, Northern Fulmar is a common breeder on most of the northern cliffs, as is Northern Gannet, a close relative of the boobies. Cormorants are common on both the Atlantic and Pacific coasts. The ancestor to the common pigeon, the Rock Pigeon, breeds primarily on coastal cliffs in western parts of the Palearctic. Both the

Eurasian Oystercatcher roosts on rocks but feeds on beaches and mudflats.
© IAIN CAMPBELL, TROPICAL BIRDING TOURS

Left: **Crested Auklet is an alcid that breeds across the n. Pacific Ocean on both sides of the Bering Strait.** © IAIN CAMPBELL, TROPICAL BIRDING TOURS

Below: **Atlantic Puffin is an alcid found on both sides of the n. Atlantic Ocean.** © IAIN CAMPBELL, TROPICAL BIRDING TOURS

rocky and sandy shorelines host many feeding species, including shorebirds such as Purple Sandpiper, Sanderling, Black-bellied Plover, and Eurasian Oystercatcher, and other birds such as Great Cormorant. Some species also use the beaches to breed, including Little Tern.

WHERE TO SEE: Nemuro Peninsula, Japan; Vardo, Norway; the Anglesey, Wales.

PALEARCTIC PELAGIC WATERS

The pelagic waters south of the Arabian Sea merge with those of the Afrotropics and Indo-Malaya, and those on the Atlantic and Pacific sides of the Palearctic link with those of the Nearctic. To avoid overlap Palearctic pelagic waters are treated in the other sections; see NEARCTIC PELAGIC WATERS. For this habitat in the Indian Ocean, see INDO-MALAYAN PELAGIC WATERS.

PALEARCTIC CROPLAND

IN A NUTSHELL: Familiar farmland found throughout most of Europe and Asia. **Habitat Affinities:** NEARCTIC CROPLAND AND GRAZING LAND. **Species Overlap:** There is some overlap with surrounding natural habitats.

DESCRIPTION: In the North African and Middle Eastern sections of this region, the arid climate has limited large-scale crop farming to some degree, relative to other parts of the region. As this is one of the most water-stressed areas of the planet, farming here historically has been concentrated in the wetter fertile areas, such as the Atlas Mountain valleys of Morocco and the Nile River valley and delta of Egypt. These two nations are the biggest cereal producers in n. Africa, raising wheat, barley, and grain maize as their main crops. Other important crops in this subregion include cotton, olives, grapes, and dates. Saudi Arabia is the world's largest producer of dates, while Tunisia is one of the planet's biggest suppliers of olive oil. Due to the aridity of the region, large-scale irrigation systems

Five species of cranes congregate on fields in Japan.
© SAM WOODS, TROPICAL BIRDING TOURS

have been needed in many areas to turn them into crop-producing landscapes, such as the Euphrates River valley in the east of the region. In 2018, it was estimated that only 5% of the region was suitable for crop production; as irrigation technologies move forward, croplands will spread into more areas previously considered unsuitable.

The countries of the European Union have had a long history of conversion of native habitats to croplands. Over 40% of the land area, or more than 433 million acres (175 million ha), is classified as agricultural areas. Europe has a diverse crop matrix, dominated by cereals (e.g., wheat, barley, grain maize, rice) but also including root crops like potatoes and sugar beet; pulses, such as beans and peas; and oil seeds, like rape. Within the member states, France and Germany are the largest producers; the United Kingdom's output is also significant.

In the east of the region, in Asia, the largest crop by far is rice (often occupying between a quarter and a half of all arable land in a nation), but wheat, barley, and soybeans are also significant. China, Uzbekistan, and Japan are the biggest producers of wheat in the region; China, Russia, and Ukraine are some of the world's largest soybean producers; and Russia, Ukraine, and Turkey are all among the largest barley producers. In the rice paddies of e. Asia (see PADDY FIELDS AND OTHER CROPLAND in the Indo-Malayan chapter), the avifauna is depauperate compared to native wetlands, although egrets, bitterns, rails, and shorebirds may use these on occasion, depending on the water levels and the amount of human disturbance.

WILDLIFE: Many birds that were historically associated with farms have seen drastic declines in recent decades with the change in farming intensity resulting from the EU's Common Agricultural Policy. This has been due to habitat clearance, such as the elimination of hedgerows from large-scale farms, and also the increased use of agrochemicals, which decimate both insect populations and plants that produce seeds on which farmland birds fed previously. Some of the species most seriously affected have been Corn Crake, Gray Partridge, Northern Lapwing, Western Yellow Wagtail, Yellowhammer, Corn Bunting, Eurasian Linnet, Eurasian Skylark, and Eurasian Tree Sparrow, while the flag bearer has been European Turtle-Dove, which has undergone a decline of more than 90%. The strongest populations of these species survive in agricultural lands of e. Europe, where farming practices have historically been less intense and smaller-scale properties still occur widely. Bird species that have persisted in large-scale farming areas, and have sometimes increased in abundance, include Stock Dove, Common Wood-Pigeon, Greater Whitethroat, European Greenfinch, European Goldfinch, Rook, and Eurasian Jackdaw.

DISTRIBUTION: Ubiquitous in most areas except Arctic, desert, and very mountainous terrain.

White Storks follow industrial harvesters in crop fields, Spain.
© IAIN CAMPBELL, TROPICAL BIRDING TOURS

PALEARCTIC GRAZING LAND

IN A NUTSHELL: Grazed lands that mimic grassland habitats. **Habitat Affinities:** NEARCTIC GRAZING LAND. **Species Overlap:** WESTERN FLOWER STEPPE.

DESCRIPTION: Grazing lands are extensive across the region; nearly 50% of European land is agricultural, which includes abundant grazing lands for livestock. The main livestock in Europe in order of magnitude are cattle (over 50% of livestock in more than 14 countries), pigs, sheep, and goats. In recent years, the Netherlands, Belgium, and Malta have recorded the highest density of livestock, while France and Germany have the highest numbers of individual units. Sheep are a strong component of the livestock in Greece and the British Isles, and Greece and Cyprus have a significant proportion of goat grazing relative to other EU states. North Africa and the Middle East were centers of domestication of cattle, sheep, and goats, and so these represent the main livestock there also.

The long presence of vast areas of grazing lands has undoubtedly led to clearance of much native habitat, particularly grasslands in upland areas and wetlands, but also woodlands, heathlands, and forests. While there can be loss of biodiversity from this reappropriation, it can conversely increase diversity in some cases, when the semi-natural habitat is more diverse than the climax vegetation. After all, much of what exists in terms of bird habitats in Europe has undergone some form of modification. Indeed, a variety of habitats are often now managed, for conservation concerns, with well-managed grazing regimes. This includes heaths and grasslands among others, where grazers are required to halt succession and maintain these habitats. These so-called semi-natural habitats, having existed alongside humans for so long, are now partly or wholly dependent on grazing. The management of this grazing, however, is crucial to maintaining such habitats, which are threatened if overgrazing occurs or chemicals are used too intensely.

Most Eurasian grassland habitats are maintained through grazing. Grazed wooded grasslands cover vast areas of the region, and many ground-nesting birds benefit from the grazing that occurs there. For example, thinning of Cork Oak forests on the Iberian Peninsula for grazing by cattle, sheep, and pigs has led to the creation of a vast human-made habitat, sometimes referred to as *dehesa*, which resembles a savanna rather than the forest it replaced long ago. As these areas combine a mosaic of habitats, with some trees remaining but plentiful grass, they can be more diverse than the forests they replaced, attracting both open- and closed-forest species. The latter include larks, shrikes, wagtails, stonechats, and buntings, and these areas often retain populations of tree-nesting species too, like tits, nuthatches, and treecreepers, due to the continued presence of pockets of trees.

A large threat to these grazing lands is conversion into croplands to meet growing food needs; these croplands are often of higher economic value but are much more destructive to wildlife. They are also susceptible to abandonment when natural vegetation growth driven by succession proceeds and then replaces this habitat with another, one no longer suitable for the current bird population. The change in density and therefore intensification of livestock grazing can also adversely affect the birds found there, for example if the animals crop some of the grasses shorter than breeding birds require.

WILDLIFE: Some of the species that can be found on grazing lands include a variety of raptors such as Lesser Spotted and Booted Eagles, as well as Common Quail, Gray and Red-legged Partridges, White Stork, Great and Little Bustards, European Roller, Eurasian Skylark, Meadow Pipit, and Western Yellow Wagtail. Eurasian Griffon is also a species of importance, as it is dependent on scavenging on the carcasses of livestock.

DISTRIBUTION: Ubiquitous in most areas except Arctic, desert, and very mountainous terrain.

HABITATS OF
THE NEARCTIC

(NORTH AMERICA)

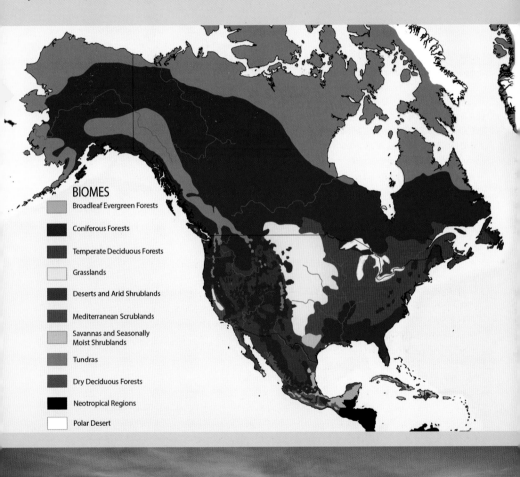

BIOMES

- Broadleaf Evergreen Forests
- Coniferous Forests
- Temperate Deciduous Forests
- Grasslands
- Deserts and Arid Shrublands
- Mediterranean Scrublands
- Savannas and Seasonally Moist Shrublands
- Tundras
- Dry Deciduous Forests
- Neotropical Regions
- Polar Desert

NEARCTIC SPRUCE-FIR TAIGA

IN A NUTSHELL: The endless forests of conifers that people imagine when they think of boreal Canada and interior Alaska. **Habitat Affinities:** EURASIAN SPRUCE-FIR TAIGA. **Species Overlap:** NEARCTIC MONTANE SPRUCE-FIR FOREST; NEARCTIC TEMPERATE MIXED FOREST; ASPEN FOREST AND PARKLAND.

DESCRIPTION: This is a North American extension of EURASIAN SPRUCE-FIR TAIGA; see the Palearctic chapter for a detailed description.

WILDLIFE: The Nearctic spruce-fir taiga is home to a tough and well-adapted set of resident mammals. Although the mammals are generally sparse in this wild country, there are some exciting possibilities. Iconic predators like Wolverine, Fisher, Canada Lynx, Gray Wolf, and American Black Bear are all found here. Moose and Caribou both make the taiga their home, and Snowshoe Hare and American Red Squirrel are common throughout.

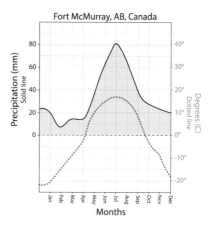

The birdlife here consists of a small set of year-round residents and a large number of breeding birds that visit for a short time during the summer. Residents include Spruce Grouse, Great Gray Owl, Northern Hawk Owl, Black-backed Woodpecker, Common Raven, Boreal Chickadee (IS), Pine Grosbeak, and White-winged Crossbill. In the summer, the forest is alive with insects that fuel a new generation of Neotropical migrants. Olive-sided Flycatcher, Yellow-bellied Flycatcher (IS), Gray-cheeked Thrush (IS), Swainson's Thrush, Cape May Warbler, Connecticut Warbler, Blackpoll Warbler (IS), and Canada Warbler all add to the astonishing spectacle that is the boreal summer.

Few reptiles and amphibians are found here, though Mink Frog and Wood Frog both make their homes in the many wetlands that dot this habitat. Wood Frogs occur farther north than any other amphibian, and their tissue can survive freezing solid for months at a time.

DISTRIBUTION: The Nearctic spruce-fir taiga is one of the most expansive habitats in the Nearctic, covering a giant swath of Canada and Alaska, and dipping into the n. United States. Reaching its eastern extent at the Atlantic Ocean in Newfoundland, the Nearctic spruce-fir taiga spans

The black form of Gray Wolf in the n. United States.
© ANDREW SPENCER

Left: Bald Eagle, the national bird of the United States, sitting in spruce-fir taiga habitat during winter. © BEN KNOOT, TROPICAL BIRDING TOURS

Above: Cape May Warbler is one of many birds that migrate to this habitat during the spring to breed. © IAIN CAMPBELL, TROPICAL BIRDING TOURS

westward to w. Alaska, though it doesn't reach coastal areas there. At the southern limits of the range, the taiga reaches n. Minnesota and Michigan, where it transitions to NEARCTIC TEMPERATE MIXED FOREST. In the north, throughout Arctic Canada and Alaska, as the stature and density of the forest decrease, and tundra (NEARCTIC ROCKY TUNDRA and BOGGY TUNDRA) appears between conifer stands, this habitat looks more like BERINGIAN TAIGA SAVANNA of the Palearctic. At high elevations in the west, Nearctic spruce-fir taiga is replaced by NEARCTIC ALPINE TUNDRA.

In addition to the Nearctic spruce-fir taiga of the far north, there are small, isolated pockets of spruce-fir in the Appalachian Mountains of the e. United States. This habitat is a relic of recent ice ages and occupies an area of less than 100 sq. mi. (160km^2). Due to limited area and heavy isolation, these pockets generally do not contain boreal fauna like the extensive western tracts do. The dominant tree species here are Red Spruce, Balsam Fir, and Fraser Fir.

WHERE TO SEE: Sax-Zim Bog, Minnesota, US; Algonquin Provincial Park, Ontario, Canada; Yukon-Charley Rivers National Preserve, Alaska, US.

NEARCTIC MONTANE SPRUCE-FIR FOREST

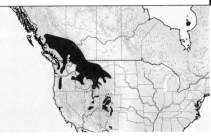

IN A NUTSHELL: A dense uniform forest of spruce and fir trees that grows at high elevations. **Habitat Affinities:** EURASIAN MONTANE CONIFER FOREST; EURASIAN SPRUCE-FIR TAIGA. **Species Overlap:** NEARCTIC SPRUCE-FIR TAIGA; MONTANE MIXED-CONIFER FOREST; HIGH-ELEVATION PINE WOODLAND.

DESCRIPTION: This high-elevation habitat of w. Canada and the United States is in many ways a southern equivalent of the more northerly NEARCTIC SPRUCE-FIR TAIGA. While the steep slope and rocky soil precludes the development of the muskegs and bogs found in the taiga, the structure and fauna are quite similar. Expansive and uniform, these forests form the seemingly endless dark green vistas that are typical of the high-elevation mountainous West.

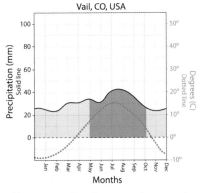

This habitat is cold year-round and has snow cover beginning as early as September and often persisting until midsummer. Annual precipitation, the vast majority of which falls as snow, generally totals 20–30 in. (500–800mm). These are relatively simple forests dominated by two species of narrow, slow-growing conifers: Engelmann Spruce and Subalpine Fir. These two species form co-dominant or monotypic stands that rarely include other trees. Douglas-fir, Lodgepole Pine, and Quaking Aspen do occur in small numbers, especially near areas of recent disturbance. These trees grow with even spacing and are typically 30–80 ft. (10–25m) tall, though individual trees can reach heights of 130 ft. (40m). The canopy is dense, its coverage ranging from 60% to 90%, and often precludes the development of shrub and herbaceous layers, which are typically sparse or absent. Shrubs that do occur include Cascade Azalea, serviceberries, Five-leaved Bramble, gooseberries, and willows. In areas that have experienced avalanches, blowdowns, or other major disturbance, grassy meadows can be found.

Nearctic montane spruce-fir forest is slow-growing and, unlike most western conifer habitats, fire-intolerant. With warmer temperatures brought by climate change, montane spruce-fir stands have suffered massive die-offs due to Mountain Pine Beetle, a pest that historically died back in winter. These beetle-caused die-offs create large swaths of forest that are then highly susceptible to sweeping wildfires, which take centuries to recover from.

WILDLIFE: The wildlife is a blend of species from the Nearctic taiga and species of adjacent montane conifer habitats. The mammals include most of the classic boreal species, such as Moose, Snowshoe

Above: Dramatic geological features can be seen alongside this habitat in Yellowstone National Park, Wyoming, US. © GEORGE LIN, TROPICAL BIRDING TOURS

Below: American Marten in spruce-fir taiga habitat during winter in Algonquin Provincial Park in Ontario, Canada. © SAM WOODS, TROPICAL BIRDING TOURS

Canada Jay (formerly Gray Jay) is familiar to many who have camped within this habitat in North American parks like Yellowstone, where it can be tame and bold, earning it the nickname "camp robber."
© SAM WOODS, TROPICAL BIRDING TOURS

Montane spruce-fir forest is prime habitat for the irruptive Evening Grosbeak. © PHIL CHAON, TROPICAL BIRDING TOURS

Hare, American Red Squirrel, Red-backed Vole, Short-tailed Weasel, American Marten, Fisher, and the elusive Canada Lynx. The lynx ranges as far south as Colorado, where populations were reestablished in the 1990s.

Many species of boreal birds are found here, including Spruce Grouse, Boreal Owl (IS), American Three-toed and Black-backed Woodpeckers, Canada Jay, Black-capped Chickadee, Golden-crowned Kinglet, Pine (IS) and Evening Grosbeaks, and Red and White-winged Crossbills. These occur alongside more typical Rocky Mountain species like Northern Pygmy-Owl, Steller's Jay, Clark's Nutcracker, Mountain Chickadee, and Cassin's Finch. In winter, Gray-crowned, Black, and Brown-capped Rosy-Finches come down to this elevation from coniferous forests above, especially during inclement weather. Flocks of hundreds can sometimes be found at bird feeders after a winter storm.

This area can have snow on the ground during any month of the year, and many locations experience frosts throughout the summer. For that reason, amphibians and reptiles are almost completely absent.

DISTRIBUTION: This is a high-elevation habitat found in w. Canada and the United States. It occurs throughout the Rocky Mountains and the Cascade Range, from British Columbia and Alberta south to n. New Mexico. It occurs as far east as c. Montana in montane islands, and as far west as the Olympic Mountains in w. Washington. As is typical of montane ecosystems, the elevation at which Nearctic montane spruce-fir occurs increases as you get closer to the equator. At the northern end of their range in British Columbia, these forests occur as low as 3,300 ft. (1,000m), while at the southern limit in New Mexico this habitat occurs as high as 11,000 ft. (3,350m). Upslope this habitat is replaced by HIGH-ELEVATION PINE WOODLAND or NEARCTIC ALPINE TUNDRA, and downslope it is usually replaced by MONTANE MIXED-CONIFER FOREST. In the Olympic range, this habitat occurs only on drier east-facing slopes and on west-facing slopes is replaced by NEARCTIC TEMPERATE RAINFOREST.

WHERE TO SEE: Yellowstone National Park, Wyoming, US; Waterton Lakes National Park, Alberta, Canada; Cameron Pass, Colorado, US.

MONTANE MIXED-CONIFER FOREST

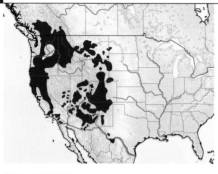

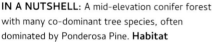

IN A NUTSHELL: A mid-elevation conifer forest with many co-dominant tree species, often dominated by Ponderosa Pine. **Habitat Affinities:** EURASIAN MONTANE CONIFER FOREST; INDO-MALAYAN PINE FOREST. **Species Overlap:** NEARCTIC MONTANE SPRUCE-FIR FOREST; PINYON-JUNIPER WOODLAND; LODGEPOLE PINE FOREST; ASPEN FOREST AND PARKLAND.

DESCRIPTION: A widely distributed and highly variable habitat of w. North America, montane mixed-conifer forest is the quintessential western forest of towering pines, with the smell of vanilla coming off sun-warmed Ponderosa Pine and the distant drumming of woodpeckers. Anyone who has spent time hiking in the mountains of the w. Nearctic region has enjoyed these vast and inviting forests. The canopy here varies from open and savanna-like at the lowest elevations to closed farther upslope. While the height of the canopy is generally 65–120 ft. (20–35m), some of the larger trees can reach heights over 200 ft. (60m). With a sparse mid-story and a variable shrub layer, these forests are quite open and easily traversed, making them a pleasant place for hiking and wildlife observation.

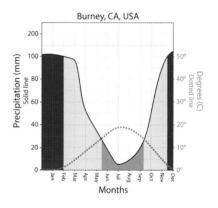

Burney, CA, USA

One of the hallmarks of the montane mixed-conifer forest is the diversity of conifer species present. Despite this variability, Ponderosa Pine is almost always a key component. Douglas-fir is the second-most widespread component of the canopy, found at all but the lowest elevations. In the Rocky Mountains, other important co-dominant trees are Lodgepole Pine, Grand Fir, White Fir, and Western Red Cedar. The Sierra Nevada and Cascades have a different set of co-dominant trees, with Sugar Pine, Jeffrey Pine, Western White Pine, Incense Cedar, and White Fir all occurring widely. The Giant Sequoia is found in the montane mixed-conifer forests of the c. Sierra Nevada. This most massive of trees changes the structure of the forests, as little understory and large canopy gaps surround these living giants.

In montane mixed-conifer forest, the mid-story is typically very sparse and comprises smaller individuals of the dominant tree types. Depending on frequency of fire and canopy density, the

Above: **Ponderosa Pines in the Black Hills of South Dakota, US.** © IAIN CAMPBELL, TROPICAL BIRDING TOURS

Right: **Pygmy Nuthatch typically occurs in Ponderosa Pine–dominated habitats.** © BEN KNOOT, TROPICAL BIRDING TOURS

shrub layer can be absent to dense. The shrub layer is highly diverse over this habitat's range, with well over 100 species represented. Shrubs typically found throughout the range include Big Sagebrush, manzanitas, ceanothuses, ninebarks, dogwoods, Huckleberry Oak, and Gambel Oak. In more open, Ponderosa Pine–dominated stands, the understory can be quite grassy, with a variety of fescues, wheatgrasses, and buffalograsses dominating the ground cover. More closed canopy stands can completely lack herbaceous ground cover, with little but fallen conifer needles covering the forest floor.

WILDLIFE: Bordering a variety of habitats that all share aspects of vertebrate communities, montane mixed-conifer forest is home to the majority of forest-dwelling mammal species of the West. Widespread large mammals include Elk, White-tailed and Mule Deer, American Black Bear, Puma (aka Mountain Lion), and Gray Wolf. Other small predators such as Fisher, Gray Fox, Long-tailed Weasel, and Bobcat are all found locally. While mice and rats are present, the most noticeable rodents in this habitat are squirrels and chipmunks, which feed heavily on the variety of cones readily found here. Abert's, Western Gray, American Red, Douglas's, and Northern Flying (IS) Squirrels; and Least and Yellow-pine Chipmunks are all common.

The bird communities are similarly diverse. Game birds like Mountain Quail, Dusky Grouse, and Sooty Grouse are readily found. Northern Goshawk is the top avian predator in this habitat. Common resident birds like Brown Creeper, Pygmy and Red-breasted Nuthatches, Mountain Chickadee, and Steller's Jay are readily seen around campgrounds and picnic areas. This forest has a high diversity of woodpeckers: White-headed (IS), Hairy, Downy, Pileated, and Black-backed Woodpeckers, along with Red-breasted, Red-naped, and Williamson's Sapsuckers, can be found exploiting mixed-aged and burned forests. In the summer, inundated with breeding migrant birds, montane mixed-conifer forest becomes alive with song. Olive-sided (IS), Hammond's (IS), and Cordilleran Flycatchers; Golden-crowned Kinglet; Cassin's Vireo; Yellow-rumped, Townsend's, and Hermit Warblers; Western Tanager; and Black-headed Grosbeak all bring an extra splash of life and color.

Reptiles and amphibians are generally sparse, as is typical of montane environments in the West. However, Rubber Boa, California Mountain Kingsnake, Sharp-tailed Snake, Western Fence Lizard, and alligator lizards (*Elgaria*) all occur.

DISTRIBUTION: Montane mixed-conifer forests are distributed widely throughout the temperate regions of the w. Nearctic. Ranging as far north as s. British Columbia, Canada, and south through

the w. United States to n. Sonora and Chihuahua, Mexico, these forests can be found in all but the driest mountains and wettest coastal ranges. They reach their eastern extent in isolated patches in the Black Hills of South Dakota and are bounded to the west by the Pacific Ocean. Montane mixed-conifer forest typically grows at elevations between 2,000 ft. and 6,000 ft. (600–1,800m). At the upper limits it transitions to NEARCTIC MONTANE SPRUCE-FIR FOREST, LODGEPOLE PINE FOREST, or HIGH-ELEVATION PINE WOODLAND. At its lower elevational limit, it is bordered by PINYON-JUNIPER WOODLAND, NEARCTIC OAK SAVANNA, PACIFIC CHAPARRAL, and NEARCTIC TEMPERATE RAINFOREST. There are often broad ecotones at these elevational boundaries.

WHERE TO SEE: Yosemite National Park, California, US; Rocky Mountain National Park, Colorado US; Black Hills, South Dakota, US; South Hills, Idaho, US.

Williamson's Sapsucker is a specialist indicator species of montane mixed-conifer forest in w. North America. © BEN KNOOT, TROPICAL BIRDING TOURS

LONGLEAF PINE SAVANNA

IN A NUTSHELL: An open, grassy pine woodland or savanna with poorly drained soils occurring on flat coastal plains and in low-lying areas of the Mississippi River valley. **Habitat Affinities:** PALEARCTIC SUBTROPICAL SAVANNA. **Species Overlap:** NEARCTIC TEMPERATE DECIDUOUS FOREST; FLORIDA OAK SCRUB.

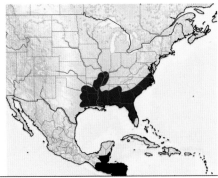

DESCRIPTION: The Longleaf Pine savanna is an open-canopied woodland or savanna occurring on the low-lying coastal plains and in the Mississippi River valley of the se. United States. The tall, narrow-trunked pine that dominates this habitat generally grows to a height of 70 ft. (22m) though occasionally as tall as 130 ft. (40m). Under historical fire regimes, trees are widely spaced (~100 ft./30m apart), and there is an open understory of grasses and small shrubs. However, with fire-suppression practices, dense shrubby undergrowth encroaches, and it can make moving through unmanaged savannas difficult. Additionally, because the soils are often poorly drained, wet grasslands and bogs are common. In the range of Longleaf Pine savanna, winters are mild, and the temperature rarely drops below freezing. Summers are hot and humid, with daily highs around 90°F (32°C). Most of the rain falls in the spring and summer months, with 43–68 in. (1,090–1,750mm) accumulating annually.

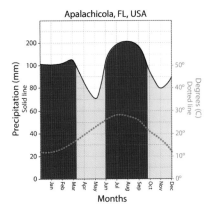

Apalachicola, FL, USA

The dominant canopy tree is Longleaf Pine, a fire-tolerant and slow-growing species. Slash Pine, Sand Pine, Pond Pine, and Virginia Pine are all also present, though rarely co-dominant. Loblolly Pine is a fast-growing but fire-susceptible species that often outcompetes and displaces Longleaf Pine in places where fire has been suppressed. Loblolly is also common around the wetter, boggier sections of Longleaf Pine savanna. The mid-story is generally sparse and includes Sweetbay Magnolia, Wax Myrtle, Sweet Gum, and Red Maple. The shrub layer, on the other hand, can be quite dense, especially after a few years without fire. Saw Palmetto is almost always present and is often joined by Fetterbush, Gallberry, and the dwarf form of live oak, among many other plants.

The herbaceous layer of the Longleaf Pine savanna is by far the most diverse vegetative component. In frequently burned areas, the ground cover is dominated by three-awns and Little Bluestem, muhly grasses, dropseeds, and a variety of sedges and forbs, including many orchid

An example of the Longleaf Pine savanna habitat type in Angelina National Forest, e. Texas, US.
© SAM WOODS, TROPICAL BIRDING TOURS

species. The poorly drained, acidic soils of the Longleaf Pine savanna are perfect growing conditions for a wide array of carnivorous plants, and the vast majority of Nearctic carnivorous plant species are found here, including many pitcher plants, sundews, and bladderworts. The famous Venus Flytrap is found nowhere else.

The Longleaf Pine savanna has been extensively damaged through widespread logging, development, and fire suppression. Today, only 5% of this habitat remains intact. Active and intensive management is underway in many areas, especially those with the endangered Red-cockaded Woodpecker, and old fire regimes are being restored through regular prescribed burns.

WILDLIFE: In much of the e. Nearctic, the mammal communities, especially of large and conspicuous mammals, are fairly similar. Virginia Opossum, White-tailed Deer, Bobcat, American Black Bear, Striped Skunk, and Common Raccoon are all abundant and noticeable in Longleaf Pine savanna. The less widespread Nine-banded Armadillo favors this habitat, especially areas with well-drained, sandy soils. Invasive Feral Hogs are a common feature and cause great damage to sensitive understory plants and terrestrial fauna.

The bird community is more distinctive. Among the common e. Nearctic birds, Wild Turkey, Mourning Dove, Barred Owl, Red-bellied and Pileated Woodpeckers, White-breasted Nuthatch, Eastern Bluebird, Eastern Towhee, and Eastern Meadowlark are all particularly abundant. This habitat is one of the few areas still containing stable populations of Northern Bobwhite. In summer, returning migrants make up a significant percentage of the avifauna. The abundant insect life provides food for Swallow-tailed and Mississippi Kites, Common Nighthawk, Chuck-will's-widow, Eastern Wood-Pewee,

Great Crested Flycatcher, White-eyed Vireo, Blue-gray Gnatcatcher, Common Yellowthroat, Northern Parula, Prairie and Pine Warblers, Indigo Bunting, Blue Grosbeak, and Summer Tanager.

Red-cockaded Woodpecker (IS), Brown-headed Nuthatch (IS), and Bachman's Sparrow (IS) are three flatwoods specialists that are rare to absent in other habitats. The endangered Red-cockaded Woodpecker nests only in mature Longleaf Pine stands, where it excavates a nest cavity in a living tree. Under these nest cavities, the woodpecker drills a number of sap wells, the sticky pitch from which helps deter snakes and other nest predators.

Reptiles and amphibians are perhaps the most exciting vertebrates found in this habitat. The endangered Gopher Tortoise is found mostly in Longleaf Pine savanna, where it excavates large burrows, which are utilized by more than 300 other species of reptiles, mammals, and invertebrates. Even Bachman's Sparrows have been observed disappearing into Gopher Tortoise burrows when escaping predators. Among the more than 30 species of snakes in this habitat, the spectacular Scarlet Snake, imposing Eastern Diamondback Rattlesnake, dainty Pygmy Rattlesnake, and endemic Pine Woods Snake are all regularly seen. Eastern Indigo Snake, the largest snake in the Nearctic, is also found here, though it is endangered. Eastern and Slender Glass Lizards, a pair of bizarre, legless, snakelike lizards, can be seen hunting for insects among the dry pine needles. There are dozens of amphibian species, with Flatwoods Salamander, Striped Newt, Ornate Chorus Frog, and Squirrel Tree Frog among the specialists. In 2018, a new species of aquatic salamander was described that lives in boggy flatwoods of sw. Alabama and Florida; at over 24 in. (60cm) in length, the Reticulated Siren is one of the largest salamanders in the world.

DISTRIBUTION: Longleaf Pine savanna is found in low-lying areas of the se. US coastal plain, from Virginia south into much of peninsular Florida, west along the Gulf of Mexico to e. Texas, and north along the Mississippi River valley through Arkansas to far s. Illinois. It is bounded to the north by NEARCTIC TEMPERATE DECIDUOUS FOREST and to the west by TALLGRASS PRAIRIE.

WHERE TO SEE: Apalachicola National Forest, Florida, US; De Soto National Forest, Mississippi, US; Croatan National Forest, North Carolina, US.

Red-Cockaded Woodpecker is found almost exclusively in Longleaf Pine savanna (photo: se. Texas, US). © SAM WOODS, TROPICAL BIRDING TOURS

Flatwoods salamanders are amphibious specialists of Longleaf Pine savanna.
© USFWS, PUBLIC DOMAIN

HIGH-ELEVATION PINE WOODLAND

IN A NUTSHELL: Open high-elevation coniferous forests growing on dry, rocky soils in the w. Nearctic.
Habitat Affinities: Upper levels of EURASIAN MONTANE CONIFER FOREST. **Species Overlap:** MONTANE MIXED-CONIFER FOREST; NEARCTIC MONTANE SPRUCE-FIR FOREST; NEARCTIC ALPINE TUNDRA.

DESCRIPTION: These sparse woodlands are found near the tree line on dry, rocky ridges and slopes in the mountains of the w. Nearctic and are populated by only the hardiest of trees. The trees are short in stature, rarely exceeding 35 ft. (11m) in height, and in persistent and intense winds are often reduced to the stunted form known as krummholz (see sidebar 2.3). In the zone where these woodlands are found, winters are long, and temperatures regularly go as low as -10°F (-23°C). In the hot, dry summer, daytime temperatures reach 90°F (32°C), though nighttime temperatures can still dip below freezing. Precipitation is generally scarce, with average rainfall ranging from 10 in. (250mm) in Great Basin Bristlecone Pine woodland in Nevada to 35 in. (900mm) in Foxtail Pine woodland in California. This habitat is a stark and emblematic feature of the high-mountain West, and spending time here usually requires a hike through some stunning mountain scenery. While the vegetation presents no obstacles, the steep slopes and loose, rocky soil can make exploration difficult.

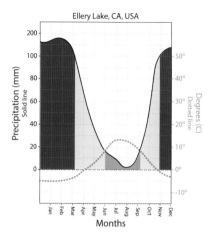

Ellery Lake, CA, USA

High-elevation pine woodlands are open stands composed of five species in the white (five-needled) pine group that are often referred to as the "high five": Great Basin Bristlecone Pine, Rocky Mountain Bristlecone Pine, Foxtail Pine, Whitebark Pine, and Limber Pine. With increasing elevation, tree size diminishes and distance between trees increases. These woodlands are generally monotypic, as each species has a disjunct range, though Limber Pine is occasionally mixed with the more restricted species. There is no notable mid-story, and ground cover is mostly bare rock, with only 5–25% vegetative ground cover on average. The sparse ground cover comprises small woody shrubs such as manzanitas, mountain mahoganies, junipers, gooseberries and currants, and bitterbrush. Herbaceous ground cover is diverse, due to the wide elevational and geographic ranges covered by this habitat, but usually includes a few grasses, especially fescues.

High-elevation pine woodlands at Targhee National Forest in Idaho, US.
© IAIN CAMPBELL, TROPICAL BIRDING TOURS

The high five are all considered keystone and foundational species that heavily influence the structure, diversity, and stability of high-montane communities and are major sources of food for alpine animals. Some of these pines are impressively ancient—all high five species have specimens known to be over 1,000 years old. An individual Great Basin Bristlecone Pine, at a staggering ~4,900 years old, is considered the oldest single organism on the planet. These slow-growing, long-lived woodlands are heavily threatened, despite their perceived toughness. Non-native White Pine Blister Rust has been decimating this habitat throughout the w. Nearctic, as have large outbreaks of Mountain Pine Beetle. Climate change and the continued march of lower-elevation habitats upslope also threaten to displace or subsume these ridgetop species.

Clark's Nutcracker is an important avian seed disperser for the trees of high-elevation pine woodlands. © BEN KNOOT, TROPICAL BIRDING TOURS

WILDLIFE: A sparse and rocky habitat, high-elevation pine woodland is home to relatively few animals. American Pikas live among the boulder-piled slopes, as do Yellow-bellied Marmots. Golden-mantled Ground Squirrel, American Red Squirrel, and Least Chipmunk all feed on the sizable seeds of Whitebark and Limber Pines. Throughout the year, squirrels create large middens of pine nuts as a winter food source. In the fall, these stockpiles are often raided by American Black Bear and Brown (aka Grizzly) Bear, both of which reap a large caloric windfall at the squirrels' expense.

The bird communities up here are made almost entirely of granivorous (seed-eating) species that survive on the ample pine nuts. Pine Grosbeak, Cassin's Finch, Red Crossbill, Gray-crowned Rosy-Finch, and Dark-eyed Junco are all regularly found in this habitat. Clark's Nutcracker is a major seed disperser in high-elevation pine woodlands, and the survival of Whitebark Pine is especially dependent on it. Nutcrackers fervently cache upwards of 90,000 seeds in a single year, often in far-flung locations. While they remember the location of most seeds, forgotten pine nuts germinate to form woodlands on isolated peaks and after fires.

DISTRIBUTION: High-elevation pine woodlands are scattered across the w. Nearctic, largely in the United States. The ranges of individual tree species vary; the highly restricted Foxtail Pine is found in a few disjunct sites in California, while the widespread Whitebark Pine occurs in the Rocky Mountains, Cascade Range, Sierra Nevada, and the Great Basin ranges, from British Columbia south to Arizona and New Mexico. These forests grow only at higher elevations, occurring as low as 6,000 ft. (1,800m) and as high as 12,000 ft. (3,650m). At their lower elevational limit they are replaced by NEARCTIC MONTANE SPRUCE-FIR FOREST, Ponderosa Pine–dominated MONTANE MIXED-CONIFER FOREST, or PINYON-JUNIPER WOODLAND. At their upper elevational limit, they transition to bare, rocky slopes, permanent snow, or NEARCTIC ALPINE TUNDRA.

WHERE TO SEE: White Mountains, California, US; Lincoln National Forest, New Mexico, US; Frank Church River of No Return Wilderness, Idaho, US.

As human beings, we love superlatives. While the differences between a record holder and second place may be almost indistinguishable, we will go to great lengths to measure that difference regardless. Superlative organisms are bizarre, impressive, inspiring, and show us the limits that living creatures can reach. Few regions hold as many hyperbolic organisms as the various coniferous habitats of California.

WORLD'S TALLEST TREE: Coast Redwood "Hyperion," 380 ft. (116m)

WORLD'S MOST MASSIVE TREE: Giant Sequoia "General Sherman," 4.2 million lb. (1.9 million kg), 52,000 cubic ft. (15,000m³)

WORLD'S OLDEST TREE: Great Basin Bristlecone Pine, ~4,900 years old

WORLD'S HEAVIEST CONIFER CONE: Coulter Pine, 11 lb. (5kg)

WORLD'S LONGEST CONIFER CONE: Sugar Pine, 23 in. (58cm)

LODGEPOLE PINE FOREST

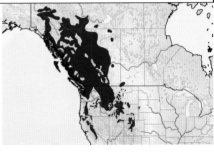

IN A NUTSHELL: A successional forest composed of uniformly aged Lodgepole Pines found at middle elevations in the mountainous West. **Habitat Affinities:** EURASIAN MONTANE CONIFER FOREST; SIBERIAN LARCH FOREST. **Species Overlap:** MONTANE MIXED-CONIFER FOREST; NEARCTIC MONTANE SPRUCE-FIR FOREST; ASPEN FOREST AND PARKLAND.

DESCRIPTION: Lodgepole Pine forest is a common habitat in mountainous areas of the w. Nearctic. It generally occurs at 8,500–10,000 ft. (2,600–3,000m), in areas where the climate is wet and cool with a very brief frost-free period in summer. Lodgepole Pine forest is a near monoculture that colonizes after major fires or other large-scale disturbances. Characterized by dense stands of same-age Lodgepole Pines, this forest has few other canopy trees. If soil conditions are favorable, small pockets of other early successional trees like Quaking Aspen may be found. The shrub and forb layers are also sparse and depauperate. Manzanitas, huckleberries, ceanothuses, and meadowsweets are the most common shrubs and can be abundant in early successional stages of this forest.

Despite the relatively sparse understory, this habitat can be surprisingly difficult to move through. Exposed to frequent fire and high winds, the forest floor is often a maze of fallen trunks.

Younger generations of Lodgepole Pines grow on top of old burns. Short forests with the burned trunks of previous generations still intact and standing are a common landscape feature. Lodgepole Pine has fire-adapted serotinous cones, which open and release seeds only after exposure to fire or other heat. If left undisturbed, Lodgepole Pine forest is replaced by NEARCTIC MONTANE SPRUCE-FIR FOREST or Ponderosa Pine–dominated MONTANE MIXED-CONIFER FOREST. But with the increased frequency and scale of fires in the w. Nearctic, the extent of Lodgepole Pine forest has increased.

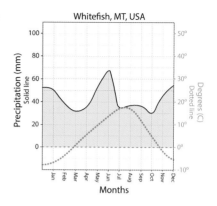

Whitefish, MT, USA

WILDLIFE: The high elevation, lack of floristic diversity, poor understory, and intense fire associated with this habitat mean it is relatively wildlife-poor. There is enough food to support a variety of small mammals, especially during years with significant cone production. Conspicuous small mammals include American Red and Douglas's Squirrels, Least Chipmunk, Golden-mantled Ground Squirrel, and Snowshoe Hare. Elk, Mule Deer, and Bighorn Sheep will use Lodgepole Pine forest for cover but spend significant amounts of time here only in areas with developed understory. The same is true of American Black Bear. Smaller predators like American Marten and Long-tailed Weasel make use of this habitat throughout the year, while Wolverine uses it for hunting and denning in winter.

Dense Lodgepole Pine forest in Yellowstone National Park, Wyoming, US.
© BEN KNOOT, TROPICAL BIRDING TOURS

Golden-mantled Ground Squirrel is a conspicuous mammal in rocky and talus slopes found within clearings in Lodgepole Pine forests. © BEN KNOOT, TROPICAL BIRDING TOURS

Red-breasted Nuthatch breeds in a variety of North American pine forests, including Lodgepole Pine forest. © PHIL CHAON, TROPICAL BIRDING TOURS

The bird communities of Lodgepole Pine are similar to those of surrounding habitats but lack species that need dense understory or larger trees. Widespread species like American Robin, Hermit Thrush, Western Wood-Pewee, Canada Jay, Ruby-crowned Kinglet, Yellow-rumped Warbler, Dark-eyed Junco, and Pine Siskin are all found. Northern Goshawk nests locally, and high-elevation species like Clark's Nutcracker and Pine Grosbeak also occur.

One notable feature of Lodgepole Pine forest is the abundance of standing dead trees, which provide excellent habitat for cavity-nesting birds. With frequent fire and large die-offs due to Mountain Bark Beetle, this habitat is heavily used by American Three-toed (IS) and Black-backed Woodpeckers. Other woodpecker species are also common, and secondary cavity nesters like Mountain Bluebird, Black-capped Chickadee, Merlin, and even Northern Hawk Owl take advantage of available cavities, a normally scarce resource.

DISTRIBUTION: This habitat occurs at middle elevations in the Rocky Mountains of Canada and the United States, from Alberta and British Columbia south through Colorado, and in the e. Cascades south through Oregon, along the Sierra Nevada, and on isolated mountaintops to s. California and Nevada.

WHERE TO SEE: Glacier National Park, Montana, US; Yellowstone National Park, Wyoming, US.

NEARCTIC TEMPERATE RAINFOREST

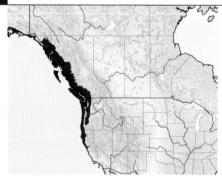

IN A NUTSHELL: An extremely wet and mossy Pacific coastal coniferous forest with towering trees. **Habitat Affinities:** NEOTROPICAL MAGELLANIC RAINFOREST; AUSTRALASIAN GONDWANAN CONIFER RAINFOREST; NEOTROPICAL VALDIVIAN RAINFOREST. **Species Overlap:** MONTANE MIXED-CONIFER FOREST; NEARCTIC MONTANE SPRUCE-FIR FOREST.

DESCRIPTION: Nearctic temperate rainforest, a towering forest laden with epiphytes, hugs a narrow strip along the Pacific coast of North America from c. California to Alaska. It is dominated by a few massive conifer species and has little mid-story and a thick understory layer of ferns, mosses, and evergreen shrubs. Temperate rainforest grows in a wet, stable climate, rarely colder than 32°F (0°C) and only occasionally warmer than 75°F (25°C) in a given year. Average precipitation is around 80 in. (2,000 mm) per year, but some areas, such as the Olympic Peninsula of Washington, receive

A characteristic abundance of fallen trees makes temperate rainforest often difficult to walk through (photo: Sol Duc Falls, Olympic Peninsula, Washington, US). © IAIN CAMPBELL, TROPICAL BIRDING TOURS

upwards of 170 in. (4,300 mm) annually. This region has two distinct seasons—a long, wet rainy season from October to May, and a short, dry, foggy summer from June to September. During the dry season, this forest receives 7–12 in. (170–300mm) of precipitation from fog alone. Despite the heavy fog, this region is best visited from April to September, outside the worst of the winter rains.

The temperate rainforest is perhaps the most dramatic of the Nearctic habitats. The multilayered canopy, regularly soaring upwards of 300 ft. (90m), is dominated by Douglas-fir, Sitka Spruce, Western Hemlock, and Western Red Cedar. From the Oregon border south, Coast Redwood is co-dominant with Douglas-fir. The massive Coast Redwood can reach 380 ft. (115.8m) in height and 29.2 ft. (8.9m) in diameter, placing it among the largest trees on earth. The

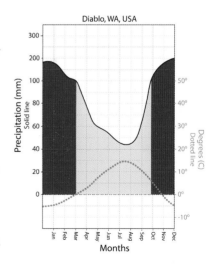

towering canopy trees are often laden with epiphytic mosses, lichens, and ferns. Underneath is a sparse mid-story layer made up of conifer saplings and smaller shade-tolerant deciduous trees like Bigleaf Maple, Vine Maple, and dogwoods. The forest floor usually supports a dense assemblage of Sword Fern, Lady Fern, rhododendrons, Salmonberry, Evergreen Huckleberry, Devil's Club, Salal, mosses, and fallen logs. This thick, often saturated understory is mostly 3–6 ft. (1–2m) tall and can be difficult to walk through.

The massive timber of the temperate rainforest made it a valuable commodity during the 19th and most of the 20th century, when huge tracts (especially Coast Redwood forests) were lost to logging. Since the latter half of the 20th century, there has been a concerted public effort to conserve the temperate rainforest, and most of the remaining tracts of old-growth rainforest are found on protected public lands. Some of these massive tracts, including the Tongass National Forest of se. Alaska and the Great Bear Rainforest of British Columbia, cover millions of acres. Apart from logging, these forests are particularly susceptible to drought, and climate-change-driven variations in average rainfall and maximum average temperature pose a major threat.

WILDLIFE: The temperate rainforest is home to a large array of the Nearctic's charismatic megafauna. The diminutive Columbian Black-tailed Deer is common throughout the range, and Roosevelt Elk is found in large herds in the habitat's southern extent. Puma, Bobcat, and American Black Bear are common throughout, and Gray Wolf and Brown (aka Grizzly) Bear are still common in British Columbia and Alaska. The temperate rainforest is home to the Kermode Bear, a large subspecies of American Black Bear famous for having a spectacular white color morph. The region also contains the massive Kodiak subspecies of Brown Bear, capable of reaching 10 ft. (3m) in length and upwards of 1,500 lb. (680kg). Watching these bears fatten themselves on salmon along the Frazer River on Kodiak Island, Alaska, is one of the great North American wildlife spectacles. Apart from the megafauna, the temperate rainforest is also home to smaller predators like Fisher, American Marten, Gray Fox, and Short-tailed Weasel. Other common mammals include Humboldt's Flying Squirrel, American Red and Douglas's Squirrels, Townsend's Chipmunk, Common Raccoon, and numerous vole species, including the Red Tree Vole (IS), which spends its entire life in the canopy eating Douglas-fir needles.

The temperate rainforest's relatively limited set of birds includes a few specialties and some widespread but uncommon species. Many of the smaller songbirds utilize the upper stratum of the canopy and are best detected by ear. Band-tailed Pigeon, Brown Creeper, Golden-crowned Kinglet, Townsend's and Hermit Warblers, Red Crossbill, and Pine Siskin can all be found feeding hundreds of feet up in the massive canopy and are best looked for along the forest edge, where they will often venture to

Douglas's Squirrel foraging on fallen timber within the temperate rainforest on the Olympic Peninsula, Washington, US. © IAIN CAMPBELL, TROPICAL BIRDING TOURS

The haunting song of Varied Thrush is a characteristic of the summertime soundtrack of the temperate rainforest of coastal nw. United States. © BEN KNOOT, TROPICAL BIRDING TOURS

lower levels. Lower down within the forest is a nice variety of raptors, woodpeckers, corvids, and songbirds, including Northern Goshawk, Merlin, Red-breasted Sapsucker, Pileated and Hairy Woodpeckers, Northern Flicker, Pacific-slope Flycatcher, Gray and Steller's Jays, Common Raven, Northwestern Crow, and Chestnut-backed Chickadee. Ruffed Grouse, Dusky Grouse, Pacific Wren, and Hermit, Swainson's, and Varied Thrushes are all common in the understory and perhaps more than any other set of birds contribute to the unique and haunting soundscape of this habitat. The two most famous avian residents are the endangered Northern Spotted Owl and Marbled Murrelet (IS), which are icons of the campaign to protect this habitat in the early 1990s. The Marbled Murrelet, the last species of bird in the United States to have its nest discovered, was found nesting on broad limbs high in the redwood canopy during the 1980s. These diminutive and fascinating seabirds nest exclusively in old-growth temperate rainforest but are most easily observed out on the open ocean.

The region is also notable for the high diversity of endemic salamanders, including 19 species that are found almost exclusively in temperate rainforest. The Wandering Salamander is commonly found on the forest floor but may also live its entire life high up in the trees. One individual was spotted 200 ft. (60m) up in a Coast Redwood, walking alongside a Marbled Murrelet nest.

Endemism: There are several small areas that contain endemic small mammals, amphibians, plants, and insects. The Olympic Peninsula is home to the Olympic Marmot, Olympic Torrent Salamander, Olympic Mudminnow, and a dozen or so endemic insects. Many of the coastal islands in British Columbia and Alaska have endemic subspecies. Marbled Murrelet is an endemic breeder to this habitat.

DISTRIBUTION: The temperate rainforest stretches for nearly 2,000 mi. (3,200km) along the Pacific coast of North America, from c. California through coastal Oregon, Washington, and British Columbia to the eastern end of Alaska's Kodiak archipelago. Found in a series of Pacific coastal mountain ranges, it is bounded to the west by the Pacific Ocean and to the east by the Ponderosa Pine–dominated MONTANE MIXED-CONIFER FOREST. Water is the key limiting factor in this habitat, and temperate rainforest does not occur in any areas without sufficient rain and fog or with unfavorably high average temperatures. Temperate rainforest is a fairly contiguous habitat where it occurs, only occasionally interrupted by intertidal salt marshes, riparian forests, or rocky coast or dune habitats.

WHERE TO SEE: Humboldt Redwoods State Park/Redwoods National Park, California, US; Olympic National Park, Washington, US; Great Bear Rainforest, British Columbia, Canada; Tongass National Forest, Alaska, US.

PINYON-JUNIPER WOODLAND

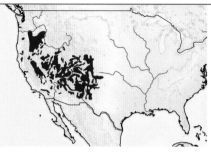

IN A NUTSHELL: An open woodland of pinyons (pines) and junipers found in arid habitats; includes elements of grassland and xeric shrub communities. **Habitat Affinities:** MEDITERRANEAN AND DRY CONIFER FOREST; PALEARCTIC SUBTROPICAL SAVANNA. **Species Overlap:** SAGEBRUSH SHRUBLAND; MONTANE MIXED-CONIFER FOREST; NEARCTIC DESERT GRASSLAND; CHIHUAHUAN DESERT SHRUBLAND.

DESCRIPTION: Pinyon-juniper woodland is one of the major habitats of the Great Basin and broader intermountain West of the United States (and into Mexico). Occurring in a narrow elevational band at 5,000–8,000 ft. (1,500–2,400m) in dry mountains and foothills, this habitat experiences an extreme range of temperatures and receives little rainfall, just 12–16 in. (300–400mm) annually. This a short, shrublike woodland, with trees rarely taller than 25 ft. (7.5m). Tree density is variable, and canopy cover ranges from about 50% in the northwest to a savanna-like 15% in the southeast. Regardless, this habitat is open enough to move through easily and is readily explored.

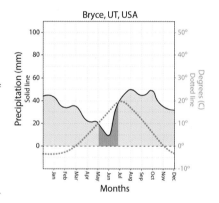

The tree component of pinyon-juniper woodland consists almost entirely of pinyon pines and junipers. The species composition and structure vary throughout the range (mostly on a northwest–southeast gradient), but key tree species include Western, Utah, Single-seeded, Alligator, and Rocky Mountain Junipers in association with Single-leaf or Two-needled Pinyons. In general, juniper species are dominant at lower elevations, while at the upper end of the elevational range the majority of the trees are pinyons.

The understory component of pinyon-juniper woodland also varies along a northwest–southeast gradient and is strongly influenced by adjacent habitats. In the northwest, pinyon-juniper woodland takes on characteristics of SAGEBRUSH SHRUBLAND, with a dense understory of Big Sagebrush, bitterbrush, rabbitbrush, and mountain mahoganies (*Cercocarpus*), with scattered perennial tussock grasses. In the southeast, the understory is composed of warm-season grasses characteristic of SHORTGRASS PRAIRIE and NEARCTIC DESERT GRASSLAND, such as Blue, Black, Hairy, and Side-oats Gramas. In the s. Rockies and on the Colorado Plateau, montane shrubs like Gambel Oak constitute a significant portion of the ground cover.

As a result of overgrazing and the suppression of high-frequency, low-intensity fires, pinyon-juniper woodlands have expanded rapidly over the past 150 years. This expansion is a threat to more restricted habitats; in particular it is detrimental to Greater Sage-Grouse, and pinyon-juniper removal is actively taking place in many protected areas.

Shrubby juniper woodland in n. Arizona, US.
© BEN KNOOT, TROPICAL BIRDING TOURS

WILDLIFE: The wildlife of pinyon-juniper woodlands is heavily influenced by adjacent habitats and holds many of the mammal species characteristic of the dry intermountain West. Mountain and Desert Cottontails, Black-tailed Jackrabbit, Mule Deer, Desert Bighorn Sheep, and Elk all feed on the understory vegetation. Pinyon species produce large and nutritious seeds (pine nuts), feeding a wide array of squirrels, chipmunks, mice, and woodrats. Pinyon Mouse is a specialist of this habitat. Dense junipers provide good cover not available in adjacent areas, and large herbivores of grassland and SAGEBRUSH SHRUBLAND habitats frequently seek shelter among them.

Pinyon-juniper woodlands hold an interesting set of bird species, from high-mountain to desert specialists, occurring here depending on time of year. Pinyons and junipers are both important food sources, and in the winter large flocks of frugivorous birds can be found gorging on juniper berries. American Robin; Eastern, Western, and Mountain Bluebirds; Townsend's Solitaire; and Cedar and Bohemian Waxwings are all species that take advantage of this berry bonanza. The large and nutritious pinyon seeds are animal-dispersed, unlike

In the w. United States, Pinyon Jay forages on pinyon seeds and is one of their principal dispersers and is therefore tied to this habitat.
© KEN BEHRENS, TROPICAL BIRDING TOURS

the winged, wind-dispersed pine seeds of many species. Pinyon pine nuts are important food for Woodhouse's Scrub-Jay, Steller's Jay, and Clark's Nutcracker, but it is the Pinyon Jay (IS) that truly specializes in this food source. Arriving in flocks that sometimes number in the hundreds, Pinyon Jays gorge on pine nuts and cache the rest for later. Forgotten pine nuts later germinate, and the movements of Pinyon Jay flocks has been shown to directly influence landscape-level genetics of pinyon forests. While the bird communities of pinyon-juniper woodland vary in close association with adjacent habitats, there are several species closely tied to this habitat. Mountain Quail, Black-chinned Hummingbird, Ferruginous Hawk, Ash-throated Flycatcher, Black-throated Gray Warbler, and Scott's Oriole use pinyon-juniper woodland preferentially over other habitats. Virginia's Warbler and Gray Flycatcher (IS) are near specialists, and Gray Vireo (IS) and Juniper Titmouse (IS) are true specialists, breeding only in pinyon-juniper woodlands.

As pinyon-juniper woodland has a very dry climate with exceedingly cold winters, very few amphibians are found here. The reptile communities vary widely throughout the range, but common and conspicuous reptiles include Western Whiptail, Gilbert's Skink, Sagebrush Lizard, Common Side-blotched Lizard, Greater Short-horned Lizard, Gopher Snake, and Prairie Rattlesnake.

DISTRIBUTION: Pinyon-juniper woodlands occupy areas of the Great Basin, the Colorado Plateau, the Rocky Mountains, and the Sonoran and Chihuahuan Deserts of the United States and Mexico, where it extends southward to 18°N in the states of Jalisco and Puebla. The habitat is bounded by coniferous forest (usually MONTANE MIXED-CONIFER FOREST) at its upper elevational limit, and at its lower elevational limit it transitions to SAGEBRUSH SHRUBLAND, NEARCTIC DESERT GRASSLAND, SHORTGRASS PRAIRIE, or CHIHUAHUAN DESERT SHRUBLAND. An extension of the pinyon-juniper woodlands is found outside of mountainous areas on limestone breaks in the Great Plains.

WHERE TO SEE: Lava Beds National Monument, California, US; Grand Canyon National Park, Arizona, US; Colorado National Monument, Colorado, US.

Pinyon-juniper among the red rocks of Arizona in wintertime.
© BEN KNOOT, TROPICAL BIRDING TOURS

MADREAN PINE-OAK WOODLAND

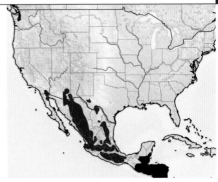

IN A NUTSHELL: A montane pine- and oak-dominated woodland with a variable canopy found from the far sw. United States through Mexico and Guatemala. **Habitat Affinities:** NEOTROPICAL PINE-OAK WOODLAND; NEOTROPICAL MIXED CONIFER FOREST; INDO-MALAYAN PINE FOREST. **Species Overlap:** MONTANE MIXED-CONIFER FOREST; NEOTROPICAL CLOUD FOREST; NEOTROPICAL DRY DECIDUOUS FOREST.

DESCRIPTION: Madrean pine-oak woodland is the dominant habitat in the mountains of the far sw. United States, w. Mexico, and Guatemala. It consists of pines and evergreen broadleaf trees. This habitat generally occurs at 5,000–9000 ft. (1,500–2,750m). The climate is temperate to subtropical, with winter lows near 40°F (4°C) and summer highs close to 90°F (32°C). The amount of precipitation is highly dependent on elevation and slope aspect, and Madrean pine-oak woodland receives anywhere from 16 in. (500mm) to 100 in. (2,500mm) of rain per year.

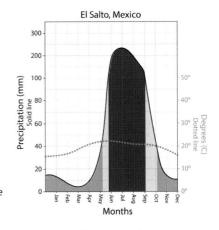

El Salto, Mexico

Structurally, Madrean pine-oak woodland varies from open to moderately dense, with canopy coverage in the range of 10–40%. The height of the canopy is highly variable and dependent on dominant species, reaching 50–100 ft. (15–30m). The mid-canopy and tall shrub layers are fairly dense, but the understory is usually open, and this is a pleasant habitat for hiking, especially in shady canyons.

Botanically, this habitat is incredibly diverse, containing over 5,000 species, more than a quarter of the plant species of Mexico. There is high diversity not only in herbaceous plants but among canopy trees as well—over 20 conifer species and nearly 50 oak species are present. No single canopy species is found throughout the entire range, but several of the most widespread and common canopy trees are Apache Pine, Chihuahuan Pine, Montezuma Pine, Hartweg's Pine, Engelmann Oak, and Uricua Oak. At higher elevations, the canopy is pine-dominated, and occasionally fir trees are present. The mid-story is open to moderately dense and usually comprises smaller oaks and madrones. The understory varies significantly with rainfall and canopy density. Woodlands with a more open canopy have understories dominated by perennial grasses like muhly grasses, lovegrasses, and Feathergrass. Small shrubby oaks are also a common feature of the understory, along with manzanitas and silktassels.

Above: **An example of Madrean pine-oak woodland in Arizona, sw. US.** © BEN KNOOT, TROPICAL BIRDING TOURS

Left: **An impressive diversity of carnivores occupies these woodlands, including this Bobcat in the United States but also Jaguar and Jaguarundi in Mexico and further south.** © PETER KNOOT

WILDLIFE: The mammalian fauna has a strong Nearctic influence, though some Neotropical species are found here as well. Carnivores include Puma, American Black Bear, Jaguarundi, Bobcat, and even Jaguar—and some Jaguars still occasionally cross into the United States. White-nosed Coati is an abundant and conspicuous omnivore, along with Ringtail, and several species of skunks. The diminutive and incredibly rare Volcano Rabbit, the only member of its genus, is endemic to a few high peaks in the Trans-Mexican Volcanic Belt.

Over 600 species of birds are found in this mountain habitat, and there are many pockets of endemism scattered throughout. The Madrean pine-oak woodland avian fauna has a strong Neotropical influence, and many tropical families including trogons, woodcreepers, euphonias, and motmots reach their northern

In Mexico during winter, Red-faced Warbler can be found roaming Madrean pine-oak woodland within diverse, mixed warbler flocks. © PHIL CHAON, TROPICAL BIRDING TOURS

extent here, as does Gray Silky-flycatcher. Hummingbirds, sparrows, jays, and tyrant flycatchers are all well represented and show a large degree of endemism. In winter, large flocks of migrant warblers contain Red, Red-faced (IS), Hermit, Black-throated Gray, Grace's, and Golden-cheeked Warblers, among others. Thick-billed and Maroon-fronted Parrots, the sole members of the genus *Rhynchopsitta*, both thrive in this habitat on a specialized diet of pine nuts. Olive Warbler is another taxonomic oddity of the Madrean pine-oak forest and the sole member of its family. Its Spanish common name, Ocotero, refers to its association with Ocote, or Montezuma Pine, one of the habitat's key species.

Reptile diversity is high in this habitat, and nearly 400 species are present. The spiny lizards (*Sceloporus*), anoles, and constrictors are particularly well represented, comprising more than half of the species found here. Amphibian diversity is also high, with over 200 species found, especially in the moister forests at high elevations. The most notable feature of the amphibian communities is the remarkable degree of endemism among salamanders. More than 15% of the world's salamander species are found in Madrean pine-oak woodland, and 40 species are found nowhere else. Isolated moist refugia scattered throughout the mountains have contributed to the level of endemism, and new species are still regularly being discovered.

Endemism: Patches of Madrean pine-oak woodland are geographically disjunct on several major mountain ranges and have been further separated during recent ice ages. There are multiple EBAs (see "Endemic Bird Areas" in the Introduction), including the woodlands of Oaxaca, the Trans-Mexican Volcanic Belt, Sierra Madre del Sur, and the Chiapas-Guatemalan Highlands.

DISTRIBUTION: Madrean pine-oak woodland is a montane ecosystem common through the mountains of the Sierra Madre Occidental, Oriental, del Sur, and del Guatemala, as far south as El Salvador. It is found in isolated pockets as far north as se. Arizona and w. Texas in the United States. Below 5,000 ft. (1,500m) these woodlands are replaced by NEARCTIC DESERT GRASSLAND, NEOTROPICAL DRY DECIDUOUS FOREST, and COLUMNAR CACTUS DESERT, among other habitats.

WHERE TO SEE: Chiricahua Mountains, Arizona, US; La Cumbre, Oaxaca, Mexico; Todos Santos Cuchumatan, Huehuetenango, Guatemala.

CHIHUAHUAN DESERT SHRUBLAND

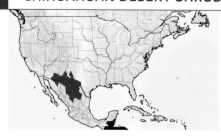

IN A NUTSHELL: A Creosote- and mesquite-dominated desert shrubland with summer monsoon rains and cold, dry winters. **Habitat Affinities:** PALEARCTIC SEMIDESERT THORNSCRUB; PALEARCTIC HOT SHRUB DESERT; AUSTRALASIAN MULGA WOODLAND AND ACACIA SHRUBLAND. **Species Overlap:** NEARCTIC DESERT GRASSLAND; COLUMNAR CACTUS DESERT; SHORTGRASS PRAIRIE.

DESCRIPTION: Along with the NEARCTIC DESERT GRASSLAND, this is one of the two main habitats of the vast Chihuahuan Desert of the sw. United States and n. Mexico. The abundance of desert grasses and lack of large cacti produce a landscape vastly different from that of the neighboring Sonoran Desert. In this short and open shrubland, with only tall spikes of flowering agaves breaking the uniform expanse of Creosote bushes, you can often see for miles.

Climatically, the combination of cold, dry winters and intense summer monsoons separates the Chihuahuan Desert from other Nearctic deserts. While occurring at the same latitudes as the

Yuccas and agaves are prominent members of the plant community within Chihuahuan Desert shrubland in Arizona, US. © BEN KNOOT, TROPICAL BIRDING TOURS

El Paso, TX, USA

(climate chart: Precipitation (mm) on left axis, Degrees (C) on right axis, Months Jan–Dec on x-axis; solid line = precipitation, dotted line = temperature)

The abundance of local reptiles means plentiful food for Greater Roadrunner. © BEN KNOOT, TROPICAL BIRDING TOURS

COLUMNAR CACTUS DESERT of Sonora and Baja California, the higher elevation and distance from the ocean means the Chihuahuan Desert shrublands have comparatively frigid winters. The region experiences temperatures below freezing nearly one-third of the year, and winter lows regularly reach 20°F (-6°C). In summer, temperatures are normally around 95–104°F (35–40°C). The most vibrant time of year is July to September, when monsoons deposit 90% of the region's annual precipitation. Cooler daytime temperatures, rapid plant growth, and renewed animal activity make this the best time to visit the region.

As in many of the s. Nearctic desert shrublands, the dominant feature on the landscape is the omnipresent Creosote. Covering huge swaths of land, this resinous, odoriferous shrub is incredibly efficient at securing water, often to the point of precluding the growth of nearby plants. Its ability to dominate the floristic landscape has earned it the common name *gobernadora*, or "governess," in Mexico. Other members of the moderately dense shrub layer of Chihuahuan Desert shrubland are mesquites, acacias, and American Tarbush.

Spiky forbs like yuccas, Sotol, and Lechuguilla are prominent on the landscape, especially when they bloom, producing flowering stalks as tall as 12 ft. (3.7m). While cacti are not as prominent here as in the columnar cactus desert to the west, the Chihuahuan Desert shrublands are nevertheless home to over 25% of the world's cactus species. Botanists theorize that cacti originated in this region and radiated throughout the Americas from this epicenter of diversity. The understory shares many grasses with the adjacent Nearctic desert grassland, including Black and Blue Gramas, Bush Muhly, dropseeds, and three-awns.

This is a resilient habitat with few threats to its conservation. In fact, overgrazing, fire suppression, and other human disturbances favor the spread of Creosote and mesquite. In many cases Chihuahuan Desert shrubland has inundated and replaced Nearctic desert grasslands.

WILDLIFE: As in many desert habitats, large mammals are scarce here, and Pronghorn and Collared Peccary are the only two found regularly. The bulk of the mammalian fauna comprises small seed-eating mammals like Antelope Jackrabbit, Desert Cottontail, White-throated Woodrat, Yellow-nosed Cotton Rat, Ord's Kangaroo Rat, and Spotted Ground Squirrel. The primary mammalian predators of these rabbits and rodents are Gray Fox, Kit Fox, and Coyote.

Western Diamondback Rattlesnake is characteristic of this habitat in the sw. United States.
© BEN KNOOT, TROPICAL BIRDING TOURS

The bird communities here have relatively low diversity that changes little throughout the year. Terrestrial birds like Gambel's Quail, Scaled Quail, Mourning Dove, and Greater Roadrunner are common. The shrubs often hold Pyrrhuloxia, Black-tailed Gnatcatcher, Black-throated Sparrow, Verdin, Cactus Wren, and Curve-billed Thrasher (IS). The tall flowering stalks of Lechuguilla and other agaves are a great place to look for perching flycatchers and birds of prey. Swainson's and Red-tailed Hawks, American Kestrel, Loggerhead Shrike, Chihuahuan Raven (IS), Say's Phoebe, and Cassin's and Western Kingbirds all regularly hunt from these high perches. In the early morning, Lechuguilla stalks are also a preferred display perch for many songbirds.

Reptiles are abundant in this habitat, especially small lizards. Greater Earless Lizard, Marbled Whiptail, Common Side-blotched Lizard, and various spiny lizards (*Sceloporus*) are common. While the majority of reptiles endemic to this habitat are found in its southern reaches in Durango and Coahuila, Mexico, the diminutive Round-tailed Horned Lizard (IS) is present throughout the Chihuahuan Desert shrublands. Common snakes include Black-necked Garter Snake, Desert Night Snake, Coachwhip, Trans-Pecos Rat Snake, Glossy Snake, and Western Diamondback Rattlesnake, among many others. On cool summer nights, snakes drawn to warm pavement are readily encountered by anyone cruising along slowly and carefully searching the roads.

DISTRIBUTION: The Chihuahuan Desert shrubland is found over a broad area of the sw. United States and n. Mexico. This area extends from c. Texas westward to se. Arizona and south over large portions of Chihuahua, Coahuila, and Durango to as far south as Zacatecas. Chihuahuan Desert shrubland lies in the rain shadow of the Sierra Madre Occidental and the Sierra Madre Oriental and is largely bounded by these two ranges. Throughout its range, Chihuahuan Desert shrubland is dotted with NEARCTIC DESERT GRASSLAND, ribbons of WESTERN RIPARIAN WOODLAND, and islands of MADREAN PINE-OAK WOODLAND in the smaller mountain ranges. The northern extent of the Chihuahuan Desert shrubland is a wide transitional zone with SHORTGRASS PRAIRIE, where elements of both habitats may be present.

WHERE TO SEE: Big Bend National Park, Texas, US.

COLUMNAR CACTUS DESERT

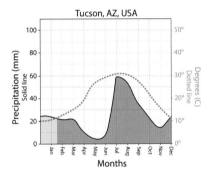

IN A NUTSHELL: A warm semidesert shrubland characterized by giant columnar cacti and late summer monsoon rains. **Habitat Affinities:** NEOTROPICAL CAATINGA; AFROTROPICAL MALAGASY SPINY FOREST; DRAGON BLOOD SEMIDESERT. **Species Overlap:** CHIHUAHUAN DESERT SHRUBLAND; PACIFIC CHAPARRAL; NEARCTIC DESERT GRASSLAND; NEOTROPICAL DRY DECIDUOUS FOREST.

DESCRIPTION: The columnar cactus desert is the lushest and most iconic of Nearctic desert landscapes. This botanically diverse desert is home to some of the strangest flora on the continent, including the emblematic Saguaro and colossal Mexican Giant Cardón cacti and the otherworldly Boojum Tree. The spectacular flora and fauna here are best seen in April–May or in the monsoon period of late summer, when animals are active and the landscape is green. This habitat is open enough that it is easily walked through, though spiny plants and midday heat should be avoided.

The columnar cactus desert is the hottest of the Nearctic desert regions, with mild winters and scorching summers, when temperatures approach 120°F (49°C). There is some relief from the summer heat when monsoon rains in July and August cool things off by roughly 10°F (5°C). The monsoons drop about half of the region's annual rain in intense and unpredictable late afternoon thunderstorms. The rest of the region's rain falls more consistently in December and January, for an annual total of 3–20 in. (75–500mm).

The most notable botanical feature here is the array of giant cacti, which are often the tallest plants. Saguaro grows to heights of 50 ft. (15m), Organ Pipe Cactus to 20 ft. (6m), and Mexican Giant Cardón reaches 60 ft. (18m). These widely spaced cacti live for centuries and serve the same function as trees for cavity-nesting birds. Along with these giants, there is a large variety of common, smaller succulents, including agaves, yuccas, barrel cacti, pricklypears, and chollas. Apart from the succulents, the other common plants are thorny shrubs and small trees, which are the dominant vegetation by area, if not by mass. The most prominent of these shrubs is Creosote, the most common plant in this and many other Nearctic desert shrublands. Paloverdes, mesquites, Catclaw Acacia, and Desert Ironwood are also abundant. In areas with higher amounts of available water, mesquites can form impenetrable thickets.

This is a resilient habitat with few major threats and relatively little development and destruction.

Conspicuous Saguaro cacti in columnar cactus desert near Tucson, Arizona, US.
© BEN KNOOT, TROPICAL BIRDING TOURS

WILDLIFE: Columnar cactus desert is the lushest and most diverse of the Nearctic desert habitats. Like many desert environments, it is relatively poor in big mammals. Collared Peccary (aka Javelina), is the most common large mammal and can often be found feeding on pricklypear fruits in season. Desert Bighorn Sheep, Coyote, Bobcat, and Puma are among the other large mammals. In the Baja California portions of the desert, there is a critically endangered subspecies of Pronghorn. Medium-size and small mammals are more abundant. Driving through the desert at night, one is likely to see Antelope Jackrabbit, Desert Cottontail, several species of kangaroo rats, and occasionally Kit Fox. After the monsoons, when Saguaro are in bloom, swarms of nectivorous Lesser Long-nosed and Mexican Long-tongued Bats move through the area. While taking advantage of the sugary feast, they often visit hummingbird feeders as well, and in the late summer can drain them overnight.

Common birds include Gambel's Quail, White-winged Dove, Harris's Hawk, Greater Roadrunner, Turkey Vulture, Pyrrhuloxia, Curve-billed Thrasher, Cactus Wren, Verdin, Black-tailed Gnatcatcher, Black-throated Sparrow, and Abert's Towhee. Gila Woodpecker and the endemic Gilded Flicker (IS) act as ecosystem engineers by constructing nesting cavities in the trunks of large cacti. These cavities allow birds that normally require trees for nesting to live in this environment; beneficiaries include Elf Owl, Ferruginous Pygmy-Owl, Purple Martin, and Brown-crested Flycatcher.

The herpetofauna of this desert landscape is dominated by snakes and lizards. Whiptails, spiny lizards (*Sceloporus*), Regal Horned Lizard, and Zebra-tailed Lizard are all commonly encountered during the day. The large Chuckwalla is often seen basking in the early morning but disappears into rocky crevices in the heat of the day. Gila Monster, a slow-moving, rotund black-and-orange lizard, is the habitat's most famous reptilian resident and one of only two venomous lizards in the world. There are more than 10 species of rattlesnakes, including Mojave, Prairie, Speckled, Baja California, and Tiger Rattlesnakes. Most of the other snakes here are rarely seen, being small, fossorial, and/or nocturnal. Amphibians are almost absent, though large numbers of Sonoran Desert Toads and Couch's Spadefoots emerge to breed after summer monsoons.

DISTRIBUTION: The columnar cactus desert is found at low elevations, below 3,600 ft. (1,100m), from sw. Arizona and se. California in the United States to ne. and c. Baja California and w. Sonora, Mexico. The northern extent of this habitat is limited by the Rocky Mountains and PINYON-JUNIPER WOODLAND. The western limit is formed by the San Gabriel Mountains and the Peninsular Ranges. Most of the eastern boundary is formed by the Sierra Madre Occidental, where the habitat transitions to NEARCTIC OAK SAVANNA and MADREAN PINE-OAK WOODLAND. South of Sonora, the habitat transitions to NEOTROPICAL DRY DECIDUOUS FOREST.

WHERE TO SEE: Saguaro National Park, Arizona, US; Organ Pipe Cactus National Monument, Arizona, US; El Vizcaíno Biosphere Reserve, Baja California, Mexico.

Harris's Hawk taking off from a giant cactus that characterizes this habitat, in Arizona, US. © BEN KNOOT, TROPICAL BIRDING TOURS

Gila Monster is the largest native lizard species in the United States and is characteristic of southwestern deserts, which include areas of impressive columnar cactus desert.
© BEN KNOOT, TROPICAL BIRDING TOURS

SALT DESERT SHRUBLAND

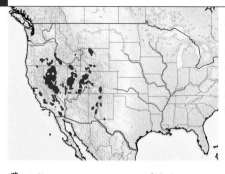

IN A NUTSHELL: A dry, sparse, and open shrubland typically found along valley floors on alkaline and saline soils. **Habitat Affinities:** EAST ASIAN COLD DESERT; AFROTROPICAL KAROO; AUSTRALASIAN CHENOPOD AND SAMPHIRE SHRUBLAND. **Species Overlap:** SAGEBRUSH SHRUBLAND; NEARCTIC DESERT GRASSLAND; CHIHUAHUAN DESERT SHRUBLAND.

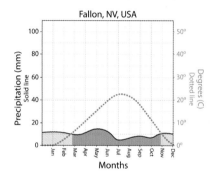

Fallon, NV, USA

Salt desert shrubland at Great Salt Lake, **Utah, US.** © IAIN CAMPBELL, TROPICAL BIRDING TOURS

DESCRIPTION: Salt desert shrubland is one of the most barren habitats of the Nearctic. It is found along valley floors, alkali flats, plateaus, washes, and gentle rocky slopes of the interior w. United States. In this sparse shrubland with expanses of barren ground, vegetation is rarely more than 3 ft. (1m) high and is usually closer to 1 ft. (0.3m). Salt desert shrubland generally grows on poor salt or alkaline soils. These soils generally form in areas that receive little precipitation and where pooling snowmelt from nearby mountains leaves large mineral deposits. Salt desert shrubland can also form near marine shale deposits, where erosion has released the salts contained in the rock. The areas

around salt desert shrubland often contain large playas and alkali flats that function as wetlands or mudflats for part of the year and are completely devoid of vegetation for the rest. This habitat is flat, open, and easily explored on foot. Visibility is obscured only by the mountains that typically encircle this habitat.

Salt desert shrubland is a cold desert that experiences hot summers and cold winters. Winter lows approach 0°F (-23°C), especially when the strong winds are taken into account. In the summer, temperatures approach 110°F (43°C), and temperature swings of 45°F (25°C) in the space of a single day are not uncommon. The area receives 5–13 in. (125–330mm) of precipitation annually, though total precipitation is usually less than 8 in. (200m).

The plants that grow in salt desert shrubland are halophytes, species that are specially adapted to growing in saline conditions. The habitat supports two distinct plant communities, shaped by the availability of water. In both cases, the landscape is covered in a mixture of small shrubs and matted subshrubs, with a sparse to moderate grassy layer. In areas with a low water table and relatively little available water, the landscape is dominated by saltbushes and Budsage. Spiny Hopsage and Winterfat are also common, usually growing as matted subshrubs. The upland areas tend to be grassier, and Squirreltail, Indian Ricegrass, and Alkali Sacaton are common native grasses. The invasive Cheatgrass and Russian Thistle, or Tumbleweed, are also abundant. In lowland sites with a high water table, the shrubs tend to be taller, and the understory is less grassy. Greasewood is by far the most dominant plant in these areas and can form a near monoculture over large areas. The few low-growing subshrubs are Iodine Bush, Utah Pickleweed, and Iodine Weed. Saltgrass is the only grass, and no other herbaceous growth occurs regularly.

WILDLIFE: Wildlife is very scarce, and this habitat can often appear lifeless. Upon closer examination, the scant halophytic vegetation provides forage for a number of small mammals. Chisel-toothed Kangaroo Rat is the only leaf-eating member of its genus, and its diet consists mostly of the highly

saline leaves of saltbushes. In order to survive on saltbush, this kangaroo rat uses its uniquely shaped teeth to scrape the salty epidermis off the leaves, leaving only the water-rich mesophyll. A specially shaped upper lip closes off the mouth during this process and prevents salt ingestion. Ord's Kangaroo Rat, Dark Kangaroo Mouse, Desert Cottontail, Black-tailed Jackrabbit, and White-tailed and Nelson's Antelope Squirrels are other common herbivores. Even mammals as large as Pronghorn are resident, though fawn mortality is high in this habitat, and populations are small compared to those in other shrublands.

Among the few resident birds, Mourning Dove, Horned Lark, Loggerhead Shrike, Rock Wren, Brewer's Sparrow, Black-throated Sparrow, and Western Meadowlark are the most evident.

Black-throated Sparrow lives within desert scrub, favoring areas where there are plentiful open areas between shrubs, such as salt desert shrubland.
© BEN KNOOT, TROPICAL BIRDING TOURS

In taller, more developed Greasewood stands, sagebrush birds like Sage Thrasher and Sagebrush Sparrow are found. In the Mojave Desert of se. California and s. Nevada and the Central Valley of California, LeConte's Thrasher is associated with this habitat. With a high abundance of rodents, this habitat can have surprisingly high numbers of raptors, especially during the winter months. Golden Eagle, Swainson's Hawk, Ferruginous Hawk, Northern Harrier, and Prairie Falcon are all regularly seen.

Common reptiles include Sagebrush Lizard, Common Side-blotched Lizard, Western Whiptail, Striped Whipsnake, and Glossy, Gopher, and Bull Snakes. The most notable resident reptile is the endangered Blunt-nosed Leopard Lizard. Salt desert shrubland makes up the majority of this species' small range in the Central Valley.

DISTRIBUTION: Salt desert shrublands occur throughout the arid mountainous w. United States, mostly in the Great Basin and the Mojave Desert. Salt desert can be found from se. Oregon east to c. Wyoming, south to s. New Mexico, and west to the Central Valley of California. Favorable growing conditions occur in intermontane valleys and basins at elevations of 5,000–7,000 ft. (1,500–2,250m), occasionally lower. Salt deserts often form a mosaic with other desert shrublands, especially Nearctic SAGEBRUSH SHRUBLAND and COLUMNAR CACTUS DESERT.

WHERE TO SEE: Carrizo Plains, California, US; Black Rock Desert Wilderness, Nevada, US; Bonneville Salt Flats, Utah, US.

NEARCTIC DESERT GRASSLAND

IN A NUTSHELL: A sparse and open grassland dominated by bunchgrasses and dotted with yuccas and cacti. **Habitat Affinities:** AUSTRALASIAN TUSSOCK GRASSLAND; AFROTROPICAL TROPICAL GRASSLAND. **Species Overlap:** CHIHUAHUAN DESERT SHRUBLAND; SHORTGRASS PRAIRIE.

DESCRIPTION: This sparse and open grassland habitat is found in the valleys and basins of the Chihuahuan Desert in the sw. United States and n. and c. Mexico. It is dominated by multiple species of warm-season bunchgrasses of varying heights, scattered among a mixture of bare earth and low-growing shrubs, cacti, yuccas, and forbs. In many places, human degradation has increased the number of shrubs and turned these grasslands into shrub savannas. Nearctic desert grassland occurs in a region with warm, humid summers and cool, dry

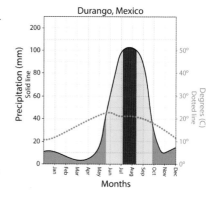

winters. The grasslands receive 9–24 in. (240–600mm) of rain in a typical year, with the majority of the rain falling in late summer in intense monsoon thunderstorms. The climate here is strongly seasonal—summer highs can regularly reach 105°F (40°C), and winter lows of 0°F (-18°C) are not unusual.

Dozens of species of bunchgrasses can be found in Nearctic desert grassland, but the most widespread and dominant species are Black and Blue Gramas, Bush Muhly, dropseeds, three-awns, Giant Sacaton, and Tubosa. While the majority of these grasses reach only 1–2 ft. (0.3–0.6m) in height, Giant Sacaton can form large bunches up to 6 ft. (2m) tall. Among the grasses are a scattering of small shrubs and succulents, including mesquites, Creosote, American Tarbush, Winterfat, pricklypears, barrel cacti, agaves, yuccas, chollas, and Ocotillo. These shrubs and succulents, which may reach heights of 10 ft. (3m), provide important cover for wildlife.

While shrubs are a natural part of this landscape, Nearctic desert grasslands have undergone widespread transition to shrublands over the past two centuries. This is largely due to overgrazing, soil loss, conversion to agriculture, and fire suppression. Only an estimated 15% of Nearctic desert grasslands remain healthy and intact.

WILDLIFE: This habitat does not support many large animals but does hold a diverse assemblage of reptiles, birds, and small mammals. Pronghorn is the most successful large mammal in the Nearctic desert grassland, but Mule Deer, Collared Peccary, and Bighorn Sheep can also be found in small numbers. The primary mammalian predators here are Bobcat and Coyote, though a few Puma occur here as well as small carnivores like Kit Fox, American Badger, and American Hog-nosed Skunk. The bulk of mammalian diversity is found among rodents and rabbits, which can reach astronomical numbers under favorable conditions. Desert Cottontail, Black-tailed and White-sided Jackrabbits, numerous species of kangaroo rats, Chihuahuan Desert Pocket Mouse, Spotted Ground Squirrel, and Mexican Prairie Dog are all common small mammals in the Nearctic desert grassland. Black-tailed Prairie Dogs are common in the northern half of the habitat's range, where they form massive dog towns, though many have been exterminated by cattle ranchers.

Nearctic desert grassland at Sulphur Springs Valley, Arizona, US, showing the extensive open grasslands area, only intermittently broken up by sparsely distributed small shrubs. © BEN KNOOT TROPICAL BIRDING TOURS

In the northern reaches of this habitat, Black-tailed Prairie Dogs are abundant.
© GABRIEL CAMPBELL, TROPICAL BIRDING TOURS

For birds, Nearctic desert grasslands are at their most diverse in winter. It is the only habitat where many grassland specialties can be seen apart from their northern breeding grounds. Common birds include Turkey Vulture, Northern Harrier, Mourning Dove, Inca Dove, American Kestrel, Cassin's Kingbird, Loggerhead Shrike, Chihuahuan Raven, Horned Lark, and Western and Eastern Meadowlarks. This area is particularly important to wintering birds like Golden Eagle; Burrowing and Short-eared Owls; Mountain Plover; Say's Phoebe; Sprague's Pipit; Baird's (IS), Grasshopper, Botteri's, Cassin's, and Vesper Sparrows; Lark Bunting; and Chestnut-collared and McCown's Longspurs. This is the primary habitat for the endangered Chihuahuan population of Aplomado Falcon and the critically endangered Worthen's Sparrow (IS), which numbers fewer than 150 individuals.

The reptile community here is generally a smaller subset of the species found in CHIHUAHUAN DESERT SHRUBLAND and SHORTGRASS PRAIRIE. Common species include Prairie Rattlesnake, Coachwhip, Desert Grassland Whiptail, and Desert Spiny Lizard.

DISTRIBUTION: Nearctic desert grassland is patchily distributed throughout the Chihuahuan Desert. From its northern limits in the United States, in se. Arizona, s. New Mexico, and w. Texas, it extends south into Mexico, across n. Sonora and through the states of Chihuahua and Coahuila. Within the Chihuahuan Desert, it occurs sporadically in MADREAN PINE-OAK WOODLAND and CHIHUAHUAN DESERT SHRUBLAND, and on the southern and eastern edges of its range, it transitions into the latter. Within its broader distribution, Nearctic desert grassland occurs mostly in valleys and basins between mountains and in lower foothills close to valley floors.

WHERE TO SEE: Sulphur Springs Valley grassland, Arizona, US; Animas Valley, New Mexico, US; Altiplano Mexicano Nordoriental, Zacatecas, Mexico.

SAGEBRUSH SHRUBLAND

IN A NUTSHELL: A short steppe or shrubland dominated by sagebrush. **Habitat Affinities:** AFROTROPICAL KAROO; PALEARCTIC SEMIDESERT THORNSCRUB. **Species Overlap:** PINYON-JUNIPER WOODLAND; SALT DESERT SHRUBLAND; SHORTGRASS PRAIRIE.

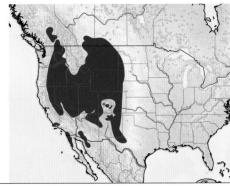

DESCRIPTION: An arid to semiarid mosaic of shrubs and grasses, this habitat occurs in the flat basins and plateaus of the intermountain region of w. North America. Mostly present in cold semideserts, sagebrush country experiences hot, dry summers with persistent wind and bitterly cold winters. The majority of the scant (10 in./250mm) annual precipitation falls as snow during the boreal winter. Growth of grasses and forbs is dependent on spring snowmelt.

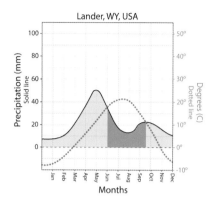

Lander, WY, USA

Based on elevation and seasonal timing of precipitation, the structure of sagebrush shrubland varies between steppe dominated by wheatgrasses and fescues and dense shrublands composed of sagebrush, saltbushes, rabbitbrush, Greasewood, and Winterfat. The understory often supports a diverse array of forbs that are important forage for herbivores incapable of digesting sagebrush. Shrub communities are typically dominated by a single sagebrush species, often Big Sagebrush. The associated faunal assemblage depends more on habitat structure than on the particular species of sagebrush present. From the Great Basin southward, the prevalence of perennial grasses is greatly diminished, and the density of shrub species decreases as well. In these areas, cryptobiotic crust, composed of blue-green algae, fungi, mosses, and lichens, is an important part of the biological community; it stabilizes barren soil, retains moisture, and aids in seed establishment. While this is typically a short and stunted shrub community (about 3 ft./1m in height), Big Sagebrush is historically capable of reaching heights over 10 ft. (3m), particularly in the eastern part of the range.

Pronghorns are abundant grazers within sagebrush.
© BEN KNOOT, TROPICAL BIRDING TOURS

Fire is an important driver in the mosaic structure of this habitat, and healthy sagebrush communities have small, infrequent, stand-replacing fires resulting in a patchwork of plant ages and density. Human activities including grazing, fire suppression, and introduction of non-native grasses have changed this regime, resulting in larger, more frequent, and higher-intensity fires.

WILDLIFE: Sagebrush shrubland is home to an abundance of native mammals, including several endemic species. Pronghorn is widespread and common in the region. Along with the endangered Pygmy Rabbit, Pronghorns are among the few species capable of digesting sagebrush. Great Basin Pocket Mouse and Sagebrush Vole are less conspicuous but widespread sagebrush obligates. The habitat is important as a winter grazing area for Mule Deer and holds the continent's largest wintering herds of Elk, which spend summers at higher elevations. Other widespread mammals include White-tailed Deer, Black-tailed Jackrabbit, Desert and Mountain Cottontails, White-tailed Prairie Dog, Coyote, Bobcat, and American Badger. Widely extirpated, large predators like Brown Bear and Gray Wolf frequently feed and forage in this habitat, though they are not resident.

Sagebrush steppe has relatively limited bird diversity with few species resident through the harsh winters. Greater (IS) and Gunnison (IS) Sage-Grouse are iconic residents, present year-round and capable of eating sagebrush. Denser stands of sagebrush support most of the true sagebrush specialists, including Sagebrush Sparrow (IS), Brewer's Sparrow, and Sage Thrasher (IS), as well as Blue-gray Gnatcatcher, Black-throated Sparrow, Gray Flycatcher, and Green-tailed Towhee. More open areas of sagebrush support a mixture of more widespread grassland and shrubland species, including Sharp-tailed Grouse, Long-billed Curlew, Mourning Dove, Swainson's Hawk, Burrowing Owl, Common Nighthawk, Loggerhead Shrike, American Kestrel, Western Meadowlark, Say's Phoebe,

Greater Sage-Grouse performs impressively ridiculous springtime displays within more open areas of sagebrush shrubland, like this one in Colorado, US. © NICK ATHANAS, TROPICAL BIRDING TOURS

Western Kingbird, and Vesper Sparrow. In winter, the abundant small mammal populations provide food for large numbers of raptors, including Red-tailed Hawk, Ferruginous Hawk, Golden Eagle, Northern Harrier, Prairie Falcon, and Short-eared Owl.

Sagebrush Lizard, Greater Short-horned Lizard, Prairie Rattlesnake, and Gopher Snake are all common and conspicuous residents throughout the habitat.

DISTRIBUTION: The entire sagebrush region covers approximately 243,000 sq. mi (630,000km²) of the interior w. United States. Almost entirely contained within the rain shadow of the Sierra Nevada, Cascade Range, and Rocky Mountains, sagebrush is prevalent in much of the Great Basin and the Wyoming Basin, and reaches into the Snake River plain, Columbia Basin, sw. Montana, the Colorado Plateau, sw. Colorado, and n. New Mexico, with incursions into sw. Canada and nw. Mexico. The northern part of this region is sagebrush steppe, where shrubs and perennial grasses are co-dominant. From the Great Basin southward, in the much drier Great Basin sagebrush vegetation type, sagebrush is dominant, and grasses are few and sparse. The entire region is dotted with small mountain ranges and experiences influence from coniferous forests throughout. Along the eastern slope of the Sierra Nevada in ne. California and Nevada and the Cascades in c. Oregon, the habitat transitions from PINYON-JUNIPER WOODLAND to true sagebrush. The encroachment of conifers into areas of historical sagebrush has been exacerbated by fire suppression.

WHERE TO SEE: Yellowstone and Grand Teton National Parks, Wyoming, US; Arapahoe National Wildlife Refuge, Colorado, US; Malheur National Wildlife Refuge, Oregon, US.

MESQUITE BRUSHLAND AND THORNSCRUB

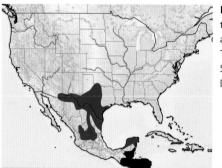

IN A NUTSHELL: A low, scrubby habitat made up of thorny bushes with small leaves, small spiny trees, and abundant cacti. **Habitat Affinities:** NEOTROPICAL THORNSCRUB; AFROTROPICAL DRY THORN SAVANNA. **Species Overlap:** NEARCTIC TROPICAL DRY FOREST; CHIHUAHUAN DESERT SHRUBLAND.

DESCRIPTION: This is an extension of NEOTROPICAL THORNSCRUB (see that habitat account for a detailed description), and though it takes many forms, most Nearctic mesquite brushland and thornscrub appears similar to that habitat. While highly variable in structure, this habitat is characterized by a short and relatively open canopy and a dense, brushy understory.

The major departure from the typical Neotropical thornscrub occurs at the northern limit of this habitat in coastal Tamaulipas, Mexico, and adjacent s. Texas, United States. Also known as mesquite shrubland or *mezquital*, this portion of the habitat has a mesquite-dominated canopy that reaches heights up to 20 ft. (6m). Mesquites are joined by Huisache, Texas Ebony, Sugar Hackberry and Granjeno. This is the most diverse of the arid shrublands found in the United States, and the shrub layer can contain dozens of species. In more pristine areas, the understory is composed of species such as Blackbrush, Amargosa, Lindheimer Pricklypear, Berlandier Wolfberry, Coyotillo, Brasil, Desert Olive, Lotebush, and Tasajillo, among others. This botanically rich shrub layer increases in both density and diversity with increasing rainfall from west to east.

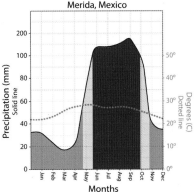

In more open areas, the understory is grassy and was historically dominated by Curly Mesquite Grass, Hooded Finger Grass, and various grama grasses. Open, grassy mezquital has become rare with the alteration of fire and grazing regimes. In areas like the King Ranch of s. Texas, intensive and conscientious management has preserved some prime examples of mesquite grasslands that are preferred by many game species of birds and mammals.

Mesquite brushland and thornscrub is a dynamic habitat that is subjected to frequent droughts, tropical storms, and fire. As a result, a single area can vary dramatically in appearance from year to year.

Patches of mesquite brushland extend into se. Arizona, sw. US (pictured). However, more extensive examples of the habitat, and its associated distinctive bird communities, are found in the Rio Grande Valley of Texas, US, and in Mexico. © PETER KNOOT

WILDLIFE: There are few endemic mammals in this habitat, though widespread species like White-nosed Coati, Bobcat, Collared Peccary, and White-tailed Deer all occur in the mainland sections. Within the United States, mesquite brushland and thornscrub habitat is critical to the survival of the Ocelot. Despite the presence of high-quality habitat, only 50 or so individuals remain in the Lower Rio Grande Valley of Texas. Strong conservation measures are underway to save not only the Ocelot but also the Jaguarundi, absent from Texas for decades. Plans to reintroduce this shy tropical feline are currently being developed, with a goal of 500-plus individuals by 2050. There are no notable mammals in this habitat in the Caribbean.

Bird communities vary slightly over the range of this habitat but tend to include skulking species that favor dense undergrowth. Thicket Tinamou, Northern and Black-throated Bobwhites, White-tipped Dove, Yucatan Wren, Green Jay, Scrub Euphonia, Audubon's Oriole, Painted Bunting, and Crimson-collared Grosbeak are typical birds. In the Lower Rio Grande Valley, the northern reaches of this habitat provide refuge for a variety of typically tropical birds. These impenetrable thickets are essentially the only habitat home to Olive Sparrow, Long-billed Thrasher, Plain Chachalaca, Altamira Oriole, and Common Pauraque within the United States and make the region they occupy one of the premier birding areas in the country. Red-crowned Parrots are found in riparian and palm microhabitats within this broader habitat range. Dense shrubland communities often have particularly loud and spectacular bouts of dawn chorus. The raucous songs of tropical birds give the s. Texas mesquite shrublands a soundscape unlike anywhere else in the United States.

Collared Peccary is another example of a more southerly species that just reaches into the United States in the far south, where habitats like mesquite shrubland reach their northern limit (photo: se. Arizona, US).
© BEN KNOOT, TROPICAL BIRDING TOURS

Green Jay is a noisy, abundant, and conspicuous representative of mesquite shrubland in the Rio Grande Valley, s. Texas, US. © IAIN CAMPBELL, TROPICAL BIRDING TOURS

This area holds few amphibian species, but reptiles are prominent. The large Texas Tortoise can be found slowly making its way through the Tamaulipan region of s. Texas and ne. Mexico. The gorgeous Texas Indigo Snake is another gem of the northern reaches of the mezquital. In s. Texas, indigo snakes are well known for their fondness for rattlesnakes as prey items and are considered an ally by many ranchers and farmers. The Yucatán region of the thornscrub is home to another slow-moving species, Black-beaded Lizard, one of only two venomous lizards in the world. Other more common reptiles include Central American Boa, Neotropical Whipsnake, Northeastern Spiny-tailed Iguana, Eastern Spiny Lizard, and Brown Anole.

Endemism: Because this habitat is found on many Caribbean islands, it has more EBAs than other similar habitats, but this endemism is more driven by isolation than by the distinctiveness of the habitat, with most birds being generalists that live in multiple habitats (see sidebar 6.2). The coastal thornscrub in Yucatán has over a dozen endemic birds, including Mexican Sheartail, Yucatán Wren (IS), Black-throated Bobwhite, and Yucatán Gnatcatcher. The North-east Mexican Gulf Slope EBA also has a few endemics, including Tamaulipas Crow, Crimson-collared Grosbeak (IS), and Tawny-collared Nightjar.

DISTRIBUTION: Found largely in dry coastal environments and montane rain shadows in Mexico and the Caribbean, Nearctic mesquite brushland and thornscrub has a patchy but significant range in the region. The northern coast of the Yucatán Peninsula, coastal Tamaulipas, and the northern sections of the Central Mexican Plateau all have large swaths of this habitat. Reaching its northern extent in the sw. United States, this habitat extends as far north as Corpus Christi, Texas, and also occurs in a very patchy distribution to se. Arizona.

Additionally, the dry southern coastal areas of Puerto Rico, Jamaica, Hispaniola, and Cuba all have heavy succulent thornscrub regions. This habitat is particularly prevalent in se. Cuba. In the Caribbean, including Mexico's east coast, this is the driest habitat in the area. As moisture and elevation increase in these areas, the thornscrub gradually transitions to NEARCTIC TROPICAL DRY FOREST. In c. Mexico, where the habitat becomes lower and drier, this habitat transitions to CHIHUAHUAN DESERT SHRUBLAND and NEARCTIC DESERT GRASSLAND.

WHERE TO SEE: Santa Ana National Wildlife Refuge, Texas, US; King Ranch, Texas, US; Merida, Yucatán, Mexico; Naranjo, Tamaulipas, Mexico; Santiago de Cuba, Cuba.

The Greater and Lesser Antilles are an area of high endemism, high diversity, and varied landscape. These islands contain everything from semidesert scrub to montane pine forest and stunted cloud forest, but the variation in habitat has little to do with the high number of endemic birds. As with many island systems, the evolution of unique species is driven more by isolation than any habitat specialization. Most Caribbean island birds are generalists. You can break them into subgroups—moist habitat birds or dry habitat birds, forest birds or scrub birds—but by and large these species are found across the majority of the habitats on the islands they inhabit.

The habitats found in the Caribbean match up with a handful of Nearctic/Neotropical habitats. The most common habitats found on Caribbean islands include NEOTROPICAL SEMI-EVERGREEN FOREST, NEOTROPICAL DRY DECIDUOUS FOREST, NEOTROPICAL THORNSCRUB, NEARCTIC MANGROVE, and NEARCTIC/NEOTROPICAL CLOUD FOREST. The very highest elevations on the wetter islands of Jamaica and Puerto Rico have ELFIN AND STUNTED CLOUD FOREST. The drier islands of Cuba and Hispaniola contain subtropical pine forests similar to MADREAN PINE-OAK WOODLAND of the s. Nearctic. These forests contain some of the few habitat specialists on the islands, including Olive-capped Warbler and Hispaniolan Crossbill.

This does not mean the islands are not worth visiting from a wildlife standpoint. With over 150 endemic species of birds, multiple endemic families, diverse herpetofauna, and even a few strange mammals, the Greater Antilles are a fantastic area to explore in a close trip from many areas of North America.

BALD CYPRESS–TUPELO FOREST

IN A NUTSHELL: Tall, flooded riverine forests of Bald Cypress and several species of broadleaf trees. **Habitat Affinities:** This is a globally unique habitat. **Species Overlap:** NEARCTIC TEMPERATE DECIDUOUS FOREST; NEARCTIC FRESHWATER WETLAND.

DESCRIPTION: One of the most visually distinctive habitats in the Nearctic, Bald Cypress–tupelo forest is often the first image that comes to mind when one pictures the se. United States. Dominated by tall, broad-based, epiphyte-laden trees, the canopy is typically dense and reaches heights of 60–100 ft. (18–30m), though Bald Cypress can be as tall as 145 ft. (44m). Bald Cypress–tupelo forests grow in hot, humid environments that typically receive 35–65 in. (900–1,650mm) of rainfall per year. This habitat occurs along floodplains or on cypress domes in the middle of open wetlands. The ground is permanently or seasonally flooded, and the maze of back channels is best explored by canoe or boardwalk.

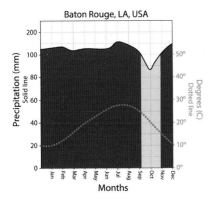

Baton Rouge, LA, USA

This is an entirely deciduous forest, where even the conifers lose their needles in winter. The majority of the canopy is composed of Bald Cypress and either Water Tupelo or Swamp Tupelo, though Water Hickory is occasionally co-dominant. Bald Cypress and Water Tupelo are both extremely broad-based trees with tapered trunks, and Bald Cypress can be as stout as 39 ft. (12m) in diameter. The mid-story in these forests is generally sparse and consists of young tupelos and smaller, shade-tolerant trees like Red Maple and Pop Ash. Vines are a common feature of the mid-story and may be abundant.

The understory is the most variable element and is dependent mostly on depth and variability of water levels. Frequently, the understory is entirely flooded, and apart from some aquatic vegetation, the only noticeable feature is the protruding "cypress knees," woody root projections whose purpose is not well understood. In seasonally dry forests, the understory can be extremely lush, dominated by rapidly growing plants tolerant of wet soil. Giant Sedge, False Nettle, Lizard's Tail, and Greater Marsh St.-John's-wort grow in areas that are flooded for much of the year, while relatively drier areas have Swamp Fern, False Pimpernel, and Camphorweed and woody shrubs like Buttonbush, willows, and Virginia Willow.

There are few surviving tracts of old-growth Bald Cypress–tupelo forest in the e. Nearctic. The inaccessible nature of these forests protected them longer than many adjacent forests, but heavy

Bald Cypresses in Cache River State Natural Area, Illinois, US.
© MICHAEL JEFFORDS AND SUE POST

logging in the early 20th century eliminated much of the remaining ancient forest. Agriculture- and flood-control-related changes to hydrology, saltwater intrusion due to rising sea levels, and increased pollution have damaged and diminished existing tracts. Bald Cypresses are capable of reaching staggering ages, and several known specimens are over 2,000 years old. A 3,500-year-old Bald Cypress was, sadly, destroyed by arson in 2012.

WILDLIFE: Bald Cypress–tupelo forests have a complex matrix of flooded lands and upland areas that are home to a mixture of both forest and wetland species. Swamp Rabbit, White-tailed Deer, American Black Bear, Common Raccoon, Bobcat, and the endangered Florida Panther, a subspecies of Puma, all utilize better-drained areas. The flooded channels are home to North American River Otter, Common Muskrat, and the introduced Nutria.

This habitat is a critical nesting area for many large wetland birds; large rookeries of egrets, herons, cormorants, and Wood Storks are a notable feature. Barred Owls are abundant, readily adapting to a diet of frogs,

Prothonotary Warbler breeds within cavities alongside swamps in this habitat in the s. United States, and often winters in Neotropical mangroves in Central America. © BEN KNOOT, TROPICAL BIRDING TOURS

salamanders, and other aquatic vertebrates. Red-headed and Pileated Woodpeckers are also more common here than in many adjacent habitats. Many species of dabbling duck spend the winter here, and Wood Duck is a ubiquitous resident. In the summer, there are many breeding neotropical migrants. Yellow-billed Cuckoo, Acadian Flycatcher, Great Crested Flycatcher, Red-eyed and Yellow-throated Vireos, Yellow-throated Warbler, and Northern Parula are all abundant breeding species. Prothonotary Warbler (IS), formerly known as the Golden Swamp Warbler, is the quintessential summer resident of Bald Cypress–tupelo forest.

Reptiles and amphibians are both common and diverse in this habitat. American Alligator is

perhaps the most famous resident—an apex predator capable of reaching 15 ft. (4.5m) in length. Alligators can be seen cruising the forested channels with their eyes barely breaking the water's surface. Bald Cypress–tupelo forests are also home to nearly 30 species of turtles, from the 180 lb. (80kg) Alligator Snapping Turtle to the diminutive Loggerhead Musk Turtle. This habitat comes alive with the sound of frogs at night.

A Cottonmouth in the Cache River State Natural Area, Illinois, US. © IAIN CAMPBELL, TROPICAL BIRDING TOURS

Southern Leopard, Pig, and River Frogs; American Bullfrog; Green and Barking Tree Frogs; and Southern Cricket Frog are among the common species here. Fully aquatic salamanders including Greater Siren and Three-toed Amphiuma are found in the murky waters, while rare species like Flatwoods Salamander make use of drier areas. Cottonmouth, Banded and Brown Water Snakes, Rough Green Snake, and Eastern Rat Snake are all common. The spectacular black-and-crimson Mud Snake is a habitat specialist.

DISTRIBUTION: Bald Cypress–tupelo forest is primarily associated with low-lying wet areas of the Atlantic and Gulf of Mexico coastal plains of the se. United States. It occurs from Delaware to s. Florida, west to e. Texas and se. Oklahoma, and north along the Mississippi and Ohio River drainages to s. Illinois and extreme se. Indiana.

WHERE TO SEE: Corkscrew Swamp Sanctuary, Florida, US; Congaree National Park, South Carolina, US; Cache River State Natural Area, Illinois, US.

NEARCTIC TEMPERATE DECIDUOUS FOREST

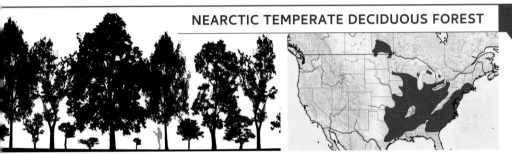

IN A NUTSHELL: A diverse, closed-canopy deciduous forest of the e. Nearctic dominated by beech, maple, oak, and hickory. **Habitat Affinities:** EUROPEAN TEMPERATE DECIDUOUS FOREST; EAST ASIAN TEMPERATE DECIDUOUS FOREST. **Species Overlap:** NEARCTIC TEMPERATE MIXED FOREST; WESTERN RIPARIAN WOODLAND; BALD CYPRESS–TUPELO FOREST.

DESCRIPTION: South of the taiga (or boreal) zone, this is the most widespread habitat in the e. Nearctic and familiar to almost everyone who has spent time outdoors east of the Mississippi River. Vast broadleaf forests have historically blanketed the e. Nearctic, and many large swaths of forest are still intact. The Nearctic temperate deciduous forest experiences distinct seasons with cold winters, hot and humid summers, and significant rainfall. Winter lows vary from 30°F to -15°F (-1 to -26°C), while summer highs are in the range of 80–95°F (27–35°C). Precipitation falls consistently year-round as either rain or snow and totals 30–60 in. (750–1,500mm).

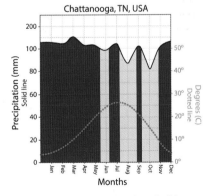

Chattanooga, TN, USA

Nearctic temperate deciduous forest typically has a closed canopy and well-developed mid-story, shrub layer, and understory. The canopy height is typically 50–80 ft. (15–25m), the trees closely spaced. The understory and shrub layer can be quite thick and difficult to navigate off trails, except in rare old-growth stands. During much of spring and early summer, the ground can be flooded or dotted

Above: **Nearctic temperate deciduous forest in the depths of winter, when the trees are largely devoid of leaves, in Maryland, US.** © IAIN CAMPBELL, TROPICAL BIRDING TOURS

Below: **Temperate deciduous forest provides both migration stopover and breeding sites for popular birds like warblers, which can create quite a bit of excitement when they arrive in springtime, at sites such as Magee Marsh, Ohio, US.** © IAIN CAMPBELL, TROPICAL BIRDING TOURS

with ephemeral pools. Thankfully, nature reserves in this habitat are well developed and plentiful, and extensive trail networks make exploring the habitat quite easy.

This forest has a high diversity of canopy trees compared to other Nearctic habitats, and species composition varies throughout the range. In general, drier, more westerly forests and south-facing slopes are dominated by a wide range of oak and hickory species, while wetter forests and those on cooler north-facing slopes predominantly consist of maples, American Beech, basswoods, and ashes. Tulip-tree is common throughout.

Nearctic temperate deciduous forest is the best-preserved example of temperate broadleaf forests in the world. However, all but a tiny fraction of this habitat was logged before 1930, and what remains is principally second growth. Additionally, American Chestnut, a keystone tree species that once comprised 25% of the individual trees of this habitat was virtually eliminated by the fungus-caused chestnut blight by about 1940. Other tree species that have succumbed to pathogens include American Elm and its congeners and many ash species, which have been decimated by Dutch elm disease and Emerald Ash Borer, respectively. Despite these losses, forest cover in the e. United States has been steadily increasing since 1920, and Nearctic temperate deciduous forest now occupies 70% of its original range.

WILDLIFE: Most large mammals in Nearctic temperate deciduous forest were eliminated by the beginning of the 20th century through overhunting and persecution. The most notable large mammal is White-tailed Deer, which is abundant to the point of being destructive. In many forests there is a clear browse line, beneath which all understory vegetation has been eliminated by deer. American Black Bear is common in many areas and increasing in much of its range. Other common mammals include Southern Flying Squirrel, Eastern Fox and Eastern Gray Squirrels, Eastern Chipmunk, Eastern Cottontail, Common Raccoon, Virginia Opossum, Striped Skunk, Red Fox, Coyote, and Bobcat.

The bird communities include many Neotropical migrants. In larger, unbroken forest tracts, the diversity of resident species is relatively low; Wild Turkey, Cooper's and Red-tailed Hawks, Barred Owl, Eastern Screech-Owl, Red-bellied and Downy Woodpeckers, Blue Jay, American Crow, Carolina Chickadee, Tufted Titmouse, White-breasted Nuthatch, and Carolina Wren are common year-round. In spring and summer, the bird diversity in these forests more than doubles, as dozens of species of songbirds return to breed. Eastern Wood-Pewee; Great Crested and Acadian Flycatchers; Red-eyed and Yellow-throated (IS) Vireos; Blue-gray Gnatcatcher; Wood Thrush; Worm-eating, Kentucky (IS), Hooded (IS), and Cerulean Warblers; and Scarlet Tanager fill the forest with song. The months of May and June are fantastic for observing birds in this habitat.

This habitat also holds large numbers of endemic reptiles and amphibians. Gray Tree Frog, Wood Frog, American Toad, Black Rat Snake, Eastern Garter Snake,

Striped Skunk is a widespread North American mammal that occurs within this habitat. © BEN KNOOT, TROPICAL BIRDING TOURS

Temperate deciduous forests in spring see the arrival of many colorful breeding warblers that have spent the winter in the Neotropics. Among them is Cerulean Warbler, which winters in the Andes of n. South America and breeds in this habitat in e. North America. © IAIN CAMPBELL, TROPICAL BIRDING TOURS

Copperhead, Broad-headed Skink, and Eastern Box Turtle are common. The salamander diversity is particularly high; Nearctic temperate deciduous forest and NEARCTIC TEMPERATE MIXED FOREST of the Appalachian Mountains are the global center for salamander diversity, with more than 20% of the world's species. The Great Smoky Mountains in particular have exceedingly high amphibian diversity and levels of endemism.

DISTRIBUTION: Nearctic temperate deciduous forest is found over most of the e. United States and small sections of s. Ontario, Canada. At its far northern boundaries in c. Wisconsin and Massachusetts, it transitions to NEARCTIC TEMPERATE MIXED FOREST. Nearctic temperate deciduous forest is also replaced by Nearctic temperate mixed forest at higher elevations in the Appalachian Mountains. Along its western boundary, from Texas north to Minnesota, this forest was historically bordered by NEARCTIC OAK SAVANNA and TALLGRASS PRAIRIE, but these habitats have largely been converted to agriculture. In the south, from s. Georgia to coastal Texas and along the coastal plain north to Virginia, Nearctic temperate deciduous forest is supplanted by LONGLEAF PINE SAVANNA and BALD CYPRESS–TUPELO FOREST.

WHERE TO SEE: Shawnee State Forest, Ohio, US; Mammoth Cave National Park, Kentucky, US; George Washington and Jefferson National Forests, Virginia, US.

SIDEBAR 6.3 ▸ HABITAT USE BY MIGRATORY BIRDS

Every year, birds that have spent the past several months in the tropics return to northern parts of the globe to breed. There are also some austral migrants, which head south during the southern summer. While some species are quite generalized, a large number are highly selective when choosing a breeding site. As an example, when Louisiana Waterthrush returns to the NEARCTIC TEMPERATE MIXED FOREST to breed, it seeks out steep ravines with clear, fast-flowing streams and establishes a territory. Within that territory, it will choose a nest site with upturned trees, large fallen logs, or cut banks, where it constructs a nest in a small cavity or hollow. Come migration time, all that deliberation goes out the window. While traveling thousands of miles under unpredictable conditions, birds can't be picky. In migration, Louisiana Waterthrushes may be observed along streams but also stagnant pools, lakeshores, scrubby thickets, parks, gardens, and even lawns. Upon arriving on their Central American wintering grounds, Louisiana Waterthrushes will seek out streams for wintering but are also regularly found along lowland rivers, marshes, and even coastal mangroves. There is some deliberation in habitat selection, but the winter habitat is much more generalized than the hyperspecific breeding requirements. This same pattern is true across most long-distance migrants—very specific breeding habitat selection, moderately specific wintering habitat selection, and anywhere they can find along the migration route.

NEARCTIC TEMPERATE MIXED FOREST

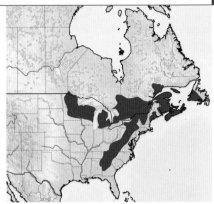

IN A NUTSHELL: A closed-canopy forest dominated by deciduous maples, beeches, and birches, along with coniferous Eastern Hemlock and Eastern White Pine. **Habitat Affinities:** EUROPEAN MOIST MIXED FOREST. **Species Overlap:** NEARCTIC TEMPERATE DECIDUOUS FOREST; NEARCTIC SPRUCE-FIR TAIGA; WESTERN RIPARIAN WOODLAND.

DESCRIPTION: Found in the e. United States and se. Canada, Nearctic temperate mixed forest is shady, moist, mossy forest crisscrossed by hemlock-lined ravines. This habitat walks the line between NEARCTIC TEMPERATE DECIDUOUS FOREST and NEARCTIC SPRUCE-FIR TAIGA while remaining distinct from both. The climate is characterized by long, cold winters and warm summers. Snow is often on the ground for five months of the year and contributes a significant portion of the 30–50 in. (750–1,250mm) of precipitation received annually.

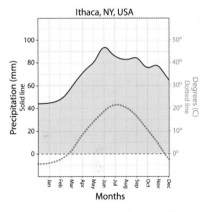

Nearctic temperate mixed forest has a very dense canopy that is typically 50–65 ft. (15–20m) tall. The poorly lit area under the canopy is home to only the most shade-tolerant plants and is often very open. Primary canopy species include American Beech, Sugar Maple, Red Maple, Black Cherry, and Eastern White Pine. The mid-story is dominated by Eastern Hemlock, which adds a distinctly northern feel. Smaller maples as well as Sweet Birch, Yellow Birch, and dogwoods make up most of the remaining mid-story. The shrub layer is often quite open. One major exception to this is encountered in the impenetrable thickets of rhododendrons and Mountain Laurel that grow along streams and ravines. The understory is also sparse and typically includes mosses, ferns, and a few herbaceous plants like violets and May-apple.

One major conservation concern is the rapid loss of Eastern Hemlock throughout this habitat. This defining tree is of great importance to breeding birds. Major die-offs caused by the Hemlock Woolly Adelgid, an introduced insect, have decimated hemlocks, especially in the southern part of their range.

WILDLIFE: The wildlife here shares features with both the NEARCTIC TEMPERATE DECIDUOUS FOREST and NEARCTIC SPRUCE-FIR TAIGA. Widespread species like White-tailed Deer, American

Nearctic temperate mixed forest in the Ozark Mountains, sc. United States.
© IAIN CAMPBELL, TROPICAL BIRDING TOURS

Black Bear, Common Raccoon, Red Fox, Eastern Cottontail, Eastern Gray Squirrel, American Red Squirrel, and Eastern Chipmunk are all apparent. The Appalachian Cottontail (IS) is endemic to the Appalachian portion of this habitat, as is Allegheny Woodrat. Typically boreal mammals like North American Porcupine, Least Weasel, Southern Bog Lemming, and Snowshoe Hare reach the southern extent of the easterly portion of their range in these forests.

Many birds also reach the southern extent of their breeding range in Nearctic temperate mixed forest. Migratory species like Alder Flycatcher; Blue-headed (IS) and Red-eyed Vireos; Brown Creeper; Winter Wren; Hermit Thrush; Black-throated Blue (IS), Black-throated Green, and Blackburnian (IS) Warblers; and Dark-eyed Junco breed here during the summer months. Open secondary growth associated with this habitat is also the primary breeding area for Golden-winged Warbler. Dense rhododendron thickets are an important breeding microhabitat for Swainson's Warbler within the Appalachians.

Relatively few reptiles are present here, but amphibian diversity is high, especially among salamanders. Mossy stream edges are home to nearly a dozen species of dusky salamanders (*Desmognathus*), and the surrounding forests hold many more. In spring, snowmelt produces ephemeral pools, free of predatory fish, which serve as critical breeding habitat for Blue-spotted and Jefferson Salamanders, Wood Frog, and Spring Peeper. Wood Turtles can also be seen, slowly making their way through the forest, often far from water.

DISTRIBUTION: This habitat of the e. United States and se. Canada is generally found to the north of NEARCTIC TEMPERATE DECIDUOUS FOREST but reaches as far south as n. Georgia at higher

Above: **An American Black Bear cub in Nearctic temperate mixed forest.** © SAM WOODS, TROPICAL BIRDING TOURS

Right: **Magnolia Warbler is one member of an attractive set of breeding warblers that visit Nearctic temperate mixed forest during the short, explosive summer breeding season (photo: Ohio, US).** © IAIN CAMPBELL, TROPICAL BIRDING TOURS

elevations in the Appalachian Mountains. It transitions to TALLGRASS PRAIRIE at its western limit in Minnesota and stretches east as far as s. Newfoundland. In se. Canada, Nearctic temperate mixed forest is replaced by NEARCTIC SPRUCE-FIR TAIGA.

WHERE TO SEE: Great Smoky Mountains National Park, Tennessee, US; Allegheny National Forest, Pennsylvania, US.

Kirtland's Warbler breeds exclusively in Jack Pine stands, found in the ecotone between Nearctic temperate mixed forest and spruce-fir taiga, making it difficult to allocate this species to a specific habitat (photo: Michigan, US). © IAIN CAMPBELL, TROPICAL BIRDING TOURS

SIDEBAR 6.4 ▸ MOUNTAINS CAN BE TWO-FACED

In the Northern Hemisphere between 30° and 50° latitude, north-facing slopes receive significantly less direct sunlight throughout the year. This results in a cooler and generally wetter microclimate on these slopes, which affects plant communities and their associated fauna. In the dry mountainous West, many localized and endemic salamanders, like the Jemez Mountains Salamander of New Mexico, are restricted to moist refugia on north-facing slopes. In the NEARCTIC TEMPERATE DECIDUOUS FOREST of the e. Nearctic, north-facing slopes often have trees more characteristic of NEARCTIC TEMPERATE MIXED FOREST, especially Eastern Hemlock. These cool microclimates also host more northerly breeding birds like Dark-eyed Junco, Black-throated Green Warbler, and Winter Wren, which might be absent from the southern slope of the same hillside.

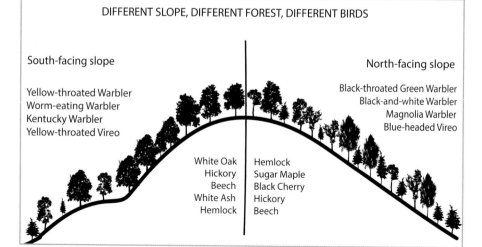

DIFFERENT SLOPE, DIFFERENT FOREST, DIFFERENT BIRDS

South-facing slope

Yellow-throated Warbler
Worm-eating Warbler
Kentucky Warbler
Yellow-throated Vireo

North-facing slope

Black-throated Green Warbler
Black-and-white Warbler
Magnolia Warbler
Blue-headed Vireo

White Oak	Hemlock
Hickory	Sugar Maple
Beech	Black Cherry
White Ash	Hickory
Hemlock	Beech

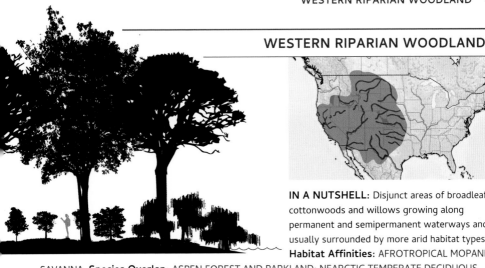

WESTERN RIPARIAN WOODLAND

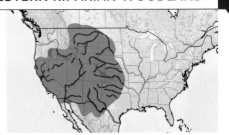

IN A NUTSHELL: Disjunct areas of broadleaf cottonwoods and willows growing along permanent and semipermanent waterways and usually surrounded by more arid habitat types. **Habitat Affinities:** AFROTROPICAL MOPANE SAVANNA. **Species Overlap:** ASPEN FOREST AND PARKLAND; NEARCTIC TEMPERATE DECIDUOUS FOREST; MONTANE MIXED-CONIFER FOREST; NEARCTIC FRESHWATER WETLAND.

DESCRIPTION: In the open expanses of the Great Plains and western shrublands and deserts, this habitat often appears as an oasis in the distance: isolated ribbons of impressively tall trees growing along the banks of rivers and streams as they traverse the landscape. This habitat spans a huge area of the United States, from the cold prairies of Montana to the sunbaked landscapes of New Mexico and California, and extends into s. Canada and n. Mexico. Despite its large range, the structure and nature of western riparian woodland remain consistent throughout—islands of dense, distinctly layered, deciduous broadleaf woodlands in a sea of other habitats. Taller forests occur in flatter areas with a wider floodplain. Steep, narrow or largely dry waterways may have few or no large trees and include only the shrub layer.

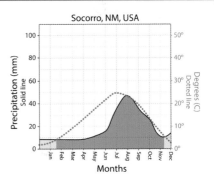

Although there are few unifying climatic characteristics, the plant composition is relatively consistent throughout. The largest and most dominant species in the canopy are Fremont and Plains Cottonwoods, which are separated by the Rocky Mountains. These trees form a dense canopy, usually greater than 60% coverage, and grow to a height of 65–80 ft. (20–25m). Under the dense canopy, the woodland can be open and park-like, with little understory. Along with these primary trees, Arizona Walnut, Goodding's Willow, and California and Arizona Sycamores are present and sometimes co-dominant in the southern half of the range. In the northern parts of the range, alders, Black Cottonwood, Box Elder, and Quaking Aspen contribute to the canopy.

The shrub layer is diverse, dense, and variable. Generally, it forms a dense thicket at the water's edge and thins as it is shaded out by canopy trees. In areas where there are no large cottonwoods, these thickets can form a dense woodland and grow up to 30 ft. (10m) tall, though heights of 5–15 ft. (1.5–4.5m) are more typical. Willow (*Salix*) is almost always the primary component or at

Above:
Yellow-breasted Chat, representing a monotypic family, is common within western riparian woodland. © BEN KNOOT, TROPICAL BIRDING TOURS

Left: **A Moose rests in western riparian woodland in the Tetons of Wyoming, US. Moose are often encountered around water, so this is a good habitat in which to seek them.** © BEN KNOOT, TROPICAL BIRDING TOURS

least co-dominant. There is a large array of other shrubs, including elderberries, serviceberries, dogwoods, poison oaks, mesquites, and various cherry species. The herbaceous layer is strongly dependent on distance from water, successional stage, and geographic range. Some grass is almost always present, especially close to the water's edge or on most recently exposed banks. In particularly wet areas, there is NEARCTIC FRESHWATER WETLAND habitat.

This is a dynamic ecosystem that is dependent on disturbance. The structure and continued survival of western riparian woodland depends on the waterways themselves. Changes in peak and minimum flows, the duration for which surface water is present, and the period between floods drastically affect the species composition and health of this habitat. Regular floods are important for creating successional zones, depositing fresh nutrients, and germinating dormant seeds. If floods are too intense, the entire area can be scoured clean and in the process robbed of nutrient-rich soils that would normally allow for regeneration.

This is one of the most threatened habitats in the West due to complex issues related to water. Flood controls have strongly altered spring flooding regimes. Water diversion for agriculture has reduced flow and killed vast swaths of woodland. Climate change has altered the length of droughts and the intensity of flooding, damaging riparian zones on both extremes of the precipitation spectrum. Additionally, non-native woody species such as tamarisk and Russian Olive are replacing willow and cottonwood stands and depleting valuable water, especially in desert ecosystems.

WILDLIFE: While western riparian woodlands generally lack unique fauna, they do have a much higher diversity and abundance of wildlife than surrounding habitats. Riparian zones serve as both magnets and corridors for nearby animals, especially those not comfortable in open environments. Common mammals include White-tailed Deer, Common Raccoon, Virginia Opossum, various squirrels, North American River Otter, North American Porcupine, woodrats, Gray Fox, and Bobcat. North American Beaver is abundant and is a very important ecosystem engineer. Dams built by beavers create wetlands and increase shrub diversity in the area. Beaver-created wetlands also protect riparian zones during periods of heavy flooding.

The bird communities of riparian zones are distinct from those of surrounding habitats, though they also include species from these adjacent habitats. The structure provided by large trees is important for nesting birds, especially raptors. Bald Eagle and Great Horned Owl are common in riparian zones throughout the range. In the Southwest, Gray Hawk and Common Black Hawk prefer to nest in riparian woodlands. Woodpeckers are common, Red-headed and Lewis's Woodpeckers, Red-naped Sapsucker, and Northern Flicker especially so. Secondary cavity nesters, including Eastern Bluebird and smaller owls like Western Screech-Owl and Elf Owl, take advantage of the many available woodpecker cavities. The dense willow thickets are full of birds. Yellow Warbler is omnipresent and perhaps defines this habitat more than any other species. Song Sparrow, Blue Grosbeak, Indigo and Lazuli Buntings, Bullock's Oriole, and Yellow-breasted Chat are also common. In the desert Southwest, endangered subspecies of Willow Flycatcher and Yellow-billed Cuckoo are found exclusively in dense riparian habitat.

Riparian zones have fairly simple reptile communities that are often a subset of generalist species from surrounding areas. The importance of western riparian woodlands to amphibians, on the other hand, cannot be overstated. Throughout much of the West, this habitat and associated wetlands provide the sole breeding grounds for frogs and toads in drier climates. Great Basin and Couch's Spadefoots; Western and Red-spotted Toads; Plains, Northern, and Relict Leopard Frogs; Canyon Tree Frog; and Northern Cricket Frog are just a few of the species that rely on this habitat for survival.

DISTRIBUTION: Western riparian woodland is found along permanent and semipermanent waterways, springs, and seeps throughout the w. United States. Tall, cottonwood-dominated woodlands are most common at low elevations, in flat areas, and along rivers with wide beds. As defined here, this habitat is found from the Great Plains west to the Central Valley of California. It occurs as far north as s. Saskatchewan and Alberta, Canada, and as far south as n. Chihuahua and Sonora, Mexico.

WHERE TO SEE: Bosque del Apache National Wildlife Refuge, New Mexico, US; Cosumnes River Preserve, California, US; Missouri Headwaters State Park, Montana, US.

ASPEN FOREST AND PARKLAND

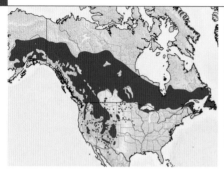

IN A NUTSHELL: Stands of Quaking Aspen, often interspersed with meadows, typically occurring within patches of coniferous forest or at the transition between grassland and coniferous forest. **Habitat Affinities:** No real affinities. **Species Overlap:** NEARCTIC SPRUCE-FIR TAIGA; NEARCTIC MONTANE SPRUCE-FIR FOREST; LODGEPOLE PINE FOREST; NEARCTIC TEMPERATE MIXED FOREST.

DESCRIPTION: Occurring in the mountainous West and throughout the taiga zone, aspen forest and parkland provides a welcome bit of variety within vast conifer-dominated areas. Aspen habitat has high biological productivity compared to surrounding areas and in summer is teeming with life. In the fall, aspen forest is clearly visible as a golden blaze amid a sea of dark green. As the name would suggest, aspen forest and parkland is dominated by the Quaking Aspen, with few other tree species present. Dense, often uniform-age stands of aspen grow to a height of 30–70 ft. (10–22m). In the mountainous West, firs, spruces, and Douglas-fir co-occur and if undisturbed will eventually replace the short-lived and shade-averse aspen. In the aspen forest and parkland of the n. Great Plains and the sub-boreal zone, Paper Birch, Balsam Poplar, and Red Maple are notable co-occurring canopy trees.

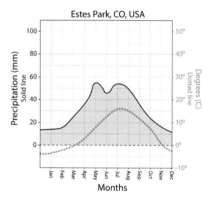

Estes Park, CO, USA

Aspen forest and parkland has an open canopy that allows for a diverse and lush understory of shrubs, grasses, and forbs. This habitat's association with high disturbance and grassland transitional zones means that small meadows and glades are an important component. Standing patiently along the edge of these glades can be an excellent strategy for wildlife-watching.

This habitat forms most commonly in areas of major disturbance—fire, landslide, avalanche, and insect kills—or in transitional zones between grassland and forest. In areas with frequent disturbance, Quaking Aspen is destroyed aboveground but quickly regenerates clonal colonies spread through it roots. One such colony in Utah, known as Pando, weighs 13 million lb. (6 million kg) and is an estimated 80,000 years old—making it a contender for the planet's heaviest and oldest organism.

This habitat is dominated by the distinctive Quaking Aspen, easily recognized by its slender, black-marked whitish trunk (photo: Teton Range, Wyoming, US). © BEN KNOOT, TROPICAL BIRDING TOURS

WILDLIFE: Aspen forests and parklands are prime mammalian grazing and browsing areas, especially compared with the surrounding coniferous forest. This results in a high concentration of ungulates, such as White-tailed Deer, Elk, Bighorn Sheep, and Moose, along with small herbivores like ground squirrels, pocket gophers, and Snowshoe Hare. With the presence of so many prey animals, these areas are attractive to Gray Wolf, Coyote, Red Fox, and American Mink. The abundant forbs include many species with edible bulbs and tubers—a favorite food for both American Black Bear and Brown Bear. Young Quaking Aspen is a preferred food tree for North American Beaver, and the older trees are often utilized in dam building. While the disturbance is helpful in maintaining aspen forest and parkland communities, flooding from dams can create conditions too wet for Quaking Aspen to grow.

Red-naped Sapsucker is a cavity nester that breeds within aspen forest. © BEN KNOOT, TROPICAL BIRDING TOURS

The bird communities here have similarities with both WESTERN RIPARIAN WOODLAND and the surrounding coniferous habitats, though with higher diversity than either of the two. Aspen trees are easily excavated and host an

abundance of cavity-nesting birds, including Mountain and Eastern Bluebirds, Yellow-bellied (IS) and Red-naped (IS) Sapsuckers, Merlin, House Wren, Northern Pygmy-Owl, and even Boreal Owl. Warbling Vireo, MacGillivray's and Wilson's Warblers, Lincoln's Sparrow, and Western Tanager are more common in aspen forest than in surrounding habitats in the West. In the East, American Woodcock likes the wet soil and clearings associated with this habitat, and the lush understory is critical breeding habitat for Ruffed Grouse. Veery, Mourning and Connecticut Warblers, Ovenbird, and Dark-eyed Junco are other ground-nesters that take advantage of the increased leaf litter and ground cover for nesting.

As in many montane and boreal habitats, the reptile and amphibian diversity is negligible.

DISTRIBUTION: Aspen forest and parkland occurs in mid- to upper elevations (generally 5,000–10,000 ft. (1,500–3,000m) in the mountainous West, from British Columbia, Canada, south into the United States through the Cascades and Sierra Nevada to c. California and south through the Rocky Mountains to c. Arizona and New Mexico. Aspen parklands are common from North Dakota west through the Canadian Prairie Provinces to c. Alberta and occur at the transition zone between TALLGRASS PRAIRIE and NEARCTIC SPRUCE-FIR TAIGA. Additionally, this habitat is found patchily in well-drained soils in the taiga zone from Maine and the Canadian Maritimes west to c. Alaska.

WHERE TO SEE: Yellowstone National Park, Wyoming, US; Rocky Mountain National Park, Colorado, US; Sax-Zim Bog, Minnesota, US.

NEARCTIC CLOUD FOREST

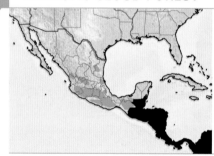

IN A NUTSHELL: Very wet, broadleaf-evergreen forest found in Mexican highlands, in which trees are laden with mosses and bromeliads, and the landscape is often inundated with fog and rain. **Habitat Affinities:** NEOTROPICAL CLOUD FOREST; AUSTRALASIAN SUBTROPICAL AND MONTANE RAINFOREST; AFROTROPICAL MONTANE FOREST. **Species Overlap:** MADREAN PINE-OAK WOODLAND; NEOTROPICAL CLOUD FOREST; NEOTROPICAL SEMI-EVERGREEN FOREST.

DESCRIPTION: This is an extension of NEOTROPICAL CLOUD FOREST; for complete description and habitat photo, see that habitat account in the Neotropics chapter.

WILDLIFE: Nearctic cloud forests are home to many widespread Neotropical mammals as well as a large array of endemic small mammals. Red Brocket Deer, Mexican Hairy Dwarf Porcupine, Northern Tamandua, White-nosed Coati, and Mexican Gray Squirrel are all common in the cloud forest. Puma, Jaguar, and Jaguarundi can be found as well but are present only in the most undisturbed and extensive forest tracts.

Nearctic cloud forests are home to an astonishing array of birds for such a small area. Well over 350 resident species can be found in the cloud forests of Mexico alone, and it is an important site for wintering migrants as well. Regularly encountered species include Red-billed Pigeon, Mottled Owl, Mountain Trogon, Northern Emerald-Toucanet (IS), Golden-olive and Ivory-billed Woodcreepers, Blue-headed Vireo, Happy Wren, Black-throated Green Warbler, and Red-headed Tanager. This habitat is also home to a wide array of hummingbirds, many of which are endemic. Long-tailed Sabrewing,

Blue-capped Hummingbird, White-tailed Hummingbird, Amethyst-throated Mountain-gem, and Garnet-throated Hummingbird are all restricted to the cloud forests of Mexico and adjacent Guatemala. In Chiapas, Mexico, and Guatemala the bizarre and spectacular Horned Guan inhabits Nearctic cloud forests on high volcanic peaks. This is also the habitat of the legendary Resplendent Quetzal. Widely considered one of the most beautiful birds on the planet, this glittering viridian and crimson jewel is an important seed disperser within the cloud forest.

The Nearctic cloud forest also hosts a wide array of reptiles and amphibians. The isolated nature of cloud forests in Mexico and the generally poor dispersal abilities of herpetofauna

Right: **Red-headed Tanager.** © PHIL CHAON, TROPICAL BIRDING TOURS

Below: **Horned Guan is a distinctive, rare, highly sought-after species of Nearctic cloud forests, typically inhabiting more remote, upper elevations of this habitat in volcanic areas. The most accessible sites are found in Guatemala, where this photo was taken.** © DANIEL ALDANA SCHUMANN, TROPICAL BIRDING TOURS

means a huge percentage of the species are found nowhere else. Some of these endemics include Smith's Arboreal Alligator Lizard, Godman's Mountain Pit Viper, and Franklin's Mushroom-tongued Salamander. The region still has incredible potential for discovery, and new species are described annually.

Endemism: With a disjunct distribution spread across many mountain ranges, Nearctic cloud forest is home to many unique species of vertebrates and contains a large number of EBAs. Notable regions of bird endemism include Los Tuxtlas in Veracruz, Mexico; the Northern Central American Highlands, of Chiapas, Mexico, and Guatemala; and the Sierra Madre del Sur in Oaxaca and Guerrero, Mexico. Nearctic cloud forest is also an important component of several Caribbean EBAs, especially those encompassing Jamaica and Puerto Rico.

DISTRIBUTION: Patchily distributed, Nearctic cloud forest occupies slightly more than 1% of the land in Mexico. Spread across 20 different states, these small patches of cloud forest occur at elevations between 2,000 ft. and 10,000 ft. (600–3,000m). Most are found in the upper parts of the Sierra Madre Oriental (Sierra de Juárez), Sierra Madre de Chiapas, Sierra Madre del Sur (Guerrero and Oaxaca), and the highlands of Jalisco. At lower elevations the Nearctic cloud forest is replaced by NEOTROPICAL SEMI-EVERGREEN FOREST and at higher elevations it is replaced by MADREAN PINE-OAK WOODLAND. Nearctic cloud forest also occurs at high elevations in Jamaica, Hispaniola, and Puerto Rico.

WHERE TO SEE: El Triunfo, Chiapas, Mexico; Blue Mountains, Jamaica; Pluma Hidalgo, Oaxaca, Mexico.

NEARCTIC TROPICAL DRY FOREST

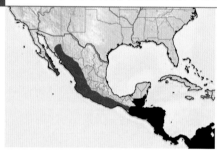

IN A NUTSHELL: Dry tropical forest that appears lush in the wet season but stark in the dry season, when the canopy loses many of its leaves. **Habitat Affinities:** NEOTROPICAL DRY DECIDUOUS FOREST; INDO-MALAYAN DRY DECIDUOUS FOREST. **Species Overlap:** MESQUITE BRUSHLAND AND THORNSCRUB; NEOTROPICAL SEMI-EVERGREEN FOREST.

DESCRIPTION: This habitat is an extension of NEOTROPICAL DRY DECIDUOUS FOREST; for complete description and habitat photos, see that habitat account in the Neotropics chapter.

WILDLIFE: This habitat is one of the world's biodiversity hotspots, though it may not seem like it at first glance. It is home to a rich array of mammals found nowhere else in the world. In the dry forests alone of Jalisco, Mexico, there are 10 endemic mammalian genera that include such species as Mexican Shrew, Trumpet-nosed Bat, Michoacan Deer Mouse, and Chamela and Magdalena Rats. Widespread species like White-tailed Deer, Collared Peccary, Virginia Opossum, Nine-banded

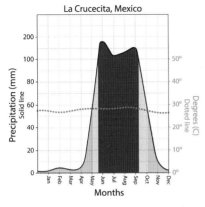

La Crucecita, Mexico

Armadillo, Mexican Gray Squirrel, Northern Tamandua, Gray Fox, Jaguarundi, Ocelot, Kinkajou, and White-nosed Coati occur as well.

The birdlife is just as divergent. Many widespread dry forest specialists occur, including Russet-crowned Motmot (IS), Streak-backed Oriole, Yellow Grosbeak, Lesser Ground-Cuckoo (IS), Lesser Roadrunner, Nutting's Flycatcher (IS), White-lored Gnatcatcher, Orange-breasted Bunting, Red-breasted Chat, and Citreoline Trogon. These species are joined by an array of localized endemics. The Balsas River basin of Michoacán and Guerrero and the interior Oaxaca region have Banded Quail, Beautiful Hummingbird, and Black-chested Sparrow. The area around Nayarit is home to Mexican Parrotlet, Mexican Woodnymph, and San Blas Jay. The narrow and windy Isthmus of Tehuantepec holds one of the Nearctic's most beautiful birds—the Rose-bellied Bunting, the plumage of which reflects every color of the setting sun from vibrant orange to those last dusky blues.

Reptiles are conspicuous and abundant here. Common species include Long-tailed Spiny Lizard, Western Spiny-tailed Iguana, Black-nosed Lizard, Least Gecko, Pacific Patchnose Snake, Western Lyre Snake, Central American Indigo Snake, and Pacific Parrot Snake. After the rains, amphibians

Tropical dry forest in Sumidero National Park in Chiapas, Mexico, during the dry season, before full leaf cover has returned. Vegetation cover is governed not by the traditional four seasons but by two seasons, dry and wet, with the forest turning green after wetter periods of the year. © NICK ATHANAS, TROPICAL BIRDING TOURS

seem to appear from nowhere—Giant Mexican Tree Frog, Marbled Toad, Sabinal Frog, and Mexican Burrowing Toad all emerge after a dormant dry season.

Endemism: This region encompasses several EBAs and is home to nearly one-third of Mexico's 120-plus endemic bird species. Major areas of endemism include Balsas River basin and interior Oaxaca, Isthmus of Tehuantepec, and the nw. Mexican Pacific slope. This habitat is also present on most Caribbean islands and includes at least some endemic birds.

DISTRIBUTION: Nearctic tropical dry forest is one of the most widely distributed habitats in w. Mexico, stretching from s. Sonora to the border of Guatemala in a wide band adjacent to the Pacific coast. These forests are also found in low inland valleys along the Balsas River and the Chiapas Depression.

Orange-breasted Bunting is a wide-ranging specialist of tropical dry forest (photo: Oaxaca, Mexico).
© PHIL CHAON, TROPICAL BIRDING TOURS

WHERE TO SEE: Huatulco, Oaxaca, Mexico; Xochicalco, Morelos, Mexico; San Blas, Nayarit, Mexico.

CEDAR SAVANNA

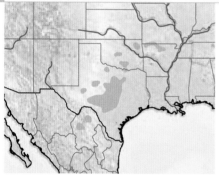

IN A NUTSHELL: An open, dry savanna of small oaks and Ashe Juniper. **Habitat Affinities:** AFROTROPICAL MIOMBO WOODLAND; AUSTRALASIAN BRIGALOW AND *CALLITRIS* WOODLANDS; MEDITERRANEAN OAK FOREST. **Species Overlap:** NEARCTIC OAK SAVANNA; CHIHUAHUAN DESERT SHRUBLAND; SHORTGRASS PRAIRIE.

DESCRIPTION: Cedar savanna is a dry and patchy, oak and juniper savanna located on steep and rocky limestone or dolomitic soils of the Texas Hill Country and a few other outlying pockets in the sc. United States and n. Mexico. The structure of the habitat varies with topography but comprises short grasses with scattered small trees, low shrubs, and rocky outcroppings. Occurring in a semiarid subtropical zone, the region has a highly seasonal climate, with moderately cold winters and hot summers. Typical winter low temperatures are 25–35°F (–4–2°C), and average summer highs

Cedar savanna in spring, in Kerr Wildlife Management Area, Texas, US.
© SAM WOODS, TROPICAL BIRDING TOURS

are around 90–95°F (29–32°C). The region receives about 30 in. (750mm) of annual precipitation, with peak precipitation falling in May and June. The Ozark subregion receives significantly more precipitation, about 45 in. (1,150mm).

Ashe Juniper, Eastern Red Cedar, and a variety of oaks, including Post, Blackjack, and Texas Oaks and Plateau Live Oak, are the primary canopy trees. Most canopy trees grow to a height of only 30–40 ft. (10–12m), while Ashe Juniper rarely surpasses 20 ft. (6m). The mid-story of these savannas is a smattering of sapling canopy trees and small shrubs, including Texas Persimmon, Agarito, Texas Mountain-laurel,

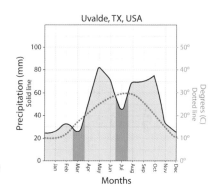

Honey Mesquite, and pricklypears. While generally sparse, the density of the shrub layer (particularly Honey Mesquite) can increase with overgrazing, choking out the normally dominant understory of Little Bluestem, Side-oats Grama, Texas Wintergrass, Curly Mesquite Grass, and buffalograsses.

Within this savanna landscape, small patches of dense forest occur on steep slopes and in valleys. These oak mottes and cedar brakes are areas of denser cover that lack a grassy understory and often harbor shade-tolerant sedges. Often small in area, cedar savannas were historically maintained through both natural and prescribed fires. As fire suppression has become a regular practice and grazing has intensified, many cedar savannas have been inundated with trees and shrubs, and have shrunk in size. This is especially true of the balds and dolomite glades in the Ozark and Ouachita Mountains, where this habitat was already relict and isolated. These pockets of this habitat are a

Cedar savanna is the primary breeding habitat for Black-capped Vireo (photo: Texas, US).
© SAM WOODS, TROPICAL BIRDING TOURS

unique subset restricted to shallow soils on the high points of hills and ridges. These remnants are relicts from a particularly hot and dry period 4,000–8,000 years ago, when this habitat was much more widespread. Without active maintenance and intervention by humans, these ancient vestiges would likely disappear.

WILDLIFE: Located at a biotic crossroads between NEARCTIC TEMPERATE DECIDUOUS FOREST, TALLGRASS PRAIRIE, and CHIHUAHUAN DESERT SHRUBLAND, the cedar savanna holds an interesting mix of mammal and bird species. Large mammals like White-tailed Deer, Collared Peccary (in Texas), and Feral Hogs are common due to the abundance of grazing land and acorn-rich oak forests. Pumas are still found in the Texas Hill Country, and a few remain in the Wichita Mountains of Oklahoma, while smaller carnivores like Bobcat, Coyote, Gray Fox, Ringtail, Common Raccoon, Striped Skunk, Western Spotted Skunk; and American Badger are more widespread and common. Other mammals found in this habitat include Black-tailed Jackrabbit, North American Porcupine, Virginia Opossum, Nine-banded Armadillo, and a large array of rodents. Perhaps the most spectacular mammal-viewing opportunities are afforded by the massive colonies of Mexican Free-tailed Bats that live in the limestone caves throughout the region. On spring and summer evenings, colonies numbering in the millions can be seen exiting these maternity caves to feed, while raptors like Red-tailed, Harris's, and Swainson's Hawks, Merlin, and Peregrine Falcon take advantage of this plentiful food source.

Common birds in the cedar savanna of Texas and the Wichita Mountains include Northern Bobwhite; Wild Turkey; White-winged and Mourning Doves; Greater Roadrunner; Black-chinned Hummingbird; Ladder-backed, Golden-fronted, and Red-headed Woodpeckers; Scissor-tailed Flycatcher; White-eyed Vireo; Black-crested and Tufted Titmice; Cliff and Cave Swallows; Bewick's Wren; Lesser Goldfinch; Field, Lark, and Rufous-crowned Sparrows; and Painted Bunting. Among nocturnal species, Common Poorwill, Common and Lesser Nighthawks, Chuck-will's-widow, and Eastern Whip-poor-will use cedar savanna—a large array of nightjars for a single habitat. The two undoubted avian stars of this habitat are the endangered Golden-cheeked Warbler (IS) and vulnerable Black-capped Vireo (IS), which are endemic and near-endemic breeding birds, respectively. Reliably found in spring and early summer, they are largely dependent on cedar savanna for food and nesting. Strips of Ashe Juniper bark are a particularly important material in nest construction, and oaks host a wide variety of caterpillars that are the primary food for chicks of these species.

A range of reptiles and amphibians can be found here. Great Plains Rat Snake, Eastern Milk Snake, Flathead Snake, Copperhead, Prairie Racerunner, Eastern Fence Lizard, and Eastern Collared Lizard are common. The areas occupied by cedar savanna also hold a large number of endemic salamanders that are associated with the limestone formations of the region but rarely with the cedar savanna itself. The underground caves and limestone aquifers of the Texas Hill Country have 14 unique species of blind aquatic salamanders in the genus *Eurycea*.

The Ozark and Ouachita Mountains of Oklahoma and Arkansas have isolated balds that, due to their restricted nature, hold a much smaller subset of the typical cedar savanna wildlife, including but not limited to Texas Mouse, Eastern Collared Lizard, Prairie Racerunner, Texas Brown Tarantula, Greater Roadrunner, and Painted Bunting.

DISTRIBUTION: In the sc. United States, cedar savanna is found mostly on the rugged limestone escarpments of the Edwards Plateau in Texas Hill Country. This core area stretches from San Antonio north and east to Waco and as far west as the edge of the Rio Grande Valley. Away from this core area, there are a few outlying pockets in n. Coahuila in Mexico, the Wichita Mountains of Oklahoma, and the Ozark and Ouachita Mountains of Missouri, Arkansas, and Oklahoma.

WHERE TO SEE: Balcones Canyonlands National Wildlife Refuge, Texas, US; Kerr Wildlife Management Area, Texas, US; Wichita Mountains National Wildlife Refuge, Oklahoma, US; White River Balds Natural Area, Missouri, US.

NEARCTIC OAK SAVANNA

IN A NUTSHELL: A widespread but disjunct habitat in which widely spaced oaks grow with an understory of grasses and forbs. **Habitat Affinities:** MEDITERRANEAN OAK FOREST. **Species Overlap:** TALLGRASS PRAIRIE; NEARCTIC DESERT GRASSLAND; CHIHUAHUAN DESERT SHRUBLAND; PACIFIC CHAPARRAL; MADREAN PINE-OAK WOODLAND; NEARCTIC TEMPERATE DECIDUOUS FOREST.

DESCRIPTION: Oak savanna occurs widely throughout the Nearctic in disjunct pockets, typically in transitional zones between grassland and forest habitats. This habitat is characterized by widely spaced broad-crowned oaks growing with an understory of grasses and forbs. These open, park-like environments are easy to walk in and are a great place to observe wildlife, especially during mast years (large seeding events).

The type and number of oak species varies by location, but oaks are always the dominant tree type. Common species include Bur, White, and Black Oaks in midwestern savannas, and Emory Oak in southwestern savannas. The Californian oak savanna is the most diverse and includes over a dozen species of oak, with Coast Live Oak, Valley Oak, California

Sierra Vista, AZ, USA

Black Oak, and Blue Oak the most prominent. The understory is usually a subset of grasses and forbs from adjacent grassland habitats. The understory of midwestern oak savanna is composed of TALLGRASS PRAIRIE species, while in southwestern oak savannas it is species from NEARCTIC DESERT GRASSLAND. In Californian oak savanna, the understory is often dominated by non-native grasses like Cheatgrass, as nearly all native grasslands have been eliminated through conversion to agriculture and introduction of invasive species.

Oak savannas occur in areas where the amount of precipitation is intermediate between that of adjacent grasslands and forests, or in areas with sandy or otherwise poor soil. Fire is crucial to the maintenance of the open, park-like nature of this habitat. In the Midwest, fire suppression has led to the near-complete disappearance of oak savanna, though active restoration efforts are underway.

WILDLIFE: While lacking a unique set of mammals, this habitat, with its open nature and abundance of food, is a great place to see wildlife. Oak savanna attracts an abundance of grazing animals, with large populations of White-tailed and Mule Deer. Squirrels and ground squirrels are common and provide a large prey base for the Coyotes, Bobcats, and American Badgers that are all readily seen here. In the western oak savanna, Ringtail is abundant relative to its presence in other habitats.

The bird life of oak savanna is strongly influenced by surrounding habitats. Wild Turkey and Lark Sparrow (IS) are found in oak savanna throughout their range. In the Midwest, Red-headed Woodpecker is particularly abundant in this habitat, while in the southwest, oak savanna is the best place to look for Montezuma Quail (IS). The oak savannas of California have a distinctive subset of birds. Oak Titmouse (IS), Nuttall's Woodpecker, and Yellow-billed Magpie are found almost exclusively in this habitat. Acorn and Lewis's Woodpeckers and Phainopepla are also especially common.

The sandy soils of oak savannas are great habitat for several species of toads, including American Toad, Oak Toad, and Eastern Spadefoot. With a large mammalian and amphibian prey base, snakes are numerous; Eastern Hognose Snake, Gopher Snake, California Kingsnake, Prairie Rattlesnake, Eastern Racer, and Ring-necked Snake are among the most common.

DISTRIBUTION: Oak savanna in the Nearctic occurs in three principal areas. The oak savanna of the midwestern United States occurs in a long strip between the TALLGRASS PRAIRIE and NEARCTIC TEMPERATE DECIDUOUS FOREST. This strip runs from Minnesota and Wisconsin south to Texas, with scattered pockets in sandy soils farther east. Most of the former oak savanna in the Midwest has become overgrown and transitioned to forest due to fire suppression.

In the Southwest, oak savanna occurs patchily as a transition between NEARCTIC DESERT GRASSLAND and MADREAN PINE-OAK WOODLAND in se. Arizona, sw. New Mexico, and w. Texas in the United States, south to Michoacán, Mexico.

Californian oak savanna forms another long strip from n. Baja California, Mexico, north up the interior of the states of California, Oregon, and Washington. It also occurs coastally and around the margins of the Central Valley in California.

WHERE TO SEE: Spenceville Wildlife Area, California, US; Oak Openings Metropark, Ohio, US; Miller Canyon, Arizona, US.

FLORIDA OAK SCRUB

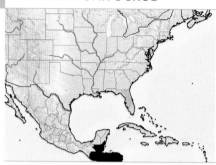

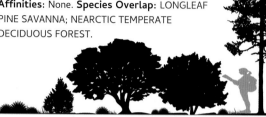

IN A NUTSHELL: A dry, open shrubland found on sandy soils in peninsular Florida. **Habitat Affinities:** None. **Species Overlap:** LONGLEAF PINE SAVANNA; NEARCTIC TEMPERATE DECIDUOUS FOREST.

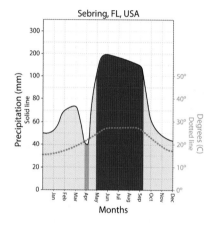

Sebring, FL, USA

DESCRIPTION: This hot, arid, open habitat is restricted to small areas of Florida, in the se. United States, and holds some unique fauna despite its small range. Florida oak scrub occurs on the sandy, nutrient-poor soils of inland ancient dune ridges as well as interior coastal dunes. The white sand soils are incredibly well drained, and despite the 50–60 in. (1,250–1,500mm) of rain received a year, this habitat remains very dry.

Florida oak scrub near St. Augustine, Florida, US, showing the dense shrub layer, sparsity of tall trees, presence of Saw Palmetto, and sandy soils characteristic of this local habitat type. © STACEY SATHER

There are rarely any trees, and the few that do occur are small and scattered Sand Pines that grow only to a height of 30 ft. (10m). The majority of the vegetation occurs in the shrub layer, which is open to moderately dense and rarely more than 6 ft. (1.8m) tall. Myrtle Oak or Scrub Oak are the two most dominant species. Sand Live Oak, Chapman's Oak, Crookedwood, pricklypears, and Florida Rosemary are usually present as well, and Saw Palmetto is often abundant.

The herbaceous layer is sparse, and the ground cover is largely exposed sand. Scattered grasses and forbs do occur, with Florida Bluestem, Gopher-apple, and Sandyfield Beaksedge the most prominent species. This area is home to a high concentration of endangered plants (15 species), including the beautiful Florida Bonamia, whose large purple-and-white flowers were once a common sight.

This is one of the most endangered habitats in the Nearctic, as over 90% has been lost to agricultural and residential development. Florida oak scrub is a fire-dependent ecosystem, relying on high-intensity but infrequent fires every 20–50 years. Natural fires burn patchily, creating a mosaic of different-aged shrub stands. Fire suppression leads to overcrowding of plants and the loss of key species. Thankfully, the majority of remaining habitat is in protected areas that are being managed for its continued survival.

WILDLIFE: The fauna is a rather limited set due to the nutrient-poor nature of this habitat and its limited range. Despite that, a large fraction of the wildlife here is unique to the area. White-tailed Deer, American Black Bear, Common Raccoon, Bobcat, and Nine-banded Armadillo all spend time foraging in Florida oak scrub, though there aren't the resources for them to live exclusively in this habitat. One of the only permanent resident mammals is the endemic and endangered Florida Mouse. The only member of its genus, this large mouse lives in burrows constructed within large Gopher Tortoise burrows.

The bird community is also sparse. The star attraction is Florida Scrub-Jay (IS), an endangered species not found outside of this habitat. Florida Scrub-Jays rely largely on the acorns found in the oak scrub for food, and cached acorns are an important method of seed dispersal. Other, common birds include Mourning Dove, Chuck-will's-widow, White-eyed Vireo, and Eastern Towhee.

The reptile communities here are the most unique aspect of this habitat. The endangered Gopher Tortoise is relatively abundant and important in creating burrows utilized by the highly local Gopher

Frog. Florida Scrub Lizard, Blue-tailed Mole Skink, and Florida Sand Skink are all endemic. Eastern Coachwhip, Mole Kingsnake, and Eastern Indigo Snake are also regularly encountered. Florida Crowned Snake and Short-tailed Snake are Florida oak scrub specialists. Florida Worm Lizard, a limbless and mostly blind reptile, is also endemic.

The locally common Florida Scrub-Jay is the ultimate indicator species of this limited habitat, never found outside of it. If you see the jay, you're in the correct habitat!
© DANIEL ALDANA SCHUMANN, TROPICAL BIRDING TOURS

This lizard is one of only two Nearctic members of a group of reptiles called amphisbaenians (the other is found on the Baja California peninsula). Also called Thunderworm, Florida Worm Lizard is usually seen aboveground only when its burrows are flooded during major storms.

DISTRIBUTION: Florida oak scrub is found on isolated sandy pockets throughout peninsular Florida, especially on Lake Wales Ridge and the Big Shrub region of Ocala National Forest.

WHERE TO SEE: Ocala National Forest, Florida, US; Merritt Island National Wildlife Refuge, Florida, US.

SHORTGRASS PRAIRIE

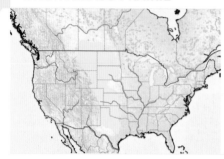

IN A NUTSHELL: A dry, open grassland with grasses rarely more than 6 in. (15cm) tall. **Habitat Affinities:** PALEARCTIC EASTERN GRASS STEPPE. **Species Overlap:** TALLGRASS PRAIRIE; NEARCTIC DESERT GRASSLAND; SAGEBRUSH SHRUBLAND.

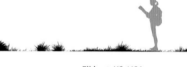

DESCRIPTION: Shortgrass prairie is a stark, low grassland occurring in the w. Great Plains in the arid rain shadow of the Rocky Mountains in c. United States and sc. Canada. This habitat is dominated by low-growing grasses and a high diversity but low density of forbs; woody vegetation makes up less than 1% of the landscape. The Great Plains shortgrass prairies experience cool winters and warm summers, with minimum temperatures ranging from 5°F (-15°C) in the north to 39°F (4°C) in the south, and maximum temperatures ranging from 64°F (18°C) in the north to 79°F (26°C) in the south. Much of the annual

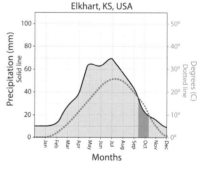

precipitation occurs during the spring and summer months, generally falling during a small number of intense events. The average rainfall for the region is 15 in. (380mm), with amounts generally increasing in the southern and eastern parts of the range. Growth of grasses is slow due to limited rainfall, and regrowth typically takes two to three years.

The shortgrass prairie is dominated by two low-growing warm season grasses: Blue Grama and Buffalograss. These two species make up the vast majority of the landcover, while a variety of forbs, succulents, and dwarf shrubs provide the majority of plant diversity in this habitat. Common shrubs and forbs include yuccas, pricklypears, Prairie Zinnia, Scarlet Globemallow, Plains Blackfoot, Slimflower Scurfpea, and Skunkbrush. In the wetter eastern sections of the shortgrass prairie, taller graminoids like Side-oats Grama and Little Bluestem can be found, though Blue Grama and Buffalograss still dominate. This broad band of mixed grass prairie is often treated as a separate grassland but is essentially an extensive ecotone between the shortgrass prairie and TALLGRASS PRAIRIE. With the

Bighorn Sheep grazing in shortgrass prairie beside the Badlands, South Dakota, US.
© GABRIEL CAMPBELL, TROPICAL BIRDING TOURS

exception of a few plants, such as yuccas, the vast majority of vegetative ground cover reaches a maximum height of 10 in. (25cm) and rarely exceeds 5 in. (13cm).

Variations in habitat structure and habitat maintenance are largely attributed to precipitation and grazing. Frequent droughts mean there is wide variation in the productivity and height of vegetation from year to year. Historically, this area was grazed by large herds of migratory herbivores, with American Bison being especially critical to the structure and annual maintenance of the shortgrass prairie. Along the habitat's southern border, changes in grazing and fire regimes have led to rapid incursion from Honey Mesquite and loss of shortgrass prairie. Unsuccessful cultivation efforts leading up to the Dust Bowl agricultural disaster of the 1930s also leave large scars on the landscape.

WILDLIFE: Shortgrass prairies were once home to the largest herds of herbivores in North America, and these animals are still a major feature of the landscape. While the vast herds of American Bison that shaped this habitat were extirpated long ago, 20th-century reintroduction efforts have established herds in protected areas throughout the shortgrass prairie's range. In addition to this landscape-altering bovid, the shortgrass prairie provides forage for Pronghorn, Mule Deer, and White-tailed Deer. Another landscape engineer, Black-tailed Prairie Dog is still an important feature of the landscape. Dog towns numbering in the millions are a thing of the past, but modern-day colonies provide prey and shelter for predators like American Badger, Coyote, the

Springtime sees Sharp-tailed Grouse use these shortgrass prairie areas for its spectacular seasonal displays, pictured here in Colorado, US. A still photo cannot give its antics full justice! © IAIN CAMPBELL, TROPICAL BIRDING TOURS

endangered Swift Fox (IS), and critically endangered Black-footed Ferret. Other common mammals on the shortgrass prairie include Black-tailed Jackrabbit, Thirteen-lined Ground Squirrel, Northern Pocket Gopher, and a variety of less-conspicuous small mammals.

The shortgrass prairie has limited bird diversity due to limited cover and resources, especially during the winter months. Only a hardy few are common year-round residents; among them are Mourning Dove, Northern Harrier, American Kestrel, Loggerhead Shrike, Horned Lark, and Western Meadowlark. Raptors such as Golden Eagle, Ferruginous Hawk, and Prairie Falcon are present year-round in small numbers. In summer, a larger variety of birds can be found, including Killdeer; Long-billed Curlew; Upland Sandpiper; Swainson's Hawk; Burrowing Owl; Common Nighthawk; Say's Phoebe; Western and Cassin's Kingbirds; Vesper, Grasshopper, and Lark Sparrows; and Lark Bunting. The region holds a variety of gallinaceous birds, including all three of the *Tympanuchus* grouse (Sharp-tailed Grouse, Lesser and Greater Prairie-Chickens), Scaled Quail, and Northern Bobwhite. Lesser Prairie-Chicken (IS) is endemic to the region, and Mountain Plover (IS), Sprague's Pipit, Baird's and Cassin's Sparrows, and McCown's Longspur (IS) are all breeding endemics or near endemics.

Prairie Racerunner, Greater Short-horned Lizard, Prairie Rattlesnake, and Gopher Snake are common and conspicuous reptile residents of this habitat.

DISTRIBUTION: Located in the Great Plains of North America, the shortgrass prairie is bounded to the west by the Rocky Mountains and extends from s. Alberta and Saskatchewan in the north to c. New Mexico and w. Texas in the south. The eastern boundary of shortgrass prairie is ill defined, as it blends with the TALLGRASS PRAIRIE in a long, dynamic, and complex ecotone, often referred to as the mixed-grass prairie. Pure shortgrass prairie occurs as far east as c. North Dakota in the north and c. Texas in the south; however, it exists as the dominant habitat type on the adjacent mixed-grass prairie.

The shortgrass prairie is the overwhelmingly dominant terrestrial habitat throughout its entire range. A scattering of prairie pothole and playa wetlands dot the region, which is also crisscrossed by narrow bands of WESTERN RIPARIAN WOODLAND. Apart from these isolated habitats, the shortgrass prairie is an unbroken habitat transitioning to SAGEBRUSH SHRUBLAND in Montana, Wyoming, and Colorado; to TALLGRASS PRAIRIE on its eastern border; and to NEARCTIC DESERT GRASSLAND of the Chihuahuan Desert in Texas and New Mexico.

WHERE TO SEE: Theodore Roosevelt National Park, North Dakota, US; Pawnee National Grassland, Colorado, US.

TALLGRASS PRAIRIE

IN A NUTSHELL: A tall, dense grassland with abundant flowering forbs. **Habitat Affinities:** PALEARCTIC WESTERN FLOWER STEPPE. **Species Overlap:** SHORTGRASS PRAIRIE; NEARCTIC OAK SAVANNA; LONGLEAF PINE SAVANNA.

DESCRIPTION: The tallgrass prairie is a dense and towering grassland habitat that historically stretched from Oklahoma in the sc. United States to Manitoba, Canada; it is the great Nearctic grassland of the e. Great Plains. A few tall species of grass dominate, producing 80% of the biomass, while a wide array of flowering plants makes up the remaining 20% of the prairie flora. While generally cooler and wetter than the SHORTGRASS PRAIRIE to its west, this habitat has such a broad latitudinal spread that it experiences a massive range of temperatures, from -31°F (-35°C) to 80°F (27°C) in the north, and 22°F (-5°C) to over 110°F (44°C) in the south. Despite these extremes, the region is overall temperate and receives a moderate amount of rainfall, 20–40 in. (500–1,000mm) annually, concentrated in intense storms in spring and early summer.

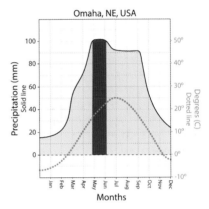

Omaha, NE, USA

Tallgrass prairie during early spring, before taller growth has occurred, in Oklahoma, US.
© IAIN CAMPBELL, TROPICAL BIRDING TOURS

Known as "the four horsemen of the prairie," Big Bluestem, Switchgrass, Indiangrass, and Little Bluestem are the core species of this habitat. These grasses are generally 2–3 ft. (0.6–0.9m) in height, though some can reach 10 ft. (3m). Growing among these core species is a bountiful collection of wildflowers, including Annual Sunflower, Leadplant, Scurfpea, Purple Coneflower, milk vetches, blazingstars, and Heath Aster, that transform the prairies into colorful patchworks during the spring and summer months. Historically rare away from permanent water, woody plants like Eastern Red Cedar, Buckbrush, Bur Oak, Common Hackberry, Plains Cottonwood, and Eastern Redbud have encroached as fire regimes have changed.

The tallgrass prairie was formerly subjected to sweeping disturbance that regularly cleansed the area and prevented the encroachment of trees. Rapid fires would tear through the grasses, recycling nutrients and renewing the habitat. The large herds of American Bison that once grazed here provided more localized disturbance and variation in vegetation height. Efforts are being made in remaining tallgrass prairie sites to restore and maintain these historical methods of disturbance, with regular burns occurring at most locations and active reintroduction of bison to larger tracts of protected land.

Unfortunately, the deep, fertile soils of the tallgrass prairie made for excellent agricultural land, and an estimated 96–99% of this habitat has been destroyed, to be replaced, largely, by soybean and corn monocultures. The largest remaining tracts exist on rocky or less arable soils in the Flint Hills of Oklahoma and Kansas and the Sandhills of Nebraska. There are major tallgrass prairie restoration projects ongoing throughout the region—perhaps the most impressive of these is the Midewin National Tallgrass Prairie in Illinois, which has revegetated an area of 20,000 acres (8,000ha) and actively hosts a herd of 50 bison.

WILDLIFE: Tallgrass prairie was once dominated by massive herds of grazing animals, but these habitat-shaping species are largely gone. While Elk and Pronghorn have been extirpated from the region, White-tailed Deer remains abundant, and American Bison is being reintroduced to numerous

Herds of burly American Bison still roam tallgrass prairies within well-protected areas.

Scissor-tailed Flycatcher is abundant and conspicuous during the boom period of a massive profusion of insect life that occurs in southern tallgrass prairies in spring and summer. Check fences and posts for the flycatcher, which habituually sallies from exposed perches. © BEN KNOOT TROPICAL BIRDING TOURS

larger tracts of protected prairie. Smaller carnivores like Coyote, Red Fox, Bobcat, American Badger, and Long-tailed Weasel remain common, feeding off abundant rodents such as Thirteen-lined Ground Squirrel, Plains Pocket Gopher, Prairie Vole, Eastern Woodrat, and Western Harvest Mouse. Larger carnivores including Gray Wolf and Puma have been largely extirpated from the prairies, but there have been sporadic individuals returning as overall population trends increase. Historically, large colonies of Black-tailed Prairie Dog were an important landscape feature, but the majority of the remaining colonies are in the SHORTGRASS PRAIRIE and mixed-grass ecotone.

The tallgrass prairie has relatively low avian diversity but very high densities of birds in the summer months, especially in the northern portion. During the winter months, the prairie is largely dominated by raptors such as Northern Harrier, American Kestrel, Red-tailed and Rough-legged Hawks, and Short-eared Owl. Northern Bobwhite and Greater Prairie-Chicken are found on the prairie year-round, and the springtime displays of Greater Prairie-Chicken are one of the great spectacles on the continent. In the summer months, the tallgrass prairie comes alive with grassland breeding specialists including Upland Sandpiper; Bobolink (IS); Sprague's Pipit; Sedge Wren; Henslow's, Grasshopper, Baird's, LeConte's, and Vesper Sparrows; Chestnut-collared Longspur; Eastern and Western Meadowlarks; and Dickcissel (IS). While prairie birds are lacking in color, their songs and flight displays are a delightful and impressive feature of the landscape. An abundance of insect prey makes Eastern and Western Kingbird and Scissor-tailed Flycatcher (in the south) ubiquitous in the warmer months.

The northern tallgrass prairie, known as the prairie pothole region, is dotted with numerous small wetlands that are critical breeding areas for North American waterfowl. Blue-winged Teal, Northern Shoveler, Gadwall, Mallard, Northern Pintail, Canvasback, Redhead, and Lesser Scaup are commonly found nesting in the cover of the upland tallgrass prairie adjacent to these wetlands.

A hyper-abundance of flowering plants makes the tallgrass prairie a very productive region for insects in the summer months. A number of rare butterflies are tallgrass prairie specialists, particularly Arogos and Dakota Skippers and Poweshiek Skipperling. Other butterflies characteristic of tallgrass prairie include Regal Fritillary, Gorgone Checkerspot, Common Ringlet, and Melissa Blue.

DISTRIBUTION: Located on the east side of the North American Great Plains from the sc. United States to sc. Canada, the tallgrass prairie stretches from n. Texas and Oklahoma in the south to Manitoba in the north, where it transitions to ASPEN FOREST AND PARKLAND or NEARCTIC SPRUCE-FIR TAIGA. The western edge of the tallgrass prairie is ill defined, as it slowly blends with the SHORTGRASS PRAIRIE in a long, dynamic, and complex ecotone often referred to as mixed-grass prairie. To the east, the prairies slowly transition to NEARCTIC OAK SAVANNA, then to forest. Historically, tallgrass prairie occurred in pockets as far east as w. Ohio and Kentucky, though little remains of these isolated eastern patches.

Other regions of the East have small, isolated pockets of prairie that have similar flora and wildlife to those of the tallgrass prairie. These small prairies were historically maintained by fire but have largely disappeared. Such small prairies were particularly common in the sc. United States and along large sections of the c. and s. Atlantic states. The Piedmont prairies are a particularly important subset of these small prairie pockets.

WHERE TO SEE: Joseph H. Williams Tallgrass Prairie Preserve, Oklahoma, US; Midewin National Tallgrass Prairie, Illinois, US; Sheyenne National Grassland, South Dakota, US.

PACIFIC CHAPARRAL

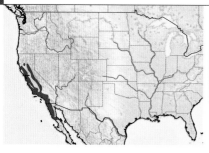

IN A NUTSHELL: A diverse and highly fire-dependent shrubland with a Mediterranean climate and largely sclerophyllous vegetation. **Habitat Affinities:** PALEARCTIC MAQUIS; PALEARCTIC GARRIGUE. **Species Overlap:** NEARCTIC OAK SAVANNA; COLUMNAR CACTUS DESERT.

DESCRIPTION: Pacific chaparral is the only truly Mediterranean habitat in the Nearctic. Chaparral experiences mild winters, when the temperature rarely drops below 32°F (0°C). The vast majority of annual precipitation falls during this period, accumulating an average of 10–17 in. (250–430mm). The summers are hot and dry, with little to no rainfall, and temperatures reach upwards of 110°F (43°C). This habitat occurs mostly on steep slopes with well-drained, rocky soils, blanketing foothills and coastal bluffs in a nearly impenetrable thicket. The woody shrubs that dominate this habitat grow 5–15 ft. (1.5–4.5m) tall and are mostly evergreen and sclerophyllous.

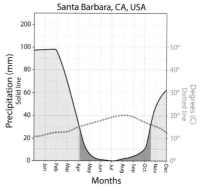

Santa Barbara, CA, USA

Pacific chaparral is an incredibly botanically diverse habitat, containing well over 1,400 documented plant species. With such high diversity, chaparral communities display a huge variation in species composition. Chamise is the most widespread and common shrub, and the closest to being a defining botanical feature. Other shrubs usually present in Pacific chaparral include manzanitas, ceanothuses, oaks, Western Poison Oak, mountain mahoganies, and Toyon. Coastal communities of Pacific chaparral regularly inundated by fog also contain areas dominated by California Sagebrush, Coyote Bush, bush lupines, and California Buckwheat.

Pacific chaparral in California, US, illustrating its floristic diversity and Mediterranean appearance.
© DORIAN ANDERSON, TROPICAL BIRDING TOURS

Pacific chaparral is famously fire-dependent. Under optimal conditions, this habitat experiences low-intensity ground fires every 5–20 years and high-intensity stand-replacing fires every 25–50 years. The characteristic flora is well adapted to these periodic burns and responds to fire in one of two ways. Re-sprouters have large subsoil burls, and the entire shrub regrows from this established root system after the destruction of aboveground vegetation. Obligate-seeders produce large quantities of seeds that remain dormant for long periods between fires. When exposed to fire and heat, these seeds germinate, replacing the parent plants consumed in the fire. A few species employ a mixture of these two strategies.

While Pacific chaparral is well adapted to fire, the human cohabitants of the area are not. This habitat surrounds some of the most densely populated areas in the Nearctic, with destructive and dangerous consequences. Fast-moving intense fires in summer and fall are often followed by heavy winter rains and rapid erosion of the now-unstable soil. This fire-flood-landslide cycle is all too familiar in s. California and n. Baja California and means that careful management of this habitat, including regular and well-timed prescribed burns, is crucial for the well-being of the human population.

WILDLIFE: Pacific chaparral is home to a high diversity of mammals, with over 50 species recorded, though none are exclusively found here. Columbian Black-tailed Deer, California Ground Squirrel, and Brush Rabbit are among the most visible mammals. There is a high diversity of small mammals; Santa Cruz Kangaroo Rat and White-eared Pocket Mouse are rare species largely restricted to chaparral environments. This habitat is a great place to see carnivores. Coyote, Bobcat, and Gray

Fox are all widespread and common, and this habitat supports remarkably high concentrations of Puma and Ringtail.

Compared to the mammalian fauna, the bird communities of Pacific chaparral are relatively species-poor. Wrentit (IS) and California Thrasher (IS) are strongly associated with Pacific chaparral and are omnipresent. Anna's Hummingbird, Greater Roadrunner, California Scrub-Jay, Bushtit, California and Spotted Towhees, and Lazuli Bunting are all very common as well. A specific subset of birds prefers recently burned chaparral, and Costa's Hummingbird, Bell's and Black-chinned Sparrows, and Lawrence's Goldfinch are readily found here for several years after a fire. California Gnatcatcher is found in the coastal, sage-heavy chaparral of s. California and n. Baja, where both the bird and the habitat are endangered. The most famous avian resident is the gargantuan, critically endangered California Condor. The majority of the world's population can be seen soaring over the chaparral-covered peaks of the coastal mountains of c. California.

A good diversity of reptiles is found in Pacific chaparral, including several specialties. Western Fence Lizard, Common Side-blotched Lizard, Western Whiptail, Gopher Snake, Pacific Rattlesnake, and Eastern Racer are all common and widespread species. Coastal Horned Lizard, Rubber Boa, Rosy Boa, and Garden Slender Salamander are all localized species occurring in Pacific chaparral.

While Pacific chaparral may be considered relatively species-poor for birds, it holds specialties, like this California Thrasher.
© DORIAN ANDERSON,
TROPICAL BIRDING TOURS

DISTRIBUTION: Pacific chaparral is found locally in the w. United States from sw. Oregon through California south to nw. Baja California, Mexico, in the Transverse and Peninsular coastal mountain ranges and in the foothills of the Sierra Nevada. Pacific chaparral dominates areas that have shallow rocky soil or have experienced major disturbance (especially fire). At higher elevations, regular frost excludes many chaparral species, and the habitat usually transitions to Ponderosa Pine–dominated MONTANE MIXED-CONIFER FOREST or SAGEBRUSH SHRUBLAND. In areas with deep, well-formed soils, Pacific chaparral is replaced by NEARCTIC OAK SAVANNA over time.

WHERE TO SEE: Pinnacles National Park, California, US; Angeles National Forest, California, US.

NEARCTIC ROCKY TUNDRA

IN A NUTSHELL: A low, mossy habitat that is under snow for much of the year. **Habitat Affinities:** EURASIAN ROCKY TUNDRA. **Species Overlap:** NEARCTIC BOGGY TUNDRA.

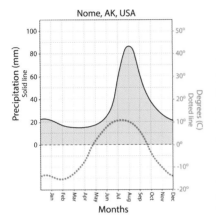

DESCRIPTION: This is an extension of EURASIAN ROCKY TUNDRA; for a full description of this habitat, see the account in the Palearctic chapter. The annual temperature in Nearctic rocky tundra is below freezing, and the average summer temperature is below 50°F (10°C).

Western Sandpiper on its breeding grounds within rocky tundra, near Nome, Alaska, US, where extensive examples of this habitat can be found. © IAIN CAMPBELL, TROPICAL BIRDING TOURS

The heaviest herbivore within rocky tundra is the Muskox (photo: near Nome, Alaska, US).
© IAIN CAMPBELL, TROPICAL BIRDING TOURS

WILDLIFE: There are few mammals hardy enough to survive the long and brutal winters in the range of Nearctic rocky tundra. Among the few that do survive are Caribou, Muskox, Arctic Fox, Arctic Hare, Arctic Ground Squirrel, collared lemmings, Brown Lemming, and voles. Lemmings and voles experience major cyclical variations in populations that drive changes in the populations of predatory birds and mammals in the region.

Apart from Common Raven and Snowy Owl, very few birds are resident in this habitat. In summer, millions of shorebirds arrive to breed here and gorge themselves on the astounding hatches of insects. While more species tend to breed in the NEARCTIC BOGGY TUNDRA, several specialize in this drier, rockier counterpart. Black-bellied Plover, American (IS) and Pacific Golden-Plovers, Western and Baird's Sandpipers, Red Knot, Surfbird, Sanderling (IS), Long-tailed Jaeger (IS), and Snowy Owl select dry and rocky tundra habitat. Snow Bunting and Lapland Longspur are abundant breeders and often the only passerines seen in this rocky habitat. On St. Matthew Island, Alaska, a large portion of breeding McKay's Buntings selects rocky sites, and in w. Alaska, Northern Wheatear also prefers this habitat.

DISTRIBUTION: Found at high latitudes from w. Alaska across the northern tier of Canadian provinces to e. Newfoundland. Most tundra is found north of 60°N, but it occurs locally as far south as 55°N in Manitoba and Alaska. Nearctic rocky tundra occurs on well-drained soils and at slightly higher elevations than the low and poorly drained NEARCTIC BOGGY TUNDRA. Rocky tundra has a very broad ecotone with NEARCTIC SPRUCE-FIR TAIGA, with small and scattered spruce trees growing in density and stature as you move south.

WHERE TO SEE: Nome, Alaska, US.

NEARCTIC BOGGY TUNDRA

IN A NUTSHELL: A low, marshy and boggy tundra with scattered meltwater pools. **Habitat Affinities:** EURASIAN BOGGY TUNDRA. **Species Overlap:** NEARCTIC ROCKY TUNDRA.

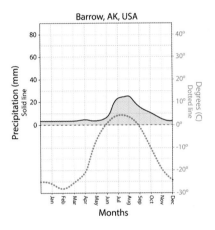

Barrow, AK, USA

DESCRIPTION: This is an extension of the EURASIAN BOGGY TUNDRA; for a full description of this habitat, see the account in the Palearctic chapter.

WILDLIFE: Wet and marshy in summer and buried in comparatively deep snows in winter, Nearctic boggy tundra has even fewer species of mammals than NEARCTIC ROCKY TUNDRA. Collared lemmings, Brown Lemming, and Red-backed Vole are the only true mammalian residents in this habitat. During the summer months, Caribou and Muskox will spend time grazing in these low areas but typically move to higher ground. Polar Bear and Arctic Fox can also be found foraging here during the summer and fall months but will return to the coast and extensive pack ice during winter and spring.

While mammal diversity is lower, the number of birds breeding in the boggy tundra is much higher than in adjacent rocky habitat. More insects, open water, and denser cover all drive the higher abundance and variety of species found here. Most Arctic breeding waterfowl use this habitat, and the boggy tundra is home to Steller's, King, and Spectacled Eiders and Long-tailed Duck (IS)—all easily among the most spectacular ducks in the world. Emperor, Ross's (IS), Snow, and Greater White-fronted Geese all nest here in clustered colonies. Ross's and Snow Geese form particularly dense and large colonies that can be heard from great distances; with their grazing, these geese alter the slow-growing tundra and can cause permanent damage and widespread loss of vegetation. In the areas surrounding these colonies, birds nesting in Nearctic boggy tundra have shown significant declines in density in response to these habitat changes. Yellow-billed (IS), Red-throated, Pacific, and Arctic Loons; Parasitic and Pomarine Jaegers; and over 25 species of shorebirds breed here as well. While shorebirds are often considered drab and skittish on migration and the wintering grounds, the breeding shorebirds of the Arctic are a completely different animal. Many obtain bright and intricately patterned alternate plumage that is utilized both in display and as camouflage. Male Buff-breasted Sandpipers perform elaborate dances to groups of potential mates; Dunlins sing complex warbling songs in flight that can last for minutes on end; and Red Knots flutter and soar in an upward spiraling display.

Above: **Once melts occur in spring, the waters of the boggy tundra open up and attract plentiful breeding opportunities for masses of shorebirds, geese, loons, and ducks, like these Spectacled Eiders in Barrow, Alaska, US. King and Steller's Eiders also breed in boggy tundra.**
© IAIN CAMPBELL, TROPICAL BIRDING TOURS

Right: **Once the waters open up in spring, masses of shorebirds descend on the boggy tundra to breed, including many Red Phalaropes, like the one pictured here in Barrow, Alaska, US. At this time, during the short period of abundant insect life and open waters available to forage within, the diversity and abundance of breeding shorebirds is high.**
© IAIN CAMPBELL, TROPICAL BIRDING TOURS

The breeding season here is incredibly short, and many shorebirds arrive, nest, and leave in less than five weeks' time. While adults will provide some preliminary care and protection, shorebird chicks are left to fend for themselves long before they are capable of flight. This leads to oddly staggered migrations and even different migratory routes between adults and juveniles. Juvenile Sharp-tailed Sandpipers (a Eurasian breeding shorebird) will often detour thousands of miles east to Alaska before heading to their wintering grounds in Australia. For this reason, juvenile Sharp-tailed Sandpipers are a regular vagrant to the United States, while adults are almost never seen.

DISTRIBUTION: The distribution of Nearctic boggy tundra mostly overlaps that of NEARCTIC ROCKY TUNDRA, stretching from w. Alaska across the northern tier of Canadian provinces to e. Newfoundland. This habitat is more common coastally and in low-lying areas.

WHERE TO SEE: Barrow, Alaska, US.

NEARCTIC ALPINE TUNDRA

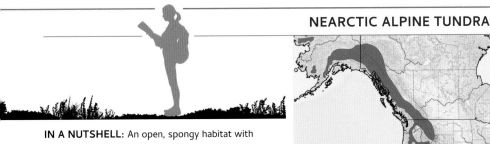

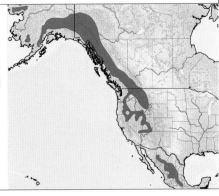

IN A NUTSHELL: An open, spongy habitat with low forbs and grasses found above tree line in temperate and subarctic regions. **Habitat Affinities:** EURASIAN ALPINE TUNDRA. **Species Overlap:** NEARCTIC ROCKY TUNDRA; HIGH-ELEVATION PINE WOODLAND.

DESCRIPTION: Alpine tundra offers a glimpse into the polar regions. After passing through the high-mountain forests and scrubby krummholz zones (see sidebar 2.3), one meets this treeless, windswept landscape, the last vegetated habitat on the tops of higher mountains. Above alpine tundra is the snow line, beyond which snow and ice persist year-round, preventing even the scant growing season experienced on the alpine tundra.

Alpine tundra is above the tree line, the exact elevation of which varies with latitude. Climatically, this is one of the harshest environments in which plants survive. A suite of adaptations allows for plant growth despite high winds, frequent disturbance, poor soils, and frigid temperatures. The

Hoary Marmot sitting within an expanse of alpine tundra in Washington, US. These large ground squirrels hibernate for many months to survive through the winter in this challenging environment.

© BEN KNOOT, TROPICAL BIRDING TOURS

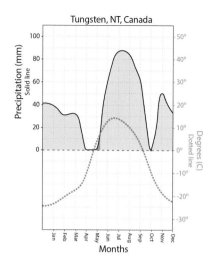

Tungsten, NT, Canada

Precipitation (mm)
Solid line

Degrees (C)
Dotted line

Months

American Pika lives on rocky slopes within alpine tundra, collecting plant materials in the growing season to sustain it through winter.
© BEN KNOOT, TROPICAL BIRDING TOURS

sedges, grasses, and forbs that dominate the landscape rarely grow more than 12 in. (30cm) tall. Many of the woody cushion plants form tight, ground-hugging mats that are an inch (2.5cm) or less above surface level. Areas sheltered from intense winds can hold small willows and taller meadow grasses but are also prone to retaining snow for longer periods, which further shortens the growing season. To counteract the truncated growing season, perennial plants keep a reserve of nutrients that allows for rapid growth under favorable conditions. Some of these plants are capable of growing, blooming, and producing seeds within a few short weeks.

In addition to their short stature, many plants of the Nearctic alpine tundra have tomentose stems and leaves—covered in dense, woolly hairs that retain moisture and trap air, protecting the stems and leaves from frost. A red pigment (anthocyanin) prevalent in many tundra plants is visible at either end of the growing season. Anthocyanin helps the plant absorb ultraviolet radiation, which is converted

The high-elevation, harsh environment of alpine tundra has a limited bird list but is home to the hardy White-tailed Ptarmigan (photo: Colorado, US).
© ANDREW SPENCER

into heat, allowing the plant to photosynthesize at very low temperatures.

For further detail on the mechanics of alpine tundra, see the EURASIAN ALPINE TUNDRA account in the Palearctic chapter.

WILDLIFE: Nearctic alpine tundra is an incredibly harsh environment with few permanent residents. The hardy few include Yellow-bellied (IS) and Hoary Marmots and Collared and American Pikas. The marmots will spend upwards of eight months of the year hibernating to escape the dire winters. The pikas spend the summer creating large hay piles of grasses and wildflowers, which they subsist on in subterranean burrows all winter long. In the summer months, the Nearctic alpine tundra is high-quality grazing land, and Mountain Goats, Elk, Bighorn Sheep, and Dall's Sheep come from lower elevations to feed.

While every year, the ROCKY TUNDRA and BOGGY TUNDRA of the Arctic serve as breeding grounds for tens of millions of migratory birds, alpine tundra does not. Surfbird is the lone shorebird that breeds on Nearctic alpine tundra, returning annually to this habitat in the North Slope of Alaska. American Pipit is the only other long-distance migrant that returns to breed on alpine tundra. Brown-capped, Black, and Gray-crowned Rosy-Finches spend time feeding on Nearctic alpine tundra in summer, preferring to nest around rocky outcrops and snowfields. Common Raven and Golden Eagle will both hunt on the open grassy elements of alpine tundra but prefer to nest on cliffs. White-tailed Ptarmigan (IS) is the only true Nearctic alpine tundra specialist, spending the entirety of its life in this habitat, throughout the Rocky Mountains and to around Anchorage. Farther north and across the top of the continent, Rock and Willow Ptarmigans are also present.

While there are no reptiles or amphibians in this habitat, there is a surprisingly high diversity of butterflies. Multiple species are endemic to isolated pockets of tundra habitat, which also hold disjunct populations of species found mostly in the Arctic.

DISTRIBUTION: Nearctic alpine tundra has a wide distribution at fairly low elevations throughout c. Alaska and Canada's Yukon and Northwest Territories. Farther south, its distribution follows the Rocky Mountains, Coast Mountains, Cascades, and Sierra Nevada. In Mexico, Nearctic alpine tundra is restricted to the heights of a few towering stratovolcanoes in the south. In terms of elevational distribution, Nearctic alpine tundra occurs near sea level at 70°N; at 4,500 ft. (1,350m) at 50°N; at 9,200 ft. (2,800m) at 40°N; and at 13,000 ft. (4,000m) at 20°N.

WHERE TO SEE: Rocky Mountain National Park, Colorado, US; Mt. Rainier National Park, Washington, US.

NEARCTIC FRESHWATER WETLAND

IN A NUTSHELL: Highly variable areas of freshwater-dominated habitat, ranging from permanent to ephemeral, with a wide array of potential hydrophilic vegetative communities. **Habitat Affinities:** PALEARCTIC TEMPERATE WETLAND. **Species Overlap:** NEARCTIC SALT MARSH; NEARCTIC BOGGY TUNDRA; NEARCTIC TIDAL MUDFLAT; TALLGRASS PRAIRIE; NEARCTIC MANGROVE; BALD CYPRESS–TUPELO FOREST.

The haunting sound of the Common Loon echoing across a still lake during springtime in the n. United States or Canada is one of the most atmospheric calls in the North American bird world. © BEN KNOOT, TROPICAL BIRDING TOURS

DESCRIPTION: Wetland communities in the Nearctic are far more varied than described here and could have a whole text written on them alone. Freshwater wetlands include everything from the massive open waters of the Great Lakes to the ephemeral playas of the Mojave Desert and all the rivers, lakes, ponds, swamps, bogs, fens, prairie potholes, cattail marshes, and inland mudflats in between. Wetland habitats that function primarily as forests (BALD CYPRESS–TUPELO FOREST, WESTERN RIPARIAN WOODLAND), NEARCTIC BOGGY TUNDRA, and several types of saltwater wetlands are addressed elsewhere in the Nearctic chapter.

WILDLIFE: Many mammals will utilize Nearctic freshwater wetlands for foraging and as a source of water, but few are expressly tied to it. Mammals with a strong affinity for wetlands include Marsh and Swamp Rabbits, Common Muskrat, North American Beaver, North American River Otter, American Mink, and Moose.

By contrast, Nearctic freshwater wetlands are one of the most productive habitats for birds on the continent. In the summer, wetlands in northern latitudes host hordes of breeding ducks, herons, grebes, rails, wrens, sparrows, and blackbirds. Trumpeter Swan, American Wigeon, Blue-winged Teal, Pacific Loon, Pied-billed and Clark's Grebes, Double-crested Cormorant, American and Least Bitterns, Yellow and King Rails, American Coot, Franklin's Gull, Black Tern, Marsh and Sedge Wrens, Common

Yellowthroat, Swamp Sparrow, and Yellow-headed and Red-winged Blackbirds are just a few wetland breeding birds restricted to the Nearctic. In winter, these same species head south to join resident wetland birds in more temperate climates. Flocks of tens of thousands of Snow Geese, Ross's Geese, Sandhill Cranes, American Wigeons, Northern Shovelers, Northern Pintails, and many more species are not uncommon in winter in the s. United States, especially in isolated wetlands in the western states.

The subtropical wetlands of the se. United States and Mexico hold their own resident avifauna, including many species shared with the Neotropics. Black-bellied Whistling-Duck, Masked Duck, Neotropic Cormorant, Anhinga, Purple Gallinule, Little Blue and Tricolored Herons, and Wood Stork are all birds more typical of southern freshwater wetlands. Many areas of the world have birds that breed in close association with swiftly flowing rivers and streams. In the Nearctic, American Dipper and Harlequin Duck fill this role.

Nearctic freshwater wetlands hold a large percentage of the continent's reptiles and amphibians. American Alligator is an iconic giant and the largest predator in many southern freshwater wetlands. Freshwater wetlands in Mexico are also home to the smaller Spectacled Caiman. Water snakes (*Nerodia*), garter snakes (*Thamnophis*), Cottonmouth, Mud Snake, Rough Green Snake, and Black Swamp Snake are all wetland specialists. Most of the Nearctic's frogs and turtles breed in wetlands, with especially high diversity in the se. United States. Newts (*Notophthalmus*), sirens (*Siren*), dwarf sirens (*Pseudobranchus*), mudpuppies (*Necturus*), and amphiumas (*Amphiuma*) are salamanders that spend most of their lives in freshwater wetlands.

There are over 1,600 species of freshwater fish endemic to the Nearctic. From frozen Arctic lakes to desert pools in the Mojave Desert, fish can be found in nearly every permanent freshwater habitat on the continent. There are some incredible species, though observing them is difficult at best. The tiny Desert Pupfish can tolerate temperatures of 113°F (45°C). The enormous American Paddlefish and Alligator Gar cruise the vast rivers of the e. United States. A rainbow multitude of darters are endemic to the tiny streams of the Appalachian Mountains.

Endemism: The Central Mexican Marshes EBA, encompassing a few wetlands near Mexico City, is home to Black-polled Yellowthroat and Aztec Rail, and once hosted the now-extinct Slender-billed Grackle. Yellowthroats are particularly sedentary, and Belding's and Altamira Yellowthroats are endemic to the Baja Peninsula and ne. Mexico, respectively. The massive Zapata Swamp in Cuba is home to the endemic Zapata Wren, Zapata Rail, Zapata Sparrow, and Cuban Crocodile, as well as the near-endemic Red-shouldered Blackbird. Many isolated marshes and river drainages have their own endemic turtles, frogs, salamanders, and fish. The small streams of the Appalachians have

A male Barrow's Goldeneye calls during the breeding season in the nw. United States.
© BEN KNOOT, TROPICAL BIRDING TOURS

high levels of endemism among darters (*Etheostoma*). Subterranean limestone aquifers in Texas and Mexico are home to dozens of endemic salamander species.

DISTRIBUTION: Nearctic freshwater wetlands are found patchily throughout the continent, interspersed among a huge array of habitats. These wetlands principally occur in low-lying and flat areas, coastal plains, near large rivers, and around the Great Lakes. Areas of the Nearctic with a large percentage of freshwater wetlands include the coast of the Gulf of Mexico, the s. Mississippi River, the prairie pothole region of the n. Great Plains, and vast expanses of NEARCTIC SPRUCE-FIR TAIGA and NEARCTIC BOGGY TUNDRA. Many artificial wetlands have been created, some by the National Wildlife Refuge System. These areas provide recreation areas for waterfowl hunters and important habitat for hundreds of nongame species as well. A self-guided auto tour through a national wildlife refuge is a great way to see a huge abundance and diversity of wildlife, particularly birds.

WHERE TO SEE: Ottawa National Wildlife Refuge, Ohio, US; Aransas National Wildlife Refuge, Texas, US; Lake Andes National Wildlife Refuge, South Dakota, US.

NEARCTIC MANGROVE

IN A NUTSHELL: Forest of salt-tolerant trees that grow coastally in the intertidal zone. **Habitat Affinities:** AUSTRALASIAN MANGROVE; INDO-MALAYAN MANGROVE FOREST. **Species Overlap:** NEARCTIC FRESHWATER WETLAND; NEARCTIC TIDAL MUDFLAT; NEARCTIC SALT MARSH.

DESCRIPTION: Growing in dense tangled forests in shallow coastal waters, mangroves are a pantropical feature of intertidal zones. The term *mangrove* refers to the particular life history of certain trees rather than any direct taxonomic group—certain salt-tolerant (halophytic) trees and shrubs of multiple lineages are called "mangroves." In some regions, mangrove forest can exhibit impressive floristic diversity, but in the Nearctic, it is relatively species-poor. While there are six mangrove species in the region, only four are common and widespread: Red Mangrove (*Rhizophora mangle*), Black Mangrove (*Avicennia germinans*), White Mangrove, and Buttonwood. Nearctic mangrove forest is typically short, 7–35 ft. (2–10m) in height, but may reach a maximum of 100 ft. (30m) in s. Mexico. These hot and humid forests are a dense maze of exposed roots emerging from soft, frequently inundated mud. Boats and boardwalks provide comfortable access to mangroves in many areas.

Mangroves serve as a critical barrier for both natural coastal habitats and human settlements. Absorbing damaging effects of strong storms, preventing saltwater inundation, and filtering nutrients and pollutants are just some of the important functions of mangroves. Unfortunately, mangrove areas are also desirable areas for human activity, and a large portion of Nearctic mangrove forest has been lost to road and housing development, shrimp farming, and salt production.

WILDLIFE: Mangroves, like many tidal environments, do not host a large array of mammals. The semiaquatic rice rats (*Oryzomys*) are the most abundant small mammal. Common Raccoons frequently forage and roost in mangroves, while the critically endangered Pygmy Raccoon of Mexico's Cozumel Island is found primarily in this habitat. Common Bottlenose Dolphin and West Indian Manatee visit deeper channels at high tide. In the Caribbean, many species of hutia (a type of rodent), including Cabrera's, Prehensile-tailed, and Desmarest's Hutia, can be found in mangroves, though not exclusively so.

Nearctic mangrove is generally poor in mammalian fauna but is a critical habitat for the rare Pygmy Raccoon, which is confined to the island of Cozumel in Mexico.

© PHIL CHAON, TROPICAL BIRDING TOURS

Functioning both as wetland and forest, Nearctic mangrove forest is home to a variety of herons, egrets, cormorants, and at low tide, shorebirds. During the boreal winter, this insect-rich environment plays host to large numbers of wintering migrant passerines including American Redstart; Prairie, Black-and-white, and Prothonotary Warblers; and Northern Waterthrush. In many areas of the world, mangrove forests have a small set of specialist birds alongside the more generalist wetland and coastal species. In the Nearctic mangrove forest, Mangrove Cuckoo (IS), Mangrove Vireo, Rufous-necked Wood-Rail, Mangrove subspecies of Yellow Warbler (IS), and American Flamingo are the species most closely tied to this habitat.

Nearctic mangrove forest is the preferred habitat of the American Crocodile, the top predator in this area. Several species of fish-eating snakes inhabit the mangroves; Banded Water Snake, Saltmarsh Snake, and Cottonmouth are the most common. During high tide several species of sea turtles will feed in the mangroves, and Kemp's Ridley Sea Turtle shows a particular fondness for mangrove roots and associated vegetation.

DISTRIBUTION: Nearctic mangrove forest is found in warm and shallow coastal environments in the s. Nearctic. Specifically, mangroves grow on the coasts of peninsular Florida, s. Texas, Mexico's e. coast from Veracruz southward, Mexico's Pacific coast in Baja Sur and from Sinaloa southward, and throughout coastal areas of the Caribbean.

WHERE TO SEE: Hatiguanico River, Cuba; Everglades National Park, Florida, US; Rio La Tovara, Nayarit, Mexico.

NEARCTIC TIDAL MUDFLAT

IN A NUTSHELL: Coastal areas of biologically rich soil that are exposed only at low tide or under favorable wind conditions. **Habitat Affinities:** AFROTROPICAL TIDAL MUDFLAT; AFROTROPICAL SALT PAN; PALEARCTIC TIDAL FLAT. **Species Overlap:** NEARCTIC FRESHWATER WETLAND; NEARCTIC SALT MARSH; NEARCTIC ROCKY COASTLINE AND SANDY BEACH.

DESCRIPTION: Principally occurring at estuaries, coastal mudflats are the result of rich organic material, sand, silt, and clay deposited by rivers. These mudflats are one of the most biologically productive habitats on the planet, as the constant flood of nutrients produces a massive amount of life, especially when compared to relatively sterile sand beaches and rocky coastlines. The unvegetated mudflats are exposed during low tides or when favorable winds result in the shifting of shallow waters. Once exposed, these open spaces, teeming with benthic invertebrates, become a crucial feeding area for masses of migratory birds and, to a lesser degree, resident coastal fauna.

WILDLIFE: Overall, coastal mudflats have little in terms of mammals, though in British Columbia and Alaska, Brown Bears are known to dig for the abundant mollusks found below the surface. Coastal mudflats are incredibly important feeding areas for migratory, wintering, and resident birds. At specific times of year, tens of thousands of waders congregate at these sites, producing a diverse

Tidal flats attract staggering congregations of migrant coastal birds, as shown here at Bolivar Flats, Texas, US, where co-author Phil Chaon is making the most of the abundant photo opportunities. Phil is surrounded by American Avocets, Piping Plovers, and a variety of sandpipers, including dowitchers, yellowlegs, and "peeps." © SAM WOODS, TROPICAL BIRDING TOURS

and spectacular congregation of species. The vast majority of North American shorebirds can be seen on Atlantic, Gulf, and Pacific coastal mudflats, though each coast holds a few species not shared with the other. Common birds at coastal mudflats include Black-bellied Plover; Willet; Least, Semipalmated, and Western Sandpipers; Marbled Godwit; Snowy Egret, and many more. The Atlantic coast holds huge concentrations of Red Knots at several key estuaries. Along the Gulf of Mexico and the Mexican Pacific coast, Reddish Egret is a mudflat specialist. In more temperate areas, exposed eelgrass beds are a popular grazing spot for Brant.

Apart from providing critical feeding grounds for migratory birds, the soft soils of coastal mudflats also serve as important spawning and nursery grounds for a wide array of fish when submerged.

DISTRIBUTION: Found locally along coastlines throughout the Nearctic.

WHERE TO SEE: Delaware Bay, Delaware and New Jersey, US; Humboldt Bay, California, US; Ensenada, Baja California, Mexico.

NEARCTIC SALT MARSH

IN A NUTSHELL: Dense grassy marshes found in coastal lowlands, especially in sheltered areas. **Habitat Affinities:** AFROTROPICAL TIDAL MUDFLAT AND SALT MARSH; AUSTRALASIAN TIDAL MUDFLAT AND SALT MARSH. **Species Overlap:** NEARCTIC TIDAL MUDFLAT; NEARCTIC MANGROVE; NEARCTIC FRESHWATER WETLAND; TALLGRASS PRAIRIE.

DESCRIPTION: Salt marshes are found in low and sheltered coastal areas that receive regular tidal inundation. In the Nearctic, salt marshes are dominated by halophytic grasses, especially cordgrasses. Height of the vegetation typically varies with distance from the waterline, with grasses increasing in height from 0.5–1 ft. (15–30cm) tall far from water to 4 ft. (120cm) near the water's edge. The high marsh also includes low-growing herbaceous plants like pickleweeds and seepweeds. These saltmarshes provide roosting habitat for many birds that feed on tidal mudflats and also are home to a small but unique subset of specialist birds and mammals.

WILDLIFE: This is a very mammal-poor environment due to the regular tidal inundation, though a very few species, such as Common Raccoon and North American River Otter, will forage here. The Salt Marsh Harvest Mouse of San Francisco Bay is one of the only resident mammals and an endangered saltmarsh endemic that forages almost exclusively on pickleweed.

Best explored by boat, the salt marsh can be teeming with waterbirds. Snowy and Great Egrets, and Little Blue, Tricolored, and Great Blue Herons are all abundant here within their ranges. Shorebirds from adjacent mudflat zones roost and forage in marshes during higher tides, and large, tightly clustered groups of Long-billed Curlew, Willet, Marbled Godwit, Western Sandpiper, Short-billed Dowitcher, and Greater and Lesser Yellowlegs are a common sight in their wintering ranges.

Salt marsh also holds a few specialized birds, though finding them in the dense, muddy grasses can be a challenge. On the Atlantic and Gulf coasts, it is the habitat of Seaside and Saltmarsh Sparrows, which can be found singing loudly from tall cordgrass stalks in spring before vanishing into the impenetrable depths. In winter they are joined by Nelson's Sparrow. The habitat is also home to Clapper Rail (IS) in the East and Ridgway's Rail in the West. These rails are best looked for walking along the edge of the grass at low tide or by listening for their raucous calls. Additionally, this is the primary habitat of the tiny Black Rail along the Atlantic and Gulf coasts. In winter, the salt marshes of the Texas Gulf coast are home to most of the world's population of the towering

Clapper Rail is a common and noisy inhabitant of salt marshes around High Island, Texas, US.

and spectacular Whooping Crane, and regular boat tours at Aransas National Wildlife Refuge are a great place to get a glimpse of this legendary bird.

DISTRIBUTION: Widespread coastally but sporadic. More common on the s. Atlantic and Gulf coasts due to gentler topography and calmer seas. On other coasts, more common where well protected, such as in Chesapeake Bay.

WHERE TO SEE: Aransas National Wildlife Refuge, Texas, US; Tijuana River Slough, California, US; Baja California (Mexico).

NEARCTIC ROCKY COASTLINE AND SANDY BEACH

IN A NUTSHELL: Coastal areas with poor rocky or sandy soil, usually lacking vegetation. **Habitat Affinities:** PALEARCTIC ROCKY COASTLINE AND SANDY BEACH; AFROTROPICAL ROCKY COASTLINE AND SANDY BEACH; INDO-MALAYAN ROCKY COASTLINE AND SANDY BEACH. **Species Overlap:** NEARCTIC TIDAL MUDFLAT; NEARCTIC SALT MARSH; NEARCTIC OFFSHORE ISLANDS; NEARCTIC PELAGIC WATERS.

DESCRIPTION: The most abundant of the Nearctic coastal habitats, this encompasses everything from the towering rocky cliffs of Maine and the Pacific Northwest to the scenic sandy beaches of Florida and the Yucatán. Coastlines of the n. Atlantic and n. Pacific coasts tend to be rockier, while the s. Atlantic and Gulf coasts tend toward sandy beaches. Sandy and rocky coastal areas tend to be nutrient- and wildlife-poor, though in some cases can be home to huge numbers of animals seasonally.

WILDLIFE: Mammals from nearby environments will use the coasts for foraging though are rarely resident. Pinnipeds such as Northern Fur Seal, Northern Elephant Seal, California and Steller's Sea Lions, and Harbor and Gray Seals use specific locales for massive haul-outs and as major pupping grounds where young are raised. Watching the daily lives of the multi-ton Northern Elephant Seals along the Pacific coast is a sight not to be missed.

These environments are great places to look for roosting flocks of terns, gulls, and shorebirds. No one species is common throughout, though Laughing Gull, Royal Tern, and Sanderling are all widespread. The rocky coasts of the Pacific hold a unique set of "rock-pipers" like Black Turnstone, Surfbird, Rock Sandpiper, and Wandering Tattler, which forage along boulders exposed at low tide.

Bar-tailed Godwit often forages on beaches.
© IAIN CAMPBELL, TROPICAL BIRDING TOURS

A Harlequin Duck resting on a rocky coastline in Washington, US. © BEN KNOOT, TROPICAL BIRDING TOURS

Surfbird is a regular sight on rocky coastlines along the west coast of North America.
© IAIN CAMPBELL, TROPICAL BIRDING TOURS

Tall cliffs along northern coastlines can be home to spectacular colonies of nesting seabirds. On the Atlantic coast, Northern Gannet, Common Murre, Razorbill, Black-legged Kittiwake, and Northern Fulmar can all be seen by the thousands, while the Pacific is home to Rhinoceros Auklet, Ancient Murrelet, Pigeon Guillemot, Tufted and Horned Puffins, and Brandt's and Pelagic Cormorants, among others. Snowy and Piping Plovers and Least Tern are all sensitive species nesting on sandy beaches and dunes. Their breeding grounds often conflict with human beach use, and large exclusionary areas are set up while nesting occurs. It wouldn't be a day at the beach without gulls, and among the more widespread species, the Yellow-footed Gull is found almost exclusively along the coast of the Baja Peninsula. Very few passerines use this environment, and among this group crows and ravens are undoubtedly the most successful. Northwestern Crow is perhaps the only rocky-coast-specific passerine in the Nearctic and may be found scavenging along the water's edge on the n. Pacific coast.

Flat rocky areas can form spectacular tide pools at low tide, and exploring these tiny marine havens can reveal all manner of crustaceans, fish, echinoderms, nudibranchs (sea slugs) and even the occasional octopus.

DISTRIBUTION: Coastal areas throughout the Nearctic.

WHERE TO SEE: Bolivar Flats, Texas, US; Big Sur, California, US; Acadia National Park, Maine, US.

NEARCTIC PELAGIC WATERS

IN A NUTSHELL: Deepwater marine environments, generally beyond the continental shelf or along major deepwater canyons. **Habitat Affinities:** Pelagic waters all over the world.

DESCRIPTION: The Nearctic region is surrounded by the Atlantic Ocean to the east, the Arctic Ocean to the north, the Pacific Ocean to the west, and the Gulf of Mexico to the south. Deep pelagic waters vary widely in accessibility and proximity throughout the region. In places like Monterey Bay, California, deep canyons allow access to pelagic zones only a few miles offshore, while in other locations, this habitat may be well over 100 mi. (160km) offshore. The pelagic and marine environments in the Nearctic are among the most well-studied in the world but remain relatively unknown compared to other environments.

Pelagic environments vary wildly throughout the Nearctic, with some holding shearwater flocks numbering in the hundreds of thousands and others being relatively barren. In these dynamic environments the presence of marine life is driven by major ocean currents, cold-water upwellings, and seasonal migrations. That being said, the northern pelagic environments hold the largest concentrations of marine life in the Nearctic. The Gulf of Mexico is generally wildlife-poor, while canyons along the Pacific coast tend to be the richest pelagic environments. The warm waters of the Gulf Stream in the Atlantic are also very productive both for feeding and as a migration route for birds, mammals, and fish.

WILDLIFE: There are major divides in the marine life of the three oceans and between warm and cold waters. Some widespread marine mammals like Humpback Whale, Fin Whale, Blue Whale, Common Bottlenose Dolphin, Killer Whale, Risso's Dolphin, and Sperm Whale are found up and

down the Atlantic and Pacific coasts. Regularly encountered mammals in Atlantic pelagic waters include Common Bottlenose Dolphin, Atlantic Spotted Dolphin, Short-finned Pilot Whale, Gervais Beaked Whale, and Risso's Dolphin. Some regularly seen Pacific specialties include Gray Whale, Northern Right Whale Dolphin, Dall's Porpoise, and Pacific White-sided Dolphin. The Arctic Ocean and far n. Atlantic have many fewer marine mammals but some great specialties, including the

Bermuda Petrel photographed during one of the Atlantic pelagic-birding boat trips off Cape Hatteras, North Carolina, US. ©
CHRIS SLOAN

A Wilson's Storm-Petrel foraging in pelagic waters off North Carolina, US. © CHRIS SLOAN

near-mythical Narwhal, Beluga Whale, Bowhead Whale, and pinnipeds like Walrus and Bearded, Harp, and Ribbon Seals. The best time for whale-watching in all three areas is the boreal summer and fall (June–September) though the Gray Whale calving season in late winter in the shallower waters around Baja California is well worth the visit.

The principal divide in pelagic bird species lies between the Pacific and Atlantic Oceans. The Arctic Ocean is principally a summer feeding ground for birds of both the n. Atlantic and the n. Pacific. All Pomarine, Parasitic, and Long-tailed Jaegers and South Polar Skua are most readily seen on pelagic waters in both oceans, though the jaegers can also be seen on tundra while breeding and on large bodies of water inland during migration. The Pacific Ocean has large seabird migrations in the fall, and August–October are the best times for pelagic trips, when typical species like Sooty (IS), Pink-footed, and Black-vented Shearwaters; Fork-tailed and Ashy Storm-Petrels; Northern Fulmar (IS); and Black-footed Albatross are joined by a wide array of less common species. Boat trips in Alaska produce the spectacular Short-tailed Albatross more and more regularly as numbers increase, and the Short-tailed Shearwater migration can number in the millions of birds. Another impressive migration occurs in s. California and off the coast of Mexico, where huge rafts of Black and Least Storm-Petrels occur in the fall and can number in the tens of thousands. The Pacific coast of Mexico has several localized endemics including Guadalupe Murrelet, Ainley's and Townsend's Storm-Petrels, and Townsend's Shearwater. Additionally, this area, especially off the coast of Oaxaca, is a good potential area to see a range of tropical petrels, including Hawaiian, Kermadec, Tahiti, and Juan Fernandez Petrels, as well as Galápagos and Christmas Shearwaters. Off the Atlantic coast, spring and summer trips out into the warm waters of the Gulf Stream produce the highest diversity of species, and Cory's (IS), Great, and Audubon's Shearwaters are regularly seen alongside Black-capped Petrel (IS) and Band-rumped, Leach's, and Wilson's Storm-Petrels. The Gulf Stream, especially around North Carolina, is the best place to see the critically endangered Bermuda Petrel away from its breeding grounds. Farther north in the Atlantic, Northern Fulmar and Manx Shearwater are common pelagic species.

WHERE TO SEE: There are a number of good options throughout the Nearctic with regular pelagic boat trips. Some of the most diverse launch from: Cape Hatteras, North Carolina, US; Monterey Bay/Half Moon Bay, California, US; Puerto Lopez, Oaxaca, Mexico.

NEARCTIC OFFSHORE ISLANDS

IN A NUTSHELL: Small offshore islands that are important breeding grounds for oceanic wildlife and a small set of other animals. Often rocky, sandy, or poorly vegetated. **Habitat Affinities:** INDO-MALAYAN OFFSHORE ISLANDS; AFROTROPICAL OFFSHORE ISLANDS; AUSTRALASIAN SANDY CAYS. **Species Overlap:** NEARCTIC ROCKY COASTLINE AND SANDY BEACH; NEARCTIC PELAGIC WATERS; NEARCTIC TIDAL MUDFLAT.

DESCRIPTION: Off the coasts of the Nearctic region lie an impressive number of small, isolated landmasses. In the northerly latitudes of the Nearctic zone these islands are volcanic (e.g., the Aleutian archipelago in Alaska) or unsubmerged sections of the continental shelf, while farther south many of the islands are formed from exposed ancient coral reefs or large accumulations of sand. Regardless of origin, one characteristic these small landmasses have in common is isolation. Often free of predators and with little disturbance from humans, Nearctic offshore islands serve as critical refuges for many pelagic species. Seasonally, many of these islands teem with thousands of seabirds and seals along with a few hardy resident animal species. Many of these islands have an array of hardy endemic plants, and a select few have endemic fauna as well.

The small landmass, concentration of life, and relative isolation mean that Nearctic offshore islands are particularly susceptible to human exploitation and destruction. Historically, many seabird nesting colonies were targeted for commercial harvest of eggs and guano, while seal colonies were decimated for furs. Introduced animals have destroyed habitat and preyed on ecologically naive species. In the Nearctic, goats, pigs, cats, rats, and mice have been the most detrimental of the introduced fauna and directly tied to several extinctions in the region. In recent years, valiant and often quite successful efforts have been made to remove these invasive species from Nearctic offshore islands. The restoration efforts on Santa Cruz Island off the coast of California have resulted in the astonishing recovery of the endangered Island Fox from a low of 135 individuals in 2000 to over 1,800 individuals today.

WILDLIFE: Offshore islands can host truly staggering numbers of wildlife during the breeding season. While mammals are less well represented on these islands than more mobile taxa, important rookery islands can hold pinniped colonies numbering in the hundreds of thousands. The Northern Fur Seal colony on St. Paul Island in Alaska's Pribilof chain once numbered over 1 million individuals. Apart from pinnipeds, the mammalian fauna on offshore islands is mostly limited to small mammals, including relatively small carnivores—Arctic Foxes have had great success colonizing northern islands, and Island Fox is endemic to the Channel Islands of California.

Least Auklet on St. Paul in the Pribilof Islands of Alaska, US, an important island for nesting seabirds such as puffins, fulmars, gulls, and cormorants.
© IAIN CAMPBELL, TROPICAL BIRDING TOURS

In far northern latitudes, millions of seabirds breed on offshore islands, and in the summer months these are fantastic locations to see a wide array of pelagic species on land. Razorbill, Dovekie, Black-legged Kittiwake, Atlantic Puffin, Common Murre, Northern Fulmar, Arctic and Common Terns, Leach's Storm-Petrel, and Northern Gannet are all abundant breeders in the n. Atlantic. Over 3 million Leach's Storm-Petrels breed on Canada's Baccalieu Island alone. In the n. Pacific, an even wider diversity of breeding alcids can be found—Tufted and Horned Puffins; Pigeon Guillemot; and Rhinoceros, Cassin's, Crested, Whiskered, and Parakeet Auklets are joined by a diverse assemblage of gulls, terns, cormorants, and tubenoses.

Offshore islands in the cool waters of Mexico's Pacific coast are home to several range-restricted endemics: Townsend's Shearwater, Townsend's Storm-Petrel, Ainley's Storm-Petrel, and Guadalupe Murrelet all have breeding ranges restricted to Guadalupe Island and the Socorro Islands. Atolls and islets off the US and Mexico in the warm waters of the Gulf of Mexico hold a different set of nesting birds, including many pantropical species—Sooty and Bridled Terns, Black and Brown Noddies, Masked and Brown Boobies—that are found in similar waters around the globe.

The herpetofauna of Nearctic offshore islands is fairly limited. Large numbers of sea turtles, including Green, Loggerhead, Kemp's Ridley, and Hawksbill nest on warm-water islands around the Nearctic. Some of the Nearctic offshore islands near Mexico's Baja Peninsula have endemic lizards and snakes, including the rattle-less Santa Catalina Rattlesnake.

DISTRIBUTION: Nearctic offshore islands are widely distributed throughout the region, in the Pacific from the Bering Sea to s. Mexico, in the Atlantic from Nova Scotia to the Caribbean, as well as in the Gulf of Mexico. The largest numbers are found along the Pacific coast (from Alaska to Mexico) and the n. Atlantic, while many small offshore islands also occur throughout the Caribbean.

WHERE TO SEE: St. Paul Island, Alaska, US; Witless Bay, Newfoundland, Canada; Channel Islands National Park, California, US.

Tufted Puffin resting on a sea-cliff breeding site on the island of St. Paul, Pribilof Islands, Alaska, US. © IAIN CAMPBELL, TROPICAL BIRDING TOURS

NEARCTIC CROPLAND AND GRAZING LAND

IN A NUTSHELL: Areas used primarily for the production of human-planted crops. **Habitat Affinities:** PALEARCTIC CROPLAND; AFROTROPICAL CROPLAND; INDO-MALAYAN PADDY FIELDS AND OTHER CROPLAND; NEOTROPICAL CROPLAND. **Species Overlap:** TALLGRASS PRAIRIE; SHORTGRASS PRAIRIE; NEARCTIC FRESHWATER WETLAND; NEARCTIC TIDAL MUDFLAT.

Hayfields provide ample sites for handsome male Bobolinks to set up breeding territories.
© PHIL CHAON, TROPICAL BIRDING TOURS

DESCRIPTION: In many parts of the world, land used for human cultivation can be highly productive areas for small mammals, birds, and reptiles. With a few exceptions, this is largely not the case in the Nearctic. Massive, industrial monocultures of row crops such as corn, soybean, and wheat are essentially biological deserts supporting little to no life. Heavy use of pesticides, protective netting, and other pest control methods further discourage use by wildlife. These few crops dominate a huge amount of the surface area, particularly in regions that were formerly grassland. This type of cultivation is most productive for wildlife after harvest or when fallow. Alfalfa and cultivated grass fields are a more natural type of monoculture that mirror native grasslands and attract a subset of native grassland birds. However, due to early harvesting, these hayfields can be traps for many nesting birds, and nests, eggs, and chicks are destroyed in the harvest.

Small-scale, mixed agriculture is a rare sight in most of the Nearctic. Smaller farms containing hedgerows, scattered trees, and a variety of crops are found mostly in Mexico, where the co-plantation of corn, squash, and beans has been occurring for thousands of years. In the United States, the farms in Amish communities tend to be much more natural habitats because of practices that included the limited application of pesticides and the use of more hedgerows and fallow areas than typical American farming.

Tree plantations generally consist of monoculture stands of native trees. While less diverse than surrounding forest, these tree stands do hold a significant subset of the wildlife of surrounding forests and are functional habitats.

Most of the grazing lands in the Nearctic are treated in the prairie and shrubland habitats.

WILDLIFE: Most large mammals use agricultural areas only fleetingly, and even this is increasingly rare. Rodent populations are often high around areas with cereal crops, sugarcane, and vineyards.

During cultivation, hayfields can be a good area to look for such bird species as Bobolink, Upland Sandpiper, Eastern and Western Meadowlarks, and Savanna and Grasshopper Sparrows. In California, large hayfields host a significant percentage of the western endemic breeding Tricolored Blackbird, and conservation groups often raise money to delay the harvest of these fields during the breeding season. Post-harvest fields for corn and cereal crops can host large flocks of geese, Sandhill Crane, American Pipit, blackbirds, Snow Bunting, and Lapland Longspur foraging for spilled grain. Additionally, the elevated rodent populations mean these areas host many wintering raptors like Rough-legged, Ferruginous, and Red-tailed Hawks; Northern Harrier; American Kestrel; and even Snowy Owl. During the harvest of rice fields in the se. United States, Virginia Rail, Sora, and Yellow Rail are readily spotted, and bird festivals have been established specifically for the viewing of the lattermost species at this time. After harvest, rice fields can be great areas to look for wintering waterfowl.

Many vineyards, particularly in the w. United States, have adopted the practice of providing nest structures for Barn Owls as a highly successful form of biological rodent control. This practice is also widespread on Amish farms. Coffee plantations in tropical and subtropical areas are very bird-poor when the coffee is grown as a sun crop. However, areas where coffee is grown as a shaded understory crop can be tremendously productive for warblers, wintering migrants, and a subset of cloud-forest species.

Tree plantations are the form of cultivation most clearly mirroring natural habitats within the Nearctic. While lacking tree species diversity, mixed-age structure, and cavities or dead snags, tree plantations can host a significant subset of local forest species including large mammals.

DISTRIBUTION: Widespread throughout the continent, areas where cultivation dominates nearly 100% of the landscape include the Great Plains of North America and the Central Valley of California. Amish farming areas are concentrated in the e. United States, and small-scale farming is common in Mexico.

Dramatic winter flocks of Sandhill Cranes and geese utilize North American grazing lands (photo: New Mexico, US). © BEN KNOOT, TROPICAL BIRDING TOURS

PLANT NAMES

The common plant names used throughout the guide can be cross-referenced to their scientific names in this list. Plants are listed by common name. Region indicates the region in the book where the plant is mentioned, not the plant's complete geographical range. Listed common names may indicate a species (Alligator Juniper, *Juniperus deppeana*), a genus (ant plant, *Myrmecodia*), or a family (aster, Asteracea).

COMMON NAME	SCIENTIFIC NAME	REGION	COMMON NAME	SCIENTIFIC NAME	REGION
Abura	*Mitragyna ciliata*	Afrotropics	Antelope Bitterbrush	*Purshia tridentata*	Nearctic
Acacia, Catclaw	*Acacia greggii*	Nearctic	Antelope Grass	*Echinochloa pyramidalis*	Afrotropics
Acacia, Flat-topped	*Acacia abyssinia*	Afrotropics	Antique Spurge	*Euphorbia antiquorum*	Indo-Malaysia
Acacia, Gum	*Senegalia senegal*	Palearctic	ant plant	*Myrmecodia*	Indo-Malaysia
Acacia, Lahai	*Acacia lahai*	Afrotropics	Apache Pine	*Pinus engelmannii*	Nearctic
Acacia, White-bark	*Vachellia leucophloea*	Indo-Malaysia	Apple, Gopher-	*Licania michauxii*	Nearctic
African Alpine Bamboo	*Yushania alpina*	Afrotropics	Apple, Mangrove	*Sonneratia alba*	Afrotropics
African Blackwood	*Dalbergia melanoxylon*	Afrotropics	Arizona Sycamore	*Platanus wrightii*	Nearctic
African Cherry	*Prunus africana*	Afrotropics	Arizona Walnut	*Juglans major*	Nearctic
African Locust Bean	*Parkia biglobosa*	Afrotropics	artemisia	*Artemisia*	Palearctic
African Mahogany	*Khaya africana*	Afrotropics	ash	*Fraxinus*	Nearctic
African Oil Palm	*Elaeis guineensis*	Afrotropics	Ash, Cape	*Ekebergia capensis*	Afrotropics
African Star-Chestnut	*Sterculia africana*	Afrotropics	Ash, Pop	*Fraxinus caroliniana*	Nearctic
African Teak	*Pericopsis elata*	Afrotropics	Ashe Juniper	*Juniperus ashei*	Nearctic
African Teak	*Pterocarpus angolensis*	Afrotropics	asparagus, sea	*Salicornia*	Australasia
African Walnut	*Lovoa trichilioides*	Afrotropics	Aspen, Eurasian	*Populus tremula*	Palearctic
Agarito	*Mahonia trifoliolata*	Nearctic	Aspen, Quaking	*Populus tremuloides*	Nearctic
Agba	*Prioria balsamifera*	Afrotropics	aster	*Asteraceae*	Afrotropics
Alan Bunga	*Shorea albida*	Indo-Malaysia	Aster, Heath	*Aster ericoides*	Nearctic
Albizia, Bitter	*Albizia amara*	Indo-Malaysia	Atlas Cedar	*Cedrus atlantica*	Palearctic
alder	*Alnus*	Nearctic, Palearctic	Australian bottlebrush	*Callistemon*	Afrotropics
			Axlewood	*Anogeissus latifolia*	Indo-Malaysia
Alder, Andean	*Alnus acuminata*	Neotropics	Ayan Spruce	*Picea jezoensis*	Palearctic
Alder, European	*Alnus glutinosa*	Palearctic	Azalea, Cascade	*Rhododendron albiflorum*	Nearctic
Aleppo Pine	*Pinus halepensis*	Palearctic	Babul	*Vachellia nilotica*	Palearctic
Alkali Sacaton	*Sporobolus airoides*	Nearctic	Baily's Cypress-pine	*Callitris baileyi*	Australasia
Alligator Juniper	*Juniperus deppeana*	Nearctic	Balau	*Shorea obtusa*	Indo-Malaysia
aloe	*Aloe*	Afrotropics	Bald Cypress	*Taxodium distichum*	Nearctic
Amargosa	*Castela erecta*	Nearctic	Balsam Fir	*Abies balsamea*	Nearctic
American Beech	*Fagus grandifolia*	Nearctic	Balsam Poplar	*Populus balsamifera*	Nearctic
American Chestnut	*Castanea dentata*	Nearctic	bamboo (China)	*Bambusoideae*	Palearctic
American Elm	*Ulmus americana*	Nearctic	bamboo	*Chusquea*	Neotropics
American Tarbush	*Flourensia cernua*	Nearctic	Bamboo, African Alpine	*Yushania alpina*	Afrotropics
Andean Alder	*Alnus acuminata*	Neotropics	Bamboo, Japanese	*Fallopia japonica*	Palearctic
Anjan	*Hardwickia binata*	Indo-Malaysia	Bangalow Palm	*Archontophoenix cunninghamiana*	Australasia
Annual Sunflower	*Helianthus annuus*	Nearctic			
Antarctic Beech	*Nothofagus moorei*	Australasia	Banksia, Heath-leaved	*Banksia ericifolia*	Australasia
Antarctic Hairgrass	*Deschampsia antarctica*	Antarctica	Banksia, Saw	*Banksia serrata*	Australasia
Antarctic Pearlwort	*Colobanthus quitensis*	Antarctica	baobab	*Adansonia*	Afrotropics

COMMON NAME	SCIENTIFIC NAME	REGION	COMMON NAME	SCIENTIFIC NAME	REGION
Baobab, Grandidier's	*Adansonia grandidieri*	Afrotropics	Blackjack Oak	*Quercus marilandica*	Nearctic
barrel cactus	*Ferocactus*	Nearctic	Black Kauri	*Agathis atropurpurea*	Australasia
Baruwa	*Saccharum bengalense*	Indo-Malaysia	Black Mangrove	*Avicennia germinans*	Afrotropics, Indo-Malaysia, Nearctic
basswood	*Tilia*	Nearctic, Palearctic			
Bauhinia, Kalahari	*Bauhinia petersiana*	Afrotropics	Black Mangrove	*Bruguiera gymnorrhiza*	Afrotropics
Beach Sheoak	*Casuarina equisetifolia*	Australasia	Black mangrove, White-flowered	*Lumnitzera racemosa*	Afrotropics
Beaksedge, Sandyfield	*Rhynchospora megalocarpa*	Nearctic	Black Oak	*Quercus velutina*	Nearctic
Bean, African Locust	*Parkia biglobosa*	Afrotropics	Black Saksaul	*Haloxylon ammodendron*	Palearctic
Bean-caper	*Tetraena simplex*	Palearctic	Blackwood, African	*Dalbergia melanoxylon*	Afrotropics
Beech, American	*Fagus grandifolia*	Nearctic	bladderwort	*Utricularia*	Indo-Malaysia
Beech, Antarctic	*Nothofagus moorei*	Australasia	blazingstar	*Liatris*	Nearctic
Beech, Black	*Nothofagus solandri*	Australasia	Bloodwood, Melville Island	*Eucalyptus cnesophila*	Australasia
Beech, Cape	*Rapanea melanophloeos*	Afrotropics	Blue Grama	*Bouteloua gracilis*	Nearctic
Beech, Common	*Fagus sylvatica*	Palearctic	Bluegrass, Bulbous	*Poa bulbosa*	Palearctic
Beech, Hard	*Nothofagus truncata*	Australasia	Blue Oak	*Quercus douglasii*	Nearctic
Beech, Lenga	*Nothofagus pumilio*	Neotropics	Blue Oak, Japanese	*Quercus glauca*	Indo-Malaysia
Beech, Mountain	*Nothofagus solandri*	Australasia	Blue Panic Grass	*Panicum antidotale*	Indo-Malaysia
Beech, Myrtle	*Nothofagus cunninghamii*	Australasia	Bluestem, Big	*Andropogon gerardii*	Nearctic
Beech, Oriental	*Fagus orientalis*	Palearctic	Bluestem, Florida	*Andropogon floridanus*	Nearctic
Beech, Red	*Nothofagus fusca*	Australasia	Bluestem, Little	*Schizachyrium scoparium*	Nearctic
Beech, Silver	*Nothofagus menziesii*	Australasia	Boarwood	*Symphonia globulifera*	Afrotropics
Benguet Pine	*Pinus kesiya*	Indo-Malaysia	Bokapi	*Staudtia stipitata*	Afrotropics
Berlandier Wolfberry	*Lycium berlandieri*	Nearctic	Boldo	*Peumus boldus*	Neotropics
Berry, Milk	*Chiococca alba*	Neotropics	Bonamia, Florida	*Bonamia grandiflora*	Nearctic
Big Bluestem	*Andropogon gerardii*	Nearctic	Boojum Tree	*Fouquieria columnaris*	Nearctic
Bigleaf Maple	*Acer macrophyllum*	Nearctic	bottlebrush, Australian	*Callistemon*	Afrotropics
Big Sagebrush	*Artemisia tridentata*	Nearctic	Bottle tree, Brazilian	*Ceiba speciosa*	Neotropics
Bilberry	*Vaccinium myrtillus*	Palearctic	box	*Eucalyptus*	Australasia
birch	*Betula*	Palearctic	Box, Poplar	*Eucalyptus populnea*	Australasia
Birch, Paper	*Betula papyrifera*	Nearctic	Box Elder	*Acer negundo*	Nearctic
Birch, Sweet	*Betula lenta*	Nearctic	boxwood	*Buxus*	Palearctic
Birch, Yellow	*Betula alleghenensis*	Nearctic	bracken	*Pteridium*	Palearctic
Bitter Albizia	*Albizia amara*	Indo-Malaysia	bramble	*Rubus*	Afrotropics
bitterbrush	*Purshia*	Nearctic	Bramble, Five-leaved	*Rubus pedatus*	Nearctic
Bitterbrush, Antelope	*Purshia tridentata*	Nearctic	Brasil	*Condalia hookeri*	Nearctic
Blackbean	*Castanospermum australe*	Australasia	Brazilian Bottle Tree	*Ceiba speciosa*	Neotropics
Black Beech	*Nothofagus solandri*	Australasia	Brigalow	*Acacia harpophylla*	Australasia
Blackbrush	*Coleogyne ramosissima*	Nearctic	Broad-leaved Paperbark	*Melaleuca quinquenervia*	Australasia
Black-butt	*Eucalyptus cambageana*	Australasia	Buckbrush	*Symphoricarpos orbiculatus*	Nearctic
Black Cherry	*Prunus serotina*	Nearctic			
Black Cottonwood	*Populus trichocarpa*	Nearctic	buckwheat	*Polygonaceae*	Palearctic
Black Cypress-Pine	*Callitris endlicheri*	Australasia	Buckwheat, California	*Eriogonum fasciculatum*	Nearctic
Blackfoot, Plains	*Melampodium leucanthum*	Nearctic	Budsage	*Picrothamnus desertorum*	Nearctic
Black Grama	*Bouteloua eriopoda*	Nearctic	buffalograss	*Bouteloua*	Nearctic

COMMON NAME	SCIENTIFIC NAME	REGION	COMMON NAME	SCIENTIFIC NAME	REGION
Buffalograss	Bouteloua dactyloides	Nearctic	Cedar, Cuban	Cedrela odorata	Neotropics
Bulbous Bluegrass	Poa bulbosa	Palearctic	Cedar, Eastern Red	Juniperus virginiana	Nearctic
Bunga, Alan	Shorea albida	Indo-Malaysia	Cedar, Incense	Calocedrus decurrens	Nearctic
Burma Padauk	Pterocarpus macrocarpus	Indo-Malaysia	Cedar, Japanese	Cryptomeria japonica	Palearctic
Bur Oak	Quercus macrocarpa	Nearctic	Cedar, Lebanon	Cedrus libani	Palearctic
bush lupine	Lupinus	Nearctic	Cedar, Red	Toona ciliata	Australasia
Bush Muhly	Muhlenbergia porteri	Nearctic	Cedar, Salt	Tamarix ramosissima	Palearctic
bushwillow	Combretum	Afrotropics	Cedar, Western Red	Thuja plicata	Nearctic
buttercup	Ranunculus	Afrotropics	Ceylon Satinwood	Chloroxylon swietenia	Indo-Malaysia
Buttonbush	Cephalanthus occidentalis	Nearctic	Chamise	Adenostema fasciculatum	Nearctic
Buttonwood	Conocarpus erectus	Nearctic	Chaparro	Castela texana	Nearctic
Cabbage, Water	Pistia stratiotes	Afrotropics	Chapman's Oak	Quercus chapmanii	Nearctic
cactus, barrel	Ferocactus	Nearctic	Cheatgrass	Bromus tectorum	Nearctic
Cactus, Lava	Brachycereus nesioticus	Neotropics	cherry	Prunus	Nearctic
Cactus, Organ Pipe	Stenocereus thurberi	Nearctic	Cherry, African	Prunus africana	Afrotropics
California Black Oak	Quercus kelloggii	Nearctic	Cherry, Black	Prunus serotina	Nearctic
California Buckwheat	Eriogonum fasciculatum	Nearctic	Cherry, Native	Exocarpos cupressiformis	Australasia
California Sagebrush	Artemisia californica	Nearctic	Cherry Laurel	Prunus laurocerasus	Palearctic
California Sycamore	Platanus racemosa	Nearctic	chestnut	Castanopsis hystrix	Indo-Malaysia
Cambodian Diospyros	Diospyros kaki	Indo-Malaysia	Chestnut, African Star-	Sterculia africana	Afrotropics
Camel Thorn	Vachellia erioloba	Afrotropics	Chestnut, American	Castanea dentata	Nearctic
Camphorweed	Pluchea camphorata	Nearctic	Chestnut, Chinese	Castanea mollissima	Palearctic
canegrass	Eragrostis	Australasia	Chestnut, Indian	Castanopsis indica	Indo-Malaysia
Cannonball Mangrove	Xylocarpus granatum	Afrotropics	Chestnut, Sweet	Castanea sativa	Palearctic
Cape Ash	Ekebergia capensis	Afrotropics	chickweed, mouse-ear	Cerastium	Afrotropics
Cape Beech	Rapanea melanophloeos	Afrotropics	Chihuahuan Pine	Pinus leiophylla	Nearctic
Cardón	Echinopsis atacamensis	Neotropics	Chinese Chestnut	Castanea mollissima	Palearctic
Cardón, Mexican Giant	Pachycereus pringlei	Nearctic	Chinese Cryptocarya	Cryptocarya chinensis	Indo-Malaysia
Carob	Ceratonia siliqua	Palearctic	chinquapin	Castanopsis	Indo-Malaysia
Cascade Azalea	Rhododendron albiflorum	Nearctic	Chir Pine	Pinus roxburghii	Indo-Malaysia
casuarina	Casuarina	Afrotropics	cholla	Cylindropuntia	Nearctic
Casuarina	Gymnostoma nobile	Indo-Malaysia	Christ's Thorn Jujube	Ziziphus spina-christi	Afrotropics
Catclaw Acacia	Acacia greggii	Nearctic	Churqui	Oxalis gigantea	Neotropics
Cat's Claw	Zanthoxylum fagara	Neotropics	cinnamon tree	Cinnamomum	Indo-Malaysia
cattail	Typha	Afrotropics	Cinnamon tree	Cinnamomum zeylanicum	Indo-Malaysia
Caucasian Walnut	Pterocarya fraxinifolia	Palearctic	Clover, Prickly	Trifolium echinatum	Afrotropics
ceanothus	Ceanothus	Nearctic	Clusterleaf, Kalahari	Terminalia brachystemma	Afrotropics
cecropia (Brazil)	Cecropia pachystachya	Neotropics	Coastal Red Milkwood	Mimusops caffra	Afrotropics
cecropia (Ecuador)	Cecropia pastasana	Neotropics	Coast Live Oak	Quercus agrifolia	Nearctic
cecropia (Central America)	Cecropia peltata	Neotropics	Coast Redwood	Sequoia sempervirens	Nearctic
cedar	Cedrus	Palearctic	Coconut, Sea	Lodoicea maldivica	Afrotropics
cedar	Juniperus	Afrotropics	coffee	Rubiaceae	Afrotropics
cedar	Widdringtonia	Afrotropics	Coffin Tree	Taiwania cryptomeriodes	Indo-Malaysia
Cedar, Atlas	Cedrus atlantica	Palearctic	Coihue, Magellanic	Nothofagus dombeyi	Neotropics
			commiphora	Commiphora	Afrotropics

COMMON NAME	SCIENTIFIC NAME	REGION	COMMON NAME	SCIENTIFIC NAME	REGION
Common Beech	*Fagus sylvatica*	Palearctic	Dark Red Meranti	*Shorea siamensis*	Indo-Malaysia
Common Hackberry	*Celtis occidentalis*	Nearctic	Darwin Cotton	*Gossypium darwinii*	Neotropics
Common Heather	*Calluna vulgaris*	Palearctic	Darwin Stringybark	*Eucalyptus tetrodonta*	Australasia
Common Hornbeam	*Carpinus betulus*	Palearctic	Date, Desert	*Balanites aegyptiaca*	Afrotropics, Palearctic
Common Lantana	*Lantana camara*	Afrotropics			
Common Maple	*Acer pseudoplatanus*	Palearctic	Date Palm, Wild	*Phoenix reclinata*	Afrotropics
Common Oak	*Quercus robur*	Palearctic	deerbrush	*Ceanothus*	Nearctic
common reed	*Phragmites*	Afrotropics	Desert Date	*Balanites aegyptiaca*	Afrotropics, Palearctic
Common Water Hyacinth	*Eichhornia crassipes*	Afrotropics			
			Desert Ironwood	*Olneya tesota*	Nearctic
Common Yew	*Taxus baccata*	Palearctic	Desert Olive	*Forestiera pubescens*	Nearctic
Coneflower, Purple	*Echinacea atrorubens*	Nearctic	Desert Sheoak	*Allocasuarina decaisneana*	Australasia
Coolabah	*Eucalyptus microtheca*	Australasia			
Copal, Zanzibar	*Hymenaea verrucosa*	Afrotropics	Devil's Club	*Oplopanax horridus*	Nearctic
cordgrass	*Spartina*	Nearctic	Diospyros, Cambodian	*Diospyros kaki*	Indo-Malaysia
Cork Oak	*Quercus suber*	Palearctic	dipterocarp	Dipterocarpaceae	Afrotropics
Cotton, Darwin	*Gossypium darwinii*	Neotropics	dipterocarp	*Dipterocarpus obtusifolius*	Indo-Malaysia
cottongrass (Arctic)	*Eriophorum*	Palearctic	dogbane	Apocynaceae	Afrotropics
Cottonwood, Black	*Populus trichocarpa*	Nearctic	dogwood	*Cornus*	Nearctic, Palearctic
Cottonwood, Fremont	*Populus fremontii*	Nearctic			
Cottonwood, Plains	*Populus deltoides*	Nearctic	Doka	*Isoberlinia doka*	Afrotropics
Coyote Bush	*Baccharis pilularis*	Nearctic	Douglas-fir	*Pseudotsuga menziesii*	Nearctic
Coyotillo	*Karwinskia humboldtiana*	Nearctic	Dragon Blood Tree	*Dracaena cinnabari*	Afrotropics
Cranberry	*Viburnum opulus*	Palearctic	dragon tree	*Dracaena*	Afrotropics
Creeping pine	*Pinus mugo*	Neotropics	dropseed	*Sporobolus*	Nearctic
Creosote	*Larrea tridentata*	Nearctic	Dropseed, Florida	*Sporobolus floridanus*	Nearctic
Crookedwood	*Lyonia ferruginea*	Nearctic	Dwarf Siberian Pine	*Pinus pumila*	Palearctic
croton	*Croton*	Afrotropics	Dwarf Willow	*Salix herbacea*	Palearctic
Cryptocarya, Chinese	*Cryptocarya chinensis*	Indo-Malaysia	East African Mahogany	*Khaya anthotheca*	Afrotropics
Cuban Cedar	*Cedrela odorata*	Neotropics	Eastern Hemlock	*Tsuga canadensis*	Nearctic
Cucumber Tree	*Dendrosicyos socotranus*	Afrotropics	Eastern Pricklypear	*Opuntia engelmannii*	Nearctic
cumbungi	*Typha*	Australasia	Eastern Red Cedar	*Juniperus virginiana*	Nearctic
Curly Mesquite	*Hilaria belangeri*	Nearctic	Eastern Redbud	*Cercis canadensis*	Nearctic
currant	*Ribes*	Nearctic	Eastern White Oak	*Quercus alba*	Nearctic
Cutover Muhly	*Muhlenbergia expansa*	Nearctic	Eastern White Pine	*Pinus strobus*	Nearctic
cypress	*Cupressus*	Palearctic	ebony	Ebenaceae	Afrotropics
Cypress, Bald	*Taxodium distichum*	Nearctic	Ebony, Mountain	*Bauhinia variegata*	Indo-Malaysia
Cypress, Japanese	*Chamaecyparis obtusa*	Palearctic	Ebony, Texas	*Pithecellobium flexicaule*	Nearctic
Cypress, Patagonian	*Fitzroya cupressoides*	Neotropics	Elder, Box	*Acer negundo*	Nearctic
cypress-pine	*Callitris*	Australasia	elderberry	*Sambucus*	Nearctic
Cypress-Pine, Baily's	*Callitris baileyi*	Australasia	Elephant Grass	*Miscanthus giganteus*	Indo-Malaysia
Cypress-Pine, Black	*Callitris endlicheri*	Australasia	elm	*Ulmus*	Palearctic
Cypress-Pine, White	*Callitris columellaris*	Australasia	Elm, American	*Ulmus americana*	Nearctic
Dahurian Larch	*Larix dahurica*	Palearctic	Emory Oak	*Quercus emoryi*	Nearctic
Dalat Pine	*Pinus dalatensis*	Indo-Malaysia	Engelmann Oak	*Quercus engelmannii*	Nearctic
			Engelmann Spruce	*Picea engelmannii*	Nearctic

COMMON NAME	SCIENTIFIC NAME	REGION	COMMON NAME	SCIENTIFIC NAME	REGION
erica	Ericaceae	Afrotropics	Frankincense	Boswellia sacra	Afrotropics
eucalyptus	Eucalyptus	Afrotropics	Fraser Fir	Abies fraseri	Nearctic
euphorbia	Euphorbia	Afrotropics	Fremont Cottonwood	Populus fremontii	Nearctic
Eurasian Aspen	Populus tremula	Palearctic	Freshwater Mangrove	Barringtonia acutangula	Indo-Malaysia
European Alder	Alnus glutinosa	Palearctic	Fruit-ridged Mallee	Eucalyptus incrassata	Australasia
Evergreen Huckleberry	Vaccinium ovatum	Nearctic	Gabon Nut	Coula edulis	Afrotropics
False-Mopane, Large	Guibourtia coleosperma	Afrotropics	Galápagos Lantana	Lantana involucrata	Neotropics
False Nettle	Boehmeria cylindrica	Nearctic	Gallberry	Ilex glabra	Nearctic
False Pimpernel	Lindernia dubia	Nearctic	Gambel Oak	Quercus gambelii	Nearctic
feathergrass	Stipa	Palearctic	Ghaf	Prosopis cineraria	Palearctic
Feathergrass	Stipa ichu	Nearctic	giant groundsel	Dendrosenecio	Afrotropics
felwort	Swertia	Afrotropics	giant lobelia	Lobelia	Afrotropics
Fern, Lady	Athyrium filix-femina	Nearctic	Giant Sacaton	Sporobolus wrightii	Nearctic
Fern, Mother Shield	Polystichum proliferum	Australasia	Giant Sedge	Carex gigantea	Nearctic
Fern, Swamp	Blechnum serratum	Nearctic	Giant Sequoia	Sequoia giganteum	Nearctic
Fern, Sword	Polystichum munitum	Nearctic	Gidgee	Acacia cambagei	Australasia
fescue	Festuca	Afrotropics, Nearctic	ginger	Zingiberaceae	Afrotropics
			Globemallow, Scarlet	Sphaeralcea coccinea	Nearctic
Fetterbush	Lyonia lucida	Nearctic	Golden Shower	Cassia fistula	Indo-Malaysia
Fever Tree	Vachellia xanthophloea	Afrotropics	Golden Wattle	Acacia pycnantha	Australasia
fig	Ficus	Afrotropics, Indo-Malaysia, Palearctic	Goodding's Willow	Salix gooddingi	Nearctic
			gooseberry	Ribes	Nearctic
Fig, Strangler	Ficus watkinsiania	Australasia	Gopher-apple	Licania michauxii	Nearctic
Finger Grass, Hooded	Chloris cucullata	Nearctic	gorse	Ulex	Palearctic
Fir, Balsam	Abies balsamea	Nearctic	Grama, Black	Bouteloua eriopoda	Nearctic
Fir, Fraser	Abies fraseri	Nearctic	Grama, Blue	Bouteloua gracilis	Nearctic
Fir, Grand	Abies grandis	Nearctic	Grama, Hairy	Bouteloua hirsuta	Nearctic
Fir, Korean	Abies koreana	Palearctic	Grama, Side-oats	Bouteloua curtipendula	Nearctic
Fir, Siberian	Abies sibirica	Palearctic	Grand Fir	Abies grandis	Nearctic
Fir, Subalpine	Abies lasiocarpa	Nearctic	Grandidier's Baobab	Adansonia grandidieri	Afrotropics
Fir, White	Abies concolor	Nearctic	Granjeno	Celtis llida	Nearctic
Five-leaved Bramble	Rubus pedatus	Nearctic	grass	Apluda aristata	Indo-Malaysia
flame tree	Delonix	Afrotropics	grass	Arundinella ciliata	Indo-Malaysia
Flatsedge, Smooth	Cyperus laevigatus	Afrotropics	grass	Arundinella villosa	Indo-Malaysia
Flat-topped Acacia	Acacia abyssinia	Afrotropics	grass	Chrysopogon zeylanicus	Indo-Malaysia
Florida Bluestem	Andropogon floridanus	Nearctic	grass	Cymbopogon flexuosus	Indo-Malaysia
Florida Bonamia	Bonamia grandiflora	Nearctic	grass	Dichanthium annulatum	Indo-Malaysia
Florida Dropseed	Sporobolus floridanus	Nearctic	Grass, Antarctic Hair-	Deschampsia antarctica	Antarctica
Florida Rosemary	Ceratiola ericoides	Nearctic	Grass, Antelope	Echinochloa pyramidalis	Afrotropics
Flytrap, Venus	Dionea muscipula	Nearctic	Grass, Elephant	Miscanthus giganteus	Indo-Malaysia
fountaingrass	Pennisetum	Afrotropics	Grass, Hippo	Vossia cuspidata	Afrotropics
foxtail	Setaria	Afrotropics	Grass, Kans	Saccharum spontaneum	Indo-Malaysia
Foxtail Pine	Pinus balfouriana	Nearctic	Grass, Mitchell	Astrebla lappacea	Australasia
frailejones (Ecuador)	Coespeletia moritziana	Neotropics	Grass, Red Oat	Themeda triandra	Afrotropics
frailejones (Venezuela)	Coespeletia palustris	Neotropics	Grass, Rhodes	Chloris gayana	Afrotropics

COMMON NAME	SCIENTIFIC NAME	REGION	COMMON NAME	SCIENTIFIC NAME	REGION
Grass, Salt	*Sporobolus spicatus*	Afrotropics	Hemlock, Western	*Tsuga heterophylla*	Nearctic
Grass, Teff	*Eragrostis tef*	Afrotropics	hickory	*Carya*	Nearctic
grass, thatching	*Hyparrhenia*	Afrotropics	Hickory, Water	*Carya aquatica*	Nearctic
grass, trident	*Tristachya*	Afrotropics	Hippo Grass	*Vossia cuspidata*	Afrotropics
grass tree	*Xanthorrhoea*	Australasia	holly	*Ilex*	Afrotropics, Palearctic
Gray Mallee	*Eucalyptus socialis*	Australasia			
Gray Mangrove	*Avicennia marina*	Afrotropics	Holly Oak	*Quercus ilex*	Palearctic
Greasewood	*Sarcobatus vermiculatus*	Nearctic	Holm Oak	*Quercus ilex*	Palearctic
Great Basin Bristlecone Pine	*Pinus longaeva*	Nearctic	Honduran Mahogany	*Swietenia macrophylla*	Neotropics
			Honey Mesquite	*Prosopis glandulosa*	Nearctic
Greater Marsh St.-John's-Wort	*Hypericum walteri*	Nearctic	Hooded Finger Grass	*Chloris cucullata*	Nearctic
			Hoop Pine	*Araucaria cunninghamii*	Australasia
groundsel, giant	*Dendrosenecio*	Afrotropics	Hopsage, Spiny	*Grayia spinosa*	Nearctic
Guarri, Natal	*Euclea natalensis*	Afrotropics	hornbeam	*Carpinus*	Palearctic
Guava, Strawberry	*Psidium cattleyanum*	Afrotropics	Hornbeam, Common	*Carpinus betulus*	Palearctic
Gum, Red	*Eucalyptus camaldulensis*	Australasia	huckleberry	*Vaccinium*	Nearctic
Gum, River Red	*Eucalyptus camaldulensis*	Australasia	Huckleberry, Evergreen	*Vaccinium ovatum*	Nearctic
Gum, Rose	*Eucalyptus grandis*	Australasia	Huckleberry Oak	*Quercus vaccinifolia*	Nearctic
Gum, Salmon White	*Eucalyptus lane-poolei*	Australasia	Huisache	*Acacia farnesiana*	Nearctic
Gum, Snow	*Eucalyptus pauciflora*	Australasia	Hungarian Oak	*Quercus farnetto*	Palearctic
Gum, Spotted	*Corymbia maculata*	Australasia	Hyacinth, Common Water	*Eichhornia crassipes*	Afrotropics
Gum, Sweet (China)	*Liquidambar formosana*	Palearctic			
Gum, Sweet (North America)	*Liquidambar styraciflua*	Nearctic	ice plant	*Aizoaceae*	Afrotropics
			Incense Cedar	*Calocedrus decurrens*	Nearctic
Gum, Sydney Blue	*Eucalyptus saligna*	Australasia	Indian Chestnut	*Castanopsis indica*	Indo-Malaysia
Gum, Timor Mountain	*Eucalyptus urophylla*	Australasia	Indiangrass	*Sorghastrum nutans*	Nearctic
Gum Acacia	*Senegalia senegal*	Palearctic	Indian Laurel	*Ficus microcarpa*	Indo-Malaysia
Gurjun Tree	*Dipterocarpus tuberculatus*	Indo-Malaysia	Indian Laurel	*Terminalia tomentosa*	Indo-Malaysia
			Indian Mangrove	*Ceriops tagal*	Afrotropics
Hackberry	*Prunus padus*	Palearctic	Indian Ricegrass	*Oryzopsis hymenoides*	Nearctic
Hackberry, Common	*Celtis occidentalis*	Nearctic	Indian Rosewood	*Dalbergia latifolia*	Indo-Malaysia
Hackberry, Sugar	*Celtis laevigata*	Nearctic	Indian Wild Date	*Phoenix sylvestris*	Indo-Malaysia
Hackenkopf	*Calligonum azel*	Palearctic	Iodine Bush	*Allenrolfea occidentalis*	Nearctic
Hairgrass, Antarctic	*Deschampsia antarctica*	Antarctica	Iodine Weed	*Suaeda torreyana*	Nearctic
Hairgrass, Silver	*Aira caryophyllea*	Afrotropics	ironbark	*Eucalyptus*	Australasia
Hairy Grama	*Bouteloua hirsuta*	Nearctic	Ironbark, Narrow-leaved	*Eucalyptus crebra*	Australasia
Hard Beech	*Nothofagus truncata*	Australasia	Ironbark, Silver-leaved	*Eucalyptus melanophloia*	Australasia
Hartweg's Pine	*Pinus rudis*	Nearctic	ironwood	*Olneya*	Nearctic
hawthorn	*Crataegus*	Palearctic	Ironwood, Desert	*Olneya tesota*	Nearctic
hazel	*Corylus*	Palearctic	Jackalberry	*Diospyros mespiliformis*	Afrotropics
Heath Aster	*Aster ericoides*	Nearctic	Japanese Bamboo	*Fallopia japonica*	Palearctic
Heather, Common	*Calluna vulgaris*	Palearctic	Japanese Blue Oak	*Quercus glauca*	Indo-Malaysia
Heath-leaved Banksia	*Banksia ericifolia*	Australasia	Japanese Cedar	*Cryptomeria japonica*	Palearctic
heliconia	*Heliconia*	Nearctic	Japanese Cypress	*Chamaecyparis obtusa*	Palearctic
heliotrope	*Heliotropium*	Palearctic	Jaragua	*Hyparrhenia rufa*	Afrotropics
Hemlock, Eastern	*Tsuga canadensis*	Nearctic			

COMMON NAME	SCIENTIFIC NAME	REGION	COMMON NAME	SCIENTIFIC NAME	REGION
Jarrah	Eucalyptus marginata	Australasia	Larch, Dahurian	Larix dahurica	Palearctic
Jeffrey Pine	Pinus jeffreyi	Nearctic	Large False-Mopane	Guibourtia coleosperma	Afrotropics
Joint Pine	Ephedra fragilis	Palearctic	laurel	Cinnamomum	Indo-Malaysia
Jujube, Christ's Thorn	Ziziphus spina-christi	Afrotropics	laurel	Lauraceae	Afrotropics
Jujube, Wild	Ziziphus spina-christi	Palearctic	Laurel, Cherry	Prunus laurocerasus	Palearctic
Junegrasses	Koeleria	Afrotropics	Laurel, Indian	Ficus microcarpa	Indo-Malaysia
Junglesop	Anonidium mannii	Afrotropics	Laurel, Indian	Terminalia tomentosa	Indo-Malaysia
juniper	Juniperus	Nearctic, Palearctic	Laurel, Mountain	Kalmia latifolia	Nearctic
			Lava Cactus	Brachycereus nesioticus	Neotropics
Juniper, Alligator	Juniperus deppeana	Nearctic	lavender	Lavandula	Palearctic
Juniper, Ashe	Juniperus ashei	Nearctic	Leadplant	Amorpha canescens	Nearctic
Juniper, Rocky Mountain	Juniperus scopulorum	Nearctic	Lebanon Cedar	Cedrus libani	Palearctic
Juniper, Single-seeded	Juniperus monosperma	Nearctic	Lebombo Wattle	Newtonia hildebrandtii	Afrotropics
Juniper, Utah	Juniperus osteosperma	Nearctic	Lechuguilla	Agave lechuguilla	Nearctic
Juniper, Western	Juniperus occidentalis	Nearctic	legume	Fabaceae	Afrotropics
Kadam	Haldina cordifolia	Indo-Malaysia	Lenga Beech	Nothofagus pumilio	Neotropics
Kahikatea	Dacrycarpus dacrydioides	Australasia	Lily, Mangrove	Crinum pedunculatum	Australasia
Kalahari Bauhinia	Bauhinia petersiana	Afrotropics	Lily, Water	Nymphaeaceae	Afrotropics
Kalahari Clusterleaf	Terminalia brachystemma	Afrotropics	Limbali	Gilbertiodendron dewevrei	Afrotropics
Kalahari Podberry	Dialium englerianum	Afrotropics			
kalanchoe	Kalanchoe	Afrotropics	Limber Pine	Pinus flexilis	Nearctic
Kalopanax	Kalopanax septemlobus	Palearctic	linden	Tilia	Nearctic, Palearctic
Kamahi	Weinmannia racemosa	Australasia			
Kans Grass	Saccharum spontaneum	Indo-Malaysia	Linden, Small-leaved	Tilia cordata	Palearctic
kapok	Ceiba	Neotropical	Lindheimer Pricklypear	Opuntia engelmannii var. Lindheimeri	Nearctic
Kapok	Ceiba pentandra	Nearctic			
Karee	Searsia lancea	Afrotropics	Lingonberry	Vaccinium vitis-idaea	Palearctic
Karri	Eucalyptus diversicolor	Australasia	Little Bluestem	Schizachyrium scoparium	Nearctic
katrafay	Cedrelopsis	Afrotropics	Live Oak	Quercus virginiana	Nearctic
kauri	Agathis	Australasia, Indo-Malaysia	Lizard's Tail	Saururus cernuus	Nearctic
			lobelia, giant	Lobelia	Afrotropics
Kauri, Black	Agathis atropurpurea	Australasia	Loblolly Pine	Pinus taeda	Nearctic
Kauri, New Zealand	Agathis australis	Australasia	Locust Bean, African	Parkia biglobosa	Afrotropics
Kauri, Sarawak	Agathis orbicula	Indo-Malaysia	Lodgepole Pine	Pinus contorta	Nearctic
Khaya	Khaya senegalensis	Afrotropics	Longleaf Pine	Pinus palustris	Nearctic
Klinki Pine	Araucaria hunsteinii	Australasia	Lotebush	Ziziphus obtusifolia	Nearctic
Korean Fir	Abies koreana	Palearctic	lovegrass	Eragrostis	Afrotropics, Nearctic
Korean Pine	Pinus koraiensis	Palearctic			
Koroch Tree	Millettia pinnata	Indo-Malaysia	lupine, bush	Lupinus	Nearctic
Kumeraho	Pomaderris kumeraho	Australasia	madrone	Arbutus	Nearctic
Lady Fern	Athyrium filix-femina	Nearctic	magellanic Coihue	Nothofagus dombeyi	Neotropics
lady's mantle	Alchemilla	Afrotropics	Magnolia, Sweetbay	Magnolia virginiana	Nearctic
Lahai Acacia	Acacia lahai	Afrotropics	Mahogany, African	Khaya africana	Afrotropics
Lancewood	Acacia shirleyi	Australasia	Mahogany, East African	Khaya anthotheca	Afrotropics
Lantana, Common	Lantana camara	Afrotropics	Mahogany, Honduran	Swietenia macrophylla	Neotropics
Lantana, Galápagos	Lantana involucrata	Neotropics	mahogany, mountain	Cercocarpus	Nearctic

COMMON NAME	SCIENTIFIC NAME	REGION	COMMON NAME	SCIENTIFIC NAME	REGION
Mahogany, Mountain	Cercocarpus betuloides	Nearctic	Melon, Nara	Acanthosicyos horridus	Afrotropics
Mahogany, Pod	Afzelia quanzensis	Afrotropics	Melville Island Bloodwood	Eucalyptus cnesophila	Australasia
Mallee, Fruit-ridged	Eucalyptus incrassata	Australasia	Meranti, Dark Red	Shorea siamensis	Indo-Malaysia
Mallee, Gray	Eucalyptus socialis	Australasia	mesquite	Prosopis	Nearctic
Mallee, Red	Eucalyptus oleosa	Australasia	Mesquite, Curly	Hilaria belangeri	Nearctic
Mallee, Ridge-fruited	Eucalyptus angulosa	Australasia	Mesquite, Honey	Prosopis glandulosa	Nearctic
Mallee, Soap	Eucalyptus diversifolia	Australasia	Mexican Giant Cardón	Pachycereus pringlei	Nearctic
Mallee, White	Eucalyptus dumosa	Australasia	Milk Berry	Chiococca alba	Neotropics
Mango	Mangifera indica	Afrotropics	milk vetch	Astragalus	Nearctic
mangrove	Bruguiera	Indo-Malaysia	Milkwood, Coastal Red	Mimusops caffra	Afrotropics
Mangrove, Black	Avicennia germinans	Afrotropics, Indo-Malaysia, Nearctic	milkwort	Polygala	Palearctic
			Mimosa	Albizia julibrissin	Palearctic
Mangrove, Black	Bruguiera gymnorrhiza	Afrotropics	miombo	Brachystegia	Afrotropics
Mangrove, Cannonball	Xylocarpus granatum	Afrotropics	miombo	Brachystegia boehmii	Afrotropics
Mangrove, Freshwater	Barringtonia acutangula	Indo-Malaysia	Mitchell Grass	Astrebla lappacea	Australasia
Mangrove, Gray	Avicennia marina	Afrotropics	Miyapok	Tetramerista glabra	Indo-Malaysia
Mangrove, Indian	Ceriops tagal	Afrotropics	Moly Redbox	Eucalyptus leptophylla	Australasia
Mangrove, Red	Rhizophora harrisonii	Afrotropics	Mongolian Oak	Quercus mongolica	Palearctic
Mangrove, Red	Rhizophora mangle	Afrotropics, Indo-Malaysia, Neotropics	Mongolian Poplar	Populus suaveolens	Palearctic
			Mongolian Wormwood	Artemisia monosperma	Palearctic
Mangrove, Red	Rhizophora mucronata	Afrotropics	monkey-orange	Strychnos	Afrotropics
Mangrove, Red	Rhizophora racemosa	Afrotropics, Nearctic	Monkey Puzzle Tree	Araucaria araucana	Neotropics
			Montezuma Pine	Pinus montezumae	Nearctic
Mangrove, White	Laguncularia racemosa	Afrotropics	Mopane	Colophospermum mopane	Afrotropics
Mangrove, White-flowered Black	Lumnitzera racemosa	Afrotropics	Mopane, Large False-	Guibourtia coleosperma	Afrotropics
			Moriche Palm	Mauritia flexuosa	Neotropics
Mangrove Apple	Sonneratia alba	Afrotropics	moringa	Moringa drouhardii	Afrotropics
Mangrove Lily	Crinum pedunculatum	Australasia	moringa	Moringa hildebrandtii	Afrotropics
Mangrove Palm	Nypa fruticans	Australasia	Morrel	Eucalyptus longicornis	Australasia
Manketti	Schinziophyton rautanenii	Afrotropics	Mother Shield Fern	Polystichum proliferum	Australasia
manzanita	Arctostaphylos	Nearctic	Mountain Beech	Nothofagus solandri	Australasia
maple	Acer	Nearctic, Palearctic	Mountain Ebony	Bauhinia variegata	Indo-Malaysia
			Mountain Laurel	Kalmia latifolia	Nearctic
Maple, Bigleaf	Acer macrophyllum	Nearctic	Mountain-laurel, Texas	Sophora secundiflora	Nearctic
Maple, Common	Acer pseudoplatanus	Palearctic	mountain mahogany	Cercocarpus	Nearctic
Maple, Red	Acer rubra	Nearctic	Mountain Mahogany	Cercocarpus betuloides	Nearctic
Maple, Sugar	Acer saccharum	Nearctic	mouse-ear chickweed	Cerastium	Afrotropics
Maple, Sycamore	Acer pseudoplatanus	Palearctic	Muchesa	Julbernardia paniculata	Afrotropics
Maple, Vine	Acer circinatum	Nearctic	muhly	Muhlenbergia	Nearctic
Maritime Pine	Pinus pinaster	Palearctic	Muhly, Bush	Muhlenbergia porteri	Nearctic
Marula	Sclerocarya birrea	Afrotropics	Muhly, Cutover	Muhlenbergia expansa	Nearctic
Mavunda	Cryptosepalum exfoliatum	Afrotropics	Mulga	Acacia aneura	Australasia
			Munondo	Julbernardia globiflora	Afrotropics
May-apple	Podophyllum peltatum	Nearctic	mustardtree	Salvadora	Afrotropics
meadowsweet	Spiraea	Nearctic	Myall	Acacia binervia	Australasia

COMMON NAME	SCIENTIFIC NAME	REGION	COMMON NAME	SCIENTIFIC NAME	REGION
myrtle	Myrtaceae	Afrotropics	Oak, Plateau Live	Quercus fusiformis	Nearctic
myrtle	Syzygium	Indo-Malaysia	Oak, Post	Quercus stellata	Nearctic
Myrtle, Wax	Myrica cerifera	Nearctic	Oak, Sand Live	Quercus geminata	Nearctic
Myrtle Beech	Nothofagus cunninghamii	Australasia	Oak, Scrub	Quercus inopina	Nearctic
Myrtle Oak	Quercus myrtifolia	Nearctic	Oak, Sessile	Quercus petraea	Palearctic
Nara Melon	Acanthosicyos horridus	Afrotropics	oak, stone	Lithofagus	Indo-Malaysia
Narrow-leaved Ironbark	Eucalyptus crebra	Australasia	Oak, Strandzha	Quercus hartwissiana	Palearctic
Natal Guarri	Euclea natalensis	Afrotropics	Oak, Texas	Quercus buckleyi	Nearctic
Native Cherry	Exocarpos cupressiformis	Australasia	Oak, Turkey	Quercus laevis	Palearctic
Needlewood Tree	Schima wallichii	Indo-Malaysia	Oak, Uricua	Quercus laurina	Nearctic
Neem	Azadirachta indica	Indo-Malaysia	Oak, Valley	Quercus lobata	Nearctic
Nettle, False	Boehmeria cylindrica	Nearctic	Oak, White	Quercus alba	Nearctic
New Guinea Oak	Cananopsis acuminatissima	Australasia	Oak, White	Quercus pubescens	Palearctic
New Zealand Kauri	Agathis australis	Australasia	Oat Grass, Red	Themeda triandra	Afrotropics
Nikau Palms	Rhopalostylis sapida	Australasia	Ocotillo	Fouquieria splendens	Nearctic
ninebark	Physocarpus	Nearctic	octopus tree	Didiereaceae	Afrotropics
Nipa Palm	Nypa fruticans	Afrotropics	Oldman Saltbush	Atriplex nummularia	Australasia
Nogales	Juglans boliviana	Neotropics	Oleander	Nerium oleander	Palearctic
Norway Spruce	Picea abies	Palearctic	olive	Oleaceae	Afrotropics
Nut, Gabon	Coula edulis	Afrotropics	Olive, Desert	Forestiera pubescens	Nearctic
Nyala Tree	Xanthocercis zambesiaca	Afrotropics	Olive, Russian	Elaeagnus angustifolia	Nearctic
oak	Quercus	Indo-Malaysia, Nearctic, Palearctic	Olive, Wild	Olea oleaster	Palearctic
			orange, monkey-	Strychnos	Afrotropics
			orchid	Orchidaceae	Afrotropics
Oak, Black	Quercus velutina	Nearctic	Organ Pipe Cactus	Stenocereus thurberi	Nearctic
Oak, Blackjack	Quercus marilandica	Nearctic	Oriental Beech	Fagus orientalis	Palearctic
Oak, Blue	Quercus douglasii	Nearctic	pachypodium	Pachypodium	Afrotropics
Oak, Bur	Quercus macrocarpa	Nearctic	Padauk, Burma	Pterocarpus macrocarpus	Indo-Malaysia
Oak, California Black	Quercus kelloggii	Nearctic	Palán-Palán	Nicotiana glauca	Neotropics
Oak, Chapman's	Quercus chapmanii	Nearctic	palm	Arecaceae	Afrotropics
Oak, Coast Live	Quercus agrifolia	Nearctic	palm	Liberbaileya	Indo-Malaysia
Oak, Common	Quercus robur	Palearctic	Palm, African Oil	Elaeis guineensis	Afrotropics
Oak, Cork	Quercus suber	Palearctic	Palm, Bangalow	Archontophoenix cunninghamiana	Australasia
Oak, Emory	Quercus emoryi	Nearctic	Palm, Mangrove	Nypa fruticans	Australasia
Oak, Engelmann	Quercus engelmannii	Nearctic	Palm, Moriche	Mauritia flexuosa	Neotropics
Oak, Gambel	Quercus gambelii	Nearctic	Palm, Nikau	Rhopalostylis sapida	Australasia
Oak, Holly	Quercus ilex	Palearctic	Palm, Nipa	Nypa fruticans	Afrotropics
Oak, Holm	Quercus ilex	Palearctic	palm, raffia	Raphia	Afrotropics
Oak, Huckleberry	Quercus vaccinifolia	Nearctic	Palm, Traveler's	Ravenala madagascariensis	Afrotropics
Oak, Hungarian	Quercus farnetto	Palearctic	Palm, Wild Date	Phoenix reclinata	Afrotropics
Oak, Live	Quercus virginiana	Nearctic	Palmetto, Saw	Serenoa repens	Nearctic
Oak, Mongolian	Quercus mongolica	Palearctic	Palo Santo	Bursera graveolens	Neotropics
Oak, Myrtle	Quercus myrtifolia	Nearctic	paloverde	Parkinsonia	Nearctic
Oak, New Guinea	Cananopsis acuminatissima	Australasia	Palo Verde	Parkinsonia aculeata	Neotropics
Oak, Pedunculate	Quercus robur	Palearctic			

COMMON NAME	SCIENTIFIC NAME	REGION	COMMON NAME	SCIENTIFIC NAME	REGION
pandanus	Pandanus	Afrotropics	Pine, Rocky Mountain Bristlecone	Pinus aristata	Nearctic
Panga Panga	Millettia stuhlmannii	Afrotropics	Pine, Sand	Pinus clausa	Nearctic
panic grass	Panicum	Palearctic	Pine, Scots	Pinus sylvestris	Palearctic
Panic Grass, Blue	Panicum antidotale	Indo-Malaysia	Pine, Single-leaf Pinyon	Pinus monophylla	Nearctic
Paperbark, Broad-leaved	Melaleuca quinquenervia	Australasia	Pine, Slash	Pinus elliottii	Nearctic
Paperbark, Weeping	Melaleuca irbyana	Australasia	Pine, Sugar	Pinus lambertiana	Nearctic
Paperbark, Weeping	Melaleuca leucadendra	Indo-Malaysia	Pine, Sumatran	Pinus merkusii	Indo-Malaysia
Paper Birch	Betula papyrifera	Nearctic	Pine, Two-needled Pinyon	Pinus edulis	Nearctic
Papyrus	Cyperus papyrus	Afrotropics			
Patagonian Cypress	Fitzroya cupressoides	Neotropics	Pine, Umbrella	Pinus pinea	Palearctic
Pearlwort, Antarctic	Colobanthus quitensis	Antarctica	Pine, Virginia	Pinus virginiana	Nearctic
Pedunculate Oak	Quercus robur	Palearctic	Pine, Western White	Pinus monticola	Nearctic
Peeling Plane	Ochna pulchra	Afrotropics	Pine, Whitebark	Pinus albicaulis	Nearctic
Persimmon, Texas	Diospyros texana	Nearctic	Pink Poui	Tabebuia rosea	Neotropics
Peumo	Cryptocarya alba	Neotropics	Pinyon Pine, Single-leaf	Pinus monophylla	Nearctic
pickleweed	Salicornia	Nearctic	Pinyon Pine, Two-needled	Pinus edulis	Nearctic
Pickleweed, Utah	Salicornia utahensis	Nearctic			
Pimpernel, False	Lindernia dubia	Nearctic	pistachio	Pistacia	Palearctic
Pindan	Acacia eriopoda	Australasia	Pistachio	Pistacia vera	Palearctic
pine	Pinus	Afrotropics	pitcher plant	Nephentes	Indo-Malaysia
Pine, Aleppo	Pinus halepensis	Palearctic	pitcher plant	Nephentes bicalcarata	Indo-Malaysia
Pine, Apache	Pinus engelmannii	Nearctic	pitcher plant	Sarracenia	Nearctic
Pine, Benguet	Pinus kesiya	Indo-Malaysia	Plains Blackfoot	Melampodium leucanthum	Nearctic
Pine, Chihuahuan	Pinus leiophylla	Nearctic	Plains Cottonwood	Populus deltoides	Nearctic
Pine, Chir	Pinus roxburghii	Indo-Malaysia	Plane, Peeling	Ochna pulchra	Afrotropics
Pine, Creeping	Pinus mugo	Neotropics	Plateau Live Oak	Quercus fusiformis	Nearctic
Pine, Dalat	Pinus dalatensis	Indo-Malaysia	Podberry, Kalahari	Dialium englerianum	Afrotropics
Pine, Dwarf Siberian	Pinus pumila	Palearctic	Pod Mahogany	Afzelia quanzensis	Afrotropics
Pine, Eastern White	Pinus strobus	Nearctic	podocarp	Dacrydium	Indo-Malaysia
Pine, Foxtail	Pinus balfouriana	Nearctic	podocarp	Podocarpus	Indo-Malaysia
Pine, Great Basin Bristlecone	Pinus longaeva	Nearctic	poison oak	Toxicodendron	Nearctic
Pine, Hoop	Araucaria cunninghamii	Australasia	Poison Oak, Western	Toxicodendron diversilobum	Nearctic
Pine, Jeffrey	Pinus jeffreyi	Nearctic	polylepis (Bolivia)	Polylepis besseri	Neotropics
Pine, Joint	Ephedra fragilis	Palearctic	polylepis (Ecuador)	Polylepis incana	Neotropics
Pine, Klinki	Araucaria hunsteinii	Australasia	Ponderosa Pine	Pinus ponderosa	Nearctic
Pine, Korean	Pinus koraiensis	Palearctic	Pond Pine	Pinus serotina	Nearctic
Pine, Limber	Pinus flexilis	Nearctic	Pop Ash	Fraxinus caroliniana	Nearctic
Pine, Loblolly	Pinus taeda	Nearctic	poplar	Populus	Palearctic
Pine, Lodgepole	Pinus contorta	Nearctic	Poplar, Balsam	Populus balsamifera	Nearctic
Pine, Longleaf	Pinus palustris	Nearctic	Poplar, Mongolian	Populus suaveolens	Palearctic
Pine, Maritime	Pinus pinaster	Palearctic	Poplar Box	Eucalyptus populnea	Australasia
Pine, Montezuma	Pinus montezumae	Nearctic	Post Oak	Quercus stellata	Nearctic
Pine, Pond	Pinus serotina	Nearctic	Poui, Pink	Tabebuia rosea	Neotropics
Pine, Ponderosa	Pinus ponderosa	Nearctic	Powderbark Wandoo	Eucalyptus accedens	Australasia

COMMON NAME	SCIENTIFIC NAME	REGION	COMMON NAME	SCIENTIFIC NAME	REGION
Prairie Zinnia	Zinnia grandiflora	Nearctic	Rocky Mountain Bristlecone Pine	Pinus aristata	Nearctic
Prickly Clover	Trifolium echinatum	Afrotropics	Rocky Mountain Juniper	Juniperus scopulorum	Nearctic
pricklypear	Opuntia	Nearctic	rose	Rosaceae	Palearctic
Pricklypear, Eastern	Opuntia engelmannii	Nearctic	Rose, Rock	Helianthemum nummularium	Palearctic
Pricklypear, Lindheimer	Opuntia engelmannii var. Lindheimeri	Nearctic	Rose Gum	Eucalyptus grandis	Australasia
protea	Proteaceae	Afrotropics	Rosemary	Salvia rosmarinus	Palearctic
Purple Coneflower	Echinacea atrorubens	Nearctic	Rosemary, Florida	Ceratiola ericoides	Nearctic
Purple-pod Terminalia	Terminalia prunioides	Afrotropics	rosewood	Dalbergia	Afrotropics
Quaking Aspen	Populus tremuloides	Nearctic	Rosewood, Indian	Dalbergia latifolia	Indo-Malaysia
rabbitbrush	Chrysothamnus	Nearctic	Russian Olive	Elaeagnus angustifolia	Nearctic
Rabbitbrush	Ericameria nauseosa	Nearctic	Russian Thistle	Echinops exaltatus	Palearctic
raffia palm	Raphia	Afrotropics	Russian Thistle	Salsola kali	Nearctic
Ragwort	Senecio jacobaea	Palearctic	Sacaton, Alkali	Sporobolus airoides	Nearctic
Ramin	Gonystylus bancanus	Indo-Malaysia	Sacaton, Giant	Sporobolus wrightii	Nearctic
rattans	Calamoideae	Afrotropics	Sage (Europe)	Salvia officinalis	Palearctic
Red Beech	Nothofagus fusca	Australasia	sagebrush	Artemisia	Nearctic
Redbox, Moly	Eucalyptus leptophylla	Australasia	Sagebrush, Big	Artemisia tridentata	Nearctic
Redbud, Eastern	Cercis canadensis	Nearctic	Sagebrush, California	Artemisia californica	Nearctic
Red Cedar	Toona ciliata	Australasia	Saguaro	Carnegeia gigantea	Nearctic
Red Gum	Eucalyptus camaldulensis	Australasia	Saksaul, Black	Haloxylon ammodendron	Palearctic
Red Mallee	Eucalyptus oleosa	Australasia	Saksaul, White	Haloxylon persicum	Palearctic
Red Mangrove	Rhizophora harrisonii	Afrotropics	Sal	Shorea robusta	Indo-Malaysia
Red Mangrove	Rhizophora mangle	Afrotropics, Indo-Malaysia, Neotropics	Salal	Gaulthoria shallon	Nearctic
Red Mangrove	Rhizophora mucronata	Afrotropics	Salmonberry	Rubus spectabilis	Nearctic
Red Mangrove	Rhizophora racemosa	Afrotropics, Nearctic	Salmon White Gum	Eucalyptus lane-poolei	Australasia
Red Maple	Acer rubra	Nearctic	saltbush	Atriplex	Nearctic
Red Oat Grass	Themeda triandra	Afrotropics	Saltbush	Atriplex centralasiatica	Palearctic
Red Spruce	Picea rubens	Nearctic	Saltbush, Oldman	Atriplex nummularia	Australasia
Redwood, Coast	Sequoia sempervirens	Nearctic	Salt Cedar	Tamarix ramosissima	Palearctic
reed, common	Phragmites	Afrotropics	Saltgrass	Distichlis spicata	Nearctic
Resin Tree	Dipterocarpus alatus	Indo-Malaysia	Salt Grass	Sporobolus spicatus	Afrotropics
restio	Restio	Afrotropics	Sand Live Oak	Quercus geminata	Nearctic
Rhodes Grass	Chloris gayana	Afrotropics	Sand Pine	Pinus clausa	Nearctic
rhododendron	Rhododendron	Nearctic	Sandyfield Beaksedge	Rhynchospora megalocarpa	Nearctic
Rice, Wild	Oryza longistaminata	Afrotropics	Sapodilla	Manilkara zapota	Neotropics
Rice, Wild	Oryza rufipogon	Indo-Malaysia	Sarawak Kauri	Agathis orbicula	Indo-Malaysia
Ricegrass, Indian	Oryzopsis hymenoides	Nearctic	Sassafras (China)	Sassafras tzumu	Palearctic
Ridge-fruited Mallee	Eucalyptus angulosa	Australasia	Sassafras, Southern	Atherosperma moschatum	Australasia
Rimu	Dacrydium cupressinum	Australasia	Satinwood, Ceylon	Chloroxylon swietenia	Indo-Malaysia
River Red Gum	Eucalyptus camaldulensis	Australasia	Sausage Tree	Kigelia africana	Afrotropics
Rock Rose	Helianthemum nummularium	Palearctic	Saw Banksia	Banksia serrata	Australasia
			Saw Palmetto	Serenoa repens	Nearctic

COMMON NAME	SCIENTIFIC NAME	REGION	COMMON NAME	SCIENTIFIC NAME	REGION
saxifrage	Saxifragaceae	Palearctic	Southern Sassafras	Atherosperma moschatum	Australasia
Scalesia	Scalesia pedunculata	Neotropics	Speargrass	Heteropogon contortus	Australasia
Scarlet Globemallow	Sphaeralcea coccinea	Nearctic	sphagnum	Sphagnum	Palearctic
Scots Pine	Pinus sylvestris	Palearctic	Spiny Hopsage	Grayia spinosa	Nearctic
Scrub Oak	Quercus inopina	Nearctic	Sponge Tree	Prosopis cineraria	Indo-Malaysia
scurfpea	Psoralea	Nearctic	Spotted Gum	Corymbia maculata	Australasia
Scurfpea, Slimflower	Psoralea tenuiflora	Nearctic	Spruce, Ayan	Picea jezoensis	Palearctic
sea asparagus	Salicornia	Australasia	Spruce, Engelmann	Picea engelmannii	Nearctic
Sea Coconut	Lodoicea maldivica	Afrotropics	Spruce, Norway	Picea abies	Palearctic
sedge (Arctic)	Eriophorum	Palearctic	Spruce, Red	Picea rubens	Nearctic
Sedge, Giant	Carex gigantea	Nearctic	Spruce, Siberian	Picea obovata	Palearctic
sedge, true	Carex	Afrotropics	Spruce, Sitka	Picea sitchensis	Nearctic
seepweed	Suaeda	Afrotropics, Nearctic	Spurge, Antique	Euphorbia antiquorum	Indo-Malaysia
			Squirreltail	Elymus elymoides	Nearctic
Sequoia, Giant	Sequoia giganteum	Nearctic	Star-Chestnut, African	Sterculia africana	Afrotropics
Seringa, Wild	Burkea africana	Afrotropics	Stinging Tree	Dendrocnide moroides	Australasia
serviceberry	Amelanchier	Nearctic	St.-John's-wort, Greater Marsh	Hypericum walteri	Nearctic
Sessile Oak	Quercus petraea	Palearctic			
Shea	Vitellaria paradoxa	Afrotropics	stonecrop	Crassulaceae	Afrotropics
Sheoak, Beach	Casuarina equisetifolia	Australasia	stone oak	Lithofagus	Indo-Malaysia
Sheoak, Desert	Allocasuarina decaisneana	Australasia	Strandzha Oak	Quercus hartwissiana	Palearctic
shepherd's tree	Boscia	Afrotropics	Strangler Fig	Ficus watkinsiana	Australasia
Shower, Golden	Cassia fistula	Indo-Malaysia	Strawberry Guava	Psidium cattleyanum	Afrotropics
Siberian Fir	Abies sibirica	Palearctic	Strawberry Tree	Arbutus unedo	Palearctic
Siberian Spruce	Picea obovata	Palearctic	Stringybark, Darwin	Eucalyptus tetrodonta	Australasia
Sicklebush	Dichrostachys cinerea	Afrotropics	Subalpine Fir	Abies lasiocarpa	Nearctic
Side-oats Grama	Bouteloua curtipendula	Nearctic	Sugar Hackberry	Celtis laevigata	Nearctic
silktassel	Garrya	Nearctic	Sugar Maple	Acer saccharum	Nearctic
Silver Beech	Nothofagus menziesii	Australasia	Sugar Pine	Pinus lambertiana	Nearctic
Silver Hairgrass	Aira caryophyllea	Afrotropics	sumac	Rhus	Palearctic
Silver-leaved Ironbark	Eucalyptus melanophloia	Australasia	Sumatran Pine	Pinus merkusii	Indo-Malaysia
Single-leaf Pinyon Pine	Pinus monophylla	Nearctic	sundew	Drosera	Indo-Malaysia, Nearctic
Single-seeded Juniper	Juniperus monosperma	Nearctic			
Sitka Spruce	Picea sitchensis	Nearctic	Sunflower, Annual	Helianthus annuus	Nearctic
Skunkbrush	Rhus aromatica	Nearctic	Swamp Fern	Blechnum serratum	Nearctic
Slash Pine	Pinus elliottii	Nearctic	Swamp Tupelo	Nyssa biflora	Nearctic
Slimflower Scurfpea	Psoralea tenuiflora	Nearctic	Sweetbay Magnolia	Magnolia virginiana	Nearctic
Slipper Orchid, White	Paphiopedilum niveum	Indo-Malaysia	Sweet Birch	Betula lenta	Nearctic
Small-leaved Linden	Tilia cordata	Palearctic	Sweet Chestnut	Castanea sativa	Palearctic
Smooth Flatsedge	Cyperus laevigatus	Afrotropics	Sweet Gum (China)	Liquidambar formosana	Palearctic
Snow Gum	Eucalyptus pauciflora	Australasia	Sweet Gum (North America)	Liquidambar styraciflua	Nearctic
Soap Mallee	Eucalyptus diversifolia	Australasia			
Soapweed Yucca	Yucca glauca	Nearctic	Sweet Thorn	Vachellia karroo	Afrotropics
Soft Tree Fern	Dicksonia antarctica	Australasia	Switchgrass	Panicum virgatum	Nearctic
Sotol	Dasylirion wheeleri	Nearctic	Sword Fern	Polystichum munitum	Nearctic
			Sycamore, Arizona	Platanus wrightii	Nearctic

COMMON NAME	SCIENTIFIC NAME	REGION	COMMON NAME	SCIENTIFIC NAME	REGION
Sycamore, California	*Platanus racemosa*	Nearctic	Tree, Fever	*Vachellia xanthophloea*	Afrotropics
Sycamore Maple	*Acer pseudoplatanus*	Palearctic	tree, flame	*Delonix*	Afrotropics
Sydney Blue Gum	*Eucalyptus saligna*	Australasia	Tree, Gurjun	*Dipterocarpus tuberculatus*	Indo-Malaysia
Tallow Tree	*Allanblackia floribunda*	Afrotropics	Tree, Koroch	*Millettia pinnata*	Indo-Malaysia
Tamarind	*Tamarindus indica*	Afrotropics	Tree, Needlewood	*Schima wallichii*	Indo-Malaysia
tamarisk	*Tamarix*	Afrotropics, Nearctic, Palearctic	Tree, Nyala	*Xanthocercis zambesiaca*	Afrotropics
			tree, octopus	Didiereaceae	Afrotropics
			Tree, Resin	*Dipterocarpus alatus*	Indo-Malaysia
Tamarisk, Wild	*Tamarix usneoides*	Afrotropics	Tree, Sausage	*Kigelia africana*	Afrotropics
Tamarugo	*Prosopis tamarugo*	Neotropics	tree, shepherd's	*Boscia*	Afrotropics
Tarbush, American	*Flourensia cernua*	Nearctic	Tree, Sponge	*Prosopis cineraria*	Indo-Malaysia
Tasajillo	*Cylindropuntia leptocaulis*	Nearctic	Tree, Stinging	*Dendrocnide moroides*	Australasia
Teak	*Tectona grandis*	Indo-Malaysia	Tree, Tallow	*Allanblackia floribunda*	Afrotropics
Teak, African	*Pericopsis elata*	Afrotropics	Tree, Tooth Brush	*Salvadora oleoides*	Indo-Malaysia
Teak, African	*Pterocarpus angolensis*	Afrotropics	Tree Fern, Soft	*Dicksonia antarctica*	Australasia
Teak, Zambezi	*Baikiaea plurijuga*	Afrotropics	trident grass	*Tristachya*	Afrotropics
Teff Grass	*Eragrostis tef*	Afrotropics	true sedge	*Carex*	Afrotropics
terminalia	*Terminalia*	Afrotropics	Tubosa	*Pleuraphis mutica*	Nearctic
Terminalia, Purple-pod	*Terminalia prunioides*	Afrotropics	Tulip-tree	*Liriodendron tulipifera*	Nearctic
Texas Ebony	*Pithecellobium flexicaule*	Nearctic	Tumbleweed	*Salsola kali*	Nearctic
Texas Mountain-laurel	*Sophora secundiflora*	Nearctic	Tupelo, Swamp	*Nyssa biflora*	Nearctic
Texas Oak	*Quercus buckleyi*	Nearctic	Tupelo, Water	*Nyssa aquatica*	Nearctic
Texas Persimmon	*Diospyros texana*	Nearctic	Turkey Oak	*Quercus laevis*	Palearctic
Texas Wintergrass	*Nassella leucotricha*	Nearctic	Two-needled Pinyon Pine	*Pinus edulis*	Nearctic
thatching grass	*Hyparrhenia*	Afrotropics			
Thistle, Russian	*Echinops exaltatus*	Palearctic	Umbrella Pine	*Pinus pinea*	Palearctic
Thistle, Russian	*Salsola kali*	Nearctic	Umbrella Thorn	*Acacia planifrons*	Indo-Malaysia
Thorn, Camel	*Vachellia erioloba*	Afrotropics	Umbrella Thorn	*Vachellia tortilis*	Afrotropics
Thorn, Sweet	*Vachellia karroo*	Afrotropics	Uricua Oak	*Quercus laurina*	Nearctic
Thorn, Umbrella	*Acacia planifrons*	Indo-Malaysia	Utah Juniper	*Juniperus osteosperma*	Nearctic
Thorn, Umbrella	*Vachellia tortilis*	Afrotropics	Utah Pickleweed	*Salicornia utahensis*	Nearctic
Thorn, Whistling	*Vachellia drepanolobium*	Afrotropics	Valley Oak	*Quercus lobata*	Nearctic
three-awn grass	*Aristida*	Nearctic	Venus Flytrap	*Dionea muscipula*	Nearctic
thyme	*Thymus*	Palearctic	Vetiver	*Chrysopogon zizanioides*	Indo-Malaysia
Timor Mountain Gum	*Eucalyptus urophylla*	Australasia	Vine Maple	*Acer circinatum*	Nearctic
Tooth Brush Tree	*Salvadora oleoides*	Indo-Malaysia	violet	*Viola*	Nearctic
Toyon	*Heteromeles arbutifolia*	Nearctic	Virginia Pine	*Pinus virginiana*	Nearctic
Traveler's Palm	*Ravenala madagascariensis*	Afrotropics	Virginia Willow	*Itea virginica*	Nearctic
			Walnut, African	*Lovoa trichilioides*	Afrotropics
Tree, Boojum	*Fouquieria columnaris*	Nearctic	Walnut, Arizona	*Juglans major*	Nearctic
tree, cinnamon	*Cinnamomum*	Indo-Malaysia	Walnut, Caucasian	*Pterocarya fraxinifolia*	Palearctic
Tree, Cinnamon	*Cinnamomum zeylanicum*	Indo-Malaysia	Wandoo	*Eucalyptus wandoo*	Australasia
Tree, Coffin	*Taiwania cryptomeriodes*	Indo-Malaysia	Wandoo, Powderbark	*Eucalyptus accedens*	Australasia
Tree, Cucumber	*Dendrosicyos socotranus*	Afrotropics	waterberry	*Syzygium*	Afrotropics
tree, dragon	*Dracaena*	Afrotropics	Water Cabbage	*Pistia stratiotes*	Afrotropics
Tree, Dragon Blood	*Dracaena cinnabari*	Afrotropics			

COMMON NAME	SCIENTIFIC NAME	REGION	COMMON NAME	SCIENTIFIC NAME	REGION
Water Hickory	*Carya aquatica*	Nearctic	White Saksaul	*Haloxylon persicum*	Palearctic
water lily	*Nymphaeaceae*	Afrotropics	White Slipper Orchid	*Paphiopedilum niveum*	Indo-Malaysia
Water Tupelo	*Nyssa aquatica*	Nearctic	Wild Date, Indian	*Phoenix sylvestris*	Indo-Malaysia
Wattle, Golden	*Acacia pycnantha*	Australasia	Wild Date Palm	*Phoenix reclinata*	Afrotropics
Wattle, Lebombo	*Newtonia hildebrandtii*	Afrotropics	Wild Jujube	*Ziziphus spina-christi*	Palearctic
Wax Myrtle	*Myrica cerifera*	Nearctic	Wild Olive	*Olea oleaster*	Palearctic
Weeping Paperbark	*Melaleuca irbyana*	Australasia	Wild Rice	*Oryza longistaminata*	Afrotropics
Weeping Paperbark	*Melaleuca leucadendra*	Indo-Malaysia	Wild Rice	*Oryza rufipogon*	Indo-Malaysia
Welwitschia	*Welwitschia mirabilis*	Afrotropics	Wild Seringa	*Burkea africana*	Afrotropics
Western Hemlock	*Tsuga heterophylla*	Nearctic	Wild Tamarisk	*Tamarix usneoides*	Afrotropics
Western Juniper	*Juniperus occidentalis*	Nearctic	willow	*Salix*	Nearctic
Western Poison Oak	*Toxicodendron diversilobum*	Nearctic	Willow, Dwarf	*Salix herbacea*	Palearctic
			Willow, Goodding's	*Salix gooddingi*	Nearctic
Western Red Cedar	*Thuja plicata*	Nearctic	Willow, Virginia	*Itea virginica*	Nearctic
Western White Pine	*Pinus monticola*	Nearctic	Winterfat	*Krascheninnikovia lanata*	Nearctic
wheatgrass	*Agropygon*	Nearctic	Wintergrass, Texas	*Nassella leucotricha*	Nearctic
wheatgrass	*Elymus*	Nearctic	Wolfberry, Berlandier	*Lycium berlandieri*	Nearctic
Whistling Thorn	*Vachellia drepanolobium*	Afrotropics	Woollybutt	*Eucalyptus miniata*	Australasia
White-bark Acacia	*Vachellia leucophloea*	Indo-Malaysia	Wormwood, Mongolian	*Artemisia monosperma*	Palearctic
Whitebark Pine	*Pinus albicaulis*	Nearctic	Yellow Birch	*Betula alleghenensis*	Nearctic
White Cypress-Pine	*Callitris columellaris*	Australasia	yellowthorn	*Rhigozum*	Afrotropics
White Fir	*Abies concolor*	Nearctic	Yew, Common	*Taxus baccata*	Palearctic
White-flowered Black Mangrove	*Lumnitzera racemosa*	Afrotropics	Yorrell	*Eucalyptus gracilis*	Australasia
White Gum	*Eucalyptus argophloia*	Australasia	Yorrell	*Eucalyptus yilgarnensis*	Australasia
White Mallee	*Eucalyptus dumosa*	Australasia	yucca	*Yucca*	Nearctic
White Mangrove	*Laguncularia racemosa*	Afrotropics, Nearctic	Yucca, Soapweed	*Yucca glauca*	Nearctic
			Zambezi Teak	*Baikiaea plurijuga*	Afrotropics
White Oak	*Quercus alba*	Nearctic	Zanzibar Copal	*Hymenaea verrucosa*	Afrotropics
White Oak	*Quercus pubescens*	Palearctic	Zinnia, Prairie	*Zinnia grandiflora*	Nearctic

INDEX

Page numbers in **bold** indicate
an illustration.

Aardvark, 248, 288, 311
Aardwolf, 248, 300, 311
acacia shrubland, 30, 57, 61, 81,
    83–85, 100, 149, 152, 292,
    296, 450
Accentor
    Alpine, 224
    Brown, 409
    Robin, 409
Adder
    Common Death, 58, 68
    Desert Death, 30
    Mud, 85
    Peringuey's, 245
    Puff, 288
    Southern, 319
Adjutant
    Greater, 226, 228, **240**
    Lesser, 226, 228
Afrotropical habitats (Sub-
    Saharan Africa)
    cities and villages, 239,
        337–38, **338**
    cropland, 237, 334–35, **335**,
        337, 525
    desert, 104, 243–45, **243**, **245**,
        246, 253, 325
    Dragon Blood semidesert,
        253–54, **254**, 453
    dry thorn savanna, 112, 243,
        245, 246, 248, 253, 281,
        282, 283, 284, 285, 288,
        289, 290, 292, 295, 296,
        296–300, **297**, **298**, **299**,
        300, 301, 303, 304, 326,
        336, 337, 354, 363, 364
    freshwater wetland, 87, 262,
        281, 302, 322–25, **322**, **323**,
        324, 329, 330
    fynbos, 66, 69, 138, 159, 246,
        247, 249, 312, 316–19, **317**,
        318, **319**, 320, 363
    grazing land, 98, 183, 336–37,
        336
    Guinea savanna, 255, 259,
        281–84, **282**, **283**, **284**, 285,
        286, 289, 296, 298, 301,
        303, 304, 308, 312, 338
    gusu woodland, 285, 286,
        288, 289–91, **290**, **291**, 292,
        296, 301, 303, 304, 385
    Indian Ocean rainforest,
        202, 249, 250, 253, 255,
        267, 273–77, **274**, **275**, **276**,
        278, 279
    inselbergs, koppies, and
        cliffs, 211, 305–7, **306**,
        307

karoo 121, 243, 244, 246–49,
    247, **248**, 312, 316, 319, 320,
    336, 456, 461
lowland rainforest, 35, 124,
    192, 255–59, **256**, **257**, **258**,
    260, 262, 263, 264, 265, 267,
    271, 273, 281, 283, 284, 288,
    302, 304, 307, 325, 327, 328
Malagasy dry deciduous
    forest, 217, 249, 250, 252,
    253, 273, 278–81, **279**, **280**
Malagasy spiny forest, 146,
    249–53, **250**, **251**, **252**, 273,
    278, 279, 281, 453
mangrove, 229, 255, 262, 264,
    322, 327–28, **327**
miombo woodland 53, 259,
    260, 261, 262, 263, 283, 284,
    285–88, **286**, **287**, **288**, 289,
    290, 291, 292, 295, 296, 301,
    303, 304, 308, 312, 488
moist mixed savanna, 50,
    262, 269, 281, 284, 285, 286,
    288, 289, 292, 295, 296,
    298, 300, 301–4, **302**, **303**,
    316, 320, 337, 338
monsoon forest, 255, 257,
    259, 260–63, **261**, **262**, **263**,
    265, 267, 271, 286, 287,
    288, 290, 302, 303, 304,
    327, 337
montane forest, 38, 41, 131,
    135, 255, 260, 263, 264,
    267–72, **268**, **269**, **270**, **271**,
    303, 304, 312, 316, 320,
    321, 484
montane grassland, 161, 223,
    246, 268, 308, 312–16, **313**,
    314, **315**, 320, 321, 334,
    335, 336
montane heath, 269, 312, 316,
    320–21, **321**
Mopane savanna 149, 285,
    288, 289, 290, 291, 292–96,
    293, **294**, **295**, 300, 301,
    303, 308, 326, 479
offshore islands, 235, 329,
    330, 331, 333–34, **333**, 523
pelagic waters, 97, 234, 329,
    330, 331–32, **331**, **332**, 333
rocky coastline, 233, 325,
    329, 330–31, 333, 519
salt marsh, 231, 322, 325, 327,
    329–30, **329**, 331, 333, 517
salt pan, 322, 325–26, **326**,
    330, 329, 516
sandy beach, 233, 325, 329,
    330–31, **330**, 333, 519
swamp forest, 225, 255, 257,
    260, 264–65, **265**, 267, 322,
    327, 328

thornscrub, 112, 253, 281, 285,
    289, 292, 296–301, **299**,
    354, 364
tidal mudflat, 231, 322, 325,
    327, 329–30, **329**, 331, 333,
    516, 517
tropical grassland, 63, 156,
    243, 246, 259, 283, 287,
    296, 308–12, **309**, **310**, **311**,
    322, 334, 335, 336, 458
Agama
    Black-necked, 288
    Common, 284
    Eritrean Rock, 245
    Ground, 291
    Namibian Rock, 300
    Tropical Spiny, 295
Agouti, 175
    Azara's, 149
Albatross
    Atlantic Yellow-nosed, **331**
    Black-browed, 166, 180
    Black-footed, 522
    Buller's, 179, 180
    Chatham, **97**
    Gray-headed, 166, 180
    Light-mantled, 166, 180
    Northern Royal, 179, 332
    Royal, 166, 180
    Salvin's, 179, 180
    Short-tailed, 522
    Sooty, 166, 180
    Wandering, 166, 180, 332
    Waved, 179, 180
Alligator, American, 470, 513
Alligator Lizard, Smith's
    Arboreal, 486
alpine heathland, Australasian,
    66, 69–71, **70**, **71**, 138,
    312, 320
Alpine Skink, Taiwan (bird), 224
alpine tundra
    Eurasian, 161, 356, 389, 390,
        407–11, **408**, **409**, 509, 511
    Nearctic, 69, 161, 389, 407,
        424, 427, 434, 436, 509–11,
        509, **510**, **511**
    paramo, 69, 114, 115, 128, 137,
        138, 161–64, **162**, **163**, 312
Amphiuma, Three-toed, 471
Anaconda, Yellow, 169
Anhinga, 513
Ani, Smooth-billed, 184
Anole, Brown, 467
Antarctic pelagic waters, 179, **180**
Antarctic tundra, 165–68
Antbird
    Bertoni's, 134
    Dot-backed, 172
    Gray-headed, 139
    Hairy-crested, 127

Lunulated, 127
Plumbeous, 172
Silvered, 172
Slender, 145
Stripe-backed, 148
White-bibbed, 134
White-cheeked, 127
White-plumed, 125, 127
White-shouldered, 172
Zeledon's, 133
Anteater, Giant, 149, 153, 158,
    169, **170**
Antelope
    Dwarf, 258
    Four-horned, 189, 218
    Roan, 283, 290, 318
    Royal, 258
    Sable, 287, **287**, 290, 318
    Saiga, 359, 396
Antelope Squirrel
    Nelson's, 457
    White-tailed, 457
Antpitta
    Crescent-faced, 139–40
    Rusty-breasted, 134, 139
    Tawny, 163
    White-throated, 136
Antshrike
    Cocha, 172
    Large-tailed, 134
    Silvery-cheeked, 148
    Western Slaty, 133
    White-bearded, 134
Antvireo, Rufous-backed, 134
Apalis, Rudd's, 262
Apostlebird, 57, 78
Aracari, Collared, 127
Arboreal Alligator Lizard,
    Smith's, 486
Arctic polar desert, 356,
    399–401, **399**, **400**, 408
Argali, 411
Armadillo
    Andean Hairy, 119
    Giant, 149, 153, 169
    Nine-banded, 149, 432,
        486–87, 490, 495
    Pink Fairy, 121
    Six-banded, **148**, 149
    Southern Three-banded, 149
Ash Borer, Emerald, 473
Asity
    Schlegel's, 279
    Yellow-bellied Sunbird-, 276
aspen forest and parkland,
    422, 428, 437, 479, 482–84,
    483, 501
Ass
    African Wild, 245
    Asiatic Wild, 221, 357, 359,
        360, 385–86, 396, 398

Attila, Citron-bellied, 172–73
Auklet
    Cassin's, 524
    Crested, **417**, 524
    Least, 416, **523**
    Parakeet, 416, 524
    Rhinoceros, 520, 524
    Whiskered, 416, 524
Australasian habitats (Australia,
    New Zealand, and New
    Guinea)
    acacia shrubland, 30, 57, 61,
        81, 83–85, 100, 149, 152,
        292, 296, 450
    alpine heathland, 66, 69–71,
        70, 71, 138, 312, 320
    Brigalow woodland, 57–58,
        58, 83, 159, 281, 285, 488
    Callitris woodland, 57–58,
        281, 285, 488
    chenopod shrubland, 30–32,
        31, 32, 63, 83, 246, 353, 456
    dry sclerophyll forest, 57, 72,
        74, 75–77, **76, 77**, 79
    dune spinifex, 28–30, **28, 29**,
        51, 63, 64, 65, 100
    Gondwanan conifer
        rainforest, 26–27, **27**, 33,
        66, 440
    large-scale farming, 98–100,
        98, 99, 100
    lowland heathland, 66–68,
        67, 68, 69, 152, 159, 316
    mallee woodland and
        scrubland, 66, 77, 81–83,
        81, 82, 85, 99, 100, 292, 358
    mangrove, 89–92, **90, 91**, 174,
        229, 327, 514
    Melaleuca savanna, 47, 49,
        56, 59–61, **59, 60, 61**
    monsoon vine forest, 43,
        45–47, **46, 47**, 48, 49, 56,
        59, 260
    montane rainforest, 35,
        38–40, **39, 40**, 199, 267,
        484
    Mulga woodland, 57, 61, 77,
        81, 83–85, **84, 85**, 100, 149,
        152, 292, 296, 450
    New Zealand beech forest,
        26, 141, 33–35, **33, 34**, 41
    open eucalypt savanna, 48,
        49, 50–53, **50, 51, 52**, 57,
        59, 289, 301, 336
    pelagic waters, 97–98, **97**,
        179, 331
    rocky coastline, 92–93, **92**,
        177, **178**, 330
    rocky spinifex, 28–30, **28, 29**,
        49, 51, 65, 100
    salt marsh, 95–96, 325,
        329, 517

samphire shrubland, 30–32,
    31, 32, 246, 353, 456
sandstone escarpments,
    92–93, **92**, 177, **178**, 330
sandy beach, 94, 95, 177, 330,
    92–93, **93**
sandy cays, 94–95, **94**,
    333, 523
sheoak woodland, 26, 141,
    33–35, **33, 34**, 41
subtropical rainforest, 26, 35,
    38–40, **39, 40**, 41, 57, 72,
    74, 267, 484
temperate eucalypt
    woodland, 69, 75, 77–80,
    78, 79, 80, 83, 85, 285, 301
temperate rainforest, 26, 33,
    41–42, **42**
temperate wetland, 86–87,
    86, 87, 322
tetrodonta woodland
    savanna, 50, 53–56, **54, 55**,
    56, 77, 289
tidal mudflat, 92, 95–96, **95**,
    96, 175, 325, 329, 517
tropical lowland rainforest,
    35–37, **36, 37**, 38, 41, 43,
    45, 89, 124, 255, 273
tropical semi-deciduous
    forest, 43–44, **44**, 143
tropical wetland, 86, 87–89,
    88, 322
tussock grassland, 28, 308,
    458, 63–65, **64, 65**
wet sclerophyll forest, 38,
    57, 72–74, **73**, 75
Avocet
    American, 506
    Pied, 326, 414
Aye-aye, 275

Babbler
    Abbot's, 197
    Arabian, 362, 365
    Bare-cheeked, 295
    Black-capped, 195
    Common, 365
    Dalat Shrike-, 201
    Eyebrowed Wren-, 201
    Gray-breasted, 208, 210
    Gray-chested, 271
    Gray-throated, 201
    Hall's, 85
    Jerdon's, 221
    Limestone Wren-, 212
    Marsh, 221
    Nonggang, 212
    Red-collared Mountain-, 271
    Rufous-vented Grass, 190
    Sooty, 212
    Streaked Wren-, 212
    Tarim, 356

White-chested, 227
White-throated, 190
    Mountain-, 272
Baboon
    Bleeding Heart, **314**
    Chacma, 248, 262
    Hamadryas, 245
    Olive, 283
    Yellow, 262, 287
Baboon, Socotra Island Blue
    (spider), 254
Badger
    American, 459, 463, 490, 493,
        497, 501
    European, 372, **378**, 379,
        384, 388
    Honey, 298, 385
    Japanese, 382
Bald Cypress–tupelo forest,
    468–71, **469, 470**, 474, 512
Bald Ibis, Northern, 393, **394**
bamboo forest, East Asian
    temperate, 133, 373–74, **374**
Bamboo Lemur
    Lac Alaotra, 323
    Western Lesser, 279
Banded Knob-tailed Gecko,
    Northern, 49
Banded Newt, Caucasian
    Northern, 376
Banded Snake, Desert, 30
Bandicoot
    Northern Brown, 36, 56
    Raffray's, 70
    Southern Brown, 68, 76
    Striped, 70
    Western Barred, 79
Banteng, 216
Barbet
    Anchieta's, 288
    Banded, 304
    Blue-throated, 216
    Coppersmith, 240
    Crested, 304
    Double-toothed, 284
    Miombo, 288
    Pied, 248, **291**
    Red-crowned, 208, 210
    Vieillot's, 283–84
    White-eared, 262
    Whyte's, 288
    Yellow-breasted, 299
    Yellow-fronted, 194
Bark Beetle, Mountain, 440
Barred Bandicoot, Western, 79
Barred Frog, Northern, 37
Bat
    Hoary, 109
    Indian Roundleaf, 190
    Lesser Long-nosed, 454
    Mexican
        Free-tailed, 490

Long-tongued, 454
New Zealand
    Lesser Short-tailed, 34
    Long-tailed, 34
Northern Blossom, 36, 52, 60
Straw-colored Fruit, 337
Trumpet-nosed, 486
Bateleur, 304
Batis
    Chinspot, 291, 304
    Pririt, 248, 300
    Woodward's, 262
beach
    Afrotropical sandy, 233, 325,
        329, 330–31, **330**, 333, 519
    Australasian sandy, 94, 95,
        177, 330, 92–93, **93**
    Indo-Malayan sandy, 231,
        233–34, **233**, 235, 330, 519
    Nearctic sandy, 330, 416, 516,
        519–20, **519, 520**, 523
    Neotropical sandy, 92, 175,
        177–78, **177**, 179
    Palearctic sandy, 414, 416–17,
        416, 417, 519
Beaded Lizard, Guatemalan, 113
Beaked Gecko, Western, 85
Beaked Whale, Gervais, 521
Bear
    American Black, 422, 430,
        432, 436, 438, 442, 448,
        470, 473, 475–76, **477**,
        483, 495
    Asiatic Black, 187, 382
    Brown, 342, **342**, 349, 360,
        372, 388, 413, 436, 442,
        463, 483, 516
    Gobi, 360
    Grizzly, 436, 442, 463,
        483, 516
    Kermode, 442
    Kodiak, 442
    Polar, 401, **404**, 406, 507
    Sloth, 214, 218
    Spectacled, 133, 163, **163**
    Sun, 226
Beaver
    Eurasian, 413
    North American, 481, 483, 512
beech forest, New Zealand, 26,
    141, 33–35, **33, 34**, 41
Bee-eater
    Black, 258
    Black-headed, 258
    Blue-headed, 258
    Böhm's, 288
    European, 386
    Green, 240
    Red-throated, 283, **284**
Beetle
    Mountain Bark, 440
    Mountain Pine, 425, 436

Beira, 245, 306
Bellbird, New Zealand, 27, 34
Beringian taiga savanna, 342,
345, 350–51, **351**, 401, 424
Bernieria, Long-billed, 279
Berrypecker, Crested, 70
Bettong
Burrowing, 79
Northern, 73
Rufous, 76
Woylie, 79
Bharal, 411
Bighorn Sheep, Desert, 445, 454
Bilby, 359
Greater, 29, 62, 65
Bison
American, 497, 500, **500**
European, 372, **372**
Bittern
American, 512
Australasian, 86
Least, 512
White-crested, 265
Black Bear
American, 422, 430, 432, 436,
438, 442, 448, 470, 473,
475–76, **477**, 483, 495
Asiatic, 187, 382
Blackbird
Eurasian, 27
Indian, 224
Red-shouldered, 513
Red-winged, 513
Saffron-cowled, 183
Scarlet-headed, 169
Tricolored, 526
Yellow-headed, 513
Blackbuck, 189, **191**, 218, 353,
363, 366
Blackcap, Eurasian, 372
Black-Cockatoo, Glossy, 57
Black Duck, Pacific, 167
Black Hawk, Common, 481
Black-headed Oriole, Ethiopian,
272
Black Snake, Red-bellied, 68
Blackstart, 362
Black-tailed Deer, Columbian,
442, 503
Blackthroat, 373
Black-throated Warbler
Blue, 476
Green, 476
Black Whipsnake, Greater, 56
Blesbok, 337
Blossom Bat, Northern, 36,
52, 60
Bluebird
Eastern, 432, 445, 481, 484
Mountain, 440, 445, 484
Western, 445
Bluebonnet, Naretha, 32

Bluebonnet Parrot, Greater, 99
Bluebuck, 318
Blue-eared Starling
Greater, 295, 304
Lesser, 284, 288
Blue Flycatcher
Mangrove, 231
Pale-chinned, 197–98
Timor, 46
White-tailed, 262
Blue Heron
Great, 517
Little, 513, 517
Blue-Magpie
Taiwan, 204
Comoro, 276
Madagascar, 275
Seychelles, 276
Blue, Melissa, 501
Blue Robin, Siberian, 342
Blue Swallow, Montane, 315
Blue-tail, Red-flanked, 344
Bluethroat, 351
Blue Tit, Eurasian, 378, **378**, 388
Blue-tongued Lizard
Centralian, 63
Western, 82
Blue-tongue, Saltbush Slender,
31
Boa
Central American, 467
Dumeril's, 252
Madagascar Ground, 280
Rosy, 504
Round Island Keel-scaled,
334
Rubber, 430, 504
Western Madagascar Tree,
280
Boar, Wild, 214, 349, 372, 379,
384, 388, 398
Boatbill, Yellow-breasted, **47**
Bobcat, 430, 432, 442, 448, **448**,
454, 459, 463, 466, 470,
473, 481, 490, 493, 495,
501, 503
Bobolink, 158, 183, 501, **525**, 526
Bobwhite
Black-throated, 466, 467
Crested, 114
Northern, 432, 466, 490,
498, 501
Bog Lemming, Southern, 476
Bokmakierie, 248, 319
Bongo, 257
Bonobo, 257
Bontebok, 318
Booby
Abbott's, 236
Brown, 94, 234, 334, 524
Masked, 234, 334, 524
Peruvian, 177, 180

Red-footed, 95, 110, 234, 334
Bottlenose Dolphin, Common,
93, 179, 234, 332, 514, 521
Boubous
Southern, 304
Tropical, 304
Bowerbird
Golden, 39
Great, 52
Regent, 39, **39**, 74
Satin, 72
Tooth-billed, 39
Box Turtle, Eastern, 474
Bracken-Warbler, Cinnamon,
320
Brambling, 371
Brant, 517
Brigalow woodland, 57–58, **58**,
83, 159, 281, 285, 488
Bright-eyed frog, Western, 280
Bristlehead, Bornean, 208,
210, **210**
Broadbill
African, 262
Banded, 198
Black-and-red, 198, 227
Black-and-yellow, **194**
Grauer's, 271
Silver-breasted, 198
Whitehead's, 201
broadleaf forest, Indo-Malayan
subtropical, 199, 201,
202–5, **203**, **204**, 380, 381
Brocket Deer, Red, 484
Brolga, **64**
Bronzewing, Flock, 64
Brown Bandicoot
Northern, 36, 56
Southern, 68, 76
Brown Howler Monkey,
Southern, 104
Brown Lemur
Common, 279
Red-fronted, 279
Sanford's, 279
Brown Snake
Ingram's, 65
Ringed, 85
Speckled, 65
Brown Tarantula, Texas, 491
Brushbird, Buff-banded, 46
Bushbuck, 262, 271, 283, 287,
290, 337
Brushfinch
White-headed, 144
Yellow-striped, 136
Brush-tailed Porcupine, African,
257
Brush-Warbler
Grand Comoro, 320
Subdesert, 252
brushland, mesquite, 112,
464–67, **465**, **466**, **467**, 486
Brushtail Possum
Common, 27, 34, 46, 73, 85

Short-eared, 39
Brushturkey, Red-billed, 90
Buffalo,
African, 271, 283, 290, 304,
311, 318, 323
Forest, 257
Asian, 221
Bulbul
Bare-faced, 212
Brown-eared, 367
Chestnut, 204
Hook-billed, 208, 210
Red-vented, 240
Swamp Palm, 265
White-throated, 197
Yellow-eared, 201
Yellow-vented, 240
Bullfinch
Eurasian, 378
Gray-headed, 188
Bullfrog
American, 471
Madagascar, 280
Bunting
Cabanis's, 288
Cinnamon-breasted, 307
Corn, 396, 419
Indigo, 433, 481
Lark, 461, 498
Lark-like, 248
Lazuli, 481, 504
McKay's, 506
Meadow, 398
Orange-breasted, 487, **488**
Ortolan, 388
Painted, 466, 490, 491
Rose-bellied, 487
Snow, 404, 506, 526
Socotra, 254
Somali, 299
Varied, 114
Yellow-breasted, 238
Burrowing Toad, Mexican, 488
Bushbird, Buff-banded, 46
Bushbuck, 262, 271, 283, 287,
290, 337
Bushchat
Pied, 224
Timor, 45, 46
Bush-Crow, Stresemann's, 337
Bushlark
Australasian, 99
Burmese, 190
Indochinese, 218
Bushpig, 262
Bushshrike
Four-colored, 262
Gorgeous, 262
Gray-headed, 304
Rosy-patched, 299
Sulphur-breasted, 304
Uluguru, 272

Bush Sparrow, Yellow-
throated, 218
Bush Squirrel
Ochre, 262
Smith's, 262, 290
Bushtit, 504
Burmese, 188
Bush Warbler
Friendly, 224
Sri Lanka, 201
Sunda, 44
Taiwan, 224
Bustard
Arabian, 245, 353
Australian, 64, **65**
Blue, 316
Buff-crested, 299
Great, 396, **396**, 420
Indian, 190
Heuglin's, 245
Houbara, 353, 362
Karoo, 248
Kori, 299, **299**
Little, 396, 420
Macqueen's, 221
Rüppell's, 245
White-bellied Barrow's, **315**
White-quilled, 319
Butcherbird, Black-backed, 53
Buttonquail
Black-rumped, 311
Painted, 75
Buzzard
Forest, 335
Jackal, 248
Macqueen's, 360
Rufous-winged, 238
Socotra, 254
Upland, 360

caatinga, 48, 129, 130, 143, 144,
146–48, **146, 147, 148**, 153,
249, 453
Cacholote, White-throated, 121
cactus desert, columnar, 249,
253, 449, 450, 451, 453–55,
454, 455, 458, 502
Cactus-Finch, Genovesa, 110
Caiman
Spectacled, 513
Yacare, 169
Caiman Lizard, Paraguay, 169
Callitris woodland, 57–58, 281,
285, 488
Camel
Bactrian, **356**, 357, 360
One-humped, 62, 363, 366
campo, 153, 156–58, **157, 158**, 168,
169, 183, 184, 308, 397
Canary
Black-headed, 248
Papyrus, 325

Yellow-fronted, 291
Canastero
Cactus, 106
Dusky-tailed, 160
Patagonian, 118
Puna, 115
Cane Rat, Greater, 257
Canvasback, 501
Capercaillie
Black-billed, 349
Western, 341
Capybara, 149, 169, **183**
Caracal, 189, 248, 293, 298, 319,
336, 385
Caracara
Chimango, 118
Striated, 118
Cardinal, Red-crested, **151**
Caribou, 351, 402, 422, 506, 507
Carpet Viper, Northeast
African, 245
Cassowary, Southern, 36, **37**
Cat
African Golden, 257
Andean Mountain, 115, 119
Black-footed, 248, 300
Geoffroy's, 117, 149
Leopard, 194, **197**, 382
Little Spotted, 149
Pallas's, 385, 398, **398**, 411
Pampas, 119, 158
Sand, 359
Catbird, Green, 39
Cave Gecko, Giant, 49
cays, Australasian sandy,
94–95, **94**, 333, 523
cedar savanna, 488–91, **489, 490**
Central Asian cold desert, 355,
357–59, **358**, 360, 363, 385
Cerrado, 131, 146, 147, 148, 150,
152–54, **153, 154**, 168, 182,
184, 213, 281, 301
Chachalaca
Chaco, 150
Plain, 466
Chaco seco, 81, 83, 120, 122, 129,
144, 147, 149–51, **150, 151**,
152, 153, 156, 158, 160, 182,
184, 189, 301
Chaffinch, Common, 349
Chameleon
Flap-necked, 263, 288
Labord's, 280
Namaqua, 245
Oustalet's, 276
Parson's, 276
Rhinoceros, 280
Seychelles Tiger, 277
Socotra, 254
Uluguru Pygmy, 272
Usambara Two-horned, 272
Warty, 252

Chamois, **347**, 348, 349, 411
Chanting-Goshawk
Dark, 291, 304
Pale, 300
chaparral, Pacific, 160, 316,
391, 430, 453, 491, 502–4,
503, 504
Chat
Arnot's, 288, 295
Buff-streaked, 316
Cape Robin-, 319
Familiar, 307
Gibber, 32
Herero, 305
Karoo, 248
Mocking Cliff-, 307
Moorland, 315, 320
Orange, 32
Red-breasted, 487
Sickle-winged, 248
Tractrac, 245
White-crowned Robin-, 284
White-fronted, 32
White-throated Robin-, 304
White-winged Cliff-, 307
Yellow-breasted, **480**, 481
Chatterer, Fulvous, 362
Chat-Tyrant, Jelski's, 140
Checkerspot, Gorgone, 501
Cheetah, 244, 283, 293, 300, 311,
318, 336
chenopod shrubland, 30–32, **31,
32**, 63, 83, 246, 353, 456
Chevrotain, Water, 258, 265
Chickadee
Black-capped, 427, 440
Boreal, 422
Carolina, 473
Chestnut-backed, 443
Mountain, 427, 430
Chicken
Greater Prairie-, **498**, 501
Lesser Prairie-, 498
Chiffchaff
Caucasian, 376
Common, 348, 378
Mountain, 376
Chihuahuan Desert shrubland,
296, 361, 444, 446, 450–52,
450, 451, 452, 453, 456, 458,
461, 464, 467, 488, 490, 491
Chilia, Crag, 160
Chilla, 106
Chimpanzee, 256, **256**, 257
Chinkara, 189, 218, 221
Chipmunk
Eastern, 473, 476
Least, 430, 436, 438
Townsend's, 442
Yellow-pine, 430
Chital, 214, 218
Chorus Frog, Ornate, 433

Chough
Red-billed, 409
White-winged, 57, 78
Chowchilla, 39
Chuckwalla, 454
Chuck-will's-widow, 432,
490, 495
Chukar, 386
Cinclodes
Buff-winged, 163
Royal, 139
Seaside, 106
Cisticola
Aberdare, 321
Desert, 311
Golden-headed, 238
Rock-loving, 307
Socotra, 254
Trilling, 288
Zitting, 311
cities and villages
Afrotropical, 239, 337–38, **338**
Indo-Malayan, 239–40,
240, 337
Civet
African, 257
Hose's Palm, 200
Masked Palm, 200
Otter, 208
Cliff-Chat
Mocking, 307
White-winged, 307
cliffs, inselbergs, and koppies,
211, 305–7, **306, 307**
cloud forest
Nearctic, 267, 468, 484–86, **485**
Neotropical, 38, 104, 124, 127,
128, 129, 131–34, **132, 133,
134**, 135, 136, 137, 138, 140,
163, 182, 199, 202, 267, 273,
447, 468, 484
stunted, 138–40, **139, 140**,
161, 468
yungas, 115, 134, 135–37, **136**,
138, 140, 150
Coachwhip, 452, 461
Eastern, 495
coastline
Afrotropical rocky, 233, 325,
329, 330–31, 333, 519
Australasian rocky, 92–93,
92, 177, **178**, 330
Indo-Malayan rocky, 231,
233–34, 330, 519
Nearctic rocky, 330, 416, 516,
519–20, **519, 520**, 523
Neotropical rocky, 92,
177–78, 179
Palearctic rocky, 414, 416–17,
416, 417, 519
Coati, White-nosed, 113, 448,
466, 484, 487

Cobra, Forest, 259, 263
Cochoa
  Green, 201
  Javan, 201
Cockatiel, **99**
Cockatoo
  Glossy Black-, 57
  Palm, 36
  Pink, 62, **84**
  Yellow-crested, 47
Cock-of-the-rocks, Andean, **133**
Colchic deciduous rainforest,
  369, 375–77, **376**
Collared-Dove, African, 299
Collared Frog, Little, 56
Collared Lizard, Eastern, 490, 491
Colobus
  Angola, 262, 271
  Guereza, 271, 337
  Niger Delta Red, 265
  Tana River Red, 262
  Western Red, 328
  Zanzibar Red, 262
columnar cactus desert, 249,
  253, 449, 450, 451, 453–55,
  454, 455, 458, 502
Comet, Red-tailed, 136
Common Dolphin, Short-
  beaked, 331
Condor
  Andean, 115
  California, 504
Conebill
  Giant, 139
  Tamarugo, 106
conifer forest
  Eurasian montane, 186, 188,
    340, 343, 346–48, **346, 347,**
    350, 366, 371, 375, 387,
    425, 428, 434, 437
  Mediterranean and dry,
    348–50, **349,** 385, 444
  montane mixed-, 348, 425,
    427, 428–30, **429, 430,** 434,
    436, 437, 438, 440, 443,
    444, 446, 447, 479, 504
  Neotropical mixed-, 26, 102–4,
    141, 447
  Gondwanan, 26–27, **27,** 33,
    66, 440
Coot, American, 512
Copperhead, 474, 490
Cormorant
  Bank, 330, 334
  Black-faced, 93
  Brandt's, 520
  Cape, 330, 334
  Crowned, 330, 334
  Double-crested, 512
  Flightless, 177
  Great, 417
  Guanay, 177, 180

Little Pied, 88
  Neotropic, 513
  Pelagic, 520
  Pied, 88
  Red-legged, 180
Cotinga, White-cheeked, 139
Cottonmouth, **470,** 471, 513, 515
Cotton Rat, Yellow-nosed, 451
Cottontail
  Andean, 163
  Appalachian, 476
  Desert, 445, 451, 454, 457,
    459, 463
  Eastern, 473, 476
  Mountain, 445, 463
Coua
  Blue, 276
  Coquerel's, 279
  Giant, 279
  Red-fronted, 276
  Running, 252
  Verreaux's, 252
Coucal
  Black, 311
  Greater, 240
  Rufous, 195
  Sunda, 231
Courser
  Bronze-winged, 288
  Burchell's, 245, 248
  Cream-colored, 353
  Double-banded, 244, 248
  Jerdon's, 190
  Temminck's, 311
Coyote, 451, 454, 459, 463,
  473, 483, 490, 493, 497,
  501, 503
Crab-Plover, 233, 330, **415,** 416
Crake
  Australian, 86
  Baillon's, 86
  Corn, 419
  Ruddy-breasted, 238
  Spotless, 86
Crane
  Black-necked, 409
  Blue, 316, **335**
  Demoiselle, 221, 384, **384**
  Hooded, 413
  Red-crowned, 413
  Sandhill, 513, 526, **526**
  Sarus, 229, **239**
  Siberian, 413
  Wattled, 324
  White-naped, 413
  Whooping, 519
Creeper
  African Spotted, 284, 288
  Brown, 430, 442, 476
Crescentchest
  Collared, 154
  Elegant, 144

Crested Tern
  Great, 95, 233, 234, 330
  Lesser, 95, 233, 330, 334
  West African, 330
Crested-Tinamou, Elegant,
  118, 119
Cricket Frog
  Northern, 481
  Southern, 471
Crocias, Gray-crowned, 201
Crocodile
  American, 515
  Cuban, 513
  Estuarine, 88
  Freshwater, 88
  Nile, 284, 323, 328
  Saltwater, 88, 90, **91,** 208, 230
  Siamese, 226
Crombec
  Cape, 248
  Northern, 299
  Red-capped, 288
cropland
  Afrotropical, 237, 334–35,
    335, 337, 525
  Nearctic, 98, 418, 525–26,
    525, 526
  Neotropical, 181–82, **181,**
    182, 525
  paddy fields and other, 219,
    227, 237–39, **237, 238, 239,**
    419, 525
  Palearctic, 418–19, **418,**
    419, 525
Crossbill
  Hispaniolan, 468
  Red, 188, 341, 348, 349, 427,
    436, 442
  White-winged, 422, 427
Crow
  American, 473
  Cape, 335
  House, 240, 338
  Northwestern, 443, 520
  Somali, 245
  Stresemann's Bush-, 337
  Tamaulipas, 467
Crowned Snake, Florida, 495
Crowned Warbler, Eastern, 368
Cuckoo
  Bornean Ground-, 226
  Common, 341–42
  Dideric, 338
  Klaas's, 338
  Lesser, 240
  Ground-, 114, 487
  Mangrove, 175, 515
  Plaintive, 240
  Sumatran Ground-, 201
  Sunda, 43
  Thick-billed, 284
  Yellow-billed, 470, 481

Cuckoo-Dove
  Slaty, 46
  Timor, 43, 46
Cuckoo-roller, 276
Cuckooshrike
  Black, 304
  Blue, 259
  Gray, 271
  Ground, 62
  Indochinese, 218
  Red-shouldered, 284
  White-breasted, 284, 285,
    295
Culpeo, 115, 119, 143, 163
Cupwing, Pygmy, 43
Curlew
  Eurasian, 414
  Far Eastern, 232, 414
  Long-billed, 463, 498, 517
Cuscus
  Common Spotted, 36
  Mountain, 39
Cutia, Vietnamese, 188, **188,** 201

Dapple-throat, 272
Dasyure
  Habbema, 70
  Speckled, 70
Day Gecko
  Madagascar Giant, **338**
  Peacock, **338**
  Round Island, 334
Death Adder
  Common, 58, 68
  Desert, 30
deciduous forest
  caatinga, 48, 129, 130, 143,
    144, 146–48, **146, 147, 148,**
    153, 249, 453
  East Asian temperate, 202,
    366, 369, 373, 378, 380–82,
    380, 381, 382, 471
  European temperate, 342,
    369, 371, 375, 377–79, **378,**
    380, 388, 471
  Indo-Malayan
    dry, 189, 191, 213, 216,
      217–19, **218, 219,** 278,
      486
    moist, 45, 196, 198, 213–16,
      214, 215, 216, 217
  Malagasy dry, 217, 249, 250,
    252, 253, 273, 278–81,
    279, 280
  Nearctic temperate, 369, 377,
    380, 381, 431, 433, 468,
    471–74, **472, 473, 474,** 475,
    476, 478, 479, 490, 491,
    493, 494
  Neotropical dry, 143–45, **144,**
    145, 146, 217, 278, 447, 449,
    453, 455, 468, 486

deciduous rainforest, Colchic,
369, 375–77, **376**
Deer
Columbian Black-tailed,
442, 503
Eld's, 218
Fallow, 34, 379, 390
Indian Hog, 214
Key, 175
Mule, 430, 438, 445, 459, 463,
493, 497
Pampas, 158, 183
Red, 349, 379, 390
Brocket, 484
Roe, 384, **384**, 390
Sika, 367, **368**, 390
White-tailed, 123, 145, 175,
430, 432, 462, 466, 470,
473, 475, 481, 483, 486,
490, 493, 495, 497, 500
Deer Mouse, Michoacan, 486
desert
Afrotropical, 104, 243–45,
243, **245**, 246, 253, 325
Arctic polar, 356, 399–401,
399, **400**, 408
Central Asian cold, 355,
357–59, **358**, 360, 363, 385
columnar cactus, 249, 253,
449, 450, 451, 453–55, **454**,
455, 458, 502
East Asian cold, 104, 355–57,
355, **356**, 359, 360, 456
Galápagos, 108–11, **109**, **110**
Himalayan montane, 161,
407–11, **410**
Neotropical desolate, 104–7,
105, **106**, 118, 202, 243, 352,
355, 357
Palearctic hot, 189, 243, 244,
245, 352–54, **352**, **354**, 356,
358, 361, 362, 364, 385
shrub, 243, 244, 245, 246,
296, 300, 352, 353,
358, 361–363, **361**, **362**,
364, 450
desert grassland, Nearctic, 308,
444, 446, 449, 450, 452,
453, 456, 458–61, **459**, **460**,
467, 491, 493, 496, 498
desert shrubland, salt, 456–58,
456, **457**, 461
desert steppe, temperate, 114,
116, 117, 355, 357, 359–60,
359, **360**, 397
desolate desert, Neotropical,
104–7, **105**, **106**, 118, 202,
243, 352, 355, 357
Devil, Thorny, **29**, 30
Dhole, 197, 214, 218
Diamondback Rattlesnake
Eastern, 433

Western, 452, **452**
Dibatag, 299
Dickcissel, 501
Dingo, 32, 52, 65, 93
Dipper
American, 513
Brown, 409
Diucon, Fire-eyed, 103
Diving-Petrel
Common, 167, 180
Peruvian, 180
Dodo, 276, 335
Dog
African Wild, 283, 287, 291,
295, **295**, 304, 305, 311
Raccoon, 367, 382
Dolphin
Atlantic Spotted, 521
Common Bottlenose, 93, 179,
234, 332, 514, 521
Ganges River, 228
Heaviside's, 332
Irrawaddy River, 228
Northern Right Whale, 521
Pacific White-sided, 521
Pantropical Spotted, 332
Pink River, 173
Risso's, 521
Short-beaked Common, 331
Spinner, 234, 332
Donacobius, Black-capped, 169
Dotterel
Black-fronted, 88
Eurasian, 409
Inland, 32, **32**, 64
Red-breasted, **95**
Red-kneed, 88
Rufous-chested, **117**
Double-collared Sunbird
Northern, **271**
Southern, 319
Douc Langur, Black-shanked, 198
Dove
African Collared-, 299
Black-backed Fruit-, 44
Black-banded Fruit-, 45, 46,
48, 49, **49**
Blue-capped Fruit-, 90
European Turtle-, 419
Galápagos, 110
Inca, 461
Laughing, 338
Mourning, 432, 452, 457, 461,
463, 490, 495, 498
Namaqua, 298, 365
Pink-headed Fruit-, 201
Red-eyed, 338
Slaty Cuckoo-, 46
Spotted, 240
Stock, 419
Timor Cuckoo-, 43, 46
Wallace's Fruit-, 90

White-tipped, 466
White-winged, 454, 490
Zebra, 240
Dovekie, 524
Dowitcher
Asian, 232, **232**, 384
Short-billed, 176, 517
Dragon
Black-collared, 63
Blue-lined, 30
Central Netted, 63
Chameleon, 56
Claypan, 31
Gibber, 32
Komodo, 44, **44**
Lined Earless, 65
Military, 30
Nullarbor Earless, 31
Pebble-mimic, 32
Rusty, 30
Smooth-snouted Earless, 32
Dragon Blood semidesert,
253–54, **254**, 453
Drill, 257
Dromedary, 62, 363, 366
Drongo
Common Square-tailed, 262
Fork-tailed, 336
Glossy-backed, 336
dry deciduous forest
caatinga, 48, 129, 130, 143,
144, 146–48, **146**, **147**, **148**,
153, 249, 453
Indo-Malayan, 189, 191,
213, 216, 217–19, **218**, **219**,
278, 486
Malagasy, 143, 217, 249,
250, 252, 253, 273, 278–81,
279, **280**
Neotropical, 143–45, **144**, **145**,
146, 217, 278, 447, 449, 453,
455, 468, 486
Dtella, Arnhem Land Spotted,
49
Duck
Black-bellied Whistling-, 513
Black Pacific, 167
Blue-billed, 86
Ferruginous, 228
Hardhead, 86
Harlequin, 513, **519**
Hartlaub's, 265, **265**
Knob-billed, 229
Lesser Whistling-, 228
Long-tailed, 507
Masked, 513
Meller's, 324
Musk, 86
Pink-eared, 86
Plumed Whistling-, 88
Wandering Whistling-, 88
White-headed, 413

White-winged, 226
Wood, 470
Dugong, 328, 329
Duiker
Blue, 262, 271
Common, 248, 283, 287, 318
Harvey's, 271
Natal Red, 262
Rwenzori Red, 321
Weyns's, 271
Dunlin, 507
Dwarf Lemur, Fat-tailed, 279
Dwarf Porcupine, Mexican
Hairy, 484

Eagle
African Fish-, 324
Bald, **424**, 481
Black-chested Snake-, 311
Booted, 376, 384, 388, 420
Changeable Hawk-, 216
Congo Serpent-, 258
Crowned, 271
Fasciated Snake-, 262
Golden, 349, 458, 461, 464,
498, 511
Gray-headed Fish-, 227
Greater Spotted, 238
Lesser Spotted, 371–72, 420
Madagascar
Fish-, 281, 324, 328
Serpent-, 276
Martial, 248
Spanish, 388
Tawny, 298
Verreaux's, 307
Wallace's Hawk-, 208
Wedge-tailed, 65
White-bellied Sea-, 233
Eagle-Owl
Cape, 320
Pharaoh, 353
Shelley's, 259
Spot-bellied, 216
Usambara, 272
Eared-Pheasant, Blue, 409
Earless Dragon
Lined, 65
Nullarbor, 31
Smooth-snouted, 32
Earless Lizard, Greater, 452
Earthcreeper
Band-tailed, 118
Buff-breasted, 119
Crag, 160
Scale-throated, 119
East Asian cold desert, 104,
355–57, **355**, **356**, 359,
360, 456
East Asian moist mixed forest,
366–68, **367**, **368**, 369, 373,
380, 381

East Asian temperate bamboo
  forest, 133, 373–74, **374**
East Asian temperate deciduous
  forest, 202, 366, 369, 373,
  378, 380–82, **380, 381,
  382,** 471
eastern grass steppe, 356, 359,
  382, 395, 396, 397–98, **397,
  398,** 407, 496
Echidna
  Eastern Long-beaked, 70
  Short-beaked, **71**
Egret
  Cattle, 336
  Chinese, 232, 413
  Great, 517
  Little, 329
  Reddish, 517
  Snowy, 517
Eider
  King, **405,** 406, 507, 508
  Spectacled, 507, **508**
  Steller's, 406, 507, 508
Elaenia, Slaty, 136
Eland
  Common, 314
  Giant, 283
Elephant
  African
    Bush, 244, 271, 283, 287,
      291, 293, **302,** 304, 305,
      318, 335
    Forest, 257, 265, 271
    Savanna, 305
  Asian, 194, 197, 207, 214, **216,**
    218, 219, 221, 226
Elephant Seal
  Northern, 519
  Southern, **166**
Elephant Shrew, Golden-
  rumped, **262**
elfin forest, 41, 115, 135, 138–40,
  139, 140, 161, 163, 468
Elk, 342, 351, 413, 430, 438, 445,
  463, 483, 500, 511
  Roosevelt, 442
Emerald-Toucanet, Northern,
  484
Emu, 100
Emuwren
  Mallee, 83
  Rufous-crowned, 29
  Southern, 67
Eremomela
  Greencap, 288, 291
  Yellow-rumped, 248
escarpments, Australasian
  sandstone, 92–93, **92,** 177,
  178, 330
espinal, 81, 83, 120, 149–51, 156,
  160, 184, 189, 301
eucalypt savanna, open, 48, 49,

50–53, **50, 51, 52,** 57, 59,
  289, 301, 336
eucalypt woodland, temperate,
  69, 75, 77–80, **78, 79, 80,**
  83, 85, 285, 301
Euphonia
  Golden-rumped, 113
  Scrub, 466
Eurasian alpine tundra, 161,
  356, 389, 390, 407–11, **408,
  409,** 509
Eurasian boggy tundra, 342,
  343, 345, 350, 399, 401,
  405–6, **405, 406,** 408, 507
Eurasian montane conifer
  forest, 186, 340, 346–48,
  346, 347, 350, 366, 371,
  375, 387, 425, 428, 434, 437
Eurasian rocky tundra, 342, 343,
  345, 350, 399, 401–4, **402,
  403, 404,** 405, 406, 407,
  408, 505
Eurasian spruce-fir taiga,
  340–42, **341, 342,** 343, 344,
  346, 350, 351, 366, 371, 377,
  379, 382, 388, 405, 422, 425
European heathland, 320,
  389–91, **390**
European moist mixed forest,
  366, 369–72, **370, 371, 372,**
  375, 377, 378, 379, 475
European moorland, 320,
  389–91
European temperate deciduous
  forest, 342, 369, 371, 375,
  377–79, **378,** 380, 388, 471

Fairywren
  Blue-breasted, 75, 78, **79**
  Purple-backed, 62
  Purple-crowned, 52, **52**
  Red-backed, **55,** 60
  Splendid, **62,** 85
  Variegated, 78
  White-winged, 31
Falanouc, Eastern, 275
Falcon
  Aplomado, 461
  Barbary, 365, 386
  Gray, 29
  Peregrine, 213, 240, 365, 490
  Prairie, 458, 464, 498
  Pygmy, 300
  Saker, 386
  White-rumped, 218
Falconet
  Collared, 216
  Spot-winged, 150
Fanaloka, 275
Fantail
  Arafura, 46
  Brown-capped, 43, 44

Friendly, 39
Mangrove, 90
Northern, 60
Rufous, 74
White-browed, 218
White-throated, 201
farming, Australasian large-
  scale, 98–100, **98, 99, 100**
Fat-tailed Mouse-Opossum,
  Elegant, 160
Fence Lizard
  Eastern, 490
  Western, 430, 504
Fernwren, 39
Ferret, Black-footed, 498
Fieldfare, **371**
Fieldwren
  Rufous, 32
  Striated, 69
File Snake, Little, 90
Finch
  Black-capped Warbling-, 150
  Black Rosy-, 427, 511
  Brown-capped Rosy-,
    427, 511
  Carbonated Sierra-, 118
  Cassin's, 427, 436
  Desert, 356
  Double-barred, 52, **60**
  Genovesa
    Cactus-, 110
    Ground-, 110
  Gouldian, **51,** 52, 100
  Gray-crowned Rosy-, 427,
    436, 511
  Gray Warbler-, 110
  Green Warbler-, 110
  Large Ground-, 110
  Long-tailed, 52
  Mangrove, 110
  Many-colored Chaco, 150
  Masked, 52
  Patagonian Sierra-, 119,
    120, 142
  Plum-headed, 57
  Raimondi's Yellow-, 106
  Ringed Warbling-, 121
  Rusty-browed Warbling-,
    136
  Sharp-beaked Ground-, 110
  Slender-billed, 106, 107
  Vegetarian, 110
  Woodpecker, 110
Finfoot, African, 265, 324
Fireback
  Crested, 195
  Crestless, 208
  Siamese, 198
Firecrest, Common, **349**
Firetail
  Diamond, 57
  Mountain, 70

Painted, 29
fir forest, Nearctic montane
  spruce-, 346, 422, 425–27,
  426, 427, 428, 430, 434,
  436, 437, 438, 440, 482
fir taiga, spruce-
  Eurasian, 340–42, **341, 342,**
    343, 344, 346, 350, 351,
    366, 371, 377, 379, 382, 388,
    405, 422, 425
  Nearctic, 340, 422–24, **423,
    424,** 425, 475, 478, 482,
    484, 501, 506, 514
Fiscal, Somali, 245
Fish-Eagle
  African, 324
  Gray-headed, 227
  Madagascar, 281, 324, 328
Fisher, 422, 427, 430, 442
Fishing-Owl
  Pel's, 262
  Rufous, 265
  Vermiculated, 265
Flameback
  Black-rumped, 218
  Common, 218
Flamecrest, 188
Flamingo
  American, 515
  Andean, 115
  Chilean, 115
  Greater, 326, **326**
  James's, 115
  Lesser, 221, 326, **326**
Flicker
  Gilded, 454
  Guatemalan, 123
  Northern, 123, 443, 481
flooded forest
  igapó, 125, 127, 169, 171–73,
    172, 173, 206, 264
  várzea, 127, 171–73, **172,
    173,** 264
flooded grassland
  Indo-Malayan seasonally,
    168, 219–22, **219, 220, 221,**
    223, 227, 237, 308
  Neotropical, 149, 153, 154,
    156, 168–70, **168, 169, 170,**
    184, 219, 227, 322
Florican
  Bengal, 221
  Lesser, 190, 221
Florida oak scrub, 431, 494–96,
  494, 495
Flowerpecker
  Black-fronted, 45
  Blood-breasted, 44, 201
  Brown-backed, 210
  Olive-crowned, 90
  Scarlet-backed, 240
  Scarlet-breasted, 210

flower steppe, western, 359,
382, 395–96, **395, 396,** 397,
398, 420, 499
Flufftail
Madagascar, 275, 320
Slender-billed, 324
Streaky-breasted, 311
Striped, 315
White-spotted, 265
Flycatcher
Acadian, 470, 473
Alder, 476
Ash-throated, 446
Atlas, 349
Black-banded, 46
Blue-throated, 216
Böhm's, 288
Broad-billed, 45
Brown-crested, 454
Cassin's, 265
Cordilleran, 430
Dull-blue, 201
European Pied, 342
Galápagos, 110
Grand Comoro, 276
Gray, 446, 463
Gray-chested Jungle-, 208
Great Crested, 433, 470, 473
Hammond's, 430
Japanese Paradise-, 368, 382
Little Pied, 43
Livingstone's, 262
Madagascar Paradise-, 276
Mangrove Blue, 231
Mariqua, 295
Mugimaki, 367–68
Narcissus, 368
Nutting's, 487
Olive-sided, 422, 430
Pacific-slope, 443
Pale, 291
Pale-chinned Blue, 197–98
Paperbark, 60
Rusty-margined, 182
Satin, 41, 72
Scissor-tailed, 490, 501, **501**
Snowy-browed, 201
Social, 182
Spotted, 342, 388
Timor Blue, 46
Ultramarine, 188
White-tailed Blue, 262
Willow, 481
Yellow, 262
Yellow-bellied, 422
Yellow-rumped, 367–68, 382
Flying Fox
Black, **47,** 52, 60, 90
Gray-headed, 90
Island, 236
Little Red, 36, 52
Madagascar, 328

Spectacled, 36, 90
Flying Lemur, Philippine, 194
Flying Squirrel
Abert's, 430
American Red, 430
Douglas's, 430
Humboldt's
Mindanao, 200
Northern, 430
Red-and-white, **188**
Southern, 473
Western Gray, 430
Foliage-gleaner, Rufous-
necked, 140
forest
Afrotropical
monsoon, 255, 257, 259,
260–63, **261, 262, 263,**
265, 267, 271, 286, 287,
288, 290, 302, 303, 304,
327, 337
montane, 38, 41, 131, 135,
255, 260, 263, 264,
267–72, **268, 269, 270,**
271, 303, 304, 312, 316,
320, 321, 484
swamp, 225, 255, 257, 260,
264–65, **265,** 267, 322,
327, 328
aspen, 422, 428, 437, 479,
482–84, **483,** 501
Australasian tropical semi-
deciduous, 43–44, **44,** 143
Bald Cypress-tupelo,
468–71, **469, 470,** 474, 512
caatinga, 48, 129, 130, 143,
144, 146–48, **146, 147, 148,**
153, 249, 453
dry sclerophyll, 57, 72, 74,
75–77, **76, 77,** 79
East Asian
moist mixed, 366–68, **367,**
368, 369, 373, 380, 381
temperate bamboo, 133,
373–74, **374**
temperate deciduous,
202, 366, 369, 373,
378, 380–82, **380, 381,**
382, 471
elfin, 41, 115, 135, 138–40, **139,**
140, 161, 163, 468
Eurasian montane conifer,
186, 188, 340, 343, 346–48,
346, 347, 350, 366, 371,
375, 387, 425, 428, 434, 437
European
moist mixed, 366, 369–72,
370, 371, 372, 375, 377,
378, 379
temperate deciduous, 342,
369, 371, 375, 377–79,
378, 380, 388, 471

igapó flooded, 125, 127, 169,
171–73, **172, 173,** 206, 264
Indo-Malayan
dry deciduous, 189, 191,
213, 216, 217–19, **218,**
219, 278, 486
freshwater swamp, 171,
206, 207, 225–27, **226,**
229, 264
limestone, 211–13, **212,** 305
mangrove, 89, 206, 208,
225, 229–31, **230, 231,**
234, 327, 514
moist deciduous, 45, 196,
198, 213–16, **214, 215,**
216, 217
peat swamp, 206–8, **207,**
208, 209, 210, 225, 227
pine, 186–88, **187, 188,** 199,
428, 447
semi-evergreen, 43, 129,
192, 196–98, **197, 198,**
211, 213, 214, 216, 260
subtropical broadleaf,
199, 201, 202–5, **203,**
204, 380, 381
lodgepole pine, 428, 430,
437–40, **438, 439,** 482
Malagasy
dry deciduous, 143, 217,
249, 250, 252, 253, 273,
278–81, **279, 280**
spiny, 146, 249–53, **250,**
251, 252, 273, 278, 279,
281, 453
matorral sclerophyll, 77, 105,
107, 159–60, **159, 160,** 316
Mediterranean and dry
conifer, 348–50, **349,**
385, 444
Mediterranean oak, 379, 385,
387–89, **388,** 391, 488, 491
montane mixed-conifer, 348,
425, 427, 428–30, **429, 430,**
434, 436, 437, 438, 440,
443, 444, 447, 479, 504
Nearctic
cloud, 267, 468, 484–86,
485
montane spruce-fir, 346,
422, 425–27, **426, 427,**
430, 434, 436, 437, 438,
440, 482
temperate deciduous,
369, 377, 380, 381, 431,
433, 468, 471–74, **472,**
473, 474, 475, 476, 478,
479, 490, 491, 493, 494
temperate mixed, 366,
422, 424, 471, 474,
475–78, **476, 477,**
478, 482

tropical dry, 464, 467,
486–88, **487, 488**
Neotropical
cloud, 38, 104, 124, 127, 128,
129, 131–34, **132, 133,**
134, 135, 136, 137, 138,
140, 163, 182, 199, 202,
267, 273, 447, 468, 484
dry deciduous, 143–45,
144, 145, 146, 217, 278,
447, 449, 453, 455,
468, 486
mixed conifer, 26, 102–4,
141, 447
semi-evergreen, 104, 108,
124, 125, 127, 128, 129–31,
130, 132, 134, 143, 148,
152, 154, 169, 182, 184,
196, 260, 468, 484, 486
New Zealand beech, 26,
33–35, **33, 34,** 41, 141
Siberian larch, 340, 342,
343–45, **344, 345,** 366–67,
402, 404, 437
stunted cloud, 138–40, **139,**
140, 161, 468
várzea flooded, 127, 171–73,
172, 173, 264
wet sclerophyll, 38, 57,
72–74, **73,** 75
yungas, 115, 134, 135–37, **136,**
138, 140, 150
Forest-Rail, Chestnut, 39
forest-steppe, Palearctic, 142,
342, 345, 382–84, **383, 384,**
385, 396
Fork-marked Lemur, Pale, 279
Fossa, 252, 275, 279, **280**
Fox
Andean, 115, 119, 163
Red, 143
Arctic, 404, 406, **406,** 413,
506, 507, 523
Bat-eared, 244, 248, 299
Cape, 248, 300, 315, 319
Corsac, 359, 385, **386,** 396
Crab-eating, **157,** 169, 175
Eastern, 473
Fennec, 353
Fuegan Red, 143
Gray, 430, 442, 451, 481, 487,
490, 503–4
Island, 523
Kit, 451, 454, 459
Pale, 245, 299
Pampas, 121
Patagonian, 106, 117
Red, 342, 348, 384, 385, 473,
476, 483, 501
Rüppell's, 245, 333
South American Gray, 106,
117, 121, **122**

Swift, 498
Tibetan Sand, 411
Francolin
Cape, 319
Chestnut-naped, 321
Chinese, 218
Crested, 304
Gray, 365
Gray-winged, 319
Hartlaub's, 300, 307
Jackson's, 316, 321
Latham's, 258
Moorland, 321
Swierstra's, 272
Free-tailed Bat, Mexican, 490
freshwater swamp forest,
Indo-Malayan, 171, 206,
207, 225–27, **226**, 229, 264
freshwater wetland
Afrotropical, 87, 262, 281,
302, 322–25, **322, 323, 324,**
329, 330
Indo-Malayan, 219, 227–29,
227, 228, 229, 237, 322
Nearctic, 322, 411, 468, 479,
480, 512–14, **512, 513,** 516,
517, 525
Friarbird
Noisy, 57
Silver-crowned, 60
Timor, 46
Frigatebird
Christmas Island, 236
Great, 95, 109, 332, 334
Frigatebird, Lesser, 95, 234,
332, 334
Magnificent, 109
Fritillary, Regal, 501
Frog
Australian Wood, 37
Barking Tree, 471
Canyon Tree, 481
Caucasian Parsley, 376
Fry's, 37
Giant Mexican Tree, 488
Gopher, 495
Gray Tree, 473
Green-eyed Tree, 37
Green Tree, 471
Little Collared, 56
Marbled, 56
Rain, 277
Masked, 49
Mink, 422
Northern
Barred, 37
Cricket, 481
Leopard, 481
Orange-thighed Tree, 37
Ornate Chorus, 433
Pig, 471
Plains Leopard, 481

Relict Leopard, 481
River, 471
Rockhole, 49
Sabinal, 488
Southern
Cricket, 471
Leopard, 471
Squirrel Tree, 433
Starry-night Reed, 277
Tawny Rocket, 56
Western Bright-eyed, 280
White-lipped Tree, 37
Wood, 422, 473, 476
Frogmouth, Marbled, 39
Fruit Bat, Straw-colored, 337
Fruit-Dove
Black-backed, 44
Black-banded, 45, 46, 48,
49, **49**
Blue-capped, 90
Pink-headed, 201
Wallace's, 90
Fulmar, Northern, 416, 520,
522, 524
Fulvetta, Golden-breasted, 373
Fur Seal
Australasian, 93
Cape, 330, 334
Northern, 519, 523
fynbos, 66, 69, 138, 159, 246, 247,
249, 312, 316–19, **317, 318,**
319, 320, 363

Gadwall, 501
Galago, Thomas's, 258
Galápagos desert and scalesia,
108–11, **109, 110**
Gallinule, Purple, 513
Gallito
Crested, 150
Sandy, 121
Gannet
Cape, **333,** 334
Northern, 416, 520, 524
Gar, Alligator, 513
garrigue, 349, 366, 389, 391,
393–94, **394,** 502
Garter Snake
Black-necked, 452
Eastern, 473
Gaur, 197, 214, 218, **219,** 221
Gazelle
Arabian Sand, 353
Dama, 299
Dorcas, 245, 299
Goitered, 353, 357, 360, 363,
366, 385
Mongolian, 398
Mountain, 353
Red-fronted, 299
Soemmerring's, 245
Speke's, 245

Thomson's, 311, 337
Gecko
Beaded, 82
Bearded, 30
Border Thick-tailed, 76
Box-patterned, 85
Burrow-plug, 85
Common
House, 240, 338, **338**
Leaf-tailed, 276
Fish-scale, **338**
Giant Cave, 49
Gibber, 32
Golden-tailed, 58
Karoo Web-footed, 245
Least, 487
Lined Leaf-tailed, 276
Madagascar Giant Day, **338**
Marbled Velvet, 49
McIlwraith, 37
Mesa, 30
Mourning, 46
Namib Web-footed, 245
Northern
Banded Knob-tailed, 49
Leaf-tailed, 37
Madagascar Velvet, **338**
Velvet, 37, 46
Ocelot, 252
Pale Knob-tailed, 30
Peacock Day, **338**
Reticulate Leaf-toed, 189
Ring-tailed, 46
Round Island Day, 334
Sakalava Madagascar Velvet,
252
Satanic Leaf-tailed, 276
Scaly, 189
Spiny-tailed, 85
Thick-tailed, 82
Tokay, 240
Tropical House, 338
Western Beaked, 85
Zigzag Velvet, 76
Gelada, **314,** 315, 321
Gemsbok, 244, 248, 295
Genet
Common, 337
Giant, 257
Miombo, 287
Rusty-spotted, 290
Gerenuk, 299, **300**
Gerygone
Dusky, 90
Golden-bellied, 230–31
Green-backed, 45–46
Mangrove, 90
Plain, 44, 46
Gharial, 228, **228**
False, 208
Giant Mouse Lemur,
Coquerel's, 279

Giant-Petrel
Northern, 97
Southern, 180
Giant Salamander, Chinese, 204
Giant Squirrel
Forest, 258, 265
Grizzled, 194
Giant Tortoise, Aldabra, 277
Gibbon
Bornean White-bearded, 208
Lar, 200
Muller's, 208, **208**
Northern White-cheeked,
204
Western Hoolock, 204
White-handed, 197
Gila Monster, 454, **455**
Giraffe, 295, 298, 304
Angolan, 244
Girdled Lizard
Black, 319
False, 319
Oelofsen's, 319
Tropical, 263
Zimbabwe, 315
Glass Lizard
Eastern, 433
Slender, 433
Glider
Feathertail, 73
Squirrel, 73, 76
Sugar, 39
Yellow-bellied, 73
Gnatcatcher
Black-tailed, 452, 454
Blue-gray, 433, 463, 473
California, 504
White-lored, 487
Yucatán, 467
Goanna
Rosenberg's, 73
Sand, 100
Goat, Mountain, 511
Godwit, 176
Bar-tailed, 329, 414, **519**
Black-tailed, 351, 414
Hudsonian, 176
Marbled, 517
Goldcrest, 348, 378
Golden Cat, African, 257
Goldeneye, Barrow's, **513**
Golden Oriole, African, 288, 291
Golden-Plover
American, 506
European, 404
Pacific, 506
Golden Sparrow, Arabian, 365
Golden Weaver, Asian, 238
Goldfinch
European, 419
Lawrence's, 504
Lesser, 490

Gondwanan conifer rainforest, 26–27, **27**, 33, 66, 440
Gonolek
  Crimson-breasted, 295, 300
  Papyrus, **323**, 324, 325
Goose
  Bar-headed, 229, 409
  Cotton Pygmy-, 88, 228
  Egyptian, 338
  Emperor, 507
  Greater White-fronted, 507
  Green Pygmy-, 87, 88
  Lesser White-fronted, 413
  Magpie, 87, 88
  Red-breasted, 413
  Ross's, 507, 513
  Ruddy-headed, 118, 184
  Snow, 507, 513
  Swan, 413
Gopher
  Northern Pocket, 498
  Plains Pocket, 501
Goral, Chinese, 200
Gorilla
  Eastern, 257
    Lowland, 257
  Mountain, **270**, 271, 315, 335
  Western, 257
    Lowland, 257
Goshawk
  Black, 335
  Dark Chanting-, 291, 304
  Northern, 341, 349, 430, 440, 443
  Pale Chanting-, 300
Grackle, Slender-billed, 513
Grass Babbler, Rufous-vented, 190
Grassbird
  Cape, 316, 319
  Little, 87
Grasscutter, 257
Grasshopper-Warbler
  Gray's, 342
  Pallas's, 344
grassland
  Afrotropical
    montane, 161, 223, 246, 268, 308, 312–16, **313**, **314**, **315**, 320, 321, 334, 335, 336
    tropical, 63, 156, 243, 246, 259, 283, 287, 296, 308–12, **309**, **310**, **311**, 322, 334, 335, 336, 458
  Australasian tussock, 28, 63–65, **64**, **65**, 308, 458
  campo, 153, 156–58, **157**, **158**, 168, 169, 183, 184, 308, 397
  Indo-Malayan
    montane, 223–25, **223**, **224**, 312

seasonally flooded, 168, 219–22, **219**, **220**, **221**, 223, 227, 237, 308
  Nearctic desert, 308, 444, 446, 449, 450, 452, 453, 456, 458–61, **459**, **460**, 467, 491, 493, 496, 498
  Neotropical flooded, 149, 153, 154, 156, 168–70, **168**, **169**, **170**, 184, 219, 227, 322
  pampas, 63, 114, 121, 149, 150, 153, 156–58, **157**, **158**, 182, 183, 184, 308, 397
  puna, 107, 114–16, **115**, **116**, 118, 135, 161, 164, 169, 312, 407
Grass-Owl, African, 324
Grassquit, Dull-colored, 182
grass steppe, eastern, 356, 359, 382, 395, 396, 397–98, **397**, **398**, 407, 496
Grasswren
  Black, 49
  Eyrean, 29
  Pilbara, 29
  Rusty, 29
  Striated, 29
  Thick-billed, 32
  White-throated, 49
Gray Flying Squirrel, Western, 430
Gray Fox, South American, 106, 117, 121, **122**
Gray Hornbill, Indian, 218
Gray Kangaroo
  Eastern, 57, 73, 79, 85, 100
  Western, 68, 82, 85, 100, **100**
Gray Mongoose, Cape, 248, 319
Gray Shrike
  Great, 360
  Iberian, 396
  Lesser, 387
  Steppe, 360
Gray Squirrel
  Eastern, 473, 476
  Mexican, 484, 487
Gray Warbler, Black-throated, 446, 449
grazing land
  Afrotropical, 98, 183, 336–37, **336**
  Nearctic, 418, 420, 525–26, **525**, **526**
  Neotropical, 183–84, **183**, **184**
  Palearctic, 420
Grebe
  Clark's, 512
  Eared, 326
  Pied-billed, 512
Greenbul
  Cabanis's, 262
  Gray-olive, 262
  Red-tailed, 258

Sombre, 262
Greenfinch
  European, 419
  Vietnamese, 188
Green-Magpie, Bornean, **201**
Green-Pigeon
  Cinnamon-headed, 208, 227
  Flores, 45
  Large, 208
  Madagascar, 279
  Timor, 45, 46
  Yellow-footed, 218
Greenshank
  Common, 414
  Nordmann's, 232, 414
Green Snake
  Rough, 471, 513
  Southeastern, 263
Green Warbler, Black-throated, 478, 484
Griffon, Eurasian, 420
Grison, Lesser, 121
Grizzly, 436, 442, 463, 483, 516
Grosbeak
  Black-headed, 430
  Blue, 433, 481
  Crimson-collared, 466, 467
  Evening, 427, **427**
  Pine, 341, 422, 427, 436, 440
  Socotra, 254
  Yellow, 487
Ground Boa, Madagascar, 280
Ground-Cuckoo
  Bornean, 226
  Lesser, 114, 487
  Sumatran, 201
Ground-Finch
  Genovesa, 110
  Large, 110
  Sharp-beaked, 110
Ground-Hornbill, Southern, **303**
Ground-Jay
  Henderson's, 356, 360
  Iranian, 353
  Mongolian, 356, **356**
  Turkestan, 358
  Xinjiang, 356
Ground-Roller
  Long-tailed, 252, **252**, 253
  Pitta-like, 275
  Rufous-headed, 275, 276
  Scaly, 275, **276**
  Short-legged, 275
Ground Skink, Round Island (reptile), 334
Ground Squirrel
  Arctic, 506
  California, 503
  Golden-mantled, 436, 438, **439**
  Spotted, 451, 459
  Thirteen-lined, 498, 501

Ground-Tyrant
  Cinnamon-bellied, 118
  Dark-faced, 106
Grouse
  Dusky, 430, 443
  Greater Sage-, 444, 463, **463**
  Gunnison Sage-, 463
  Hazel, **345**, 384
  Ruffed, 443, 484
  Sharp-tailed, 463, 498, **498**
  Sooty, 430
  Spruce, 422, 427
Grysbok
  Cape, 318
  Sharpe's, 295
Guan
  Horned, 485, **485**
  Red-faced, 136, **136**
Guanaco, 115, 117, **117**, 183
Guillemot, Pigeon, 520, 524
Guineafowl
  Crested, 262
  Helmeted, 304, 338
Guinea savanna, 255, 259, 281–84, **282**, **283**, **284**, 285, 286, 289, 296, 298, 301, 303, 304, 308, 312, 338
Gull
  Andean, 115
  Belcher's, 177
  Franklin's, 512
  Glaucous, 416
  Gray, 106
  Hartlaub's, 334
  Herring, 416
  Kelp, 334
  Laughing, 519
  Lava, 177
  Pallas's, 228
  Sabine's, 332
  Swallow-tailed, 180
  White-eyed, 334
  Yellow-footed, 520
gusu woodland, 285, 286, 288, 289–91, **290**, **291**, 292, 296, 301, 303, 304, 385

Hairy Armadillo, Andean, 119
Hairy Dwarf Porcupine, Mexican, 484
Hairy-nosed Wombat
  Northern, 57
  Southern, 82
Hamerkop, 324
Hare
  Arctic, 506
  Cape, 248, 311
  Ethiopian Highland, 315
  Hispid, 221
  Mountain, 342
  Scrub, 248

Snowshoe, 422, 425–26, 427, 438, 476, 483
Tolai, 359, 396
Hare-Wallaby, Rufous, 29
Harrier
Black, 248, 319
Cinereous, 158
Eurasian Marsh-, 229
Hen, 341, 371, 384, 388
Long-winged, 158, **158**
Montagu's, 384, 396
Northern, 458, 461, 464, 498, 501, 526
Pallid, 384
Papuan Marsh-, 70
Pied, 238
Hartebeest, Red, 248, 311
Harvest Mouse
Salt Marsh, 517
Western, 501
Hawfinch, 349
Hawk
Common Black, 481
Cooper's, 473
Ferruginous, 446, 458, 464, 498, 526
Galápagos, 110
Gray, 481
Harris's, 454, **455**, 490
Long-tailed, 258
Red-tailed, 452, 464, 473, 490, 501, 526
Roadside, 158
Rough-legged, 371, **371**, 501, 526
Rufous-tailed, 142
Swainson's, 452, 458, 463, 490, 498
White-tailed, 158
Hawk-Eagle
Changeable, 216
Wallace's, 208
Hawk Owl, Northern, 384, 422, 440
heath, Afrotropical montane, 269, 312, 316, 320–21, **321**
heath forest, kerangas, 206, 208, 209–10, **210**
heathland
Australasian
alpine, 66, 69–71, **70, 71**, 138, 312, 320
lowland, 66–68, **67, 68**, 69, 152, 159, 316
European, 320, 389–91, **390**
fynbos, 66, 69 138, 159, 246, 247, 249, 312, 316–19, **317, 318, 319**, 320, 363
Heathwren
Chestnut-rumped, 67
Shy, 83
Hedgehog, Long-eared, 359

Hedgehog Tenrec, Lesser, 252
Helmetshrike
Chestnut-fronted, 262
Retz's, 288
White, 295
Hemispingus, Piura, 139
Heron
Goliath, 324
Gray, 329
Great Blue, 517
Little Blue, 513, 517
Pacific Reef-, 95, 234
Tricolored, **176**, 513, 517
Western Reef-, 330
White-backed Night-, 324
Zigzag, 172
high-elevation pine woodland, 425, 427, 430,434–36, **435**, **436**, 509
Hillstar, Ecuadorian, 163
Himalayan montane desert, 161, 407–11, **410**
Hippopotamus, 311, 323, **324**, 337
Pygmy, 265
Hoatzin, **173**
Hobby
Eurasian, 341, 389
Oriental, 213
Hog
Feral, 432, 490
Giant Forest, 257, 271
Pygmy, 221
Red River, 257, 265
Hog Deer, Indian, 214
Hog-nosed Skunk
American, 459
Humboldt's, 117
Patagonian, 117
Hognose Snake
Blond, 280
Eastern, 493
Madagascar, 280
Honey-buzzard, European, 376
Honeyeater
Banded, 52, 53
Bar-breasted, 60
Black, 83
Black-headed, **73**
Bridled, 39
Brown-backed, 60
Crescent, 72
Fuscous, 78
Golden-backed, 52
Gray, 85
Gray-headed, 52
Gray-streaked, 70
Green-backed, 37
Indonesian, 45
Mangrove, 90, 92
New Holland, 68, **68**
Orange-cheeked, 70
Painted, 78

Purple-gaped, 83
Regent, 78, **78**
Rufous-banded, 53
Rufous-throated, 53
Smoky, 39
Spiny-cheeked, 85
Striped, 57
Tawny-crowned, 68
White-cheeked, 68
White-fronted, 83
White-lined, 49
White-naped, 75
White-throated, 74
Yellow-eared, 44
Yellow-faced, 78
Yellow-plumed, **81**, 83
Yellow-throated, 75
Yellow-tinted, 52
Honeyguide
Greater, 335
Lesser, 335
Hooded Scaly-foot, Western, 63
Hoolock Gibbon, Western, 204
Hoopoe, Eurasian, **392**
Hoopoe-Lark
Greater, 245, 353, **354**
Lesser, 245
Hornbill, 336
Black-and-white-casqued, 271
Black-casqued, 259
Blyth's, 36
Bradfield's, **290**, 291
Brown, 198
Crowned, 262
Great, 198, **198**
Indian Gray, 218
Pale-billed, 288
Rhinoceros, 195
Rufous-necked, 204
Rusty-cheeked, 198
Silvery-cheeked, 271
Southern Ground-, **303**
White-crested, 259
Wreathed, 198
Wrinkled, 208
Yellow-casqued, 259
Horned Lizard
Coastal, 504
Regal, 454
Round-tailed, 452
Hornero, Crested, 149
Horse
Feral, 62
Przewalski's, 398
Wild, 396
House Gecko
Common, 240, 338, **338**
Tropical, 338
House-Martin, Asian, 342
Howler Monkey
Black-and-gold, 170

Southern Brown, 104
Huet-huet, Black-throated, 103, 104
Hummingbird
Anna's, 504
Beautiful, 487
Black-chinned, 446, 490
Blue-capped, 485
Costa's, 504
Garnet-throated, 485
Mangrove, 175
Volcano, 164
White-tailed, 485
Hutia
Cabrera's, 514
Desmarest's, 514
Prehensile-tailed, 514
Hyena
Brown, 244, 248, 295, 300, 330
Spotted, 244, 288, 291, **291**, 295, 298, 304, 337
Striped, 245, 299, 353, 366, 385
Hylia, Green, 258
Hyliota, Yellow-bellied, 288
Hypocolius, 365, **365**
Hyrax
Eastern Tree, 271
Rock, **306**

Ibex
Alpine, 411
Nubian, 353, 363, 392
Siberian, 411
Walia, 315
Ibis
Black-headed, 229
Buff-necked, 169
Giant, 218
Hadada, 338
Madagascar, 275
Northern Bald, 393, **394**
Red-naped, 238
Spot-breasted, 265
Straw-necked, **87**
Wattled, 315, **315**
White-shouldered, 218
Ibisbill, 228, 409, **409**
Idol, Moorish, **94**
igapó flooded forest, 125, 127, 169, 171–73, **172, 173**, 206, 264
Iguana
Black, 113
Galápagos Land, **110**
Galápagos Pink Land, 109
Marine, 177, **178**
Northeastern Spiny-tailed, 467
Western Spiny-tailed, 487
Impala, 295, 304, 337

Imperial-Pigeon
Gray, 236
New Caledonian, 40
Pied, 236
Timor, 43, 44
Yellowish, 90
Indian Ocean rainforest, 202,
249, 250, 253, 255, 267,
273–77, **274, 275, 276,**
278, 279
Indigobird, Village, 338
Indigo Snake
Central American, 487
Eastern, 433, 495
Texas, 467
Indo-Malayan habitats
(Southeast Asia and India)
cities and villages, 239–40,
240, 337
dry deciduous forest, 189,
191, 213, 216, 217–19, **218,**
219, 278, 486
freshwater swamp forest,
171, 206, 207, 225–27, **226,**
229, 264
freshwater wetland, 219,
227–29, **227, 228, 229,**
237, 322
kerangas, 206, 208, 209–10,
210
limestone forest, 211–13,
212, 305
mangrove forest, 89, 206,
208, 225, 229–31, **230, 231,**
234, 327, 514
moist deciduous forest, 45,
196, 198, 213–16, **214, 215,**
216, 217
montane grassland, 223–25,
223, 224, 312
offshore islands, 94, 234,
235–36, **236,** 333, 523
paddy fields and other
cropland, 219, 227, 237–39,
237, 238, 239, 419, 525
peat swamp forest, 206–8,
207, 208, 209, 210, 225, 227
pelagic waters, 234–35, **235,**
331, 418
pine forest, 186–88, **187, 188,**
199, 428, 447
rocky coastline, 231, 233–34,
330, 519
salt pan, 231–33, **232,** 325, 329
sandy beach, 231, 233–34,
233, 235, 330, 519
seasonally flooded grassland,
168, 219–22, **219, 220, 221,**
223, 227, 237, 308
semi-evergreen forest, 43,
129, 192, 196–98, **197, 198,**
211, 213, 214, 216, 260

subtropical broadleaf forest,
199, 201, 202–5, **203, 204,**
380, 381
thornscrub, 112, 189–91, **190,**
191, 217, 363, 364
tidal mudflat, 95, 231–33, **232,**
234, 325, 329
tropical lowland rainforest,
35, 36, 124, 192–95, **193,**
194, 196, 198, 199, 200, 206,
208, 209, 211, 226, 235, 238,
255, 273
tropical montane rainforest,
138, 186, 199–201, **200, 201,**
202, 203–4, 209, 214, 216,
267, 273
Indri, 275, **275**
inselbergs, koppies, and cliffs,
211, 305–7, **306, 307**
islands
Afrotropical offshore, 235,
329, 330, 331, 333–34,
333, 523
Indo-Malayan offshore, 94,
234, 235–36, **236,** 333, 523
Nearctic offshore, 333, 519,
523–24, **523, 524**

Jabiru, 169
Jacamar, White-chinned, 173
Jacana
Bronze-winged, 229
Comb-crested, 87, **88**
Madagascar, 281, 324
Pheasant-tailed, 229, **229**
Wattled, 169
Jackal, 385
Black-backed, 244, 248, 295,
299, 314, 326, 330
Golden, 216
Side-striped, 287, 304, 323
Jackdaw, Eurasian, 419
Jackrabbit
Antelope, 451, 454
Black-tailed, 445, 457, 459,
463, 490, 498, 459
Jacky-winter, 78
Jaeger
Long-tailed, 332, 404, 506, 522
Parasitic, 234, 332, 507, 522
Pomarine, 332, 404, 507, 522
Jaguar, **130,** 136, 153, 170, 173,
175, 448, 484
Jaguarundi, 136, 448, 466,
484, 487
Javelina, 454
Jay
Blue, 473
California Scrub-, 504
Canada, 427, **427,** 440
Curl-crested, 154
Eurasian, 349

Florida Scrub-, 495, **495**
Gray, 427, **427,** 443
Green, 466, **467**
Henderson's Ground-,
356, 360
Iranian Ground-, 353
Mongolian Ground-, 356, **356**
Pinyon, **445,** 446
San Blas, 487
Siberian, 341, **342**
Steller's, 427, 430, 443, 446
Turkestan Ground-, 358
White-tailed, 145, **145**
White-throated Magpie-, 145
Woodhouse's Scrub-, 446
Xinjiang Ground-, 356
Jerboa, Thick-tailed Three-
toed, 396
Jird, Midday, 360, **360**
Junco
Dark-eyed, 436, 440, 476,
478, 484
Volcano, 164
Jungle-Flycatcher, Gray-
chested, 208
juniper woodland, pinyon-, 348,
428, 430, 436, 444–46, **445,**
446, 455, 461, 464

Kaka, 34
Kakapo, 27
Kangaroo
Bennett's Tree-, 36
Doria's Tree-, 39
Eastern Gray, 57, 73, 79,
85, 100
Lumholtz's Tree-, **40**
Musky Rat-, 36
Red, 31, 32, 62, 65, 82, 85,
85, 100
Western Gray, 68, 82, 85,
100, **100**
Kangaroo Mouse, Dark, 457
Kangaroo Rat
Chisel-toothed, 457
Ord's, 451, 457
Santa Cruz, 503
Karoo, 121, 243, 244, 246–49,
247, 248, 312, 316, 319, 320,
336, 456, 461
Kea, 27, **27,** 69
Keelback, 88
Keeled Plated Lizard,
Madagascar, 252
kerangas, 206, 208, 209–10, **210**
Kestrel
American, 452, 461, 463, 498,
501, 526
Australian, 99
Eurasian, 338, 341
Greater, 248
Lesser, 360

Rock, 338
Killdeer, 498
Killer Whale, Pygmy, 332
Kingbird
Cassin's, 452, 461, 498
Eastern, 501
Tropical, 182, 184
Western, 452, 464, 498, 501
Kingfisher
American Pygmy, 173, **175**
Beach, 90
Brown-hooded, 304
Brown-winged, 230
Buff-breasted Paradise-, 36
Chocolate-backed, 258
Cinnamon-banded, 45
Collared, 90, 328
Giant, 324
Green-and-rufous, 173
Madagascar Pygmy-, 276
Malagasy, 328
Mangrove, 328
Red-backed, 63
Stork-billed, 227
Striped, 291
White-throated, **221**
Kinglet
Golden-crowned, 427,
430, 442
Ruby-crowned, 440
King-Parrot, Australian, 73
Kingsnake
California, 493
Mountain, 430
Mole, 495
Kinkajou, 133, 487
Kipunji, 271
Kite
Black, **56,** 65, 338
Black-breasted, 65
Black-shouldered, 99
Letter-winged, 64
Mississippi, 432
Red, 371
Scissor-tailed, 299
Swallow-tailed, 432
White-tailed, 158
Kittiwake, Black-legged, 416,
520, 524
Klipspringer, 306, **307**
Knob-tailed Gecko
Northern Banded, 49
Pale, 30
Knot
Great, 232, **232,** 414
Red, 330, 414, 506, 507, 517
Koala, 73, **73,** 74, 76
Kob, 283, **283,** 311
Kodkod, 103
Koel, Asian, 240
koppies, cliffs, and inselbergs,
211, 305–7, **306, 307**

Krait, Banded Sea, 90
Kudu
    Greater, 287, 290, 295, 304
    Lesser, 299
Kusimanse, Alexander's, 265

Lammergeier, 409
Land Iguana
    Galápagos, **110**
    Pink, 109
Langur
    Black-shanked Douc, 198
    Delacour's, **212**, 213
    François', 204, 212–13
    Golden, 204
    Gray, 214
    Hatinh, 213
    Hose's, 210
    Laotian, 212
    Nilgiri, 200
    Purple-faced, 194
Lapwing
    Andean, 115, 163
    Banded, 100
    Gray-headed, 238
    Northern, 419
    River, 228
    Sociable, 360, 396
    Spot-breasted, 315
larch forest, Siberian, 340, 342,
    343–45, **344**, **345**, 366–67,
    402, 404, 437
Lark
    Bimaculated, 360
    Black, 396, **396**
    Black-Crowned Sparrow-,
        245
    Black-eared Sparrow-, 248
    Botha's, 316
    Calandra, 387, 396
    Chestnut-headed Sparrow-,
        245
    Crested, 245, 387
    Desert, 245, 353
    Dune, 245, **245**
    Dupont's, 393
    Flappet, 311
    Gray's, 245
    Greater Hoopoe-, 245,
        353, **354**
    Horned, 457, 461, 498
    Karoo, 248
    Lesser
        Hoopoe-, 245
        Short-toed, 362
    Magpie-, 90
    Masked, 245
    Mongolian, 398
    Somali Long-billed, 245
    Stark's, 245
    Temminck's, 353
    Thekla's, 396

Wood, 389
Laughingthrush
    Ashy-headed, 194
    Barred, 373
    Cambodian, 201
    Elliot's, 188
    Nilgiri, 201
    White-whiskered, 188
Leafbird, Golden-fronted, 216
Leaf-eared Mouse, Darwin's,
    106
Leaf Monkey
    Phayre's, 200
    Red, 208, 226
    Silvered, 208
Leaf-tailed Gecko
    Common, 276
    Lined, 276
    Northern, 37
    Satanic, 276
Leaf-toed Gecko, Reticulate,
    189
Leaf Warbler
    Hainan, 204
    Limestone, 212
    Mountain, 43, 44
    Timor, 44
Lechwe
    Nile, 311
    Southern, 310, 311
Leiothrix, Red-billed, 204
Lemming
    Brown, 506, 507
    Southern Bog, 476
Lemur
    Black-and-white Ruffed, 275
    Cleese's Woolly, 279
    Common Brown, 279
    Coquerel's Giant Mouse, 279
    Crowned, 279
    Fat-tailed Dwarf, 279
    Gray-brown, 252
    Gray Mouse, 252
    Lac Alaotra Bamboo, 323
    Madame Berthe's Mouse,
        279
    Mongoose, 279
    Pale Fork-marked, 279
    Petter's Sportive, 252
    Philippine Flying, 194
    Red-fronted Brown, 279
    Red Ruffed, 275
    Ring-tailed, 252, **280**
    Sanford's Brown, 279
    Western
        Lesser Bamboo, 279
        Woolly, 279
    White-footed Sportive, 252
Leopard Frog
    Northern, 481
    Plains, 481
    Relict, 481

Southern, 470–71
    Blunt-nosed, 458
Leopard, 189, 194, 197, 214, **215**,
    218, 219, 240, 245, 257, 271,
    283, 288, 291, 304, 305,
    306, 314, 319, 336, 337,
    353, 385
    Clouded, 208
    Snow, **410**, 411
Lesser Bamboo Lemur,
    Western, 279
Lesser Short-tailed Bat, New
    Zealand, 34
limestone forest, Indo-
    Malayan, 211–13, **212**, 305
Linnet, Eurasian, 419
Linsang, Spotted, 204
Liocichla, Scarlet-faced, 201
Lion, 244, 271, 283, 288, 291, 298,
    304, 306, 311, 314, 318, 336
    Asiatic, 218, 353
Lizard
    Black-beaded, 467
    Black Girdled, 319
    Black-lined Plated, 295
    Black-nosed, 487
    Blunt-nosed Leopard, 458
    Burton's Snake-, 82
    Bushveld, 291
    Cape Mountain, 319
    Centralian Blue-tongued, 63
    Coastal Horned, 504
    Common Side-blotched, 446,
        452, 458, 504
    Desert Spiny, 461
    Eastern
        Collared, 490, 491
        Fence, 490
        Glass, 433
        Spiny, 467
    False Girdled, 319
    Florida
        Scrub, 495
        Worm, 495, 496
    Four-lined Plated, 252
    Frilled, 100
    Giant Plated, 307
    Greater
        Earless, 452
        Short-horned, 446, 464,
            498
    Guatemalan Beaded, 113
    Kalahari Plated, 291
    Long-tailed Spiny, 487
    Madagascar Keeled Plated,
        252
    Oelofsen's Girdled, 319
    Ornate Sandveld, 315
    Paraguay Caiman, 169
    Regal Horned, 454
    Round-tailed Horned, 452
    Sagebrush, 446, 458, 464

Shingleback, 82, **82**
Shovel-snouted, 245
Slender Glass, 433
Smith's Arboreal Alligator,
    486
Southern Rock, 319
Stumpy-tailed, 82
Three-eyed, 252
Three-lined Plated, 252
Tropical Girdled, 263
Tsingy Plated, 280
Wedge-snouted, 245
Western
    Blue-tongued, 82
    Fence, 430, 504
    Plated, 280
Zebra-tailed, 454
Zimbabwe Girdled, 315
lodgepole pine forest, 428, 430,
    437–40, **438**, **439**, 482
Logrunner
    Australian, 39
    Southern, 74
Long-beaked Echidna, Eastern,
    70
Long-billed Lark, Somali, 245
Longclaw
    Abyssinian, 315
    Sharpe's, 316
    Yellow-throated, 311
longleaf pine savanna, 431–34,
    432, **433**, **434**, 474, 494,
    499
Long-nosed Bat, Lesser, 454
Longspur
    Chestnut-collared, 461, 501
    Lapland, 404, 506, 526
    McCown's, 461, 498
Long-tailed Bat, New Zealand,
    34
Longtail, Green, 272
Long-tongued Bat, Mexican,
    454
Loon
    Arctic, 406, 507
    Common, **512**
    Pacific, 507, 512
    Red-throated, 507
    Yellow-billed, 406, 507
Lorikeet
    Iris, 44
    Leaf, 44
    Little, 75
    Musk, 78
    Olive-headed, 44
    Red-collared, **59**, 60
    Scaly-breasted, **76**
    Varied, 52
Loris, Red Slender, 200
Lory
    Brown, 90
    Red, 90

Lovebird
 Black-collared, 258
 Black-winged, 304
Lowland Gorilla
 Eastern, 257
 Western, 257
lowland heathland,
  Australasian, 66–68, **67**,
  68, 69, 152, 159, 316
lowland rainforest
 Afrotropical, 35, 124, 192,
  255–59, **256**, **257**, **258**, 260,
  262, 263, 264, 265, 267, 271,
  273, 281, 283, 284, 288, 302,
  304, 307, 325, 327, 328
 Australasian tropical, 35–37,
  36, **37**, 38, 41, 43, 45, 89,
  124, 255, 273
 Indo-Malayan tropical, 35, 36,
  124, 192–95, **193**, **194**, 196,
  198, 199, 200, 206, 208, 209,
  211, 226, 235, 238, 255, 273
 Neotropical, 35, 103, 104,
  124–29, **124**, **125**, **126**, **128**,
  131, 132, 134, 136, 147, 149,
  169, 171, 172, 174, 175, 182,
  192, 255, 273
Lynx
 Canada, 422, 427
 Eurasian, 349, 384
 Iberian, 349, 388, **388**, 392
 Sand, 385
Lyrebird
 Albert's, 74
 Superb, 72
Lyre Snake, Western, 487

Macaque
 Assam, 200
 Assamese, 204
 Barbary, 349, 392
 Bonnet, 214
 Crab-eating, 208, 230
 Formosan Rock, 204
 Japanese, 382, **382**
 Long-tailed, 230
 Rhesus, 212, 214
 Southern Pig-tailed, 197, 308
 Stump-tailed, 197
Macaw
 Hyacinth, 170, 184, **184**
 Indigo, 147
 Red-and-green, 154
 Red-shouldered, 154
 Spix's, 147
Madagascar Swift (reptile)
 Cuvier's, 280
 Merrem's, 252
Madagascar Tree Boa,
  Western, 280
Madagascar Velvet Gecko
 Northern, *338*

Sakalava, 252
Madrean pine-oak woodland,
  123, 382, 447–49, **448**, **449**,
  452, 455, 461, 468, 484,
  486, 491, 493
Magellanic rainforest, 33, 102,
  104, 118, 141–43, **142**,
  375, 440
Magpie
 Australian, 27
 Bornean Green-, **201**
 Taiwan Blue-, 204
 Yellow-billed, 493
Magpie-Jay, White-throated, 145
Magpie-Lark, 90
Malagasy dry deciduous forest,
  143, 217, 249, 250, 252, 253,
  273, 278–81, **279**, **280**
Malagasy spiny forest, 146,
  249–53, **250**, **251**, **252**, 273,
  278, 279, 281, 453
Malkoha
 Black-bellied, 208
 Green, 262
 Scale-feathered, 201
 Sirkeer, 218
Mallard, 167, 501
Malleefowl, 83, 99
mallee woodland and
  scrubland, 66, 77, 81–83,
  81, **82**, 85, 99, 100, 292, 358
Mamba
 Black, 288
 Jameson's, 259
Mammoth, Woolly, 402
Manakin
 Araripe, 147, **147**
 Orange-crowned, 172
 Yungas, 136
Manatee
 Amazonian, 173
 West African, 323, 328, 329
 West Indian, 514
Mandrill, 257
Mangabey
 Black-crested, 265
 Golden-bellied, 265
 Gray-cheeked, 265
 Sooty, 328
mangrove
 Afrotropical, 229, 255, 262,
  264, 322, 327–28, **327**
 Australasian, 89–92, **90**, **91**,
  174, 229, 327, 514
 Nearctic, 327, 468, 512,
  514–15, **515**, 517
 Neotropical, 174–75, **174**, **175**,
  229, 327
mangrove forest, Indo-
  Malayan, 89, 206, 208,
  225, 229–31, **230**, **231**, 234,
  327, 514

Mangrove Snake, White-
  bellied, 90
Mantella, Betsileo, 280
maquis, 114, 121, 160, 316, 336,
  349, 366, 385, 387, 388,
  389, 391–93, **391**, **392**, 502
Mara, Patagonian, 121
Margay, 113, 136
Marmot
 Bobak, 396
 Hoary, **509**, 511
 Olympic, 443
 Yellow-bellied, 436, 511
Marsh-Harrier
 Eurasian, 229
 Papuan, 70
Marten
 American, **426**, 427, 438, 442
 Pine, 348, 349, 372
Martin
 Asian House-, 342
 Banded, 311
 Gray-throated, 228
 Purple, 454
 Rock, 307
 White-eyed River, 228
matorral sclerophyll forest and
  scrub, 77, 105, 107, 159–60,
  159, **160**, 316
Meadowlark
 Eastern, 432, 461, 501, 526
 Pampas, 158
 Western, 457, 461, 463, 498,
  501, 526
Mediterranean and dry conifer
  forest, 348–50, **349**,
  385, 444
Mediterranean oak forest, 379,
  385, 387–89, **388**, 391,
  488, 491
Meerkat, 248, 300, 315
Melaleuca savanna, 47, 49, 56,
  59–61, **59**, **60**, **61**
Melidectes
 Belford's, 39
 Long-bearded, 70
 Short-bearded, 70
 Sooty, 70
Merganser, Scaly-sided, 413
Merlin, 440, 443, 484, 490
Mesite
 Subdesert, 252, 253
 White-breasted, 279
mesquite brushland and
  thornscrub, 112, 464–67,
  465, **466**, **467**, 486
Metaltail
 Neblina, 140
 Viridian, 140
Mexican Tree Frog, Giant, 488
Milk Snake, Eastern, 490
Miner

Bell, 73
 Black-eared, 83
 Campo, 158
 Coastal, 106, 107
 Grayish, 106
 Slender-billed, 115
 Thick-billed, 106, 107
Minivet
 Ashy, 368, 382
 Jerdon's, 190
 Small, 218
 White-bellied, 218
Mink
 American, 483, 512
 European, 413
Minke Whale, Antarctic, 97
Miombo Sunbird
 Eastern, 288
 Western, 288
miombo woodland, 53, 259, 260,
  261, 262, 263, 283, 284,
  285–88, **286**, **287**, **288**, 289,
  290, 291, 292, 295, 296, 301,
  303, 304, 308, 312, 488
Mockingbird
 Española, **109**, 110
 Floreana, 110
moist deciduous forest, Indo-
  Malayan, 45, 196, 198,
  213–16, **214**, **215**, **216**, 217
Mole-Rat
 Big-headed African, 315
 Naked, 299
Mole Skink, Blue-tailed
  (reptile), 495
Monal, Chinese, 409
Mongoose
 Black, 300, 306
 Bushy-tailed, 262
 Cape Gray, 248, 319
 Jackson's, 271
 Marsh, 323
 Slender, 271
 Sokoke Dog, 262
 Yellow, 248
Monito del Monte, 103
Monitor
 Asian Water, 208, 230, 240
 Lace, 68
 Mangrove, 90
 Merten's Water, 88
 Nile, 324
 Pygmy Desert, 30
 Rusty, 90
 Savanna, 288
 Short-tailed, 30
 Spencer's, 65
 Spotted Tree, 56
Monjita, Rusty-backed, 118
Monkey
 Allen's Swamp, 265
 Azara's Night, 170

Barbara Brown's Titi, 148
Black-and-gold Howler, 170
Blue, 262
Brown Woolly, 128
Campbell's, 328
De Brazza's, 265
Dryas, 265
Golden Snub-nosed, 367,
   373, **374**
Green, 283
L'Hoest's, **269**, 271
Mona, 265, 328
Patas, 283, 299
Phayre's Leaf, 200
Preuss's, 271
Proboscis, 208, **231**
Red Leaf, 208, 226
Red-tailed, 257
Silvered Leaf, 208
Silvery Woolly, 128
Snow, 382, **382**
Southern Brown Howler, 104
Tantalus, 283
Vervet, 287
monsoon forest, Afrotropical,
   255, 257, 259, 260–63, **261,
   262, 263**, 265, 267, 271, 286,
   287, 288, 290, 302, 303,
   304, 327, 337
monsoon vine forest,
   Australasian, 43, 45–47,
   46, 47, 48, 49, 56, 59, 260
Monster, Gila, 454, **455**
montane conifer forest,
   Eurasian, 186, 188, 340,
   343, 346–48, **346, 347**, 350,
   366, 371, 375, 387, 425, 428,
   434, 437
montane desert, Himalayan,
   161, 407–11, **410**
montane forest, Afrotropical,
   38, 41, 131, 135, 255, 260,
   263, 264, 267–72, **268, 269,
   270, 271**, 303, 304, 312, 316,
   320, 321, 484
montane grassland
   Afrotropical, 161, 223, 246,
   268, 308, 312–16, **313, 314,
   315**, 320, 321, 334, 335, 336
   Indo-Malayan, 223–25, **223,
   224**, 312
montane heath, Afrotropical,
   269, 312, 316, 320–21, **321**
montane mixed-conifer forest,
   348, 425, 427, 428–30, **429,
   430**, 434, 436, 437, 438,
   440, 443, 444, 447, 479, 504
montane rainforest
   Australasian, 35, 38–40, **39,
   40**, 199, 267, 484
   Indo-Malayan tropical, 138,
   186, 199–201, **200, 201,**

202, 203–4, 209, 214, 216,
   267, 273
montane spruce-fir forest,
   Nearctic, 346, 422, 425–27,
   426, 427, 428, 430, 434,
   436, 437, 438, 440, 482
monte, 114, 116, 118, 120–22, **121,
   122**, 149, 150, 156, 158, 182,
   184, 336
moorland, European, 320,
   389–91
Moose, 342, 351, 413, 422, 425,
   480, 483, 512
Mopane savanna, 149, 285, 288,
   289, 290, 291, 292–96, **293,
   294, 295**, 300, 301, 303,
   308, 326, 479
Moth, Comet, 277
Motmot
   Rufous, **125**
   Russet-crowned, 114, 487
   Turquoise-browed, 145
Mountain-Babbler
   Red-collared, 271
   White-throated, 272
Mountain Cat, Andean, 115, 119
Mountain-gem, Amethyst-
   throated, 485
Mountain Kingsnake, California,
   430
Mountain Lion, 430
Mountain Pit Viper, Godman's,
   486
Mountain Squirrel
   Carruther's, 271
   Tanganyika, 271
Mountain-Tanager, Masked, 140
Mountain Zebra
   Cape, 248, 319
   Hartmann's, 244, 295
Mouse
   Chihuahuan Desert Pocket,
      459
   Dark Kangaroo, 457
   Darwin's Leaf-eared, 106
   Florida, 495
   Great Basin Pocket, 463
   Michoacan Deer, 486
   Pinyon, 445
   Salt Marsh Harvest, 517
   Texas, 491
   Western Harvest, 501
   White-eared pocket, 503
Mousebird
   Blue-naped, 299, **299**
   Speckled, 304
Mouse Lemur
   Coquerel's Giant, 279
   Gray, 252
   Madame Berthe's, 279
Mouse-Opossum, Elegant Fat-
   tailed, 160

mudflat
   Afrotropical tidal, 231, 322,
      325, 327, 329–30, **329**, 331,
      333, 516, 517
   Australasian tidal, 92, 95–96,
      95, 96, 175, 325, 329, 517
   Indo-Malayan tidal, 95,
      231–33, **232**, 234, 325, 329
   Nearctic tidal, 329, 414,
      512, 514, 516–17, **516**, 519,
      523, 525
   Neotropical tidal, 175–76,
      176, 177
Mudlark, Australian, 90
Mudminnow, Olympic, 443
Mulga woodland, 57, 61, 77, 81,
   83–85, **84, 85**, 100, 149, 152,
   292, 296, 450
Munia
   Alpine, 70
   Chestnut, 238
   Snow Mountain, 70
Muntjac
   Gongshan, 204
   Red, 216
Murre
   Common, 520, 524
   Thick-billed, 416
Murrelet
   Ancient, 520
   Guadalupe, 522, 524
   Marbled, 443
Mushroom-tongued
   Salamander, Franklin's,
   486
Muskox, 402, 506, **506**, 507
Muskrat, Common, 413, 470,
   512
Musk Turtle, Loggerhead, 470
Myna
   Apo, 201
   Bali, 216
   Common, 240, 338
Myzomela
   Red-headed, 90
   Scarlet, 60

Nail-tail Wallaby
   Bridled, 57
   Northern, 52
Narwhal, 522
Nativehen, Black-tailed, 86
Nearctic habitats (North
   America)
   alpine tundra, 69, 161, 389,
      407, 424, 427, 434, 436,
      509–11, **509, 510, 511**
   aspen forest and parkland,
      422, 428, 437, 479, 482–84,
      483, 501
   Bald Cypress–tupelo forest,
      468–71, **469, 470**, 474, 512

boggy tundra, 165, 405, 424,
   505, 506, 507–8, **508**, 511,
   512, 514
cedar savanna, 488–91,
   489, 490
Chihuahuan Desert
   shrubland, 296, 361, 444,
   450–52, **450, 451, 452**, 453,
   456, 458, 461, 464, 467,
   488, 490, 491
cloud forest, 267, 468,
   484–86, **485**
columnar cactus desert, 249,
   253, 449, 450, 451, 453–55,
   454, 455, 458, 502
cropland, 98, 418, 525–26,
   525, 526
desert grassland, 308, 444,
   446, 449, 450, 452, 453,
   456, 458–61, **459, 460**, 467,
   491, 493, 496, 498
Florida oak scrub, 431,
   494–96, **494, 495**
freshwater wetland, 322, 411,
   468, 479, 480, 512–14, **512,
   513**, 516, 517, 525
grazing land, 418, 420,
   525–26, **525, 526**
high-elevation pine
   woodland, 425, 427, 430,
   434–36, **435, 436**, 509
lodgepole pine forest, 428,
   430, 437–40, **438, 439**, 482
longleaf pine savanna, 431–34,
   432, 433, 434, 474, 494, 499
Madrean pine-oak
   woodland, 382, 447–49,
   448, 449, 452, 455, 461,
   468, 484, 486, 491, 493
mangrove, 327, 468, 512,
   514–15, **515**, 517
mesquite brushland and
   thornscrub, 112, 464–67,
   465, 466, 467, 486
montane mixed-conifer
   forest, 348, 425, 427,
   428–30, **429, 430**, 434, 436,
   437, 438, 440, 443, 444,
   447, 479, 504
montane spruce-fir forest,
   346, 422, 425–27, **426, 427**,
   428, 430, 434, 436, 437,
   438, 440, 482
oak savanna, 281, 285, 289,
   292, 385, 387, 430, 455,
   474, 488, 491–93, **492**, 499,
   502, 504
offshore islands, 333, 519,
   523–24, **523, 524**
Pacific chaparral, 160, 316,
   391, 430, 453, 491, 502–4,
   503, 504

pelagic waters, 331, 418, 519, 521–22, **521**, **522**, 523
pinyon-juniper woodland, 348, 428, 430, 436, 444–46, **445**, **446**, 455, 461, 464
rocky coastline, 330, 416, 516, 519–20, **519**, **520**, 523
rocky tundra, 165, 401, 424, 505–6, **505**, **506**, 507, 508, 509, 511
sagebrush shrubland, 246, 444, 445, 446, 456, 458, 461–64, **462**, **463**, 496, 498, 504
salt desert shrubland, 456–58, **456**, **457**, 461
salt marsh, 325, 329, 512, 514, 516, 517–19, **518**
sandy beach, 330, 416, 516, 519–20, **519**, **520**, 523
shortgrass prairie, 444, 446, 450, 452, 458, 461, 488, 496–98, **497**, **498**, 499, 501, 525
spruce-fir taiga, 340, 422–24, **423**, **424**, 425, 475, 478, 482, 484, 501, 506, 514
tallgrass prairie, 63, 156, 395, 433, 474, 478, 484, 490, 491, 493, 496, 498, 499–502, **499**, **500**, **501**, 512, 517, 525
temperate deciduous forest, 369, 377, 380, 381, 431, 433, 468, 471–74, **472**, **473**, **474**, 475, 476, 478, 479, 490, 491, 493, 494
temperate mixed forest, 366, 422, 424, 471, 474, 475–78, **476**, **477**, **478**, 482
temperate rainforest, 102, 427, 430, 440–43, **441**, **442**, **443**
tidal mudflat, 329, 414, 512, 514, 516–17, **516**, 519, 523, 525
tropical dry forest, 464, 467, 486–88, **487**, **488**
western riparian woodland, 452, 471, 475, 479–81, **480**, 483, 498, 512
Neotropical habitats (Central and South America)
Antarctic pelagic waters, 179–80, **180**
Antarctic tundra and tussock grass, 165–68, **165**, **166**, **167**
caatinga, 48, 129, 130, 143, 144, 146–48, **146**, **147**, **148**, 153, 249, 453
campo, 153, 156–58, **157**, **158**, 168, 169, 183, 184, 308, 397

Cerrado, 131, 146, 147, 148, 150, 152–54, **153**, **154**, 168, 182, 184, 213, 281, 301
Chaco seco, 81, 83, 120, 122, 129, 144, 147, 149–51, **150**, **151**, 152, 153, 156, 158, 160, 182, 184, 189, 301
cloud forest, 38, 104, 124, 127, 128, 129, 131–34, **132**, **133**, **134**, 135, 136, 137, 138, 140, 163, 182, 199, 202, 267, 273, 447, 468, 484
cropland, 181–82, **181**, **182**, 525
desolate desert, 104–7, **105**, **106**, 118, 202, 243, 352, 355, 357
dry deciduous forest, 143–45, **144**, **145**, 146, 217, 278, 447, 449, 453, 455, 468, 486
elfin forest, 41, 115, 135, 138–40, **139**, **140**, 161, 163, 468
espinal, 81, 83, 120, 149–51, 156, 160, 184, 189, 301
flooded grassland, 149, 153, 154, 156, 168–70, **168**, **169**, **170**, 184, 219, 227, 322
Galápagos desert and scalesia, 108–11, **109**, **110**
grazing land, 183–84, **183**, **184**
igapó flooded forest, 125, 127, 169, 171–73, **172**, **173**, 206, 264
lowland rainforest, 35, 103, 104, 124–29, **124**, **125**, **126**, **128**, 131, 132, 134, 136, 147, 149, 169, 171, 172, 174, 175, 182, 195, 255, 273
Magellanic rainforest, 33, 102, 104, 118, 141–43, **142**, 375, 440
mangrove, 174–75, **174**, **175**, 229, 327
matorral sclerophyll forest and scrub, 77, 105, 107, 159–60, **159**, **160**, 316
mixed conifer forest, 26, 102–4, 141, 447
monte, 114, 116, 118, 120–22, **121**, **122**, 149, 150, 156, 158, 182, 184, 336
pampas, 63, 114, 121, 149, 150, 153,156–58, **157**, **158**, 182, 183, 184, 308, 397
paramo, 69, 114, 115, 128, 137, 138, 161–64, **162**, **163**, 312
Patagonian steppe, 107, 116–18, **117**, 119, 120, 122, 141, 142, 143, 165, 167, 184, 359
pelagic waters, 179–80, **179**
pine-oak woodland, 123, **123**, 447

puna, 107, 114–16, **115**, **116**, 118, 135, 161, 164, 169, 407
rocky coastline, 92, 177–78, 179
sandy beach, 92, 175, 177–78, **177**, 179
semi-evergreen forest, 104, 108, 124, 125, 127, 128, 129–31, **130**, 132, 134, 143, 148, 152, 154, 169, 182, 184, 196, 260, 468, 484, 486
semidesert scrub, 104, 107, 112, 114, 116, 118–20, **119**, **120**, 122, 159, 246, 253
stunted cloud forest, 138–40, **139**, **140**, 161, 468
thornscrub, 108, 112–14, **112**, **113**, 118, 128, 143, 253, 296, 464, 468
tidal mudflat, 175–76, **176**, 177
Valdivian rainforest, 26, 102–4, **103**, 141, 440
várzea flooded forest, 127, 171–73, **172**, **173**, 264
wetland, 149, 153, 154, 156, 168–70, **168**, **169**, **170**, 184, 219, 227, 322
yungas, 115, 134, 135–37, **136**, 138, 140, 150
Nesomys, Western, 280
Netted Dragon, Central, 63
Newt
  Caucasian Northern Banded, 376
  Striped, 433
New Zealand beech forest, 26, 33–35, **33**, **34**, 41, 141
Nicator, Eastern, 262
Nighthawk
  Common, 432, 463, 490, 498
  Lesser, 490
Night-Heron, White-backed, 324
Nightingale, Common, 388
Nightjar
  Abyssinian, 320
  Archbold's, 70
  Bonaparte's, 210
  Egyptian, 360, 362, **362**
  Eurasian, 389
  Fiery-necked, 335
  Freckled, 307
  Large-tailed, 46
  Pennant-winged, 288
  Pygmy, 148
  Red-necked, 392
  Rwenzori, 320
  Standard-winged, 283, **284**
  Syke's, 362, 365
  Tawny-collared, 467
Night Monkey, Azara's, 170
Night Snake, Desert, 452

Nilgai, 218, **220**, 366
Noddy
  Black, 95, 334, 524
  Brown, 94, 95, 234, 236, 334, 524
  Lesser, 334
Northern Banded Newt, Caucasian, 376
Nutcracker
  Clark's, 427, 436, **436**, 440, 446
  Eurasian, 188, 348
  Kashmir, 349
Nuthatch
  Algerian, 349
  Brown-headed, 433
  Burmese, 218
  Eurasian, 388
  Giant, 188
  Indian, 218
  Krüper's, 349
  Pygmy, **429**, 430
  Red-breasted, 430, **439**
  Western Rock, 393, **394**
  White-breasted, 432, 473
  White-browed, 188
Nutria, 143, 470
Nyala, 295, 304
  Mountain, 321

oak forest, Mediterranean, 379, 385, 387–89, **388**, 391, 488, 491
oak savanna, Nearctic, 281, 285, 289, 292, 385, 387, 430, 455, 474, 488, 491–93, **492**, 499, 502, 504
oak scrub, Florida, 431, 494–96, **494**, **495**
oak woodland
  Madrean pine-, 382, 447–49, **448**, **449**, 452, 455, 461, 468, 484, 486, 491, 493
  Neotropical pine-, 123, **123**, 447
Ocelot, 113, 153, 173, **173**, 466, 487
Ocotero, 449
offshore islands
  Afrotropical, 235, 329, 330, 331, 333–34, **333**, 523
  Indo-Malayan, 94, 234, 235–36, **236**, 333, 523
  Nearctic, 333, 519, 523–24, **523**, **524**
Okapi, 257, 259
Olinguito, 133
Olivaceous Warbler, Western, 392
Onager, 385–86
Openbill, Asian, 238
Opossum
  Derby's Woolly, 123

Elegant Fat-tailed Mouse-,
160
Virginia, 432, 473, 481, 486,
490
Orangutan
Bornean, 194, 207–8, 210,
226, **226**
Sumatran, 208
Oribi, 283, 287, 311, 314
Oriole
African Golden, 288, 291
Altamira, 466
Audubon's, 466
Black-headed, 304
Black-naped, 382
Bullock's, 481
Ethiopian Black-headed, 272
Scott's, 446
Slender-billed, 188
Spot-breasted, 145
Streak-backed, 487
Orphean Warbler
Eastern, 388
Western, 388, 392
Oryx
Arabian, 353, **362**, 363
Beisa, 299
Scimitar-horned, 299
Ostrich, Common, 244, 248,
304, 311, 326
Otter
Eurasian, 413
Giant, 169, **170**, 173
North American River, 470,
481, 512, 517
Smooth-coated, 240
Ovenbird, 484
Owl
African
Grass-, 324
Wood-, 335
Austral Pygmy-, 142
Barn, 526
Barred, 432, 470, 473
Boreal, 348, 349, 427, 484
Burrowing, 119, 461, 463, 498
Cape Eagle-, 320
Desert, 353
Eastern Screech-, 473
Elf, 454, 481
Eurasian Pygmy-, 349
Ferruginous Pygmy-, 454
Flores Scops-, 43
Great
Gray, 422
Horned, 481
Little, 392
Long-eared, 349, 388
Luzon Scops-, 188
Maned, 258
Mantanani Scops-, 236
Mottled, 484

Northern
Hawk, 384, 422, 440
Pygmy-, 427, 484
Spotted, 443
Pels Fishing-, 262
Pharaoh Eagle-, 353
Pygmy-, 348
Red, 276
Rufous Fishing-, 265
Serendib Scops-, 194
Shelley's Eagle-, 259
Short-eared, 461, 464, 501
Snowy, **403**, 404, 506, 526
Socotra Scops-, 254
Sokoko Scops-, 262
Spot-bellied Eagle-, 216
Usambara Eagle-, 272
Vermiculated Fishing-, 265
Western Screech-, 481
Yungas Pygmy-, 136
Owlet, Pearl-spotted, 295
Oxpecker, **336**, 337
Oystercatcher
African, 330
Eurasian, **416**, 417
Pied, 93, **93**, 96
Sooty, 93

Pacific chaparral, 160, 316, 391,
430, 453, 491, 502–4,
503, 504
Paddlefish, American, 513
paddy fields and other
cropland, 219, 227, 237–39,
237, 238, 239, 419, 525
Pademelon
Calaby's, 39, 70
Red-legged, 36, 39, 73, 74
Red-necked, 39, 41, 74
Rufous-bellied, 41
Painted-Snipe, Greater, 238
Palearctic habitats (Europe,
Northern Asia, and North
Africa)
Arctic polar desert, 356,
399–401, **399, 400**, 408
Beringian taiga savanna, 342,
345, 350–51, **351**, 401, 424
Central Asian cold desert,
355, 357–59, **358**, 360,
363, 385
Colchic deciduous rainforest,
369, 375–77, **376**
cropland, 418–19, **418, 419**, 525
East Asian cold desert, 104,
355–57, **355, 356**, 359,
360, 456
East Asian moist mixed
forest, 366–68, **367, 368**,
369, 373, 380, 381
East Asian temperate bamboo
forest, 133, 373–74, **374**

East Asian temperate
deciduous forest, 202,
366, 369, 373, 378, 380–82,
380, 381, 382, 471
eastern grass steppe, 356,
359, 382, 395, 396, 397–98,
397, 398, 407, 496
Eurasian alpine tundra, 161,
356, 389, 390, 407–11, **408,
409**, 509, 511
Eurasian boggy tundra, 342,
343, 345, 350, 399, 401,
405–6, **405, 406**, 408, 507
Eurasian montane conifer
forest, 186, 188, 340, 343,
346–48, **346, 347**, 350, 366,
371, 375, 387, 425, 428,
434, 437
Eurasian rocky tundra, 342,
343, 345, 350, 399, 401–4,
402, 403, 404, 405, 406,
407, 408, 505
Eurasian spruce-fir taiga,
340–42, **341, 342**, 343, 344,
346, 350, 351, 366, 371,
377, 379, 382, 388, 405,
422, 425
European heathland 320,
389–91, **390**
European moist mixed
forest, 366, 369–72, **370,
371, 372**, 375, 377, 378,
379, 475
European moorland, 320,
389–91
European temperate
deciduous forest, 342,
369, 371, 375, 377–79, **378**,
380, 388, 471
forest-steppe, 142, 342, 345,
382–84, **383, 384**, 385, 396
garrigue, 349, 366, 389, 391,
393–94, **394**, 502
grazing land, 420
Himalayan montane desert,
161, 407–11, **410**
hot desert, 189, 243, 244, 245,
352–54, **352, 354**, 356, 358,
361, 362, 364, 385
hot shrub desert, 243, 244,
245, 246, 296, 300, 352,
353, 358, 361–63, **361, 362**,
364, 450
maquis, 114, 121, 160, 316, 336,
349, 366, 385, 387, 388,
389, 391–93, **391, 392**, 502
Mediterranean and dry
conifer, 348–50, **349**,
385, 444
Mediterranean oak forest,
379, 385, 387–89, **388**, 391,
488, 491

rocky coastline, 414, 416–17,
416, 417, 519
sandy beach, 414, 416–17, **416,
417**, 519
semidesert thornscrub, 30,
118, 120, 189, 296, 300, 352,
354, 361, 362, 364–66, **365**,
394, 450, 461
Siberian larch forest, 340,
342, 343–45, **344, 345**,
366–67, 402, 404, 437
subtropical savanna, 61, 357,
385–87, **386**, 431, 444
temperate desert steppe,
114, 116, 117, 355, 357,
359–60, **359, 360**, 397
temperate wetland, 86, 322,
411–13, **412, 413**, 512
tidal flat, 414–16, **414, 415**, 516
western flower steppe, 359,
382, 395–96, **395, 396**, 397,
398, 420, 499
Palm Civet
Hose's, 200
Masked, 200
Palmcreeper, Point-tailed, 127
pampas, 63, 114, 121, 149, 150,
153,156–58, **157, 158**, 182,
183, 184, 308, 397
Panda
Giant, 367, 373, **374**
Red, **203**, 373
Pangolin
Giant, 257
Ground, 288, 311
Long-tailed, 265
Panther, Florida, 470
Paradise-Flycatcher
Japanese, 368, 382
Madagascar, 276
Paradise-Kingfisher, Buff-
breasted, 36
Paradise-Whydah, Broad-
tailed, 288
Parakeet
Blossom-headed, 218
Burrowing, 121, **122**
Cactus, 148
Echo, 276
Gray-headed, 216
Green-cheeked, 136
Nanday, 150
New Caledonian, 40
Orange-fronted, 145
Red-breasted, 216
Rose-ringed, 338
Slender-billed, 103
White-eyed, 154
Yellow-chevroned, 154
Yellow-crowned, 27, 34
paramo, 69, 114, 115, 128, 137,
138, 161–64, **162, 163**, 312

Pardalote, Spotted, 73, 74
parkland, aspen, 422, 428, 437, 479, 482–84, **483**, 501
Parrot
 Australian
  King-, 73
  Ringneck, 99
 Bourke's, 85
 Brown-headed, 304
 Eclectus, 36
 Golden-shouldered, 52, **55**
 Gray, 335
 Greater
  Bluebonnet, 99
  Vasa, 276
 Ground, 67
 Hooded, 52
 Lesser Vasa, 276
 Maroon-fronted, 449
 Mascarene, 276
 Meyer's, 291
 Mulga, 85
 Night, 29, 64
 Olive-shouldered, 46
 Painted Tiger-, 70
 Princess, 62
 Red-cheeked, 45
 Red-crowned, 466
 Red-rumped, **98**, 99
 Red-spectacled, 104
 Rock, **92**
 Superb, 78, 99
 Swift, 78
 Thick-billed, 449
 Tucuman, 136
 Turquoise, 57, 78, 99
 Turquoise-fronted, 150
 Vinaceous-breasted, 104
 Yellow-fronted, 272
Parrotbill
 Black-breasted, 221
 Brown, 373
 Golden, 373
 Rusty-throated, 373
Parrotfinch
 Mount Mutis, 44
 Tricolored, 46
Parrotlet, Mexican, 487
Parrot Snake, Pacific, 487
Parsley Frog, Caucasian, 376
Partridge
 Black, 208
 Chestnut-breasted, 204
 Chestnut-headed, 201
 Crested, 194
 Daurian, 398
 Gray, 419, 420
 Hainan, 204
 Orange-necked, 198
 Red-legged, 392, 420
 Rufous-throated, 201
 Taiwan, 204

Udzungwa, 271
 White-necklaced, 204
Parula, Northern, 433, 470
Patagonian steppe, 107, 116–18, **117**, 119, 120, 122, 141, 142, 143, 165, 167, 184, 359
Patchnose Snake, Pacific, 487
Pauraque, Common, 466
Peacock, Congo, 259
Peacock-Pheasant
 Bronze-tailed, 201
 Germain's, 198
 Hainan, 204
 Mountain, 201
 Palawan, 195
peat swamp forest, Indo-Malayan, 206–8, **207**, **208**, 209, 210, 225, 227
Peccary
 Chacoan, 149
 Collared, 136, 451, 454, 459, 466, **466**, 486, 490
 White-lipped, 136
Pectinator, Speke's, 306
Peeper, Spring, 476
pelagic waters
 Afrotropical, 97, 234, 329, 330, 331–32, **331**, **332**, 333
 Antarctic, 179–80, **180**
 Australasian, 97–98, **97**, 179, 331
 Indo-Malayan, 234–35, **235**, 331, 418
 Nearctic, 331, 418, 519, 521–22, **521**, **522**, 523
 Neotropical, 179–80, **179**
 Palearctic, 418
Pelican
 Australian, 86
 Dalmatian, 229
 Great White, 229, 326
 Peruvian, 177
 Spot-billed, 226, 228, 231
Penguin
 African, **330**, 334
 Galápagos, 177
 Humboldt, 177, 179
 King, **165**
 Little, 93
 Magellanic, **177**
 Yellow-eyed, 93
Perch, Nile, 323
Petrel
 Ainley's Storm-, 522, 524
 Antarctic, 180
 Ashy Storm-, 522
 Band-rumped Storm-, 522
 Bermuda, **521**, 522
 Black-bellied Storm-, 180
 Black-capped, 522
 Black Storm-, 522
 Black-winged, 334

Blue, 167, 180
 Bulwer's, 234, 334
 Cape, 97, 180, 332
 Common Diving-, 167, 180
 Elliot's Storm-, 179, 180
 European Storm-, 332
 Fork-tailed Storm-, 522
 Galápagos, 109, 179
 Gray-backed Storm-, 180
 Great-winged, 97
 Hawaiian, 522
 Herald, 334
 Jouanin's, 234
 Juan Fernandez, 179–80, 522
 Kermadec, 334, 552
 Leach's Storm-, 332, 522, 524
 Least Storm-, 522
 Markham's Storm-, 180
 Masatierra, 180
 Northern Giant-, 97
 Peruvian Diving-, 180
 Providence, 97
 Ringed Storm-, 180
 Snow, 180, **180**
 Soft-plumaged, 167, 180
 Southern Giant-, 180
 Swinhoe's Storm-, 234
 Tahiti, 522
 Townsend's Storm-, 522, 524
 Trindade, 334
 Wedge-rumped Storm-, 179
 Westland, 180
 White-chinned, 167, 180
 White-headed, 167
 Wilson's Storm-, 234, 332, **521**, 522
Pewee
 Eastern Wood-, 432, 473
 Western Wood-, 440
Phainopepla, 493
Phalarope, Red, **508**
Pheasant
 Blue Eared-, 409
 Bronze-tailed Peacock-, 201
 Cheer, 188
 Copper, 381
 Elliot's, 204
 Germain's Peacock-, 198
 Golden, 373
 Hainan Peacock-, 204
 Hume's, 188
 Koklass, 367
 Lady Amherst's, 373
 Mountain Peacock-, 201
 Palawan Peacock-, 195
 Reeves's, 381, **381**
 Ring-necked, 381
 Swinhoe's, 204, **204**
Phoebe, Say's, 452, 461, 463, 498
Piapiac, 284, 336
Picathartes

Gray-necked, 307
 White-necked, 307
Piculet
 African, 258
 Spotted, 148
 Tawny, 148
Pied Flycatcher
 European, 342
 Little, 43
Pig-tailed Macaque, Southern, 197, 208
Pigeon
 Band-tailed, 442
 Chestnut-quilled Rock-, 48, 49
 Cinnamon-headed Green-, 208, 227
 Common Wood-, 419
 Comoro, 276
  Blue-, 276
 Flores Green-, 45
 Gray Imperial-, 236
 Large Green-, 208
 Madagascar
  Blue-, 275
  Green-, 279
 Metallic, 236
 New Caledonian Imperial-, 40
 Nicobar, 236, **236**
 Nilgiri Wood-, 201
 North American Passenger, 64
 Pale-vented, 182
 Partridge, 52
 Pied Imperial-, 236
 Rameron, 271
 Red-billed, 484
 Rock, 240, 338, 416
 Seychelles Blue-, 276
 Spinifex, 29, **29**
 Squatter, **56**
 Timor
  Green-, 45, 46
  Imperial-, 44
 White-collared, 306, 338
 White-quilled Rock-, 49
 Yellow-footed Green-, 218
 Yellowish Imperial-, 90
Pika
 American, 436, **510**, 511
 Collared, 511
 Steppe, 396
Pilotbird, 72
Pilot Whale
 Long-finned, 97
 Short-finned, 332, 521
Pine Beetle, Mountain, 425, 436
pine forest,
 Indo-Malayan, 186–88, **187**, **188**, 199, 428, 447
 lodgepole, 428, 430, 437–40, **438**, **439**, 482

pine-oak woodland
Madrean, 123, 382, 447–49, **448, 449,** 452, 455, 461, 468, 484, 486, 491, 493
Neotropical, 123, **123,** 447
pine savanna, longleaf, 431–34, **432, 433, 434,** 474, 494, 499
pine woodland, high-elevation, 425, 427, 430, 434–36, **435, 436,** 509
Pintail, Northern, 501, 413
pinyon-juniper woodland, 348, 428, 430, 436, 444–46, **445, 446,** 455, 461, 464
Pipipi, 34
Pipit
African, 311
Alpine, 70
American, 511, 526
Meadow, 420
Nilgiri, 224
Paddyfield, 238
Peruvian, 106
Richard's, 384
Sokoke, 262
South Georgia, 167
Sprague's, 461, 498, 501
Tree, 379
Yellow-breasted, 316
Pitohui, White-bellied, 90
Pitta
African, 262
Bar-bellied, **212**
Blue, 198
Blue-headed, 195
Blue-rumped, 212
Eared, 198
Elegant, 45
Fairy, 212, 382
Garnet, 195
Gurney's, 198
Mangrove, 230, **230**
Papuan, 36
Rainbow, 45, 46, **46**
Schneider's, 201
Pit Viper, Godman's Mountain, 486
Plains-wanderer, 31, **31,** 32, 100
Plated Lizard
Black-lined, 295
Four-lined, 252
Giant, 307
Kalahari, 291
Madagascar Keeled, 252
Three-lined, 252
Tsingy, 280
Western, 280
Platypus, **42**
Ploughbill, Wattled, 38
Plover
American Golden-, 506

Black-bellied, 176, **232,** 404, 414, 417, 506, 517
Chestnut-banded, 326
Common Ringed, 414
Crab-, 233, 330, **415,** 416
Diademed Sandpiper-, **115**
Egyptian, 324
European Golden-, 404
Forbes's, 311
Greater Sand-, 96, 414
Hooded, 93, **93**
Kentish, 233, 414
Kittlitz's, 326
Lesser Sand-, 96, 414, 415
Magellanic, 118
Malaysian, 233, **233**
Mountain, 461, 498
Oriental, 360, 398
Pacific Golden-, 506
Piping, 506, 520
Red-capped, 96
Snowy, 520
White-fronted, 330
Pochard
Baer's, 228, 413
Madagascar, 324
Pocket Gopher
Northern, 498
Plains, 501
Pocket Mouse
Chihuahuan Desert, 459
Great Basin, 463
White-eared, 503
polar desert, Arctic, 356, 399–401, **399, 400,** 408
Polecat
European, 413
Steppe, 396, 398
Poorwill, Common, 490
Porcupine
African Brush-tailed, 257
Brazilian, 170
Cape, 248
Mexican Hairy Dwarf, 484
North American, 476, 481, 490
Porpoise, Dall's, 521
Possum
Common
Brushtail, 27, 34, 46, 73, 85
Ringtail, 39, 73
Coppery Ringtail, 70
Eastern
Pygmy, 41, 73, 76
Ringtail, 41
Honey, 68, **68**
Long-tailed Pygmy, 70
Rock Ringtail, 46, 49
Short-eared Brushtail, 39
Striped, 36, **37**
Western Pygmy, 68
Potoroo, Gilbert's, 68
Potto, 258

prairie
shortgrass, 444, 446, 450, 452, 458, 461, 488, 496–98, **497, 498,** 499, 401, 525
tallgrass, 63, 156, 395, 433, 474, 478, 484, 490, 491, 493, 496, 498, 499–502, **499, 500, 501,** 512, 517, 525
Prairie-Chicken
Greater, 498, 501
Lesser, 498
Prairie Dog
Black-tailed, 459, **460,** 497, 501
Mexican, 459
White-tailed, 463
Pratincole
Australian, 52, 64, 88
Black-winged, 396
Collared, 387
Small, 228
Prinia
Brown, 218
Graceful, 362
Karoo, 248, 319
Prion
Antarctic, 167, 180
Slender-billed, 180
Priprites, Black-capped, 104
Pronghorn, 451, 454, 457, 459, **462,** 463, 497, 500
Ptarmigan
Rock, 409, 511
White-tailed, 511, **511**
Willow, 351, **351,** 511
Pudu, Southern, 103, 143
Puffback
Black-backed, 262, 304
Pink-footed, 271
Puffbird, Sooty-capped, 175
Puffin
Atlantic, 416, **417,** 524
Horned, 416, 520, 524
Tufted, 416, 520, 524, **524**
Puku, 287, 290, 311, 323
Puma, 115, 117, 143, 275, 430, 442, 448, 454, 459, 470, 484, 490, 501, 504
puna, 107, 114–16, **115, 116,** 118, 135, 161, 164, 169, 407
Pupfish, Desert, 513
Pygmy Chameleon, Uluguru, 272
Pygmy-Goose
Cotton, 88, 228
Green, 87, 88
Pygmy Kingfisher, American, 173, **175**
Pygmy-Kingfisher, Madagascar, 276
Pygmy-Owl, 348
Austral, 142
Eurasian, 349

Ferruginous, 454
Northern, 427, 484
Yungas, 136
Pygmy Possum
Eastern, 41, 73, 76
Long-tailed, 70
Western, 68
Pygmy Squirrel
African, 258
Whitehead's, 200
Pyrrhuloxia, 452, 454
Python
African Rock, 288, 311
Green Tree, 37
Oenpelli, 49
Olive, 49
Rough-scaled, 46
Scrub, 37
Pytilia
Green-winged, 304
Orange-winged, 288

Quagga, 248
Quail
Banded, 487
Blue, 311
Brown, 70
Common, 335, 420
Gambel's, 452, 454
Montezuma, **492,** 493
Mountain, 430, 446
Ocellated, 123
Scaled, 452, 498
Snow Mountain, 70
Quailfinch, 335
Quail-plover, 299
Quail-thrush
Chestnut, 83
Chestnut-breasted, 85
Nullarbor, 32
Spotted, 74, 75
Quetzal, Resplendent, 485
Quokka, 68
Quoll
Eastern, 41
New Guinean, 36
Northern, 56
Spotted-tailed, 76

Rabbit
Brush, 503
European, 29
Marsh, 512
Pygmy, 463
Riverine, 248
Swamp, 470, 512
Volcano, 448
Raccoon
Common, 432, 442, 470, 473, 376, 381, 490, 495, 514, 517
Crab-eating, 169, 175
Pygmy, 514, **515**

Racer
Eastern, 493, 504
Nagarjun Sagar, 189
Racerunner, Prairie, 490, 491, 498
Racket-tail, Booted, **134**
Rail
Aztec, 513
Black, 517
Chestnut, 90
Chestnut Forest-, 39
Clapper, 517, **518**
Galápagos, 110
Giant Wood-, 158
King, 512
Little Wood-, 175
Madagascar, 324
Nkulengu, 258
Okinawa, 204
Ridgeway's, 517
Rufous-necked Wood-, 515
Sakalava, 324
Tsingy Wood-, 279, 280
Virginia, 526
White-throated, 276, 328
Yellow, 512, 526
Zapata, 513
Rail-babbler, Malaysian, 36, 195
rainforest
Afrotropical lowland, 35, 124,
192, 255–59, **256, 257, 258**,
260, 262, 263, 264, 265,
267, 271, 273, 281, 283, 284,
288, 302, 304, 307, 325,
327, 328
Australasian
montane, 35, 38–40, **39**,
40, 199, 267, 484
subtropical, 26, 35, 38–40,
39, 40, 41, 57, 72, 74,
267, 484
temperate, 26, 33,
41–42, **42**
tropical lowland, 35–37,
36, 37, 38, 41, 43, 45,
89, 124, 255, 273
Colchic deciduous, 369,
375–77, **376**
Gondwanan conifer, 26–27,
27, 33, 66, 440
Indian Ocean, 202, 249, 250,
253, 255, 267, 273–77, **274**,
275, 276, 278, 279
Indo-Malayan
tropical lowland, 35, 36,
124, 192–95, **193, 194**,
196, 198, 199, 200, 206,
208, 209, 211, 226, 235,
238, 255, 273
tropical montane, 138, 186,
199–201, **200, 201**, 202,
203–4, 209, 214, 216,
267, 273

Magellanic, 33, 102, 104, 118,
141–43, **142**, 375, 440
Nearctic temperate, 102, 427,
430, 440–43, **441, 442, 443**
Neotropical lowland, 35, 103,
104, 124–29, **124, 125, 126**,
128, 131, 132, 134, 136, 147,
149, 169, 171, 172, 174, 175,
182, 195, 255, 273
Valdivian, 26, 102–4, **103**,
141, 440
Rain Frog, Marbled, 277
Rat
Australian Swamp, 73
Big-headed African Mole-,
315
Brown, 34
Chamela, 486
Chisel-toothed Kangaroo, 457
Dassie, 306
False Water, 90
Galápagos Rice, 109
Greater Cane, 257
Laotian Rock, 212
Long-haired, 64
Magdalena, 486
Malagasy Giant Jumping,
279, 280
Naked Mole-, 299
Ord's Kangaroo, 451, 457
Santa Cruz Kangaroo, 503
Subalpine Woolly, 70
Yellow-nosed Cotton, 451
Rat-Kangaroo, Musky, 36
Rat Snake
Black, 473
Eastern, 471
Great Plains, 490
Trans-Pecos, 452
Rattlesnake
Baja California, 455
Eastern Diamondback, 433
Mojave, 455
Pacific, 504
Prairie, 446, 455, 461, 464,
493, 498
Pygmy, 433
Santa Catalina, 524
Speckled, 455
Tiger, 455
Western Diamondback,
452, **452**
Raven
Brown-necked, 358, 362
Chihuahuan, 452, 461
Common, 422, 443, 506, 511
Rayadito, Thorn-tailed, 142
Razorbill, 416, 520, 524
Red Colobus
Niger Delta, 265
Tana River, 262
Western, 328

Zanzibar, 262
Red Duiker
Natal, 262
Rwenzori, 321
Red Flying Squirrel, American,
430
Red Fox
Andean, 143
Fuegan, 143
Redhead, 501
Redpoll, Hoary, 371
Redshank
Common, 351, 414
Spotted, 351, 414
Red Squirrel
American, 422, 427, 436, 438,
442, 476
Eurasian, 349
Redstart
American, 515
Black, 387
Brown-capped, 136
White-winged, 349, 409
Redthroat, 85
Reed Frog, Starry-night, 277
Reedbuck
Bohor, 311
Mountain, 314
Southern, 311
Reedling, Bearded, 412, **413**
Reed Warbler
African, 324
Australian, 87
Blyth's, 229
Clamorous, 229, 328
Reef-Heron
Pacific, 95, 234
Western, 330
Reindeer, 351, 402
Rhea
Greater, 149, 158, 169
Lesser, **116**, 118, 158
Rhebok, Gray, 315, 319
Rhinoceros
Black 244, 283, 293, 295, 298,
304, 305, 320
Indian, 197, 214, 221
Sumatran, 207, 226
White, 283, 304, 305, **310**, 311
Woolly, 402
Rice Rat, Galápagos, 109
Ridley Turtle
Kemp's, 515, 524
Olive, 234, 236, 331, 334
Riflebird
Magnificent, 36, 37
Paradise, 39, 74
Victoria's, 36
Rifleman, 34
Right Whale
Pygmy, 97
Southern, 97, 332

Right Whale Dolphin, Northern,
521
Ringed Plover, Common, 414
Ringlet, Common, 501
Ringneck Parrot, Australian, 99
Ringtail, 448, 490, 493, 504
Ringtail Possum
Common, 39, 73
Coppery, 70
Eastern, 41
Rock, 46, 49
riparian woodland, western,
452, 471, 475, 479–81, **480**,
483, 498, 512
River Dolphin
Ganges, 228
Irrawaddy, 228
Pink, 173
River Martin, White-eyed, 228
River Otter, North American,
470, 481, 512, 517
Roadrunner
Greater, **451**, 452, 454, 490,
491, 504
Lesser, 114, 487
Robin
American, 440, 445
Bearded Scrub-, 304
Black Scrub-, 299
European, 349, **378**, 379
Flame, 69, **71**
Gray-headed, 39
Hooded, 62, 78
Indian, 218
Japanese, 367
Kalahari Scrub-, 300
Karoo Scrub-, 248, 319
Mangrove, 90
Miombo Scrub-, 288, **288**
Northern Scrub-, 37
Pink, 41, **42**
Red-capped, 57, **58**
Rufous-tailed Scrub-, 358,
387, 392
Siberian Blue, 342
Snow Mountain, 39, 70
Southern Scrub-, 83
Subalpine, 70
Western Yellow, 75
White-faced, 36, 37
White-starred, 271
Robin-Chat
Cape, 319
White-crowned, 284
White-throated, 304
Rock Agama
Eritrean, 245
Namibian, 300
Rocket Frog, Tawny, 56
Rockfinch, Pale, 365
Rockfowl
Gray-necked, 307

White-necked, 307
Rockjumper, Cape, 319
Rock Lizard, Southern, 319
Rock Nuthatch, Western,
 393, **394**
Rock-Pigeon
 Chestnut-quilled, 48, 49
 White-quilled, 49
Rock Python, African, 288, 311
Rock Rat, Laotian, 212
Rockrunner, 300, 307
Rock Thrush
 Littoral, 252
 Miombo, 288
Rock-Wallaby
 Brush-tailed, 76
 Short-eared, 49
 Wilkins', **48**
Rockwren, New Zealand, 69
Roller
 Blue-bellied, 283
 European, 420
 Indian, 218
 Lilac-breasted, **303**, 304
 Long-tailed Ground-, 252,
  252, 253
 Pitta-like Ground-, 275
 Racket-tailed, 288
 Rufous-headed Ground-,
  275, 276
 Scaly Ground-, 275, 276
 Short-legged Ground-, 275
Roofed Turtle, Red-crowned,
 228
Rook, 419
Rope Squirrel, Thomas's, 265
Rosefinch
 Pallas's, 384
 Taiwan, 224
Rosella, Crimson, 73
Rosy-Finch
 Black, 427, 511
 Brown-capped, 427, 511
 Gray-crowned, 427, 436, 511
Rough-legged Hawk, 371, **371**
Rough-winged Swallow,
 Southern, 184
Roundleaf Bat, Indian, 190
Royal Albatross, Northern,
 179, 332
Rubythroat, Himalayan, 409
Ruff, 326, 404, **406**
Ruffed Lemur
 Black-and-white, 275
 Red, 275

Sabrewing, Long-tailed, 484
Saddleback, **34**
sagebrush shrubland, 246, 444,
 445, 446, 456, 458, 461–64,
 462, 463, 496, 498, 504
Sage-Grouse

Greater, 444, 463, **463**
 Gunnison, 463
Salamander
 Blue-spotted, 476
 Caucasian, 376
 Chinese Giant, 204
 Flatwoods, 433, **434**, 471
 Franklin's Mushroom-
  tongued, 486
 Garden Slender, 504
 Jefferson, 476
 Jemez Mountains, 478
 Olympic Torrent, 443
 Wandering, 443
Saltator, Black-throated, 154
salt desert shrubland, 456–58,
 456, 457, 461
salt marsh
 Afrotropical, 231, 322, 325, 327,
  329–30, **329**, 331, 333, 517
 Australasian, 95–96, 325,
  329, 517
 Nearctic, 325, 329, 512, 514,
  516, 517–19, **518**
salt pan
 Afrotropical, 322, 325–26,
  326, 330, 329, 516
 Indo-Malayan, 231–33, **232**,
  325, 329
Sambar, 187, 197, 218, 224
samphire shrubland, 30–32, **31**,
 32, 246, 353, 456
Sanderling, 93, 95, 96, 233, 331,
 414, 417, 506, 519
Sandgrouse
 Black-bellied, 362, 396
 Chestnut-bellied, 245, 362
 Lichtenstein's, 245, 353, 365
 Namaqua, 245, 248
 Pallas's, 356, 358, 360, 398
 Pin-tailed, 353, 360, **386**, 387
 Spotted, 245, 353
 Yellow-throated, 311
Sandpiper
 Baird's, 506
 Broad-billed, 414
 Buff-breasted, 507
 Curlew, 326, 351, 414
 Green, 384, 406
 Least, 517
 Purple, 351, 404, 417
 Rock, 519
 Semipalmated, 176, 517
 Sharp-tailed, 406, 408
 Solitary, 406
 Spoon-billed, 232, 351, 414
 Terek, **232**, 330
 Upland, 183, 498, 501, 526
 Western, 176, **505**, 506, 517
 White-rumped, 176
Sandpiper-Plover, Diademed,
 115

Sand-Plover
 Greater, 96, 414
 Lesser, 96, 414, 415
Sand Skink, Florida (reptile),
 495
Sand Snake, Mahafaly, 252
Sand-swimmer, Broad-banded,
 63
Sandveld Lizard, Ornate, 315
Sapsucker
 Red-breasted, 430, 443
 Red-naped, 430, 481, **483**,
  484
 Williamson's, 430, **430**
 Yellow-bellied, 484
Satinbird, Crested, 39
savanna
 Afrotropical
  dry thorn, 112, 243, 245,
   246, 248, 253, 281, 282,
   283, 284, 285, 288, 289,
   290, 292, 295, 296–300,
   297, 298, 299, 300, 301,
   303, 304, 326, 336, 337,
   354, 363, 364, 464
  moist mixed, 50, 262,
   269, 281, 284, 285, 286,
   288, 289, 292, 295, 296,
   298, 300, 301–4, **302**,
   303, 316, 320, 337,
   338
 Beringian taiga, 342, 345,
  350–51, **351**, 401, 424
 cedar, 488–91, **489, 490**
 Cerrado, 131, 146, 147, 148,
  150, 152–54, **153, 154**, 168,
  182, 184, 213, 281, 301
 Chaco seco, 81, 83, 120, 122,
  129, 144, 147, 149–51, **150**,
  151, 152, 153, 156, 158, 160,
  182, 184, 189, 301
 espinal, 81, 83, 120, 149–51,
  156, 160, 189, 301
 Guinea, 255, 259, 281–84, **282**,
  283, 284, 285, 286, 289,
  296, 298, 301, 303, 304,
  308, 312, 354
 longleaf pine, 431–34, **432**,
  433, 434, 474, 494, 499
 Melaleuca, 47, 49, 56, 59–61,
  59, 60, 61
 Mopane, 149, 285, 288, 289,
  291, 292–96, **293, 294**,
  295, 300, 301, 303, 308,
  326, 479
 Nearctic oak, 281, 285, 289,
  292, 385, 387, 430, 455,
  474, 488, 491–93, **492**, 499,
  402, 504
 open eucalypt, 48, 49, 50–53,
  50, 51, 52, 57, 59, 289,
  301, 336

Palearctic subtropical, 61,
  357, 385–87, **386**, 431,
  444
 tetrodonta woodland, 50,
  53–56, **54, 55, 56**, 77, 289
 scalesia, Galápagos desert and,
  108–11, **109, 110**
Scaly-foot
 Brigalow, 58
 Western Hooded, 63
Scaly-Tail, Cameroon, 258
Scaup, Lesser, 501
sclerophyll forest
 dry, 57, 72, 74, 75–77, **76**,
  77, 79
 matorral, 77, 105, 107, 159–60,
  159, 160, 316
 wet, 38, 57, 72–74, **73**, 75
Scops-Owl
 Flores, 43
 Luzon, 188
 Mantanani, 236
 Serendib, 194
 Socotra, 254
 Sokoke, 262
Screamer, Horned, 169
Screech-Owl
 Eastern, 473
 Western, 481
scrub
 Florida oak, 431, 494–96,
  494, 495
 matorral sclerophyll forest
  and, 77, 105, 107, 159–60,
  159, 160, 316
 Neotropical semidesert,
  104, 107, 112, 114, 116,
  118–20, **119, 120**, 122, 159,
  246, 253
Scrub-bird, Rufous, 41
Scrubfowl, Tabon, 236
Scrub-Jay
 California, 504
 Florida, 495, **495**
 Woodhouse's, 446
scrubland, mallee woodland
 and, 66, 77, 81–83, **81, 82**,
  85, 99, 100, 292, 358
Scrub-Robin
 Bearded, 304
 Black, 299
 Kalahari, 300
 Karoo, 248, 319
 Miombo, 288, **288**
 Northern, 37
 Rufous-tailed, 358, 387, 392
 Southern, 83
Scrubwren
 Atherton, 39
 White-browed, 39
 White-browed, 75
Sea-Eagle, White-bellied, 233
Sea Krait, Banded, 90

Seal
    Australasian Fur, 93
    Bearded, 522
    Cape Fur, 330, 334
    Gray, 519
    Harbor, 519
    Harp, 522
    Northern
        Elephant, 519
        Fur, 519, 523
        Ribbon, 522
        Southern Elephant, **166**
Sea Lion
    California, 519
    Steller's, 519
Secretarybird, 248, **311**
Seedeater
    Plumbeous, 154
    Variable, 182, 184
    Yellow-bellied, 182
Seedsnipe
    Gray-breasted, 119, **120**
    Least, 118
    Rufous-bellied, **162**
semi-deciduous forest,
        Australasian tropical,
        43–44, **44**, 143
semidesert, Dragon Blood,
        253–54, **254**, 453
semidesert scrub, Neotropical,
        104, 107, 112, 114, 116,
        118–20, **119**, **120**, 122, 159,
        246, 253
semidesert thornscrub,
        Palearctic, 30, 118, 120,
        189, 296, 300, 352, 354,
        361, 362, 364–66, **365**, 394,
        450, 461
semi-evergreen forest
    Indo-Malayan, 43, 129, 192,
        196–98, **197**, **198**, 211, 213,
        214, 216, 260
    Neotropical, 104, 108, 124,
        125, 127, 128, 129–31, **130**,
        132, 134, 143, 148, 152, 154,
        169, 182, 184, 196, 260, 468,
        484, 486
Seriema, Red-legged, 153,
        158, 169
Serin
    European, 372
    Syrian, 349
    Yemen, 307
Serow
    Chinese, 200, 212
    Formosan, 204
Serpent-Eagle
    Congo, 258
    Madagascar, 276
Serval, 311, 314, 323
Shag, Antarctic, 180
Sheartail, Mexican, 467

Shearwater
    Audubon's, 522
    Black-vented, 522
    Christmas, 522
    Cory's, 332, 522
    Flesh-footed, 234, 332
    Galápagos, 179, 522
    Great, 180, 522
    Manx, 522
    Pink-footed, 522
    Short-tailed, 522
    Sooty, 167, 180, 332, 522
    Streaked, 234
    Townsend's, 522, 524
    Tropical, 332, 334
    Wedge-tailed, 234, 332, 334
Sheathbill, Snowy, **167**
Sheep
    Bighorn, 438, 459, 483,
        497, 511
    Dall's, 511
    Desert Bighorn, 445, 454
Shelduck, Ruddy, 228, **412**
sheoak woodland, 61–63, **62, 63**
Shoebill, **322**, 324
Sholakili, Nilgiri, 201
shortgrass prairie, 444, 446,
        450, 452, 458, 461, 488,
        496–98, **497**, **498**, 499,
        501, 525
Short-horned Lizard, Greater,
        446, 464, 498
Short-tailed Bat, New Zealand
        Lesser, 34
Short-toed Lark, Lesser, 362
Shortwing
    Flores, 44
    Rusty-bellied, 204
    White-browed, 44
Shoveler, Northern, 501, 513
Shovel-nosed Snake, Narrow-
        banded, 30
Shrew
    Golden-rumped Elephant,
        262
    Mexican, 486
    Pen-tailed Tree, 226
Shrike
    Gray-capped, 188
    Great Gray, 360
    Iberian Gray, 396
    Lesser Gray, 387
    Loggerhead, 452, 457, 461,
        463, 498
    Long-tailed, 387
    Masked, 349, 362, 393
    Red-tailed, 362
    Steppe, 360
    White-tailed, 295, 300
    Woodchat, 392
    Yellow-billed, 284
Shrike-Babbler, Dalat, 201

Shrikejay, Crested, 195
Shrikethrush
    Bower's, 39
    Little, 228
    Sandstone, 48, 49
Shrike-tit, Northern, 53
Shrike-Tyrant, White-tailed, 113
shrub desert, Palearctic hot,
        243, 244, 245, 246, 296,
        300, 352, 353, 358, 361–63,
        361, 362, 364, 450
shrubland
    acacia, 30, 57, 61, 81, 83–85,
        100, 149, 152, 292, 296, 450
    chenopod, 30–32, **31, 32**, 63,
        83, 246, 353, 456
    Chihuahuan Desert, 296, 361,
        444, 450–52, **450**, **451**, **452**,
        453, 456, 458, 461, 464,
        467, 488, 490, 491
    garrigue, 349, 366, 389, 391,
        393–94, **394**, 502
    Karoo, 121, 243, 244, 246–49,
        247, 248, 312, 316, 319, 320,
        336, 456, 461
    maquis, 114, 121, 160, 316, 336,
        349, 366, 385, 387, 388,
        389, 391–93, **391, 392**, 502
    monte, 114, 116, 118, 120–22,
        121, 122, 149, 150, 156, 158,
        182, 184, 336
    sagebrush, 246, 444, 445,
        446, 456, 458, 461–64, **462**,
        463, 496, 498, 504
    salt desert, 456–58, **456**,
        457, 461
    samphire, 30–32, **31, 32**, 246,
        353, 456
    Siberian larch forest, 340, 342,
        343–45, **344, 345**, 366–67,
        402, 404, 437
    Side-blotched Lizard, Common,
        446, 452, 458, 504
    Sierra-Finch
        Carbonated, 118
        Patagonian, 119, **120**, 142
    Sifaka
        Coquerel's, 279
        Crowned, 279
        Golden-crowned, 279
        Perrier's, 279
        Verreaux's, **251**, 252, 279
        Von der Decken's, 279
    Silky-flycatcher, Gray, 449
    Silverbird, 304
    Siren
        Greater, 471
        Reticulated, 433
    Siskin
        Cape, 319
        Pine, 440, 442
    Sitatunga, 323

Sittella, Black, 39
Skimmer
    African, 324
    Indian, 228
Skink (reptile)
    Bayon's, 315
    Blue-tailed Mole, 495
    Broad-headed, 474
    Florida Sand, 495
    Gilbert's, 446
    Gold-spotted, 252
    Grass, 311, 315
    Great Desert, 63
    Ovambo Tree, 295
    Round Island Ground, 334
    Western, 280
    Yakka, 85
Skink, Taiwan Alpine (bird), 224
Skipper
    Arogos, 501
    Dakota, 501
Skipperling, Poweshiek, 501
Skua
    Chilean, 180
    Great, 167
    South Polar, 180, 234, 522
Skunk
    American Hog-nosed, 459
    Humboldt's Hog-nosed, 117
    Patagonian Hog-nosed, 117
    Striped, 432, 473, **473**, 490
    Western Spotted, 490
Skylark
    Eurasian, 419, 420
    Oriental, 238
Slaty Antshrike, Western, 133
Slender Blue-tongue, Saltbush,
        31
Slender Loris, Red, 200
Slender Salamander, Garden,
        504
Sloth
    Brown-throated Two-toed,
        145
    Hoffman's Two-toed, 145
Snake
    Australian Green Tree, 56, 88
    Banded Water, 471, 515
    Black
        Rat, 473
        Swamp, 513
    Black-necked Garter, 452
    Blond Hognose, 280
    Broad-headed, 73, 76
    Brown Water, 471
    Bull, 458
    Central American Indigo, 487
    Cross-barred Tree, 263
    Desert
        Banded, 30
        Night, 452
    Dunmall's, 58

Dwyer's, 85
Eastern
    Garter, 473
    Hognose, 493
    Indigo, 433, 495
    Milk, 490
    Rat, 471
Fierce, 65
Flathead, 490
Florida Crowned, 495
Forest Vine, 259
Four-lined, 280
Glossy, 452, 458
Gopher, 446, 458, 464, 493,
    498, 504
Great Plains Rat, 490
Ingram's Brown, 65
Little File, 90
Madagascar Hognose, 280
Mahafaly Sand, 252
Moon, 30
Mud, 471, 513
Mulga, 85
Narrow-banded Shovel-
    nosed, 30
Ornamental, 58
Pacific
    Parrot, 487
    Patchnose, 487
Pale-headed, 73
Pine Woods, 433
Red-bellied black, 68
Red-naped, 85
Ringed Brown, 85
Ring-necked, 493
Rough Green, 471, 513
Saltmarsh, 515
Savanna Vine, 263
Scarlet, 433
Sharp-tailed, 430
Short-tailed, 495
Southeastern Green, 263
Speckled Brown, 65
Texas Indigo, 467
Trans-Pecos Rat, 452
Velvety Swamp, 169
Western Lyre, 487
White-bellied Mangrove, 90
Yellow-naped, 85
Snake-Eagle
    Black-chested, 311
    Fasciated, 262
Snake-eye, Leschenault's, 189
Snake-Lizard, Burton's, 82
Snake-necked Turtle, Northern,
    88
Snapping Turtle
    Alligator, 470
    Northern, 88
Snipe
    Greater Painted-, 238
    Jameson's, 163

Pin-tailed, 238
Solitary, 409
Swinhoe's, 384
Snowfinch
    Blanford's, 356
    Pere David's, 360
Snub-nosed Monkey, Golden,
    367, 373, **374**
Solitaire
    Rodrigues, 276
    Townsend's, 445
Songlark
    Brown, 99
    Rufous, 99
Sooty-Woodpecker, Northern,
    195
Sora, 526
Spadefoot
    Couch's, 455, 481
    Eastern, 493
    Great Basin, 481
Sparrow
    Abd al Kuri, 254
    Arabian Golden, 365
    Bachman's, 433
    Baird's, 461, 498, 501
    Bell's, 504
    Black-chested, 487
    Black-chinned, 504
    Black-throated, 452, 454,
        457, **457**, 463
    Botteri's, 461
    Brewer's, 457, 463
    Cassin's, 461, 498
    Eurasian Tree, 240, 419
    Field, 490
    Grasshopper, 461, 498,
        501, 526
    Henslow's, 501
    House, 240, 338
    Lark, 490, 493, 498
    LeConte's, 501
    Lincoln's, 484
    Nelson's, 517
    Olive, 466
    Rufous-crowned, 490
    Sagebrush, 458, 463
    Saltmarsh, 517
    Savanna, 526
    Seaside, 517
    Sind, 190
    Socotra, 254
    Song, 481
    Spanish, 396
    Swamp, 513
    Vesper, 461, 464, 498, 501
    Worthen's, 461
    Yellow-throated Bush, 218
    Zapata, 513
    Zarudny's, 358, **358**
Sparrowhawk
    Eurasian, 371, 379

Levant, 376
    Rufous-breasted, 335
Sparrow-Lark
    Black-crowned, 245
    Black-eared, 248
    Chestnut-headed, 245
Sparrow-Weaver, Chestnut-
    backed, 288
Spectral Tarsier, Gursky's, **194**
Spiderhunter, Whitehead's, 201
Spinebill
    Eastern, 72
    Western, **67**
Spinetail
    Araucaria Tit-, 104
    Plain-mantled Tit-, 118
    Red-shouldered, 148
    Striolated Tit-, 104
spinifex, 48, 61, 81, 84
    dune, 28–30, **28**, **29**, 51, 63,
        64, 65, 100
    rocky, 28–30, **28**, **29**, 49, 51,
        65, 100
Spiny Agama, Tropical, 295
spiny forest, Malagasy, 146,
    249–53, **250**, **251**, **252**, 273,
    278, 279, 281, 453
Spiny Lizard
    Desert, 461
    Eastern, 467
    Long-tailed, 487
Spiny-tailed Iguana
    Northeastern, 467
    Western, 487
Spinytail, Ocellated, 245
Spoonbill
    Black-faced, 232, 413
    Eurasian, **228**, 229
Sportive Lemur
    Petter's, 252
    White-footed, 252
Spotted Creeper, African,
    284, 288
Spotted Cuscus, Common, 36
Spotted Dolphin
    Atlantic, 521
    Pantropical, 332
Spotted Dtella, Arnhem
    Land, 49
Spotted Eagle
    Greater, 238
    Lesser, 371–72, 420
Spotted Owl, Northern, 443
Spotted Skunk, Western, 490
Spotted Woodpecker
    Great, 378
    Lesser, 372, 378
    Middle, 388
Spot-throat, 272
Springbok, 244, 247, 248, 295
Springhare, Southern, 299
spruce-fir forest, Nearctic

montane, 346, 422, 425–27,
    426, **427**, 428, 430, 434,
    436, 437, 438, 440, 482
spruce-fir taiga
    Eurasian, 340–42, **341**, **342**,
        343, 344, 346, 350, 351,
        366, 371, 377, 379, 382, 388,
        405, 422, 425
    Nearctic, 340, 422–24, **423**,
        424, 425, 475, 478, 482,
        484, 501, 506, 514
Spurfowl, Mount Cameroon, 272
Square-tailed Drongo,
    Common, 262
Squirrel
    Abert's Flying, 430
    African Pygmy, 258
    American Red, 422, 427, 436,
        438, 442, 476
    Flying, 430
    Arctic Ground, 506
    California Ground, 503
    Carruther's Mountain, 271
    Deppe's, 123
    Douglas's, 438, 442, **442**
    Flying, 430
    Eastern Gray, 473, 476
    Eurasian Red, 349
    Forest Giant, 258, 265
    Gambian Sun, 283
    Golden-mantled Ground,
        436, 438, **439**
    Grizzled Giant, 194
    Guayaquil, 145, **145**
    Humboldt's Flying, 442
    Mexican Gray, 484, 487
    Mindanao Flying, 200
    Nelson's Antelope, 457
    Northern Flying, 430
    Ochre Bush, 262
    Red-and-white, **188**
    Red-bellied Coast, 262
    Red-legged Sun, 258
    Slender-tailed, 258
    Smith's Bush, 262, 290
    Southern Flying, 473
    Spotted Ground, 451, 459
    Tanganyika Mountain, 271
    Thirteen-lined Ground,
        498, 501
    Thomas's Rope, 265
    Variable, 240
    Variegated, 123
    Whitehead's Pygmy-, 200
    White-tailed Antelope, 457
Starling
    Black-bellied, 262
    Black-winged, 216
    Brahminy, **191**
    Bristle-crowned, 306
    Bronze-tailed, 284
    Daurian, 384

European, 338
Golden-breasted, 299
Greater Blue-eared, 295, 304
Lesser Blue-eared, 284, 288
Meves's, 295
Neumann's, 306
Purple, 284
Red-winged, 306, 338
Réunion, 276
Rosy, 396
Sharpe's, 271
Sharp-tailed, 291
Socotra, 254
Wattled, 336
White-cheeked, 384
Starthroat, Plain-capped, 114
Steenbok, 248, 290, 319
steppe
eastern grass, 356, 359, 382,
395, 396, 397–98, **397**, **398**,
407, 496
Palearctic forest-, 142, 342,
345, 382–84, **383**, **384**,
385, 396
Patagonian, 107, 116–18, **117**,
119, 120, 122, 141, 142, 143,
165, 167, 184, 359
temperate desert, 114, 116,
117, 355, 357, 359–60, **359**,
360, 397
western flower, 359, 382,
395–96, **395**, **396**, 397, 398,
420, 499
Stilt
Banded, 86
Black-winged, 329
Pied, 86
Stint
Little, 414
Red-necked, 351
Stonechat
African, 320, 335
European, 389
Réunion, 320
Siberian, 384
Stork
Abdim's, 311
Black-necked, 88, 229
Maguari, 158
Marabou, 338
Milky, 228
Oriental, 413
Painted, 228
Saddle-billed, 324
Storm's, **207**, 208, 231
White, 311, **419**, 420
Wood, 169, 470, 513
Yellow-billed, 324
Storm-Petrel
Ainley's, 522, 524
Ashy, 522
Band-rumped, 522

Black-bellied, 180
Black, 522
Elliot's, 179, 180
European, 332
Fork-tailed, 522
Gray-backed, 180
Leach's, 332, 522, 524
Least, 522
Markham's, 180
Ringed, 180
Swinhoe's, 234
Townsend's, 522, 524
Wedge-rumped, 179
Wilson's, 234, 332, **521**, 522
Stubtail, Timor, 46
subtropical broadleaf forest,
Indo-Malayan, 199, 201,
202–5, **203**, **204**, 380, 381
subtropical rainforest,
Australasian, 26, 35,
38–40, **39**, **40**, 41, 57, 72,
74, 267, 484
subtropical savanna, Palearctic,
61, 357, 385–87, **386**,
431, 444
Sugarbird, Cape, 319, **319**
Sunangel, Purple-throated, 140
Sunbeam, Shining, 140
Sunbird
Anchieta's, 288
Carmelite, 328
Congo, 265
Copper-throated, 231
Eastern
Miombo, 288
Violet-backed, 299
Fork-tailed, 204
Madagascar, 320
Malachite, 315, 319, 320
Mariqua, 300
Mouse-brown, 328
Neergaard's, 262
Nile Valley, 365
Northern Double-collared,
271
Olive, 262
Olive-backed, 240
Orange-breasted, 319, **319**
Palestine, 362
Plain-backed, 262
Purple, 218, 240
Red-tufted, 315
Scarlet-chested, 304
Socotra, 254
Southern Double-collared,
319
Tacazze, 320
Variable, 304
Western
Miombo, 288
Violet-backed, 284, 288
White-breasted, 304

Sunbird-Asity, Yellow-bellied,
276
Sunbittern, **169**
Sungazer, 315
Sungem, Horned, 154
Suni, 262
Sun Squirrel
Gambian, 283
Red-legged, 258
Surfbird, 177, 506, 511, 519, **520**
Swallow
Cave, 490
Chilean, 142
Cliff, 490
Gray-rumped, 311
Hill, 223, 224
Montane Blue, 315
Southern Rough-winged, 184
White-tailed, 337
swamp forest
Afrotropical, 225, 255, 257,
260, 264–65, **265**, 267, 322,
327, 328
Indo-Malayan
freshwater, 171, 206, 207,
225–27, **226**, 229, 264
peat, 206–8, **207**, **208**, 209,
210, 225, 227
Swamphen, Gray-headed, 228
Swamp Rat, Australian, 73
Swamp Snake
Black, 513
Velvety, 169
Swamp Warbler
Golden, 470
Madagascar, 328
White-winged, 324
Swan
Black, 86
Trumpeter, 512
Swift (bird)
Little, 338
White-rumped, 338
Swift (reptile)
Cuvier's Madagascar, 280
Merrem's Madagascar, 252
Swiftlet, Mountain, 70

Tachuri, Gray-backed, 154
Tahr
Arabian, 353, 363
Nilgiri, 224, **224**
taiga
Eurasian spruce-fir, 340–42,
341, **342**, 343, 344, 346,
350, 351, 366, 371, 377, 379,
382, 388, 405, 422, 425,
Nearctic spruce-fir, 340,
422–24, **423**, **424**, 425, 475,
478, 482, 484, 501, 506, 514
taiga savanna, Beringian, 342,
345, 350–51, **351**, 401, 424

Tailorbird
African, 272
Long-billed, 272
Mountain, 201
Taipan
Coastal, 56
Angolan, 328
Gabon, 328
tallgrass prairie, 63, 156,
395, 433, 474, 478, 484,
490, 491, 493, 496, 498,
499–502, **499**, **500**, **501**,
512, 517, 525
Tamandua
Northern, 484, 487
Southern, **150**, 170
Tamarin
Black-mantled, 128
Golden-mantled, **126**, 128
Tanager
Blue-and-yellow, 113
Blue-gray, 182
Flame-rumped, 182
Golden-crowned, 140
Gray-headed, 173
Masked Mountain-, 140
Opal-crowned, 127
Opal-rumped, 127
Palm, 182
Red-headed, 484, **485**
Rust-and-yellow, 136
Scarlet, 473
Scarlet-throated, 147
Scrub, 113, **113**
Summer, 433
Western, 430, 484
White-banded, 154
Tapaculo
Chucao, 103, **103**, 104
Ochre-flanked, 103, 104
White-browed, 136
White-throated, 160, **160**
Zimmer's, 136
Tapir
Brazilian, **172**, 175
Malayan, 194, 207, 226
Mountain, 133
Tarantula, Texas Brown, 491
Tarsier
Gursky's Spectral, **194**
Philippine, 194
Tattler, Wandering, 519
Tayra, **134**
Teal
Baikal, 413
Blue-winged, 501, 512
Cape, 326
Marbled, 413
temperate bamboo forest, East
Asian, 133, 373–74, **374**
temperate deciduous forest
East Asian, 202, 366, 369,

373, 378, 380–82, **380**, **381**, **382**, 471
European, 342, 369, 371, 375, 377–79, **378**, 380, 388, 471
Nearctic, 369, 377, 380, 381, 431, 433, 468, 471–74, **472**, **473**, **474**, 475, 476, 478, 479, 490, 491, 493, 494
temperate desert steppe, 114, 116, 117, 355, 357, 359–60, **359**, **360**, 397
temperate eucalypt woodland, 69, 75, 77–80, **78**, **79**, **80**, 83, 85, 285, 301
temperate mixed forest, Nearctic, 366, 422, 424, 471, 474, 475–78, **476**, **477**, **478**, 482
temperate rainforest
  Australasian, 26, 33, 41–42, **42**
  Nearctic, 102, 427, 430, 440–43, **441**, **442**, **443**
temperate wetland
  Australasian, 86–87, **86**, **87**, 322
  Palearctic, 86, 322, 411–13, **412**, **413**, 512
Tenrec
  Lesser Hedgehog, 252
  Tailless, 275
  Web-footed, 275
Tern
  Arctic, 524
  Black, 512
  Black-naped, 95, 233, 234
  Bridled, 95, 234, 524
  Common, 524
  Damara, 330
  Great Crested, 95, 233, 234, 330
  Inca, 177, **178**, 179, 180
  Least, 520
  Lesser Crested, 95, 233, 330, 334
  Little, 95, 233, 417
  Peruvian, 180
  River, 228
  Roseate, 95, 233, 234, 334
  Royal, 519
  Sandwich, 233, 330, 334
  Sooty, 94, 95, 234, 334, 524
  West African Crested, 330
  White, 236
  White-cheeked, 334
Tetraka, Appert's, 279
tetrodonta woodland savanna, 50, 53–56, **54**, **55**, **56**, 77, 289
Thamnornis, 252
Thick-knee
  Beach, 90, **91**, 96, 233

Eurasian, 387, 396
Great, 228
Peruvian, 106, **106**
Thick-tailed Gecko, Border, 76
thorn savanna
  Afrotropical dry, 112, 243, 245, 246, 248, 253, 281, 282, 283, 284, 285, 288, 289, 290, 292, 295, 296, 296–300, **297**, **298**, **299**, **300**, 301, 303, 304, 326, 336, 337, 354, 363, 364
  Chaco seco, 81, 83, 120, 122, 129, 144, 147, 149–51, **150**, **151**, 152, 153, 156, 158, 160, 182, 184, 189, 301
  espinal, 81, 83, 120, 149–51, 156, 160, 184, 189, 301
Thornbill
  Blue-mantled, 163–64
  Inland, 78
  Mountain, 39
  Rainbow-bearded, 140, **140**
  Slaty-backed, 85
  Slender-billed, 32
  Striated, 75
  Yellow, 57, 78
  Yellow-rumped, 78
thornscrub
  Afrotropical, 112, 253, 281, 285, 289, 292, 296–301, **299**, 354, 364
  Indo-Malayan, 112, 189–91, **190**, **191**, 217, 363, 364
  mesquite brushland and, 112, 464–67, **465**, **466**, **467**, 486
  Neotropical, 108, 112–14, **112**, **113**, 118, 128, 143, 253, 296, 464, 468
  Palearctic semidesert, 30, 118, 120, 189, 296, 300, 352, 354, 361, 362, 364–66, **365**, 394, 450, 461
Thrasher
  California, 504, **504**
  Curve-billed, 452, 454
  LeConte's, 458
  Long-billed, 466
  Sage, 458, 463
Three-banded Armadillo, Southern, 149
Three-toed Jerboa, Thick-tailed, 396
Three-toed Woodpecker
  American, 427, 440
  Eurasian, 341
Thrush
  Austral, 142
  Bassian, 41
  Chestnut-backed, 44
  Chinese, 367
  Dusky, 342

Gray-cheeked, 422
Hermit, 440, 443, 476
Island, 224
Littoral Rock, 252
Miombo Rock-, 288
Mistle, 348, 372
Orange-banded, 46
Siberian, 344, **368**
Song, 27, 379, 392
Swainson's, 422, 443
Varied, 443, **443**
Wood, 473
Thunderworm, 496
tidal flat, Palearctic, 414–16, **414**, **415**, 516
Tiger, 194, 197, 214, 230
  Bengal, 218, **218**, 219
  Siberian, 367
  Sumatran, 226
Tiger-Parrot, Painted, 70
Tinamou
  Andean, 119
  Elegant Crested-, 118, 119
  Ornate, 115
  Thicket, 466
Tinkerbird
  Moustached, 271
  Western, 271
  Yellow-fronted, 283
Tit
  Black-bibbed, 188
  Black-browed, 188
  Black-throated, 204
  Burmese, 188
  Carp's, 295
  Cinereous, 409
  Coal, 349
  Crested, 349
  Eurasian Blue, 378, **378**, 388
  Great, 358, 388
  Iriomote, 204
  Long-tailed, 384
  Miombo, 288
  Rufous-bellied, 288, 291
  Siberian, 341
  Sombre, 349
  Turkestan, 358
  Varied, 368, **368**, 382
  White-naped, 190
  Willow, 378
Tit-hylia, 259
Titi Monkey, Barbara Brown's, 148
Titmouse
  Black-crested, 490
  Juniper, 446
  Oak, 493
  Tufted, 473, 490
Tit-Spinetail
  Araucaria, 104
  Plain-mantled, 118
  Striolated, 104

Tit-Tyrant
  Ash-breasted, 139
  Black-crested, 140
  Tufted, 113
Tit-Warbler, White-browed, 409, **409**
Toad
  American, 473, 493
  Cane, 37, 56, 88
  Caucasian, 376
  Marbled, 488
  Mexican Burrowing, 488
  Oak, 493
  Red-spotted, 481
  Sonoran Desert, 455
  Western, 481
Toadlet, Mimic, 56
Tody-Tyrant, Yungas, 136
Tomtit, 27, 34
Torrent Salamander, Olympic, 443
Tortoise
  Aldabra Giant, 277
  Angulate, 319
  Gopher, 433, 495
  Parrot-beaked, 319
  Radiated, 252
  Spider, 252
  Texas, 467
Toucan
  Channel-billed, 127
  Toco, **154**
  White-throated, 127
Toucanet, Northern Emerald-, 484
Towhee
  Abert's, 454
  California, 504
  Eastern, 432, 495
  Green-tailed, 463
  Spotted, 504
Tragopan
  Cabot's, 204
  Temminck's, 373
  Western, 188
Trainbearer, Black-tailed, 113
Tree Boa, Western Madagascar, 280
Treecreeper
  Bar-tailed, 188
  Black-tailed, 52, 53
  Hume's, 188
  Red-browed, 73, 74, 75
  Short-toed, 388
  White-browed, 85
  White-throated, 75
Tree Frog
  Barking, 471
  Canyon, 481
  Giant Mexican, 488
  Gray, 473
  Green, 471

Green-eyed, 37
Orange-thighed, 37
Squirrel, 433
White-lipped, 37
Tree Hyrax, Eastern, 271
Tree-Kangaroo
Bennett's, 36
Doria's, 39
Lumholtz's, **40**
Tree Monitor, Spotted, 56
Treepie
Gray, 188
Hooded, 190
Rufous, 218
Tree Python, Green, 37
Tree-rat, Black-footed, 46
Treerunner, White-throated, 142
Tree Shrew, Pen-tailed, 226
Tree Skink, Ovambo (reptile), 295
Tree Snake
Australian Green, 56, 88
Cross-barred, 263
Tree Sparrow, Eurasian, 240, 419
Tree Vole, Red, 442
Trogon
Bar-tailed, 271
Black-headed, 145
Citreoline, 487
Mountain, 484
Narina, 262, **263**
Whitehead's, 201
tropical dry forest, Nearctic, 464, 467, 486–88, **487**, **488**
tropical grassland, Afrotropical, 63, 156, 243, 246, 259, 283, 287, 296, 308–12, **309**, **310**, **311**, 322, 334, 335, 336, 458
tropical lowland rainforest
Australasian, 35–37, **36**, **37**, 38, 41, 43, 45, 89, 124, 255, 273
Indo-Malayan, 35, 36, 124, 192–95, **193**, **194**, 196, 198, 199, 200, 206, 208, 209, 211, 226, 235, 238, 255, 273
tropical montane rainforest, Indo-Malayan, 138, 186, 199–201, **200**, **201**, 202, 203–4, 209, 214, 216, 267, 273
tropical semi-deciduous forest, Australasian, 143–44, **44**
tropical wetland, Australasian, 86, 87–89, **88**, 322
Tropicbird
Red-billed, **179**
Red-tailed, 234, 236, 334
White-tailed, 234, 236, 331
Tsessebe, Common, 311

Tuco-tuco, Magellanic, 143
Tui, 27, **27**, 34
tundra
Antarctic, 165–68
Eurasian
alpine, 161, 356, 389, 390, 407–11, **408**, **409**, 509, 511
boggy, 342, 343, 345, 350, 399, 401, 405–6, **405**, **406**, 408, 507
rocky, 342, 343, 345, 350, 399, 401–4, **402**, **403**, **404**, 405, 406, 407, 408, 505
Nearctic
alpine, 69, 161, 389, 407, 424, 427, 434, 436, 509–11, **509**, **510**, **511**
boggy, 165, 405, 424, 505, 506, 507–8, **508**, 511, 512, 514
rocky, 165, 401, 424, 505–6, **505**, **506**, 507, 508, 509, 511
paramo, 69, 114, 115, 128, 137, 138, 161–64, **162**, **163**, 312
tupelo forest, Bald Cypress–, 468–71, **469**, **470**, 474, 512
Turaco
Bannerman's, 272
Fischer's, 262
Great Blue, **257**, 258
Livingstone's, 262
Prince Ruspoli's, 304
Purple-crested, 304
Red-crested, 272
Ross's, 262
Schalow's, 262
Violet, 284
Turca, Moustached, 160
Turkey, Wild, 432, 473, 490, 493
Turnstone
Black, 519
Ruddy, 93, 95, **232**, 233, 331
Turtle
Alligator Snapping, 470
Eastern Box, 474
Green, 234, 236, 331, 334, 524
Hawksbill, 234, 236, 331, 334, 524
Kemp's Ridley, 515, 524
Leatherback, 234, 236, 331, 334
Loggerhead, 234, 236, 331, 334, 524
Loggerhead Musk, 470
Northern
Snake-necked, 88
Snapping, 88
Olive Ridley, 234, 236, 331, 334
Red-crowned Roofed, 228

Wood, 476
Turtle-Dove, European, 419
tussock grass, Antarctic tundra and, 165–68, **165**, **166**, **167**
tussock grassland, Australasian, 28, 308, 458, 63–65, **64**, **65**
Twinspot, Peters's, 262
Two-horned Chameleon, Usambara, 272
Two-toed Sloth
Brown-throated, 145
Hoffman's, 145
Tyrannulet, Minas Gerais, 145
Tyrant
Ash-breasted Tit-, 139
Black-crested Tit-, 140
Cattle, **183**
Chocolate-vented, 118, 184
Cinnamon-bellied Ground-, 118
Dark-faced Ground-, 106
Jelski's Chat-, 140
Short-tailed Field, 106
Spectacled, 158, **183**
Strange-tailed, **183**
Tufted Tit-, 113
White-tailed Shrike-, 113
Yungas Tody-, 136

Urial, 385

Valdivian rainforest, 26, 102–4, **103**, 141, 440
Vanga
Bernier's, 276
Blue, 279, **279**
Helmet, 276
Hook-billed, 279
Lafresnaye's, 252
Red-tailed, 276
Rufous, 279
Sickle-billed, 279
Van Dam's, 279–80
White-headed, 279
várzea flooded forest, 127, 171–73, **172**, **173**, 264
Vasa Parrot
Greater, 276
Lesser, 276
Velvet Gecko
Marbled, 49
Northern, 37, 46
Madagascar, **338**
Sakalava Madagascar, 252
Zigzag, 76
Verdin, 452, 454
Vicuña, 115
villages, cities and
Afrotropical, 239, 337–38, **338**
Indo-Malayan, 239–40, **240**, 337

Vine Snake
Forest, 259
Savanna, 263
Violet-backed Sunbird
Eastern, 299
Western, 284, 288
Viper
Gaboon, 259, 263
Godman's Mountain Pit, 486
Northeast African Carpet, 245
Rhinoceros, 259
Vireo
Black-capped, **490**
Blue-headed, 476, 478, 484
Cassin's, 430
Gray, 446
Mangrove, 175, 515
Red-eyed, 470, 473, 476
Warbling, 484
White-eyed, 433, 490, 495
Yellow-throated, 470, 473, 478
Viscacha, Mountain, **106**, 119
Visorbearer, Hyacinth, 154
Vole
Prairie, 501
Red-backed, 427, 507
Red Tree, 442
Sagebrush, 463
Vontsira
Grandidier's, 252
Narrow-striped, 252, 280
Ring-tailed, 275
Vulture
Bearded, 409
Cinereous, 386
Palm-nut, 262, **265**
Turkey, 454, 461

Wagtail
African Pied, 330
Eastern Yellow, 238
Mekong, 228
Western Yellow, 419, 420
Wallaby
Agile, 52, 56, 60, **61**, 88, 100
Black-striped, 76
Bridled Nail-tail, 57
Brush-tailed Rock-, 76
Northern Nail-tail, 52
Parma, 73
Pretty-faced, 74, 76, **77**
Red-necked, 57
Rufous Hare-, 29
Short-eared Rock-, 49
Swamp, 73, 79
Whiptail, 57
Wilkins' Rock, **48**
Wallaroo
Antilopine, 52, **52**, 56, 88, 100
Black, 49
Common, 65, 85
Euro, 32, 52, 62

Wallcreeper, 409
Walrus, **400**, 401, 522
Wapiti, Altai, 384
Warbler
African Reed, 324
Aquatic, 412
Arabian, 365, **365**
Arctic, 371
Australian Reed, 87
Black-and-white, 478, 515
Blackburnian, 181, **182**, 476
Blackpoll, 422
Black-throated
Blue, 476
Gray, 446, 449
Green, 476, 478, 484
Blyth's Reed, 229
Canada, 181, 422
Cape May, 422, **424**
Cerulean, 181, 473, **474**
Chestnut-vented, 319
Cinnamon Bracken-, 320
Clamorous Reed, 229, 328
Connecticut, 422, 484
Cyprus, 388, 392
Dartford, 389, **390**
Dusky, 342
Eastern
Crowned, 368
Orphean, 388
Friendly Bush, 224
Garden, 379
Golden-cheeked, 449, 490
Golden Swamp, 470
Golden-winged, 476
Goldman's, 123, **123**
Grace's, 449
Grand Comoro Brush-, 320
Grauer's, 271
Gray's Grasshopper-, 342
Green, 376
Greenish, 372
Hainan Leaf, 204
Hermit, 430, 442, 449
Hooded, 473
Kentucky, 473, 478
Kirtland's, **478**
Kopje, 248, **248**, 307
Layard's, 248
Limestone Leaf, 212
MacGillivray's, 484
Madagascar Swamp, 328
Magnolia, **477**, 478
Mangrove, 515
Marmora's, 392
Miombo Wren-, 288, 295
Mountain Leaf, 43, 44
Mourning, 181, 484
Namaqua, 248
Neumann's, 271
Olive, 123, 449
Olive-capped, 468

Olive-tree, 388
Oriole, 284
Paddyfield, 238, 412
Pallas's Grasshopper-, 344
Papyrus Yellow-, 324
Pine, 433
Pink-headed, 123
Prairie, 433, 515
Prothonotary, 470, **470**, 515
Red, 449
Red-faced, 449, **449**
Rufous-eared, 248, **248**
Sardinian, 392
Scrub, 358, 362
Socotra, 254
Speckled, 57, 393
Sri Lanka Bush, 201
Stierling's Wren-, 288
Subdesert Brush-, 252
Sunda Bush, 44
Swainson's, 476
Taiwan Bush, 224
Timor Leaf, 44
Townsend's, 430, 442
Upcher's, 387
Veery, 484
Victorin's, 319
Virginia's, 446
Western
Olivaceous, 392
Orphean, 388, 392
White-browed Tit-, 409, **409**
White-tailed, 272
White-winged Swamp, 324
Willow, 342, 348, 349
Wilson's, 484
Wood, 378, 379
Worm-eating, 473, 478
Yellow, 110, 481, 515
Yellow-breasted, 43, 44
Yellow-rumped, 123, **123**, 430, 440
Yellow-throated, 470, 478
Warbler-Finch
Gray, 110
Green, 110
Warbling-Finch
Black-capped, 150
Ringed, 121
Rusty-browed, 136
Warthog, Common, 283
Wasp
Common, 34
German, 34
Waterbuck, 283, 304, 323
Watercock, 229
Waterhen, White-breasted, 240
Water Monitor
Asian, 208, 230, 240
Merten's, 88
Water Rat, False, 90
Water Snake,

Banded, 471, 515
Brown, 471
Waterthrush
Louisiana, 474
Northern, 515
Wattlebird
Little, 68
Western, 68
Waxbill
Black-faced, 300
Black-tailed, 262
Waxwing
Bohemian, 341, 445
Cedar, 445
Japanese, 342
Weasel
Least, 476
Long-tailed, 123, 430, 438, 501
Short-tailed, 34, 427, 442
Weaver
Asian Golden, 238
Bar-winged, 288
Brown-capped, 271
Chestnut-backed Sparrow-, 288
Clarke's, 262
Forest, 262
Nelicourvi, 276
Orange, 265
Sakalava, 279
Spectacled, 304
Village, 304, 338
Web-footed Gecko
Karoo, 245
Namib, 245
Weevil, Giraffe-necked, 277
western flower steppe, 359, 382, 395–96, **395**, **396**, 397, 398, 420, 499
western riparian woodland, 452, 471, 475, 479–81, **480**, 483, 498, 512
wetland
Afrotropical freshwater, 87, 262, 281, 302, 322–25, **322**, **323**, **324**, 329, 330
Australasian
temperate, 86–87, **86**, **87**, 322
tropical, 86, 87–89, **88**, 322
Indo-Malayan freshwater, 219, 227–29, **227**, **228**, **229**, 237, 322
Nearctic freshwater, 322, 411, 468, 479, 480, 512–14, **512**, **513**, 516, 517, 525
Neotropical, 149, 153, 154, 156, 168–70, **168**, **169**, **170**, 184, 219, 227, 322
Palearctic temperate, 86, 322, 411–13, **412**, **413**, 512

Whale
Antarctic Minke, 97
Beluga, 522
Blue, 97, 180, 234, 332, 521
Bowhead, 522
Bryde's, 179
Fin, 521
Gervais Beaked, 521
Gray, 521, 522
Humpback, 180, 234, 332, 521
Killer, 332, 521
Long-finned Pilot, 97
Melon-headed, 332
Minke, 180
Pygmy
Killer, 332
Right, 97
Short-finned Pilot, 332, 521
Southern Right, 97, 332
Sperm, 179, 521
Wheatear
Black, 393
Black-eared, 393
Cyprus, 388, 392
Desert, 353, 360
Finsch's, 362, 387
Hooded, 353, 362
Isabelline, 356, 362, 398
Mourning, 353
Northern, 506
White-crowned, 353
Whimbrel, 93, 176, 329, 404
Whipbird
Eastern, **40**, 72, 74
Papuan, 39
Whip-poor-will, Eastern, 490
Whipsnake
Greater Black, 56
Neotropical, 467
Striped, 458
Whiptail
Desert Grassland, 461
Marbled, 452
Western, 446, 458, 504
Whistler
Bare-throated, 43, 44
Black-tailed, 92
Fawn-breasted, 45
Gilbert's, 57, 99
Green-backed, 188
Island, 90
Lorentz's, 39
Mangrove, 231
New Caledonian, 40
Olive, 41
Red-lored, 83
Rufous, 57, 78
White-breasted, 92
Whistling-Duck
Black-bellied, 513
Lesser, 228
Plumed, 88

Wandering, 88
White-bearded Gibbon
Bornean, 208
Northern, 204
White-eye
Australian Yellow, 90
Bonin, 204
Cape, 319
Comoro, 320
Dark-crowned, 44
Flores, 45
Javan, 208, 231
Madagascar, 320
Mountain, 44
Timor, 45, 46
White-browed, 43, 44
Whiteface, Chestnut-breasted, 32
White-fronted Goose
Greater, 507
Lesser, 413
Whitehead, 27
White Pelican, Great, 229, 326
White-sided Dolphin, Pacific, 521
Whitethroat, Greater, 419
Whydah, Broad-tailed Paradise-, 288
Wigeon, American, 512, 513
Wild Ass
African, 245
Asiatic, 357, 359, 360, 385–86, 396, 398
Wildcat
African, 337
Asiatic, 396
European, 372, 388
Iriomote, 204
Wild Dog, African, 283, 287, 291, 295, **295**, 304, 305, 311
Wildebeest
Black, 248, 315
Blue, 295, 311
Willet, 176, 517
Wolf
Ethiopian, 315, 321, **321**
Gray, 349, 372, 384, 385, 411, 422, **423**, 430, 442, 463, 483, 501
Indian, 221
Maned, 153, **153**, 169
Tibetan, 411
Wolverine, 342, **345**, 351, 422, 438
Wombat
Common, **80**, 82
Northern Hairy-nosed, 57
Southern Hairy-nosed, 82
Woodcock, American, 484
Woodcreeper
Golden-olive, 484

Ivory-billed, 484
Long-billed, 172
Moustached, 145
Scimitar-billed, 150, 172
Striped, 172
Wood Frog, Australian, 37
Woodhoopoe
Green, **294**, 295, 304
Violet, **294**, 295
woodland
Brigalow, 57–58, **58**, 83, 159, 281, 285, 488
Callitris, 57–58, 281, 285, 488
gusu, 285, 286, 288, 289–91, **290**, **291**, 292, 296, 301, 303, 304, 385
high-elevation pine, 425, 427, 430, 434–36, **435**, **436**, 509
Madrean pine-oak, 123, 382, 447–49, **448**, **449**, 455, 461, 468, 484, 486, 491, 493
mallee, 66, 67, 81–83, **81**, **82**, 98, 100, 292, 358
miombo, 53, 259, 260, 261, 262, 263, 283, 284, 285–88, **286**, **287**, **288**, 289, 290, 291, 292, 295, 296, 301, 303, 304, 308, 312, 488
Mulga, 57, 61, 77, 81, 83–85, **84**, **85**, 100, 149, 152, 292, 296, 450
Neotropical pine-oak, 123, **123**, 447
pinyon-juniper, 348, 428, 430, 436, 444–46, **445**, **446**, 455, 461, 464
sheoak, 61–63, **62**, **63**
temperate eucalypt, 69, 75, 77–80, **78**, **79**, **80**, 83, 85, 285, 301
western riparian, 452, 471, 475, 479–81, **480**, 483, 498, 512
woodland savanna, tetrodonta, 50, 53–56, **54**, **55**, **56**, 77, 289
Woodnymph, Mexican, 487
Wood-Owl, African, 335
Woodpecker
Acorn, 493
American Three-toed, 427, 440
Arabian, 365
Bearded, 304
Bennett's, 291
Black, 341, **347**, 348, 372
Black-backed, 422, 427, 430, 440
Black-bodied, 150
Black-headed, 218

Cream-colored, **128**
Downy, 430, 473
Eurasian Three-toed, 341
Fine-spotted, 284
Fulvous-breasted, 216
Gila, 454
Golden-fronted, 490
Golden-tailed, 304
Gray-headed, 372
Great Spotted, 378
Ground, 307
Hairy, 430, 443
Japanese, 382
Ladder-backed, 490
Lesser Spotted, 372, 378
Levaillant's, 349
Lewis's, 481, 493
Magellanic, 142, **142**
Middle Spotted, 388
Northern Sooty-, 195
Nuttall's, 493
Okinawa, 204
Orange-backed, 208
Pileated, 430, 432, 443, 470
Pygmy, 382
Red-bellied, 432, 473
Red-cockaded, 433, **433**
Red-headed, 470, 481, 490, 493
Sind, 365
White-backed, 367, 378, 388
White-headed, 430
Yellow-crowned, 218
Wood-Pewee
Eastern, 432, 473
Western, 440
Wood-Pigeon
Common, 419
Nilgiri, 201
Wood-Rail
Giant, 158
Little, 175
Rufous-necked, 515
Tsingy, 279, 280
Woodrat
Allegheny, 476
Eastern, 501
White-throated, 451
Woodshrike, Common, 218
Woodstar, Purple-collared, 113
Woodswallow
Black-faced, 62
White-browed, 78
Woolly Lemur
Cleese's, 279
Western, 279
Woolly Monkey
Brown, 128
Silvery, 128
Woolly Opossum, Derby's, 123
Woolly Rat, Subalpine, 70
Worm Lizard, Florida, 495, 496

Wren
Band-backed, 123
Banded, 114
Bewick's, 490
Cactus, 452, 454
Carolina, 473
Happy, 484
House, 484
Marsh, 512
Pacific, 443
Rock, 457
Sedge, 501, 512
South Island, 69
Winter, 476, 478
Yucatan, 466, 467
Zapata, 513
Wren-Babbler
Eyebrowed, 201
Limestone, 212
Streaked, 212
Wrentit, 504
Wren-Warbler
Miombo, 288, 295
Stierling's, 288
Wrybill, **96**
Wryneck, Rufous-necked, 335

Xenops, Great, 147

Yellow-Finch, Raimondi's, 106
Yellowhammer, 419
Yellowhead, 34
Yellowlegs
Greater, 517
Lesser, 517
Yellow-nosed Albatross, Atlantic, **331**
Yellow Robin, Western, 75
Yellowthroat
Altamira, 513
Belding's, 513
Black-polled, 513
Common, 433, 512–13
Olive-crowned, 184
Yellow Wagtail
Eastern, 238
Western, 419, 420
Yellow-Warbler, Papyrus, 324
Yellow White-eye, Australian, 90
Yuhina, Indochinese, 204
yungas, 115, 134, 135–37, **136**, 138, 140, 150

Zebra
Cape Mountain, 248, 319
Common, 248, 295, 304, 311, 337
Grévy's, 311
Hartmann's Mountain, 244, 295